Cephalopod Culture

Background: Newly born *Octopus vulgaris* paralarvae feeding on *Cancer pagurus* zoea (Photo by Manuel Nande) Top right corner: *Nautilus pompilius* (Fig. 10.1, Photo by Gregory Barord) Middle right side: *Euprymna hyllebergi* hatchling (Fig. 15.4, Photo by Jaruwat Nabhitabhata) Bottom right corner: *Sepiella inermis* broodstock (Fig. 13.1, Photo by Jaruwat Nabhitabhata)

José Iglesias • Lidia Fuentes • Roger Villanueva
Editors

Cephalopod Culture

Springer

Editors
José Iglesias
Centro Oceanográfico deVigo
Instituto Español de Oceanografía (IEO)
Vigo
Pontevedra
Spain

Roger Villanueva
Institut de Ciències del Mar (CSIC)
Barcelona
Spain

Lidia Fuentes
Centro Oceanográfico deVigo
Instituto Español de Oceanografía (IEO)
Vigo
Pontevedra
Spain

ISBN 978-94-017-8647-8 ISBN 978-94-017-8648-5 (eBook)
DOI 10.1007/978-94-017-8648-5
Springer New York Heidelberg Dordrecht London

Library of Congress Control Number: 2014933556

Printed on acid-free paper

Springer is part of Springer Science+Business Media (www.springer.com)

Preface

Cephalopods—octopuses, squids, cuttlefishes and *nautiluses*—are arguably the most intelligent, and perhaps charismatic, invertebrates on the planet. They occur in all the world's oceans, from the intertidal areas to the deep sea. But, why culture cephalopods? For food, is the instinctive answer—high growth rates and short life spans make cephalopods ideal candidates for commercial aquaculture since they have the potential to rapidly reach market size.

However, as '*Cephalopod Culture*' makes clear, the reasons for culturing cephalopods are more varied. Cephalopod species have been used as model organisms in neurobiology since the early twentieth century: they were, for example, the 'lab rat' for the Nobel Prize winning work of Alan L. Hodgkin and Andrew F. Huxley on the initiation and propagation of nerve impulses. Also, as their position as a neurological model was cemented, it became scientifically important in the latter half of the twentieth century to perfect small scale cephalopod culture protocols. At the same time, the world fishery potential for cephalopods was being realised, with world catches rising eight-fold as many finfish stocks declined.

However, some cephalopod stocks are now overfished, and others are at maximum potential: can large commercial scale culture of cephalopods provide a solution? '*Cephalopod Culture*' draws on the expertise of nearly 50 cephalopod researchers from across four continents to provide a thorough scientific reference on the state of the art of cephalopod culture today. Tracing the history of cephalopod culture, this book provides a wealth of information on the constraints and bottlenecks in the culturing process, paying particular attention to the problems of feeding planktonic early life stages, whose complex behaviours and nutritional requirements require the provision of live prey.

The diversity of cephalopods, and their wide variation in life history strategies (eggs of some species hatch to relatively large benthic 'miniature adults' whilst other species hatch to planktonic stages just a few millimetre in length), means that species have unique culture needs. This is where '*Cephalopod Culture*' excels. With 16 chapters dedicated to in-depth culturing methods for particular species, this book provides a laboratory manual that distils all the recently published research on each species, as well as detailing system requirements and management. These chapters, written by experts in cephalopod culture, are an essential read for students, technicians, amateur aquarists, researchers attempting to culture species as yet uncultured, researchers attempting to improve yields or reduce costs, those in industry looking to upscale cephalopod culture enterprises for food production, as well as suppliers to the aquarium trade.

'*Cephalopod Culture*' also looks to the future. Recent breakthroughs, such as the successful rearing of large octopus hatchlings on a wholly artificial diet of squid paste, are highlighted, as are areas requiring further research; for example, there is a need to understand better the nutritional requirements of planktonic hatchlings so that appropriately enriched zooplankton can be raised as prey items.

Additional economic benefits of cephalopod culture are also emphasised. These include restocking, as has been successfully achieved for octopus in the Seto Inland Sea of Japan; pharmaceutical exploitation of antibacterial and potential anticancer activities reported from squid ink; the use of modified cuttlebone in tissue engineering; and the many and varied uses of cuttlefish oil.

In all, this is a pioneering text, which draws together a vast array of knowledge on cephalopod culture and provides the foundation for future advances in this significant field.

President, Cephalopod International Advisory Council — Dr. Louise Allcock

Contents

Part II Main Cultured Cephalopods

Contributors

José Pedro Andrade CCMAR-CIMAR L.A., Centro de Ciências do Mar do Algarve, Universidade do Algarve, Campus de Gambelas, Faro, Portugal

Gregory J. Barord Biology Department, Brooklyn College, City University of NY Graduate Center, Brooklyn, NY, USA

Jennifer A. Basil Biology Department, Brooklyn College, City University of NY Graduate Center, Brooklyn, NY, USA

Sigurd von Boletzky Observatoire Océanologique, CNRS, Laboratoire Arago, Banyuls-sur-Mer, France

Claudia Caamal-Monsreal Unidad Multidisciplinaria de Docencia e Investigación, Facultad de Ciencias, Universidad Nacional Autónoma de México (UNAM), Puerto de Abrigo s/n Sisal, Yucatán, México

Jesús Cerezo Valverde IMIDA-Acuicultura, Consejería de Agricultura y Agua de la Región de Murcia, Murcia, Spain

Pedro Domingues Centro Oceanográfico de Vigo, Instituto Español de Oceanografía, Vigo, Spain

Ana Farías Instituto de Acuicultura, Sede Puerto Montt, Universidad Austral de Chile, Puerto Montt, Chile

Lidia Fuentes Oceanographic Center of Vigo, Instituto Español de Oceanografía (IEO), Vigo, Spain

Pedro Gallardo Unidad Multidisciplinaria de Docencia e Investigación, Facultad de Ciencias, Universidad Nacional Autónoma de México (UNAM), Puerto de Abrigo s/n Sisal, Yucatán, México

José García García Consejería Agricultura, Agua y Medio Ambiente de la Región de Murcia, IMIDA, Murcia, Spain

Benjamín García García IMIDA-Acuicultura, Consejería de Agricultura y Agua de la Región de Murcia, Murcia, Spain

Camino Gestal Instituto de Investigaciones Marinas (IIM), CSIC, Vigo, Spain

José Iglesias Oceanographic Center of Vigo, Instituto Español de Oceanografía (IEO), Vigo, Spain

Yuzuru Ikeda Department of Chemistry, Biology and Marine Sciences, Faculty of Science, University of the Ryukus, Nishihara, Japan

Noussithé Koueta Laboratoire de Biologie des Mollusques Marins et des Ecosystèmes Associés: BioMEA, Université de Caen, Caen Cedex, France

CNRS INEE—FRE3484 biomea Biologie des Mollusques marins et des Ecosystèmes Associés, Esplanade la Paix, Université de Caen, Caen Cedex, France

Estelle Le Bihan IVAMER, Innovation Marine, Thury-Harcourt, France

Qi Li Fisheries College, Ocean University of China, Qingdao, China

Xiang-Zhi Lin Third Institute of Oceanography, SOA, Xiamen, China

Chang Liu Fisheries College, Ocean University of China, Qingdao, China

Zhao-Sheng Liu Fisheries College, Ocean University of China, Qingdao, China

Manuel Luaces Arrecifes del Atlántico S.L., Muxía, A Coruña, Spain

Maite Mascaró Unidad Multidisciplinaria de Docencia e Investigación, Facultad de Ciencias, Universidad Nacional Autónoma de México (UNAM), Puerto de Abrigo s/n Sisal, Yucatán, México

Jennifer Mather University of Lethbridge, Lethbridge, AB, Canada

Óscar Monroig Instituto de Acuicultura Torre de la Sal (IATS-CSIC), Castellón, Spain

Institute of Aquaculture, School of Natural Sciences, University of Stirling, Scotland, UK

Jaruwat Nabhitabhata Excellence Centre for Biodiversity of Peninsular Thailand (CBIPT), Faculty of Science, Prince of Songkla University, Hatyai, Songkhla, Thailand

Juan Carlos Navarro Instituto de Acuicultura Torre de la Sal (IATS-CSIC), Castellón, Spain

Michelle K. Nishiguchi Department of Biology – MSC 3AF, New Mexico State University, Las Cruces, NM, USA

Alberto Olivares Facultad de Ciencias del Mar y de Recursos Biológicos, Universidad de Antofagasta, Antofagasta, Chile

Cristina Pascual Unidad Multidisciplinaria de Docencia e Investigación, Facultad de Ciencias, Universidad Nacional Autónoma de México (UNAM), Puerto de Abrigo s/n Sisal, Yucatán, México

Graham J. Pierce University of Aberdeen, Newburgh, Aberdeenshire, UK

CESAM & Departamento de Biologia, Universidade de Aveiro, Aveiro, Portugal

Julio Portela Centro Oceanográfico de Vigo, Instituto Español de Oceanografía (IEO), Vigo, Spain

Yao-Sen Qian Fisheries College, Ocean University of China, Qingdao, China

Ocean Fishery Station of Ganyu, Lianyungang, China

Manuel Rey-Méndez Departamento de Bioquímica y Biología Molecular, Facultad de Biología, CIBUS, Campus Vida, Universidad de Santiago de Compostela, Santiago de Compostela, A Coruña, Spain

Carlos Rosas Unidad Multidisciplinaria de Docencia e Investigación, Facultad de Ciencias, Universidad Nacional Autónoma de México (UNAM), Puerto de Abrigo s/n Sisal, Yucatán, México

Francisco Javier Sánchez Instituto Español de Oceanografía (IEO), Oceanographic Center of Vigo, Vigo, Spain

David Scheel Alaska Pacific University, Anchorage, AK, USA

Susumu Segawa Department of Ocean Sciences, Faculty of Marine Science, Tokyo University of Marine Science and Technology, Minato-ku, Tokyo, Japan

António V. Sykes CCMAR-CIMAR L.A., Centro de Ciências do Mar do Algarve, Universidade do Algarve, Campus de Gambelas, Faro, Portugal

Íker Uriarte Instituto de Acuicultura, Sede Puerto Montt, Universidad Austral de Chile, Puerto Montt, Chile

Carlos Veiga Samertolameu S. Coop. Galega, Moaña, Pontevedra, Spain

Helene Viala IVAMER, Innovation Marine, Thury-Harcourt, France

Érica A. G. Vidal Centro de Estudos do Mar (CEM), Universidade Federal do Paraná (UFPR), Cx. P. 61, Pontal do Paraná PR, 83255-976, Brazil

Center for Marine Studies, University of Parana—UFPR, Pontal do Paraná, PR, Brazil

Roger Villanueva Institut de Ciències del Mar (CSIC), Barcelona, Spain

Zhao-Kai Wang Third Institute of Oceanography, SOA, Xiamen, China

Óscar Zúñiga Facultad de Ciencias del Mar y de Recursos Biológicos, Universidad de Antofagasta, Antofagasta, Chile

Wei-Bing Zheng Third Institute of Oceanography, SOA, Xiamen, China

Xiao-Dong Zheng Fisheries College, Ocean University of China, Qingdao, China

Part I
Introduction

Chapter 1
Cephalopod Biology

Sigurd von Boletzky and Roger Villanueva

Abstract Cephalopod biology is briefly surveyed in the context of cephalopod culture, considering its promises and limits. The motivations for undertaking culture work vary greatly, both in a basic and applied science perspective. Under most circumstances, the outcome of an experimental culture remains uncertain until a second generation is achieved. Culture conditions of benthic, shallow-water species are more easily reached in captivity than the living conditions of offshore, pelagic and deep-sea cephalopods. In general terms, difficulties in launching a culture from eggs are inversely proportional to the size of the hatchlings involved. Although cephalopod hatchlings are always actively foraging predators equipped with the organ systems typical of the adults, a combination of very small size and a pelagic lifestyle represents a challenge for their culture, particularly to satisfy their nutritional requirements.

Keywords Cephalopods · Biology · Culture · Development · Reproduction

1.1 Introduction

Half a century after the first successful culture attempts (Choe and Oshima 1963; Itami et al. 1963), a brief overview of cephalopod biology—even when limited to the context of aquaculture—will necessarily be somewhat incomplete. For a recent overview of cephalopod biology, we therefore recommend reading the early chapters of Boyle and Rodhouse (2005). Cephalopods are exclusively marine molluscs (with only few species tolerating low salinities); they have separate sexes and direct development (without a truly larval stage); they are carnivores or (more rarely) scavengers;

S. von Boletzky (✉)
Observatoire Océanologique, CNRS, Laboratoire Arago,
66650 Banyuls-sur-Mer, France
e-mail: sigurd.boletzky@sfr.fr

R. Villanueva
Institut de Ciències del Mar (CSIC), Passeig Marítim de la Barceloneta 37–49,
08003 Barcelona, Spain
e-mail: roger@icm.csic.es

J. Iglesias et al. (eds.), *Cephalopod Culture*, DOI 10.1007/978-94-017-8648-5_1,

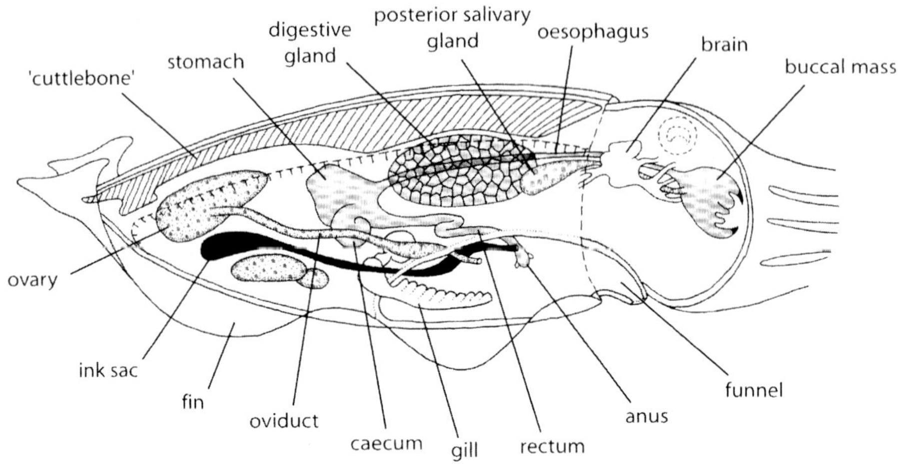

Fig. 1.1 Generalised anatomy of a cuttlefish. (After Boyle and Rodhouse 2005)

and some cephalopods constantly swim or hover in mid-water, whereas others stay close to the bottom, and some bury in sandy substrata to remain immobile for large parts of the day (thus hiding from predators along with minimizing energy expenditures). Although we view examples from different groups with various lifestyles (including some nektonic squid species that become gregarious during juvenile development), we are not able to cover the full breadth of adaptive features of cephalopod biology that might be exploited in culture work. However, one common feature of all cephalopod cultures—present and future—must be emphasized, namely the inevitably artificial character of cephalopod life in captivity.

Culture, even when achieved in very large enclosures designed to mimic a natural environment, always introduces some artificial elements into the life of captive cephalopods. Not the least important of these elements are the absence of potential predators and a general loss of sensory input. Even in an advanced culture system, cephalopod hatchlings and juveniles are thus conditioned by captivity from the outset. One should keep in mind that—no matter how small they are—newly hatched cephalopods are already accomplished organisms equipped with complex sensory organs, a very elaborate nervous system and powerful adult-like effectors (Nixon and Young 2003). Disregarding body size, the similarity between newly hatched and adult cephalopods is indeed striking. As examples, Figs. 1.1, 1.2 and 1.3 show the generalised anatomy of adult coleoid cephalopods and Fig. 1.4 illustrates some general morphological features of the hatchlings.

In many bottom-living inshore cephalopods, the conditioning by artificial elements, such as confinement or illumination, may of course be exploitable for the purposes of a culture. Some of these inshore species indeed are physiologically flexible and can adapt to artificial environments. They may live and grow on unnatural diets, become sexually mature and reproduce even if low growth rates can lead to a reduced adult size (Boletzky 1987). The reasons for such reduced

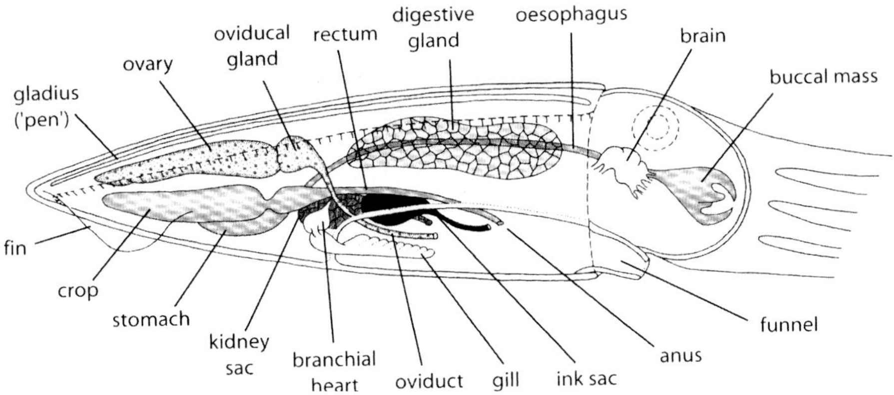

Fig. 1.2 Generalised anatomy of a loliginid squid. (After Boyle and Rodhouse 2005)

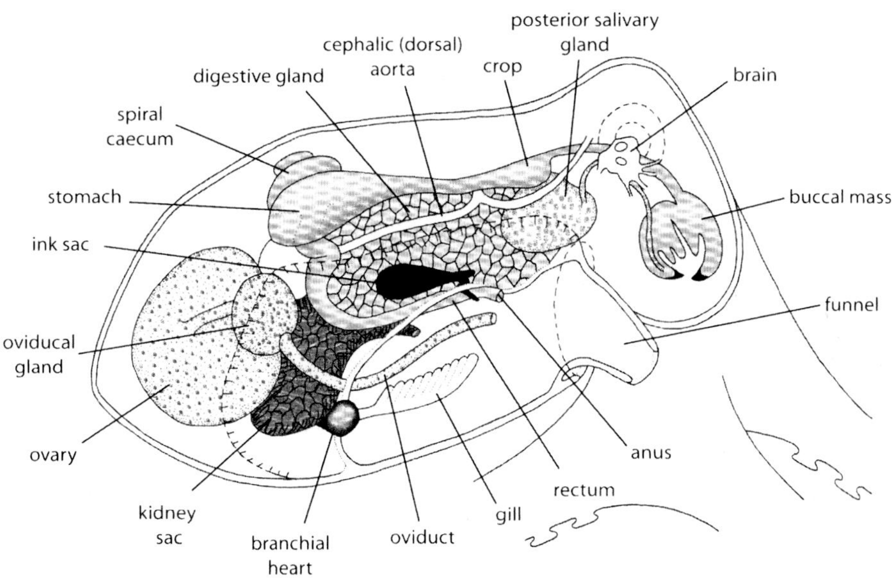

Fig. 1.3 Generalised anatomy of a female octopus. (After Boyle and Rodhouse 2005)

growth rates may be related to quality or quantity of food or stress caused by general culture conditions (tolerance to crowding, unsuitable artificial light and temperature levels, among others). With regard to food intake, a social hierarchy may become established within a group of captive individuals and may lead to substantial variations in individual food intake. And, last but not least, much like the adults, the small animals are also subject to various diseases (Hanlon and Forsythe 1990a, b; Hochberg 1990, Chap. 6).

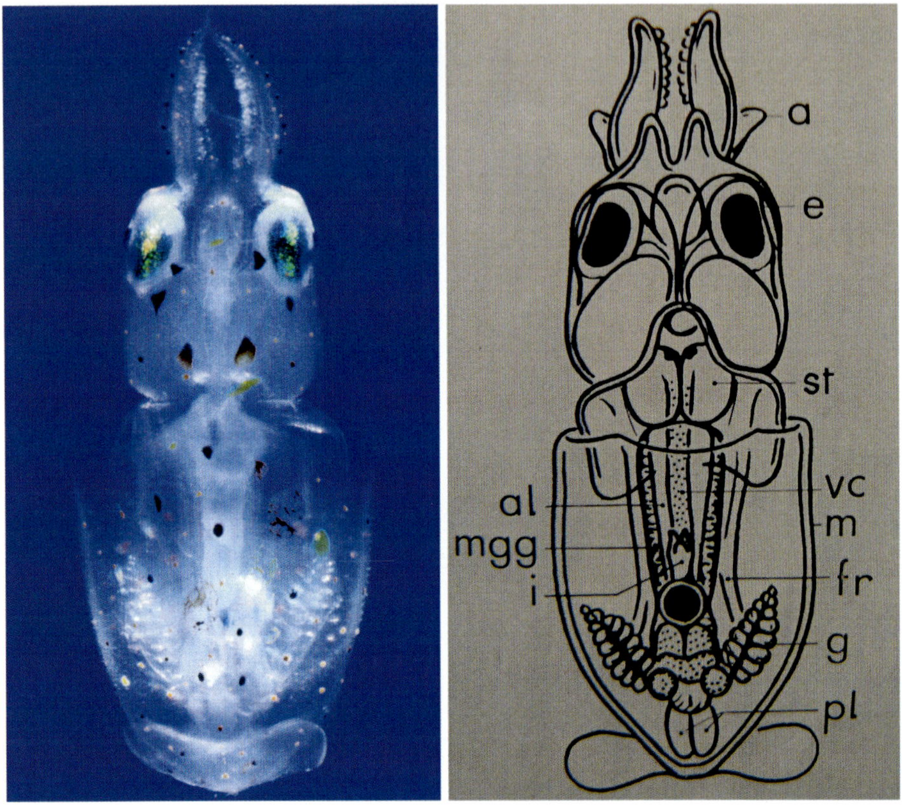

Fig. 1.4 *Left:* Live hatchling of *Loligo vulgaris*, in dorsal view. Individual photographed under anaesthesia (2–5 % ethanol), potentially causing chromatophore contraction (Villanueva, original). *Right:* Semi-schematic representation of a loliginid squid hatchling with a three-lobed inner yolk sac, in ventral view. *al* anterior lobe, *pl* posterior lobe, *a* arm, *e* eye, *fr* funnel retractor, *g* gill, *i* intestine, *m* mantle, *mgg* midgut gland, *st* statocyst, *vc* vena cephalica. (After Boletzky 2010)

Living individuals with such highly complex sensory organs and neural pathways involving multiple effectors are not automata. It is not surprising therefore that the results obtained in culture work are not always fully reproducible. The very existence of such unpredictable results may draw attention to particularly interesting biological problems related to individual variations, epigenetic effects and phenotypic plasticity. Perhaps the practical aims of a cephalopod culture may be assessed—and possibly redirected more precisely—based on the recognition of technical or conceptual limitations.

1.2 Reproduction and Growth

The starting condition for culture of any species is given by the reproduction of mature genitors. By producing fertilized eggs, the parent individuals form a potential P generation for hopefully continuing, successive offspring generations

(F1, F2, F3…). Subsequent reproduction of a few individuals brought up in the first rearing experiment may suffice for some generations, but sooner or later inbreeding may generate negative effects (Andrade et al. 2012). This will necessitate the occasional introduction of new genitors from wild populations. Cephalopods have a single period of sexual maturity, spawning single- or multiple-egg masses during a more or less protracted period of maturity. Thus, the question arises whether during culture experiments the brooding stock should be replaced by new individuals every new season or period. When live genitors (or egg masses with viable embryos) are not available, an alternative method to start a culture may be the collection of ova and spermatozoa from freshly caught, mature individuals, followed by in vitro fertilization. An essential condition for normal development of these embryos is the preparation of an oviducal jelly layer embedding the eggs (Ikeda et al. 1993). This allows the chorion membrane to expand, thus providing the necessary perivitelline space for the developing embryo. This method allowed the fertilization of 12 oceanic squid species (Villanueva et al. 2012) in view of the difficulties involved in obtaining eggs from spawning of captive oceanic squid broodstock maintained in aquaria (Bower and Sakurai 1996; O'Dor and Balch 1985).

When weighing the pros and cons of candidate species for culture, one should remember that the reproductive behaviours (Hanlon and Messenger 1996) and related features of anatomy and functional morphology (Budelmann et al. 1997) vary greatly among systematic groups of cephalopods, especially with regard to their respective ecological adaptations and lifestyles. There are numerous ways how spermatozoids are packed in spermatophores, which are produced by glands of the male duct; how these packages are transferred to a chosen female during mating; and how a spermatophore is transformed into a bubble-shaped spermatangium that may be stored by the female (in special places of the integument, or in pouches, or—in the case of octopods—in the oviduct or even in the ovary itself; Nesis 1996). Particularly complex mechanisms allow the spermatozoa to be finally released from their storage sites and activated so that they can pass through the micropyle of an egg chorion.

The ostensibly simple mating system of cephalopods (one female, one male) may be subject to considerable complications. Thus, the basic mating pattern may be modified by the actions of *sneaker* males or by other mechanisms of sperm competition (Hanlon and Messenger 1996; Huffard et al. 2008; Sato et al. 2010; Iwata et al. 2011). The observation of multiple paternity in hatchlings from a single egg mass suggests that this 'complication' is an important advantage for natural selection, though a bad one for aquaculture (Boyle et al. 2001; Buresch et al. 2003; Nesis 1996; Hoving et al. 2010; Voight 2001; Voight and Feldheim 2009; Quinteiro et al. 2011). Recent observations on *Loligo bleekeri* show that males of different sizes that employ different mating behaviours also produce two different forms of spermatozoa (Iwata et al. 2011).

Mating and spawning in captivity may generate some stress for the females if males have continuous access to them. Since females store the sperm after each mating in sufficient amounts to last for weeks or months, it is advisable to keep mature males separate from spawning females. Spawning is indeed of high energy cost for both males (Franklin et al. 2012) and females, especially if the female produces large masses of eggs with gelatinous egg case material, either continually or intermittently, over several weeks or months. Declining physiological conditions in spawning females are often recognizable by the production of more or less disorganized egg masses (Boletzky 1987). Under normal conditions, egg masses are structured at spawning in a specific way that sets the stage for a more or less complex process of modifications taking place during the development of the embryos, from the initial condition having permitted the passage of spermatozoa to the final condition allowing the young animals to hatch out (Boletzky 1998).

1.2.1 Fecundity

Depending on the species considered, average fecundities vary from a few hundred (e.g. sepiolid squids) to several thousands (e.g. sepiid cuttlefish) or hundreds of thousands (e.g. merobenthic octopods like *Octopus vulgaris* or *O. cyanea*) or around one million eggs as in the pelagic octopod *Ocythoe tuberculata* (Salman and Akalin 2012). The maximum potential fecundity estimated for the jumbo squid *Dosidicus gigas* is 32 million oocytes; so far, this is the largest number recorded for any cephalopod (Nigmatullin and Markaida 2009). When spawning is drawn out over weeks or months, there may be pauses of variable lengths, but there is no state of sexual rest (regression of reproductive organs). The terminology used with cephalopods therefore has the term *multiple spawning*. The sum total of eggs laid over a long time by a single female may be several times the instantaneous holding capacity of the ovary sac (see Sect. 1.2.2). Apart from the high number of eggs thus produced, the staggered hatching of young individuals, which results from the staggered spawning events, may open up some interesting possibilities for cephalopod culture. In particular, it would be interesting to test broad mating combinations among the offspring (e.g. involving descendants of different males).

1.2.2 Growth

No matter whether it is attained through fast or slow growth (Moltschaniwskyj 2004), reaching a large adult size is not necessary for carrying through a successful cephalopod culture. If sexual maturity is attained at a comparatively small adult size, in many bottom-living species a long spawning activity based on continued maturation of ova can still lead to a high lifetime fecundity of reared individuals (Boletzky 1988). The highest growth rates observed under wild conditions (as deduced from postmortem analysis of periodic growth structures, e.g. statoliths) can be equalled under ideal culture conditions, with varied, high-quality food, especially when the maximum amount of food ingested per day represents a tissue mass nearly 10 % of the body mass of the young cephalopod (Mangold and Boletzky 1973). Rations of 30 % of the body mass have been reported for juvenile benthic octopuses (Quintana et al. 2011). However, the actual investment of ingested energy in body mass production changes with age (Briceño et al. 2010). The observations of Pecl and

Moltschaniwskyj (1999) on the growth of wild-caught *Idiosepius* held in aquaria clearly show that under these conditions the processes of somatic growth can be modified at the cellular level, the mantle muscle growing thicker, with a greater proportion of mitochondria-rich tissue, muscle fibres with smaller mitochondrial cores and fewer small muscle fibres. For details on nutrition and growth, see Chap. 5.

1.3 Life Cycle

The whole life cycle of coleoid cephalopods varies, depending on the species, from about 3–4 months (Idiosepiidae) to several years (Boyle and Rodhouse 2005). The embryonic phase may vary from about 1 week in pelagic squids producing very small eggs (Sakurai et al. 1996) to more than 1 year in deep-water species producing very large eggs (Boletzky 1994; Laptikhovsky 1999). In the Pearly Nautilus (Saunders and Ward 1987), the whole life cycle is much longer, incorporating several years of uninterrupted reproductive activity (an extreme form of the protracted terminal spawning typical for cephalopods, as indicated in Sect. 1.2.1). The very large eggs need about 1 year to develop to hatching. This is not unique, since certain deep-water octopods have similar egg sizes and similar durations of embryonic development (Wood et al. 1998).

1.3.1 Embryos

Depending on the species considered, ovum sizes vary from less than 1 mm to about 40 mm. In spite of these enormous size differences, the basic pattern of embryogenesis is the same in all cephalopods. The ovum is fertilized by a spermatozoon having passed through the micropyle into the chorion (which, except in the incirrate Octopoda, is embedded in oviducal jelly and nidamental envelopes). The zygote undergoes partial cleavage: The cytoplasm concentrated at the animal pole is subdivided by the formation of shallow cleavage furrows, which do not cut into the underlying yolk mass. While in the centre of this cleavage zone, small blastomeres are subsequently formed, the peripheral segments (the so-called blastocones) remain in continuity with the yolk mass; they are the founders of the prospective yolk syncytium, which will break down the yolk and export the nutritive elements to the blood (Boletzky 2003). From the unilayered blastula, epibolic gastrulation starts with a progressive expansion of the central cell plate (prospective outer germ layer, or ectoderm) over the peripheral blastomeres adjacent to the ring of blastocones. This peripheral complex forms the inner germ layer (mesendoderm), which soon becomes completely covered by the centrifugally advancing edge of the ectoderm. This edge forms the incipient blastopore lip of the gastrula.

When looking at subsequent developmental stages of cephalopod embryos and hatchlings, one has to remember that the marginal blastopore lip of the early gastrula forms the entire outer yolk sac envelope with its blood lacuna and the muscular elements (derived from the mesodermal part of the inner germ layer), which spread over the yolk mass. These muscular elements form a network organized to create peristaltic waves of surface contractions generating the early embryonic blood circulation. From the outer yolk sac the blood space extends around the remaining yolk mass inside the 'embryo proper' (i.e. the embryonic body derived from the gastrula).

In embryos developed from small eggs, the outer yolk sac is smaller than, or about the same size as, the embryo proper, whereas in embryos developing from large eggs, the outer yolk sac is huge compared to the embryo proper. Whatever the size, however, the remaining yolk mass (i.e. the yolk not yet 'used up' during embryogenesis) is finally taken up by the body of the embryo before it hatches. This so-called inner yolk sac (belonging to the digestive gland) is the persistent part of the embryonic 'yolk organ' (Boletzky 2010).

To leave the egg case (i.e. the chorion and complementary envelopes), cephalopod hatchlings use a special hatching gland situated on the mantle tip. The local digestion of the envelopes by the hatching enzyme produces an opening, through which the hatchling extracts itself, generally using auxiliary hatching equipments (e.g. integumental ciliatures). The different hatching mechanisms are closely related to the structure and consistency of the egg cases, which may have undergone strong modifications during the development of the embryos (Boletzky 1998).

1.3.2 Hatchlings

All cephalopods hatch as complete little animals, which are devoid of true larval features (hence the term *paralarva* proposed by Young and Harman 1988). Remaining yolk reserves are absorbed to depletion while the young animal begins to ingest and digest food, generally live prey (e.g. crustacean larvae), which are captured using visual hunting techniques. This parallel embryonic and postembryonic nutrition may last several days or weeks. Early growth of hatchlings shows a phase with no net growth in dry weight due to the yolk consumption (Vidal et al. 2002). Once the yolk reserves are used up, hatchlings depend on active foraging and—to a limited extent—on energy sources stored in the cells of certain tissues. Depending on the actual hatching time, the young animal starts out under more or less favourable conditions of individual fitness. Premature hatching, which is frequently triggered by artificial stimuli (Villanueva and Norman 2008), may substantially reduce survival chances, since the remaining outer yolk sac is lost, and a large inner yolk sac represents a heavy ballast for an actively swimming animal, limiting the jet propulsion power by reducing the available water volume of the mantle cavity.

In most cephalopods, the hatchlings have a lifestyle that is rather similar to the adult lifestyle. Thus, the hatchlings of nektonic and macroplanktonic squids are planktonic in that they are transported by currents, but within a given water mass in the cubic metre range, they move about actively by jet propulsion (Bartol et al. 2009), approaching potential prey by forward swimming and escaping from would-be predators by backward jetting (often combined with inking). This is essentially a micronektonic way of life similar to the lifestyle of adult squid (gregarious behaviour in loliginid squids appearing only several weeks after hatching). In short, these species have a holopelagic life cycle.

Hatchlings of benthic and nektobenthic cuttlefish (Guibé and Dickel 2011) and sepiolids with the exception of *Heteroteuthis* (Hoving et al. 2008), in their turn, have the same lifestyle as their bottom-living parents, including burying in sandy substrata. In other words, they have a holobenthic (at times, partially demersal) life cycle. Two exceptions to this general rule of ontogenetic lifestyle stability are known.

The hatchlings of the idiosepiid pygmy squids are always planktonic and later take up the peculiar quasi-benthic lifestyle of the adult animals (which rest when not foraging, either close to the bottom, attaching themselves to the lower face of plants and other overhanging structures, or to floating algae). This lifestyle switching is group typical (Boletzky et al. 2005).

The hatchlings of many bottom-living octopuses (of the family Octopodidae) start out as planktonic young animals, very similar in overall aspects to the hatchlings of pelagic octopods (e.g. Argonautidae). In contrast to the lifestyle switching observed in these species (and in the Idiosepiidae), certain octopus species have a life cycle without a planktonic post-hatching phase. In other words, they are holobenthic, whereas those octopodid species passing through a planktonic phase can be called merobenthic (Villanueva and Norman 2008).

Some confusion is due to the designation *larva,* which is often used for planktonic cephalopod hatchlings. It is derived from the adoption of terms proposed for newly hatched fish (Nesis 1979), and sometimes also alludes to gastropods. As a possible way out of the confusion, Young and Harman (1988) suggested *paralarva* as a new term for young cephalopods that do not share the biotope of their adults, thus introducing an ecological criterion.

1.3.3 Juveniles and Subadults

Depending on whether or not a paralarval phase exists, the actual juvenile phase of the life cycle starts either at the end of the paralarval phase or (if no paralarval phase exists) virtually at hatching. External morphological changes can be identified in octopuses at settlement, when the paralarval phase is ending (Villanueva and Norman 2008), such as positive allometric arm growth, the addition of new suckers, chromatophore, iridophore and leucophore formation, the development of skin sculptural components and a horizontal pupillary response. At the same time, octopuses appear to lose the Kölliker organs that cover the body surface and the lateral line system analogue. These structures have not been reported for adult benthic octopuses (Budelmann et al. 1997). A minor morphological change is the loss of the oral denticles of the beaks. These transformations are reflected in changes in the relative sizes of the various lobes of the paralarval and juvenile octopus brain (Nixon and Mangold 1996). Subadults are characterized by an adult-like external morphometry in individuals that are not yet sexually mature. This is the phase of sexual differentiation, during which male and female organs grow from early rudiments to the onset of sexual maturation (as defined by different maturation stage systems).

1.3.4 Adult Life

The sexual maturity of adult individuals culminates in the formation and storage of functional gametes, all the associated organs (oviducal and nidamental glands in females, spermatophoric glands and stores in males) being ready to function.

1.3.4.1 Sexual Dimorphism

In general terms, most cephalopods have adult females and males of roughly similar sizes, sometimes with one sex distinctly larger than the other. An exception to this general rule is the group of pelagic octopods of the families Alloposidae, Tremoctopodidae, Ocythoidae and Argonautidae, all of which have dwarf males, some with very peculiar forms of pelagic life (e.g. obligate rafting), and with an extremely large copulatory arm that may autotomize. None of these or other pelagic species (of families Bolitaenidae, Amphitretidae, Vitreledonellidae) have ever been cultured. However, future attempts to culture them may provide essential biological insights.

1.3.4.2 Egg Care

True egg care is known only in incirrate octopods. In pelagic octopods, the spawning females always keep their masses of stalked eggs, either holding them in their arms or fixating clusters of stalked eggs on calcified structures secreted by special integumental glands of certain arms. In *Tremoctopus,* a pair of egg-carrying rods is formed by the spawning female. A bilaterally and nearly symmetric brood case is built (and used also as a floating device) by the female *Argonauta* (Finn and Norman 2010). *Ocythoe* is the only ovoviviparous form known; the eggs develop while slowly descending the extremely long oviducts from where they hatch.

The different forms of egg care behaviour in the only bottom-living octopod family (Octopodidae) are similar to the modes of egg care known in pelagic octopods: In some species, large eggs or clusters of small eggs are carried around by the mother, whereas in the majority of octopodid species, single eggs or clusters of eggs are glued to a solid substratum (inside a den, often represented by man-made items like pots, bottles or car tires!) in a way reminiscent of the egg-carrying devices of *Argonauta* and *Tremoctopus*.

The few examples of egg brooding in decabrachian cephalopods are known from gonatid and bathyteuthid squids (Okutani et al. 1995; Seibel et al. 2005; Bush et al. 2012). It is somewhat questionable here to use the term egg care (the way it is used for octopods, where egg care includes sucker-based manipulation and ventilation with water flushes from the funnel). Indeed, the oceanic squid female simply holds in her arms the gelatinous egg mass she has produced; the female probably starves while serving as a floating device for the egg mass.

1.3.4.3 Post-Spawning Period and Senescence

In most shallow-water cephalopod species cultured until now, the senescent period is preceded by a phase of reduced or even arrested food ingestion, starting with the first spawning events (Anderson et al. 2002). The egg-laying period is located at the very end of the individual's life in most shallow-water species and is followed by a relatively quick and degenerative metabolic process. Some exceptions in coleoid cephalopods are deep-sea cephalopods with protracted spawning periods such as cirrate octopods, in which sexual maturity probably occupies most of their life cycle and egg laying is not directly related with the end of the life cycle and senescence (Collins and Villanueva 2006). Skin lesions and some diseases could also be an early symptom of senescence, such as coccidiosis, that has been reported among the most prevalent diseases of coastal octopuses leading to a poor condition (Gestal et al. 2007; Moltschaniwskyj et al. 2007; Pascual et al. 2010, Chap. 6).

1.4 Conclusion

The most general aim of cephalopod cultures may be highlighted as a programme of the utmost comprehensive cephalopod biology. The biological requirements of a cultured species must be thoroughly understood during all the phases of its life cycle. Such a broad and deep understanding will rarely exist ready prior to a culture experiment, so in general the experimental approach is inescapable. Or—to put it in more elegant terms—the *bottom-up* view of the experimental biologist applying available bits of knowledge needs to be seconded by the *top-down* view of the experienced biologist judging from his/her comprehensive knowledge of specific lifestyles, behavioural repertoires and physiological conditions at all life stages that have to be considered. The know-how actually needed in cephalopod culture has to result from continued confrontation and combination of these complementary views.

Only such an ambitious form of know-how warrants a practice that will hopefully stand the test of existing regulations on animal welfare.

References

Anderson RC, Wood JB, Byrne RA (2002) Octopus senescence: the beginning of the end. J Appl Anim Welf Sci 4:275–283

Andrade JP, Baptista FD, Sykes AV, Porta JM, Porta J (2012) Genetic variation in consecutive generations of cultured cuttlefish, *Sepia officinalis*. Cephalopod International Advisory Council Symposium, 27 Oct—2 Nov 2012, Florianópolis, Brazil, Abstracts Book

Bartol IK, Krueger PS, Stewart WJ, Thompson JT (2009) Pulsed jet dynamics of squid hatchlings at intermediate Reynolds numbers. J Exp Biol 212:1506–1518

Boletzky SV (1987) Fecundity variation in relation to intermittent or chronic spawning in the cuttlefish, *Sepia officinalis* L. (Mollusca, Cephalopoda). Bull Mar Sci 40:382–387

Boletzky SV (1988) A new record of long-continued spawning in *Sepia officinalis* (Mollusca, Cephalopoda). Rapp Comm Int Mer Médit 31:257

Boletzky SV (1994) Embryonic development of cephalopods at low temperatures. Antarctic Sci 6:139–142

Boletzky SV (1998) Cephalopod eggs and egg masses. Ocean Mar Biol Ann Rev 36:341–371

Boletzky SV (2003) Biology of early life stages in cephalopod molluscs. Adv Mar Biol 44:143–204

Boletzky SV (2010) The 'yolk organ' of cephalopod embryos: on transient functions from crawling substratum to provisional knapsack. Ferrantia 59:14–21

Boletzky SV, Nishiguchi MK, Nabhitabhata J, Nugranad J (eds) (2005) *Idiosepius*: ecology, biology and biogeography of a mini-maximalist. Phuket Mar Biol Cent Res Bull 66:11–22

Bower JR, Sakurai Y (1996) Laboratory observations on *Todarodes pacificus* (Cephalopoda: Ommastrephidae) egg masses. Am Malacol Bull 13:65–71

Boyle P, Rodhouse P (2005) Cephalopods—Ecology and Fisheries. Blackwell Science, Blackwell Publishing, Oxford

Boyle PR, Noble L, Emery AM, Craig S, Black KD, Overnell J (2001) Development and hatching in cephalopod eggs. A model system for partitioning environmental and genetic effects on development. In: Atkinson D, Thorndyke M (eds) Environment and Animal Development. Genes, Life Histories and Plasticity 251–267. Oxford, BIOS Scientific

Briceño F, Mascaró M, Rosas C (2010) Energy demand during exponential growth of *Octopus maya*: exploring the effect of age and weight. ICES J Mar Sci 67:1501–1508

Budelmann BU, Schipp R, Boletzky SV (1997) Cephalopoda. In: Harrison F (ed) Microscopic anatomy of invertebrates, vol 6A: Mollusca II, chapter 3, pp. 119–414

Buresch KC, Boal JG, Knowles J, Debose J, Nichols A, Erwin A, Painter SD, Nagle GT, Hanlon RT (2003) Contact chemosensory cues in egg bundles elicit male-male agonistic conflicts in the squid *Loligo pealeii*. J Chem Ecol 29:547–560

Bush SL, Hoving HJT, Huffard CL, Robison BH, Zeidberg LD (2012) Brooding and sperm storage by the deep-sea squid *Bathyteuthis berryi* (Cephalopoda: Decapodiformes). J Mar Biol Ass UK 92:1629–1636

Choe S, Oshima Y (1963) Rearing of cuttlefishes and squids. Nature 197:307

Collins MA, Villanueva R (2006) Taxonomy, ecology and behaviour of the cirrate octopods. Ocean Mar Biol Ann Rev 44:277–322

Finn JK, Norman MD (2010) The argonaut shell: gas-mediated buoyancy control in a pelagic octopus. Proc Roy Soc (B) 277:2967–2971

Franklin AM, Squires ZE, Stuart-Fox D (2012) The energetic cost of mating in a promiscuous cephalopod. Biol Lett 8:754–756

Gestal C, Pascual S, Guerra A (2007) *Aggregata octopiana* (Protista: Apicomplexa): a dangerous pathogen during commercial *Octopus vulgaris* ongrowing. ICES J Mar Sci 64:1743–1748

Guibé M, Dickel L (2011) Embryonic visual experience influences post-hatching shelter preference in cuttlefish. Vie Milieu 61:243–246

Hanlon RT, Forsythe JW (1990a) Diseases of Mollusca: Cephalopoda. In: Kinne O (ed) Diseases of Marine Animals III: 23–46. Biologische Anstalt Helgoland, Hamburg

Hanlon RT, Forsythe JW (1990b) Structural Abnormalities and Neoplasia. In: Kinne O (ed) Diseases of marine animals III: 203–204. Biologische Anstalt Helgoland, Hamburg

Hanlon RT, Messenger JB (1996) Cephalopod Behaviour. Cambridge University Press.

Hochberg FG (1990) Diseases caused by Protistans and Metazoans. In: Kinne O (ed), Diseases of Marine Animals III: 47–227. Biologische Anstalt Helgoland, Hamburg

Hoving HJT, Laptikhovsky V, Piatkowski U, Önsoy B (2008) Reproduction in *Heteroteuthis dispar* (Rüppell, 1844)(Mollusca: Cephalopoda): a sepiolid reproductive adaptation to an oceanic lifestyle. Mar Biol 154:219–230

Hoving HJT, Lipinski MR, Videler JJ, Bolstad KSR (2010) Sperm storage and mating in the deep-sea squid *Taningia danae* Joubin, 1931 (Oegopsida: Octopoteuthidae). Mar Biol 157:393–400

Huffard CL, Caldwell RL, Boueka F (2008) Mating behaviour of *Abdopus aculeatus* (d'Orbigny 1834)(Cephalopoda: Octopodidae) in the wild. Mar Biol 154:353–362

Ikeda Y, Sakurai Y, Shimazaki K (1993) Fertilizing capacity of squid (*Todarodes pacificus*) spermatozoa collected from various sperm storage sites, with special reference to the role of gelatinous substance from oviducal gland in fertilization and embryonic development. Inv Reprod Develop 23:39–44

Iwata Y, Shaw P, Fujiwara E, Shiba K, Kakiuchi Y, Hirohashi N (2011) Why small males have big sperm: dimorphic squid sperm linked to alternative mating behaviours. BMC Evol Biol 11:236

Itami K, Izawa Y, Maedea S, Nakai K (1963) Notes on the laboratory culture of the octopus larvae. Bull Jap Soc Sci Fish 29:514–520

Laptikhovsky VV (1999) Improved mathematical model to study the duration of embryogenesis in cephalopod molluscs. Ruthenica 9:141–146

Mangold K, Boletzky SV (1973) New data on reproductive biology and growth of *Octopus vulgaris*. Mar Biol 19:7–12

Moltschaniwskyj NA (2004) Understanding the process of growth in cephalopods. Mar Freshw Res 55:379–386

Moltschaniwskyj NA, Hall K, Lipinski MR, Marian JEAR, Nishiguchi M, Sakai M, Shulman DJ, Sinclair B, Sinn DL, Staudinger M, Van Gelderen R, Villanueva R, Warnke K (2007) Ethical and welfare considerations when using cephalopods as experimental animals. Rev Fish Biol Fish 17:455–476

Nesis KN (1979) The larvae of cephalopods. Biol Mor 4:26–37

Nesis KN (1996) Mating, spawning and death in oceanic cephalopods: a review. Ruthenica 6:23–64

Nigmatullin CM, Markaida U (2009) Oocyte development, fecundity and spawning strategy of large sized jumbo squid *Dosidicus gigas* (Oegopsida: Ommastrephinae). J Mar Biol Ass UK 89:789–801

Nixon M, Mangold K (1996) The early life of *Octopus vulgaris* (Cephalopoda: Octopodidae) in the plankton and at settlement: a change in lifestyle. J Zool London 239:301–327

Nixon M, Young JZ (2003) The brains and lives of Cephalopods. Oxford University Press, Oxford

O'Dor RK, Balch N (1985) Properties of *Illex illecebrosus* egg masses potentially influencing larval oceanographic distribution. North Atl Fish Org Scient Coun Stud 9:69–76

Okutani T, Nakamura I, Seki K (1995) An unusual egg-brooding behavior of an oceanic squid in the Okhotsk Sea. Venus 54:237–239

Pascual S, González AF, Guerra A (2010) Coccidiosis during octopus senescence: Preparing for parasite outbreak. Fish Res 106:160–162

Pecl GT, Moltschaniwskyj NA (1999) Somatic growth processes: how are they altered in captivity? Proc R Soc Lond B 266:1133–1139

Quintana D, Rosas C, Moreno-Villegas E (2011) Relationship between nutritional and rearing parameters of *Octopus maya* juveniles fed different rations of crab paste. Aquaculture Nutrition 17e:379–388.

Quinteiro J, Baibal T, Oukhattar L, Soukri A, Seixas P, Rey-Mendez M (2011) Multiple paternity in the common octopus *Octopus vulgaris* (Cuvier, 1797), as revealed by microsatellite DNA analysis. Moll Res 31:15–20

Sakurai Y, Bower JA, Nakamura Y, Yamamoto S, Watanabe K (1996) Effect of temperature on development and survival of *Todarodes pacificus* embryos and paralarvae. Am Malacol Bull 13:89–95

Salman A, Akalin M (2012) A rare pelagic cephalopod *Ocythoe tuberculata* (Octopoda: Argonautoidea): the record fecundity for Octopoda and new data on morphometry. Turk J Fish Quat Sci 12:339–344

Sato N, Kasugai T, Ikeda Y, Munehara H (2010) Structure of the seminal receptacle and sperm storage in the Japanese pygmy squid. J Zool 282:151–156

Saunders WB, Ward PD (1987) Ecology, Distribution and Population Characteristics of *Nautilus*. In: Saunders WB, Landman NH (Eds) Nautilus: the biology and paleobiology of a living fossil. Plenum Press, New York, pp. 137–162

Seibel BA, Robison BH, Haddock SHD (2005) Post-spawning egg care by a squid. Nature 438:929

Vidal EAG, DiMarco FP, Wormuth JH, Lee PG (2002) Influence of temperature and food availability on survival, growth and yolk utilization in hatchling squid. Bull Mar Sci 71:915–931

Villanueva R, Norman MD (2008) Biology of the planktonic stages of benthic octopuses. Oceanog Mar Biol Ann Rev 46:105–202

Villanueva R, Staaf DJ, Argüelles J, Bozzano A, Camarillo-Coop S, Nigmatullin CM, Petroni G, Quintana D, Sakai M, Sakurai Y, Salinas-Zavala CA, De Silva-Dàvila R, Tafur R, Yamashiro C, Vidal EAG (2012) A laboratory guide to *in vitro* fertilization of oceanic squids. Aquaculture 342–343:125–133

Voight JR (2001) The relationship between sperm reservoir and spermatophore length in benthic octopuses (Cephalopoda: Octopodidae). J Mar Biol Ass UK 81:983 986

Voight JR, Feldheim KA (2009) Microsatellite inheritance and multiple paternity in the deep-sea octopus *Graneledone boreopacifica* (Mollusca: Cephalopoda). Invert Biol 128:26–30

Wood JB, Kenchington E, O'Dor RK (1998) Reproduction and embryonic development time of *Bathypolypus arcticus*, a deep-sea octopod (Cephalopoda: Octopoda). Malacologia 39:11–19

Young RE, Harman, RF (1988) "Larva", "Paralarva", and "Subadult" in cephalopod terminology. Malacologia 29:201–207

Chapter 2
Behaviour

Jennifer Mather and David Scheel

Abstract The coleoid cephalopods, typified by cuttlefish, squid, and octopuses, are carnivorous molluscs. Of the better-known coastal cephalopods, many live in shallow water, are short-lived, physiologically efficient, and nocturnal. The behaviour of cephalopods overall is poorly known: a basic ethogram is available for one cuttlefish and one octopus species. Cuttlefish and squid eat primarily fish and crustaceans; the octopus diet includes crustaceans and molluscs. Most cephalopods prefer live natural food and prepared diets reduce growth compared to natural food. Cephalopods are generally solitary. A semelparous life history, parental care of eggs only in the octopuses (and a few squid) and no overlap of generations restrict the opportunity for social behaviour. Cuttlefish reproductive tactics may be complex, and squid swim with conspecifics in schools. Smaller cephalopods are at risk for cannibalism from larger ones (common in 59% of species and high in 24%). Cephalopods use visual displays, and some have size-based social hierarchies in captivity. Coastal coleoid cephalopods grow rapidly (live only 1–2 years), mature at an early age, and many die shortly after laying eggs. Many species of squid and cuttlefish aggregate for spawning, while male octopuses locate receptive females by chemoreception. Most young hatch at a small size, are planktonic, and must hunt live appropriate-sized prey. Major challenges to mariculture include keeping tiny planktonic paralarvae alive, providing adequate diet for growth, and avoiding cannibalism within high density captive populations.

Keywords Behaviour · Cephalopods · Life history · Diet · Antipredator · Reproduction

J. Mather (✉)
University of Lethbridge, Lethbridge, AB, Canada
e-mail: mather@uleth.ca

D. Scheel
Alaska Pacific University, Anchorage, AK, USA
e-mail: dscheel@alaskapacific.edu

J. Iglesias et al. (eds.), *Cephalopod Culture*, DOI 10.1007/978-94-017-8648-5_2,

Table 2.1 Aspects of the behaviour of the three major cephalopod groups[a]

Behaviour	Cuttlefish	Squid	Octopus
Habitat choice	Benthic, semi-benthic	Pelagic	Benthic
Sensory	Vision, chemical	Vision, mechanical	Vision, chemical
Locomotion	Jet	Jet	Crawl/jet
Antipredator	Camouflage, hiding, ink	Jet escape, ink	Hiding, camouflage, ink
Skin displays	Camouflage, sexual	Sexual	Camouflage
Prey	Fish, crustacean	Fish	Mollusc, crustacean
Sociality	Social at reproduction	Social groups	Solitary
Care of young	None	None	Female egg brooding

[a] Many minor exceptions

2.1 Introduction

Behaviour is one of the most simple and obvious aspects of the life of any animal. But besides being easy to see, it has logical links to the ecology of animals' interactions with their surroundings and the physiology of their internal workings (Drickamer et al. 1996). Generating an environment that allows a captive cephalopod to express its normal range of behaviour, such as providing shelter for cuttlefish and octopuses, and space for squid, is important for their health and competence. But behaviour can also be used as a cue of trouble. Unusual coloration, poor muscle tone, change in feeding behaviour, or location in an atypical place or position can tell the people keeping the cephalopods that something is wrong.

There are two problems with describing the behaviour of 'the cephalopod'. The first one is that there is a large number of species, normally inhabiting and adapted to a wide variety of marine habitats (see Table 2.1). The abundant coleoid cephalopods are typified by cuttlefish, squid, and octopuses, many of which live in shallow water, are short-lived, semelparous, and physiologically efficient though not necessarily active more than 25 % of the time (Mather 1988). The few nautiloids, represented by *Nautilus* and *Allonautilus* (see Chap. 10), are long-lived iteroparous inhabitants of the cold depths.

Some cephalopods, such as the octopuses, seem 'preadapted' to confinement (Boycott 1954), occupying a sheltered 'home' for around 70 % of the time (Mather et al. 1985; Mather 1988). In contrast, the jet-propelled squid in their normal pelagic environment are typified by *Sepioteuthis sepioidea*, who make a startle jet escape at an average of eight times per hour, moving a mode of 2 m distance (Mather 2010), a difficult response in a laboratory environment.

A second problem with describing the behaviour of cephalopods is that it is poorly known, though see Hanlon and Messenger (1996) for an excellent but somewhat dated summary. A basic ethogram (description of the common behaviour of a species) can only be pieced together for *Octopus vulgaris*—see Wells' (1978) and Mather et al.'s (2010a) popular books—and *Sepia officinalis* (see Guerra 2006), though the behaviour of the latter has not been studied in the wild. Squid and cuttlefish species often gather to mate and their agonistic and sexual behaviour is fairly well known (Hall and Hanlon 2002; Hanlon et al. 2002; Jantzen and Havenhand

2003a, b) but descriptions of the rest of their behavioural repertoire are lacking. Many cephalopods are nocturnally active and their activity cycle is not flexible (Meisel et al. 2006), though again *O. vulgaris* is the exception. The behaviour of nocturnal animal is difficult to study, as shining a light on them can change their behaviour and few scientists have a night-viewing system that works in the water.

Nevertheless, the behaviour of cephalopods can reveal basic adaptations to the necessities of life. The first necessity is adapting to the physical, chemical, and biological environment around them, including predators. Although predators may be absent in a laboratory environment, antipredator reactions such as inking will still be present. The second necessity is provision of food, and although the laboratory environment may provide food easily, there are the problems of appropriate diet, food preparation, and enrichment with food provision for young. A third necessity is relationship with conspecifics; although many of the cephalopods are not social (Boal 2006), this can bring problems for confinement itself, such as cannibalism (Ibáñez and Keyl 2010). A fourth necessity is provision of opportunities for reproduction, from the stage of mating through guarding of eggs in octopuses to the provision of a proper environment for planktonic paralarvae (see Chap. 23). Understanding these problems can lead to providing an enriched environment for cephalopods in captivity. Enrichment of the captive environment is not just window-dressing for aquarium audiences (Peters et al. 2005). Cuttlefish raised in an enriched environment grow faster and learn better (Dickel et al. 2000) and habitat enrichment increases the activity and widens the range of skin patterns in mudflat octopuses (Biegel and Boal 2006). This is particularly important for development, as young cuttlefish given a complex background learn to produce skin patterns and dig in the sand more quickly (Poirier et al. 2004, 2005). Taking these aspects of biology into consideration, Moltschaniwskyj et al. (2007) have written a thorough review of ethical and welfare considerations when keeping cephalopods.

2.2 Antipredator Behaviours

Without the protection of the molluscan shell, cephalopods are at risk for predation, particularly from fish (Packard 1972). They have evolved a galaxy of antipredator responses, including skin displays mediated by the complex chromatophore system (Messenger 2001). Excellent bottom-matching camouflage (Hanlon and Messenger 1988) is a specialty of the cuttlefish and this matching has been used to investigate the ability of the cephalopod visual system to analyse its environment (e.g. Barbosa et al. 2008). Young nektonic squid such as *S. sepioidea* use a variety of patterns accompanied by postures to match features of their environment (Moynihan and Rodaniche 1977; Mather et al. 2010b). In the open water, many squid and cuttlefish use reflexive countershading for camouflage from predators above or below (Ferguson et al. 1994). On the approach of a potential predator, cephalopods of all groups use startle deimatic displays with eyespots (Langridge et al. 2007; Mather 2010; Staudinger et al. 2012). These are selectively used in the presence of visual predators (Langridge et al. 2007) and are directed towards an approaching potential predator (Mather 2010), though not to an imminent danger. Octopuses (*Octopus cyanea*) escaping from a scuba diver following them may use unpredictable pattern changes (Hanlon et al. 1999b) and eventually escape. Similarly, young cuttlefish (Hanlon and Messenger 1988), *S. sepioidea* (Mather et al. 2010b), and *Euprymna scolopes* (Anderson and Mather 1996) use a combination of an unpredictable sequence of moves to different locations, pattern and posture change and ejection of a squid-sized 'dummy' dark blob (see Hanlon and Messenger 1996 for discussion).

The primary response of octopuses is hiding (Mather 1994). The provision of even a pair of bricks in the laboratory evokes this response, and an octopus not given such material may cower arms outward in the corner of a tank. Such a shelter can also be provided in the form of an open cube or a pot, which is very useful for transport of a captive animal with the minimum amount of handling stress. Given more material, however, octopuses will manipulate it to form a shape they prefer. In the wild, octopuses can hide in shells, human discards such as beer bottles (Anderson et al. 1999), crevices, and under rocks. They do not find the 'ideal' home in terms of characteristics such as volume and aperture area, but instead modify shelter by clearing out sand and rubble, detaching algal fronds, and bringing rocks to block the aperture (Mather 1994). Many octopus species are likely limited in the wild by the availability of shelter (see Hartwick et al. 1978 for *Enteroctopus dofleini*). Cuttlefish and other sepiolids (Mather 1986; Anderson et al. 2004) bury themselves in sand if they are provided with small grain size and an adequate depth. Squid primarily use jet-propelled escape responses (O'Dor and Webber 1986).

Most cephalopods have the ability to release ink during a major threat, and do so in combination with escape and appearance change. This may deter predators by either blocking their chemoreception (Wood et al. 2010) or impeding their vision. Ink can be dispersed in the sea as a screen or held together as a 'dummy' (Anderson and Mather 1996). Ink can also become an alarm cue to conspecifics (Wood et al. 2008). Given its efficacy, it is surprisingly rarely used, but a replacement must be metabolized for future threats if ink is lost by ejection. Increased or maximal initial threat eventually causes escape responses in most cephalopods. In the laboratory, many of the display responses are harmless and seldom seen; eyespots seem to be the exception. But jet escape responses can lead to posterior mantle skin damage (Hanlon et al. 1983) and ink circulating in a restricted tank environment can damage the health of cephalopods and any other animals in the same system.

2.3 Foods and Feeding Behaviours

All cephalopods are carnivores, and most exhibit a preference for live natural foods. These factors have constrained the aquaculture of cephalopods and are part of the current challenge of improving aquaculture methods, particularly for early life stages where mortality is the highest (e.g. Vaz-Pires et al. 2004; Sykes et al. 2006; Uriarte et al. 2011). In the wild, cephalopods may forage solitarily, as cuttlefish and octopuses typically do, or in aggregate, as may squid (Neill and Cullen 1974; Boal 2006), although many schooling squid forage solitarily at night. The coleoid cephalopods are visually guided predators, although octopuses may also forage tactilely (for review, see Hanlon and Messenger 1996). For many cephalopods, hard remains of prey—from stomach contents for squid and from midden piles outside the den for octopuses—provide information about diet composition. Diet remains much less well known from deep water species, away from dens, or from soft prey whose remains are not found in middens or stomach, although that is changing as information from molecular methods becomes available (e.g. Lorrain et al. 2011). However, in such cases, molecular data have confirmed that descriptions of diet content from hard remains include the majority of cephalopod prey (e.g. Stowasser et al. 2006; Hunsicker et al. 2010; Lorrain et al. 2011), although some molecular methods may provide greater temporal detail (e.g. Hunsicker et al. 2010).

2.3.1 Cuttlefish (Order Sepiida) and Squid (Order Teuthida)

The main foods of cuttlefish and squid are shrimp, crab, and fish, but they also are known to consume other crustaceans (euphasiids, copepods, cirripedes, amphipods), some molluscs (including gastropods, bivalves, and cephalopods), as well as polychaetes (Hanlon and Messenger 1996). Knowledge of squid-feeding ecology is limited (Lorrain et al. 2011). Squid are active predators whose total dietary composition is broad and varies with growth. Among *Illex argentinus* individuals, 55–85 % included crustaceans in their diet, mainly *Themisto gaudichaudii* (order Amphipoda) and euphasiids, while the occurrence of squid was 12–13 %, and 3–29 % of squid individuals consumed fish (Ivanovich and Brunetti 1994). Similarly, diets of common Atlantic squid *I. illecebrosus* and *Doryteuthis* (*Loligo*) *pealeii* were diverse, and while that of *I. illecebrosus* was dominated by fishes, squid, and non-decapod crustaceans, the stomach contents of *D. pealeii* consisted of largely unidentified prey remains, molluscs, and copepods (Bowman et al. 2000). Cannibalism occurs commonly and is increased by crowding (Ivanovich and Brunetti 1994; Ibáñez and Keyl 2010). Many squid are harvested by jigging (e.g. Chen et al. 2008) possibly indicating attacks to a wide range of stimuli that may contribute to a broad diet.

Both cuttlefish and squid attack prey in stages consisting of attention behaviours, a positioning approach, and ambush-like strike in which the long tentacles shoot out to seize the prey (Messenger 1968; Neill and Cullen 1974). The attention behaviours involve changes in body patterning, arm positioning, and orientation of the body towards the prey (Messenger 1968). For slower moving prey (such as crabs) of both squid and cuttlefish, approach occurs only from behind the prey, and seizure occurs with partially opened arms without use of the tentacles (Duval et al. 1984).

There is little work on the behavioural ecology of squid and cuttlefish diet choice, and these groups are considered predatory generalists (e.g. Guerra 2006). However, cuttlefish have been an important invertebrate model system to understand learning in the context of the ontogeny of hunting behaviour and prey choice (e.g. Messenger

1973; Dickel et al. 2000; Darmaillacq et al. 2004; Cole and Adamo 2005; Agin et al. 2006; Guibé et al. 2012). Within 24–48 h after hatching, cuttlefish strike at small suitable prey (Messenger 1973); attack latency, success rates, and suitability of potential prey items can be influenced by learning (e.g. Hanlon and Messenger 1996; Darmaillacq et al. 2004; Cole and Adamo 2005). Ontogenetic changes in prey selection also occur in squid, where adults may occupy a higher trophic level than smaller juveniles (e.g. Hunsicker et al. 2010; Ibáñez and Keyl 2010).

Squid depend on movement of prey items and contrast with the background to locate potential prey and elicit attack (Hanlon et al. 1983). Lacking these stimuli, especially motion, squid and cuttlefish will not attack and will starve to death. Prey size range does not appear critical, with both large and small squid taking prey ranging in size from macroplanktonic to nearly the same length as the squid itself (Hanlon et al. 1983). Squid orient visually towards their prey (Hanlon and Messenger 1996), as do cuttlefish, although cuttlefish also exhibit searching behaviour for partially buried prey in the substrate, which they uncover by expelling a jet of water over the substrate which can blow away cover (Hanlon and Messenger 1996). All squid are cannibalistic, a tendency enhanced under food shortage and in the presence of smaller or injured conspecifics. Further, mating behaviours such as male courtship of females and male–male aggression can disrupt feeding and may lead to injuries to fins or other areas that contribute to cannibalism (Hanlon 1990). Thus, feeding in captivity can be promoted and cannibalism reduced by keeping squid in tanks containing all individuals of the same size and sex.

2.3.2 Octopuses

The diets of a few species of octopuses are well known. Many species use dens in shallow water and discard hard remains of prey, including shells, carapaces, and bones, in midden piles outside the dens. The diet of shallow-water benthic octopuses is dominated by crustaceans (e.g. Mather et al. 2012) and molluscs such as bivalves (e.g. Vincent et al. 1998) and snails (Ambrose 1984). Diets typically are dominated by one to a handful of prey species, but diverse other prey occur occasionally in diets (e.g. Ambrose 1984; Scheel and Anderson 2012). Such occasional items may include alternative crustaceans or bivalves, but also almost any other prey group characterized by hard parts such as gastropods, chitons, cephalopods, echinoderms, fish, and even birds (Sazima and de Almeida 2006; Nightingail 2012; Scheel and Anderson 2012). Soft items (e.g. worms) may also be eaten but are less likely to be detected in the diet. In some populations, individual octopuses appear to exhibit prey specificity (Anderson et al. 2008b), although diet composition across the population is broad; in other populations, individual specialization appears not to be the rule (Mather 2011; Mather et al. 2012; Scheel and Anderson 2012; Leite et al. in sub).

Hard shells protecting crabs, bivalves, and other prey represent a challenge to octopuses' intent on feeding on the soft tissue inside, and octopuses have several methods to surmount this challenge. Marks left by octopuses on hard remains of prey may indicate how the octopus handled the food item (*E. dofleini*: Dodge and Scheel 1999). Octopuses are known to apply strength alone to pull open prey (McQuaid 1994; Steer and Semmens 2003; Anderson and Mather 2007) so that not all remains from prey eaten by octopuses will be marked. Pulling may be the method of choice for some prey, especially bivalves (Steer and Semmens 2003), but pulling is not always successful. Octopuses are also well known for drilling through the shell of their prey, using salivary enzymes, papilla, and radula (Nixon and Maconnachie 1988). Drilling forms a small ovoid depression in the outer layer of the shell that penetrates the inner layer with holes 1.5–3.0 mm long and 0.25–2.0 mm wide (inner and outer width dimensions, respectively; Nixon and Maconnachie 1988; Dodge and Scheel 1999). Octopuses are known to drill hard shells and carapaces, as well as puncture softer tissue such as eyes (*Eledone cirrhosa*: Grisley et al. 1996). In some species of octopuses attacking bivalves, drill attempts consistently may be made in a location where the prey is vulnerable (Wodinsky 1969; Cortez et al. 1998; Anderson et al. 2008a). However, this does not appear universal and some octopus-drilled locations are variable or may be a function of prey type or size (*E. dofleini*: Dodge and Scheel 1999; Scheel et al. 2007). Alternatively, octopuses may chip prey with their beaks (Anderson 1994; Dodge and Scheel 1999) leaving characteristic marks on the edges of bivalves or breakage patterns on crab carapaces, chelae, or legs.

According to foraging theory (Pulliam 1974; Sih 1984; Stephens and Krebs 1986), diet selection by rate-maximizing foragers among spatially mixed prey types will be determined by prey energy content, handling time, and encounter rates. Spatial segregation of prey types may result in habitat selection influencing encounter rates and prey selection (Vincent et al. 1996). Alternatively, octopuses may act as risk-minimizing or time-minimizing foragers (Scheel et al. 2007; Leite et al. 2009). If so, this could result in a preference for larger prey (greater energy content) without regard for handling time. This seems particularly likely given the octopuses often consume food at a den or other shelter (Mather 1991a), and spend the majority of their time hiding (e.g. *O. vulgaris*: Mather 1988, *E. dofleini*: Scheel and Bisson 2012), which allows time to drill, chip, or pull open prey in safety (see Sect. 2.2). Scheel et al. (2007) and Scheel and Anderson (2012) found that *E. dofleini* exhibit a preference for larger prey individuals within a species, and for larger species among similar prey types (crustaceans). Preferences by this octopus species may further be influenced by detectability of the prey, possibly itself a function of prey camouflage behaviour, epiphytes, or escape responses. There is now a growing interest in examining the constraints octopuses may face in nutrient trade-offs (e.g. Lee 1994; Rigby and Sakurai 2004; Onthank and Cowles 2011) and their effects on diet, but research in this area is only beginning.

Few studies examine how octopuses choose where to forage or how they find food. *O. cyanea* conducts tactile, speculative, and saltatory foraging on shallow-water reefs (Yarnall 1969, Forsythe and Hanlon 1997), similar to speculative web-over foraging described for *E. dofleini* in shallow water (Johnson 1942, Cosgrove 2002). In both species, as well as *O. vulgaris*, foraging excursions may follow along habitat edges such as reef edges or cliff faces preferentially in habitats containing suitable dens or other shelter (Mather 1991b; Forsythe and Hanlon 1997; Scheel and Bisson 2012; Smith 2012). Different species may forage at intervals from several times per day (*E. dofleini*: Mather et al. 1985; *O. vulgaris*: Mather 1991a) to every 2–3 days (*E. dofleini*: Scheel and Bisson 2012). Foraging trip lengths are generally small, extending no more than 30–57 m from a den for large species (*O. cyanea*: Ivey 2007; *E. dofleini*: Scheel and Bisson 2012) and to only 6 m for *O. vulgaris* (Mather 1994).

Octopuses and cuttlefish will take prepared foods, and there has been some effort to create prepared diets of both natural (e.g. shrimp, squid) and terrestrial (e.g. chicken) foods for rearing cephalopods in captivity (Lee et al. 1991; Lee 1994; Garcia et al. 2011; Rosas et al. 2011), with more recent work showing increased promise for solving this difficult problem. Although both cuttlefish and octopuses grew on prepared diets, in all cases growth rates were below those on natural foods, and in many cases cuttlefish and octopuses on prepared diets did not grow (Lee et al. 1991; Domingues et al. 2007; Domingues et al. 2008; Valverde et al. 2008; Rosas et al. 2011). Cuttlefish eating prepared foods exhibited lower assimilation efficiencies than on natural foods (Rosas et al. 2007), possibly in part due to interference with digestion by binders used in prepared foods (Rosas et al. 2008; Garcia et al. 2011). Live foods, especially for early developmental stages, continue to have more success than prepared diets. However, both cuttlefish and octopuses can readily be trained to take minimally processed but nonliving marine foods, such as fresh or frozen fish, squid, or shrimp (e.g. Koueta et al. 2006). Octopus growth rates may be higher on mixed than monotypic diets (Rigby and Sakurai 2004), and nutrition ratios may be important in maximizing growth (Lee 1994; Aguila et al. 2007; Onthank and Cowles 2011) although this remains poorly understood.

2.4 Nonsexual Social Interactions

A generalization that cephalopods are solitary outside of the reproductive period would be true, with small exceptions and some variation (Boal 2006). With a semelparous life history (see below), the presence of parental care only in the octopuses and a fairly complete lack of overlap of generations, cephalopods are not likely to have been selected for social behaviour. Cooperation would be selected for by three different mechanisms (Drickamer et al. 1996). One is mutualism, the benefit of interactions to both individuals, which is unlikely unless animal live in close proximity, like the squid. A second is kin selection, again unlikely when planktonic dispersal means that most cephalopods are not living near kin. A third is reciprocity, where one individual benefits another, expecting a future benefit, and again this is unlikely in animals such as cephalopods that avoid one another much of their lifetime. Nevertheless, social recognition occurs in some cephalopods (Boal 2006) and other social behaviours may occur but have yet to be carefully studied.

Outside of the reproductive period, cuttlefish may be solitary, although field data (Corner and Moore 1980) are fragmentary, and reproductive social tactics may have been a selective force for cuttlefish intelligence (Brown et al. 2012). A size-based dominance hierarchy is true for older, though not newly hatched cuttlefish (Warnke 1994). Boal (1996) tackled the problem of social recognition through laboratory observations. She found that cuttlefish kept together in tanks did not react differently to familiar vs unfamiliar conspecifics and were not closer to familiar than unfamiliar ones. Given visual stimuli, they were more aroused (with a higher ventilation rate) by the sight of prey items than that of conspecifics (Boal and Ni 1996). When they were maintained in a small or a large tank, cuttlefish spaced themselves more widely in the large one and seldom approached to within two body lengths. In a small tank, there were more male agonistic zebra displays and more displacement of one animal by another (Boal et al. 1999). More importantly, cuttlefish in the large tank consumed 25 % more shrimp prey, suggesting that the crowding in the small (1.5 m diameter for three animals) tank was stressful and that this stress was reducing food intake. This is not good news for anyone who wishes to cultivate cuttlefish in captivity; a more complex habitat may provide visual separation and reduce this stress.

Squid are different, in that they prefer to swim with conspecifics (Hurley 1978), in approximately parallel orientation and within a body length or two in captivity, even in a very large tank, 15 m in diameter (Mather and O'Dor 1984). Because the adults do not live after egg laying and the young are planktonic, again there is little likelihood of kin recognition. Most squid gather with conspecifics, though Moynihan and Rodaniche (1982) noted *S. sepioidea* swimming with *Doryteuthis (Loligo) plei*. Groups sort by size; although *S. sepioidea* are attracted to conspecifics whatever the size, smaller animals are at risk for cannibalism from larger ones and so maintain several body lengths distance. It has been suggested that squid on the end of a line are sentinels, watching for predators and escaping first from them (Hanlon and Messenger 1996). The presence of sentinels would suggest cooperative behaviour, where individuals would assume periods of excess risk and trade off this risk for better protection outside of their sentinel time (Drickamer et al. 1996). However, Adamo and Weichelt (1999) found that such sentinel behaviour was not true for *S. lessoniana*. With predator threat (Mather 2010) or when schools are larger (Mather and O'Dor 1984), squid more closely to one another, possibly monitoring spacing through water deformation received through their lateral line analogue (Budelmann and Bleckmann 1988). Like cuttlefish, many squid have visual displays, including mostly male agonistic ones (Di Marco and Hanlon 1997). In captivity, males set up a dominance hierarchy, with larger ones dominating smaller and winning agonistic interactions. But, as for cuttlefish (Warnke 1994) and octopuses (Mather 1980), the presence of a dominance hierarchy in a crowded situation does not prove any social recognition in the wider spaces of the natural environment.

Octopuses are perhaps the most solitary of cephalopods, with their density likely dictated by a lack of predator pressure rather than any mutual attraction. Such a lack can be direct, as when predators are excluded from a specific area (Aronson 1986 for *Octopus briareus*), or indirect, when shelter is limiting (for *E. dofleini*, Hartwick et al. 1978). Octopuses observed in the wild do not defend territories (Aronson 1986; Mather et al. 1985), although they may defend their immediate surroundings, such as a sheltering home (Cigliano 1993). This lack of defence is accompanied by frequent moves from one to another small home range (Mather and O'Dor 1991). The exception may be *Abdopus aculeatus* (Huffard et al. 2008) which is known to gather at high densities in sea grass beds during reproduction. However, it is not known whether the animals are permanent or temporary occupants of these restricted areas. In the laboratory, octopuses may maintain dominance hierarchies based on size (Mather 1980). Cigliano's (1993) observations that with time, interactions may decrease, suggest recognition of this hierarchy. Tricario et al. (2011) have demonstrated the possibility of familiarity in *O. vulgaris*, but the flaws in the statistical analyses make it difficult to confirm.

The general lack of sociality in cephalopods may be one factor contributing to the frequency of cannibalism in the group. In their review of 34 species, Ibáñez and Keyl (2009) report cannibalism was common in 59 % of species reported and high in 24 %. The combination of little sociality with short lifespan, semelparous reproduction, and high metabolic rate may favour cannibalism. In the wild, it may contribute to population limitation, although for culture in the laboratory it is a major problem. In *I. illecebrosus* squid, cannibalism is found during high densities (O'Dor and Dawe 1998), and this is also true for *O. briareus* (Aronson 1983). It may be a response to a limited food supply, as in squid during migration (O'Dor and Dawe 1998), and some reports of cannibalism may be distorted due to stress during capture in fisheries. It is also size based, and where males and females are dimorphic in size, the larger sex may consume the smaller one.

Regardless of circumstances, cannibalism is a major problem for anyone keeping cephalopods in captivity. By definition, culturing animals means keeping them at high densities, which not only is stressful for them (Boal et al. 1999) but also creates a situation for consumption of animals by each other. To some extent, cannibalism can be avoided by keeping animals of the same size together, and since cannibalism is also partly dependent on food supply, it is useful to have a good supply of preferred food—which may be expensive and hard to procure (see discussion of food and feeding). Giving cephalopods a complex environment in captivity may maximize their ability to escape from conspecifics who would consume them, as well as increasing their growth and learning capacity (Dickel et al. 2000). Yet enrichment (Anderson and Wood 2001; Mather in press) is only a partial solution to the expression of normal behaviour in an unnatural environment.

2.5 Reproduction and Lifespan

2.5.1 Life History

Shallow-water coleoid cephalopods grow rapidly, mature at an early age, and are typically semelparous (but see below), dying shortly after laying eggs. Most cephalopods live only for 1–2 years (Boletzky 2003b; e.g. for *O. vulgaris*, Katsanevakis and Verriopoulos 2006b). Thus, abundance is limited by juvenile recruitment rates rather than adult survival.

Adult size and age vary greatly. The smallest ones live just a few months (e.g. 3 months for the pygmy squid *Idiosepius* spp., Boletzky 2003b), while the largest ones (e.g. *E. dofleini*: Hartwick 1983) are associated with delayed maturity in cold-water habitats and growth to large size (Wood and O'Dor 2000; Farías et al. 2009). There is a stronger relationship of cephalopod lifespan with temperature than with body size (Wood and O'Dor 2000). The giant Pacific octopus, *E. dofleini*, is a cold-water species and unusual in living up to 5 years in captivity (Hartwick 1983).

2.5.2 Movement

A thorough review of cephalopod movement behaviours is beyond the scope of this chapter (see Semmons et al. 2007) but movement ecology has the potential to affect husbandry. Long-distance movements may occur for most sepiolids. *S. officinalis* makes seasonal onshore–offshore migrations (Guerra 2006), likely due to winter cooling of the waters. Cuttlefish, with their cuttlebone-based buoyancy mechanism, can also perform daily vertical migration—upward at night for food and downward in the daytime (Webber et al. 2000). Migration may also be site focused, as *Sepia apama* in Australia makes long-distance migration to gather in restricted areas for breeding (Hall and Hanlon 2002).

Squid are the most mobile cephalopods (see possible tracking techniques, Semmens et al. 2007). Not only do many species move metres back and forth over the short term (Mather 2010), they also gather in daytime in the shallows and disperse over deeper waters at night to feed (Hanlon and Messenger 1996). But larger and more open-ocean species may make huge lifetime movements. *I. illecebrosus* gathers to feed off the Grand Banks in northeastern North America; the adults may move offshore to mate and spawn and their eggs are encased in a large gelatinous capsule and drift south to off the southeast coast. There the eggs hatch and the young begin a northward journey (O'Dor and Dawe 1998). Similarly, *Dosidicus gigas* move southwards off the coast of South America to feeding grounds, then slowly back to their equatorial spawning grounds (Nesis 1983), with the newly hatched young dispersing westwards and drifting southeast. With global warming, their range is expanding northwards. Animals with movement patterns as far-ranging and complex as these will be difficult to raise in the confines of even a large laboratory setting.

While octopuses generally are amenable to confinement within a small environment when provided with a sheltering 'home', they eventually will attempt to leave. This may be tied to the octopus' short-term occupation of small home ranges (Mather and O'Dor 1991; Scheel and Bisson 2012), as they only stay in a restricted area for days or weeks. Thus, there is a 'laboratory lore' of how to keep your octopus in the tank. Heavy weights on the lid, locking lids, and outdoor carpeting around the rim of the tank may act as deterrents. The likelihood of escape is somewhat different amongst species that have been kept in captivity, with *O. vulgaris* being the most likely one to escape (Wood and Anderson 2004). It is suggested though not proven that an enriched environment, more space, and better nutritional state may keep the octopus confined. Still, some of them just want to leave. In the longer term, octopuses may make onshore–offshore migrations, perhaps for reproduction or to access a more appropriate food supply, but this may be irrelevant to their culture unless light level affects their reproduction.

2.5.3 Semelparity

Coastal cephalopods typically mature in about a year (Boyle and Boletzky 1996) and although many are semelparous, Rocha et al. (2001) have identified five different cephalopod reproductive strategies along a gradient in environmental stability, including four varieties of multiple spawning, three of which are nonterminal. Recognizing this complexity in cephalopod spawning behaviours has blurred the sharp distinction between semelparity and iteroparity. Cuttlefish lay individual or small groups of eggs and sequester them in protected spaces. They spawn intermittently over a period of several months. The squid deposit eggs in large egg masses, some seasonally in mass synchronized spawning events, and are typified by terminal spawning although there are some examples of nonterminal spawning (Rocha et al. 2001). According to Rocha et al. (2001), octopuses span the continuum between semelparity and iteroparity: Many shallow-water octopuses are small-egged species laying eggs in festoons or small clusters of a single spawning event, while deep-water octopuses such as *Graneledone* sp. but also some shallow-water octopuses (Anderson and Wood 2012) are large-egged species depositing eggs singly, perhaps over a protracted period. *E. megalocyathus* is an intermediate example (Ortiz 2006). In either case, females brood eggs over a long period and die after hatching of the young.

2.5.4 Mating and Spawning

Male cuttlefish court females, using multiple mating strategies (Norman et al. 1999), and deceive other males via female mimicry (Norman et al. 1999; Brown et al. 2012). Mass spawning is seen in the giant Australian cuttlefish, *S. apama* (Hall and Hanlon 2002). During these mass spawnings, both sexes have multiple mates; males defended females but not egg-laying sites, and males attempted takeovers of paired females through agonistic displays and attention to opportunities for extra-pair copulations via stealth or female mimicry (Hall and Hanlon 2002). Many aspects of this interaction, but not mass spawning, are typical of Sepiida.

S. officinalis males present a striped body pattern (intense zebra display) to other males; larger and darker males are dominant in such interactions (Boal 1997). However, females show consistent preference not for the larger or darker males, but for the most recently mated males and those showing fewer zebra displays (Boal 1997). She suggested that females used chemical cues rather than visual ones in assessing males. In captivity, males may initiate copulation without obvious courtship (which females may avoid through escape responses), and male–male aggression occurs. The extent to which these behaviours are artefacts of spatial constraint in captivity is not known (Adamo et al. 2000) and may pose challenges for cultivation. Hanlon et al. (1999a) observed that males initiate mating in the head-to-head posture, and then direct jets of water on at the female's buccal membrane, likely to flush spermatangia placed there by previous mating, and then transfer their own spermatangia to the buccal membrane using the hectocotylus. The male then manipulates the spermatangium on the female to break it open so that sperm are released. Females appear to terminate mating and are then guarded briefly by the male.

Some squid species (e.g. *Loligo vulgaris*: Arnold 1990; *I. illecebrosus*: Hendrickson 2004) but not others (e.g. *D. gigas*: Nigmatullin et al. 2001) aggregate for spawning. Spawning grounds, e.g. for *L. reynaudii*, may be used repeatedly, spawning may not be synchronized within an entire population, and squid move between grounds over an extended period (Sauer et al. 2000). Nonetheless, spawning aggregations are harvested and support managed fisheries (Rodhouse 2001). In captivity, *L. vulgaris* mating behaviour and male–male aggression may be induced by the presence of a recently laid egg mass (or visually similar object) in the tank or even in the visual field (Arnold 1990), as well as by pheromones present in the egg mass (King et al. 2003; Cummins et al. 2011). Arnold (1990) observed that on detecting such an object, individual sexually mature squid investigate the object tactilely and may jet water at it (possibly an effort to flush away sperm from other males, see 'Cuttlefish' above). Males begin to dart about, display to other males, and place themselves between females and rival males. Females and males display to each other, accentuating oviducal gland and testes, respectively (*L. reynaudii*, Hanlon et al. 2002). A male will swim alongside a female and raise one or two medial arm in an S-shaped curved display posture. Dark bands or patches also feature in this display, especially in competition with rival males, who may also be chased (Byrne et al. 2003; Mather 2004). Male–male contests may also include physical contact such as fin beating (Hanlon et al. 2002). Social hierarchies determined via dominance in agnostic displays develop in captivity (Arnold 1990), while paired males have an advantage over intruder males in the wild (Hanlon et al. 2002), an effect due to female choice, as females jet to avoid unwanted male mating attempts. Copulation of paired *L. vulgaris*, described by Arnold (1990) and of *L. reynaudii* described by Hanlon et al. (2002), is preceded by the male positioning himself alongside but slightly below the female, flashing chromatophores. The male then grabs the female and positions his arms close to her mantle opening. He reaches into his mantle with his hectocotylus and picks up spermatophores, which are quickly ejaculated and cemented to the inside of the female mantle near the opening of the oviduct. The male then releases the female. Copulatory behaviour may be interspersed with egg laying, and newly released sperm may be observed on just-laid egg masses. Female choice may operate on several levels, including female manipulation of sex ratios, avoidance of mating attempts, and selection of stored sperm to fertilize eggs (Hanlon et al. 2002), leading to multiple paternity within egg strings (Shaw and Sauer 2004). Copulation alternated with egg laying will continue until both sexes are exhausted. Squid die within hours. Sneaker males of *L. reynaudii* (Hanlon et al. 2002), as well as other squid (e.g. *D. gigas*: Nigmatullin et al. 2001) and cuttlefish (Hanlon et al. 1999a) mate in the head-to-head position. Mating behaviour may be variable and dependent on context (Hanlon et al. 2002; Jantzen and Havenhand 2003a). Thus, there are opportunities to manipulate the onset of spawning under cultivation, although courtship and mating behaviours may be inseparable from male–male aggression.

Male octopuses may use chemoreception to locate receptive females (Voight 1991a; Di Cristo et al. 2005). Populations may have a single (Ambrose 1988) or two (Katsanevakis and Verriopoulos 2006b) spawning (and hence recruitment) peaks per year, with post-settlement mortality rates >50% within 3 months. Octopuses are not territorial and males apparently do not guard receptive females (Voight 1991b; but for exception see Huffard et al. 2010). During precopulatory behaviours, the male approaches the female and extends the hectocotylus towards her. If not rebuffed, males touch the female with the ligula (which females may use in assessment of male maturity), and if receptive, females remain motionless (e.g. *Paroctopus digueti*: Voight 1991b). In some species, females also may approach males (in captivity, Cheng and Caldwell 2000), and males may not always distinguish between male and female potential mates, or even among species (Lutz and Voight 1994; Cheng and Caldwell 2000). Copulation then occurs in positions varying from at arm's length to the male engulfing the female beneath his web. The male inserts the hectocotylus into the female mantle cavity. Mating may last up to several hours and is typically terminated by the female (e.g. *Octopus joubini*: Mather 1978; *P. digueti*: Voight 1991b; *Hapalochlaena lunulata*: Cheng and Caldwell 2000). In at least some species, more complex male mating strategies may involve sneaker males, mate guarding, and competition (Huffard et al. 2008). Thus, aspects of mating, including male initiation of mating and mating technique, may depend on crowding or familiarity with territory (Mather 1978), and so mating under cultivation may differ from in the wild.

2.5.5 Egg-Care Behaviours

Cuttlefish selectively deposit their eggs on support, such as in crevices among appropriate corals (Boletzky 1998). Squid deposit their eggs, often en masse, on or under appropriate bottom substrate, such as in sand, under rocks, or on sessile organisms, anchor lines or other support, or in pelagic balloons (Oegopsids). Except for family Gonatidae and Bathyteuthida squid (Bush et al. 2012), neither squid nor cuttlefish are known to exhibit egg-brooding behaviours.

Most small-egged octopus species cement festoons of eggs to sheltered hard substrate, such as inside the roof of a den. Eggs are laid over several days. Females typically stop eating just before or about the time of egg laying, although this is not invariant. Females then tend eggs, manipulating each strand with arm tips and blowing jets of water across them. Dead eggs are removed, as may be any showing signs of fungi or algae (Batham 1957; Gabe 1975; Anderson and Wood 2012).

Females brood the eggs continuously and become increasingly pale during this period. When the eggs begin to hatch, the female may aid the process by manipulating the eggs. After paralarvae hatch, or in some cases before, when the female is completely spent, she will crawl out of her den and a short distance away to die. When a female dies prior to paralarvae hatching, her eggs may be quickly predated or overcome by algae and fungi.

2.5.6 Hatching and Paralarvae

Cuttlefish hatch as juveniles, without distinct paralarvae. Immediately after hatching, cuttlefish begin to learn to capture size-appropriate prey in a fashion similar to adults (e.g. Hanlon and Messenger 1988; Guibé et al. 2012). Cuttlefish hatch with fully differentiated tentacles and use these for prey capture from the start. In contrast, Teuthid squid also have no true larval form but some species do have an ecological and morphological paralarvae stage wherein they do not yet assume the adult form and lifestyle (Boletzky 2003a). Social interactions, such as schooling of adult Loliginid squid, do not occur in newly hatched cephalopods. Loliginid squid begin schooling several weeks after hatching and this continues into adult life (Boletzky 2003a). Like cuttlefish, Loliginid squid begin learning attack behaviours by experience, foraging immediately post hatching (Boletzky 2003a; Darmaillacq et al. 2008) but prey choice is partially cued prior to hatching. However, because Loliginid squid do not hatch with fully formed tentacles, their prey capture methods must change during ontogeny. For both of these cephalopod groups, hatchlings learn to position themselves appropriately to attack prey from a direction most likely to yield success, and to adopt different attack behaviours depending on the availability and density of both prey and competitors (Boletzky 2003a).

Octopus hatchlings will still have a remnant of the yolk sac attached that may sustain them for the first few days out of the egg. As for squid, hatchlings that do not yet have the adult form and lifestyle are termed paralarvae (Young and Harman 1988). These are planktonic, positively phototaxic (causing them to swim upwards in the water column), and neustonic (orienting towards the water–air interface); at the ocean's surface they hunt planktonic crustaceans and molluscs appropriate for their size. Prey are located visually (Villanueva et al. 1996) and must be provided at sufficient density and frequency to facilitate growth. Adequate nutrition for paralarvae of meroplanktonic octopus species poses one of the greatest challenges to cultivation. Paralarvae grow rapidly and move about by jetting (Villanueva et al. 1995). As they approach the size at which they settle, they may cling to floating objects (Nixon and Mangold 1996). Recruitment of meroplanktonic species to the benthic population is heavily influenced by oceanographic parameters including temperature influencing growth rate, and current patterns influencing retention and recruitment to the source population (Katsanevakis and Verriopoulos 2006a). Octopus hatchlings with adult form and lifestyle are termed juveniles, and these occur in the large-egged forms (Voight and Grehan 2000; Anderson and Wood 2012), many of which are found in deep sea. They adopt a

Table 2.2 Signs of problems with cephalopods in captivity

Cuttlefish	Semi-solitary
(Provide substrate for digging)	Not feeding
(Hiding in sand, daytime)	Lack of skin colour
	Not seeking shelter
	Abrasions on posterior, fin edges
	Poor posture
	Unresponsive to startle
	Buoyancy problems (floating)
Squid	Social
(Watch for escape response, jet collisions with tank)	Not feeding
	Lack of skin colour
	Separation from group
	Settled on substrate
	Abrasions on posterior, fins
	Unresponsive to startle
	Jerky, hesitant swimming
Octopod	Solitary
(Manipulation of substrate is normal)	Not feeding
(Prevent escape)	Lack of skin colour (except in sleep)
	Poor muscle tone (except in sleep)
	End of arms missing (not when previously acquired)
	Not seeking shelter (provide it)
	Unresponsive to startle

benthic lifestyle immediately, but also grow rapidly, feeding on crustaceans and other small prey (Forsythe 1984).

2.6 Conclusions

This chapter reveals how complex and poorly understood are the behaviours of cephalopods. Yet, despite this, knowledge of this natural behaviour can serve as a foundation for successful culture of different cephalopod species. Even this partial account gives details of limitations, suggestions for how to assure that animals thrive in captivity, and, perhaps most importantly, an indication of which behaviours to monitor to assure oneself that the animals are doing well or warning signs that they are in trouble (see Table 2.2). The major obstacles to successful cultivation remain keeping planktonic paralarvae alive, providing adequate diet for growth, and avoiding cannibalism within high density captive populations. Progress has been made in providing adequate diet, but no satisfactory solution is yet available on even this minimum requirement for successful mariculture.

References

Adamo SA, Weichelt KJ (1999) Field observations of schooling in the oval squid, *Sepioteuthis lessoniana* (Lesson, 1830). J Moll Stud 65:377–380

Adamo SA, Brown WM, King AJ et al (2000) Agonistic and reproductive behaviours of the cuttlefish *Sepia officinalis* in a semi-natural environment. J Moll Stud 66:417–419

Agin V, Chichery R, Chichery M-P et al (2006) Behavioural plasticity and neural correlates in adult cuttlefish. Vie Milieu 56(2):81–87

Aguila J, Cuzon G, Pascual C et al (2007) The effects of fish hydrolysate (CPSP) level on *Octopus maya* (Voss and Solis) diet: digestive enzyme activity, blood metabolites, and energy balance. Aquaculture 273(4):641–655. doi:10.1016/j.aquaculture.2007.07.010

Ambrose RF (1984) Food preferences, prey availability, and the diet of *Octopus bimaculatus* Verrill. J Exp Mar Biol Ecol 77:29–44

Ambrose RF (1988) Population dynamics of *Octopus bimaculatus*: influence of life history patterns, synchronous reproduction and recruitment. Malacologia 29(1):23–39

Anderson RC (1994) Octopus bites clam. Festivus 25(5):58–59

Anderson RC, Mather JA (1996) Escape responses of *Euprymna scolopes* Berry 1912 (Cephalopoda: Sepiolidae). J Moll Stud 62:543–545

Anderson RC, Mather JA (2007) The packaging problem: bivalve prey selection and prey entry techniques of the octopus *Enteroctopus dofleini*. J Comp Psychol 121(3):300–305

Anderson RC, Wood JB (2001) Enrichment for giant Pacific octopuses: happy as a clam? J Appl Anim Welf Sci 4:157–168

Anderson RC, Wood JB (2012) Raising baby octopuses. Drum Croaker 43:34–40

Anderson RC, Hughes PC, Mather JA et al (1999) Determination of the diet of *Octopus rubescens* Berry, 1953 (Cephalopods, Octopodidae) through examination of its beer bottle dens in Puget Sound. Malacologia 41:455–460

Anderson RC, Mather JA, Steele CW (2004) Burying and associated behaviors of *Rossia pacifica* (Cephalopoda: Sepiolidae). Vie Milieu 54:13–19

Anderson RC, Sinn DL, Mather JA (2008a) Drilling localization on bivalve prey by *Octopus rubescens* Berry, 1953 (Cephalopoda: Octopodidae). Veliger 50(4):326–328

Anderson RC, Wood JB, Mather JA (2008b) *Octopus vulgaris* in the Caribbean is a specializing generalist. Mar Ecol Prog Ser 371:199–202

Arnold JM (1990) Squid mating behavior. In: Gilbert DL, Adelman WJJ, Arnold JM (eds) Squid as experimental animals. Plenum Press, New York, pp 65–76

Aronson RB (1986) Life history and den ecology of *Octopus briareus* Robson in a marine lake. J Exp Mar Biol Ecol 95:37–56

Barbosa A, Mäthger LM, Hanlon RT (2008) Cuttlefish camouflage: the effects of substrate contrast and size in evoking uniform, mottle or disruptive body patterns. Vis Res 48:1242–1253

Batham EJ (1957) Care of eggs by *Octopus maorum*. T Roy Soc NZ 84:629–638

Biegel M, Boal JG (2006) The effect of habitat enrichment on the mudflat octopus. Shape of Enrich 15:3–6

Boal JG (1996) Absence of social recognition in laboratory-reared cuttlefish, *Sepia officinalis* L. (Mollusca: Cephalopoda). Anim Behav 52:529–537

Boal JG (1997) Female choice of males in cuttlefish (Mollusca: Cephalopoda). Behaviour 134(13/14):975–988

Boal JG (2006) Social recognition: a top down view of cephalopod behaviour. Vie Milieu 56(2):69–79

Boal JG, Ni JN (1996) Ventilation rate of cuttlefish, *Sepia officinalis*, in response to visual stimuli. Veliger 39:342–347

Boal JG, Hylton RA, Gonzalez SA, Hanlon RT (1999) Effects of crowding on the social behavior of cuttlefish (*Sepia officinalis*). Contemp Top Lab Anim Sci 38:49–55

Boletzky S (1998) Cephalopod eggs and egg masses. Oceanogr Mar Biol Annu Rev 36:341–371

Boletzky S (2003a) Biology of early life stages in cephalopod molluscs. Adv Mar Biol 44:143–203

Boletzky S (2003b) A lower limit to adult size in coleoid cephalopods: elements of a discussion. Berliner Palaobiol Abh 3:19–28

Bowman RE, Stillwell CE, Micheals WL et al (2000) Food of northwest Atlantic fishes and two common species of squid. In: NOAA Technical Memorandum NMFS-NE-155. National Oceanic and Atmospheric Administration, Woods Hole

Boycott BB (1954) Learning in *Octopus vulgaris* and other cephalopods. Pubbl Staz Zool Nap 25:67–93

Boyle P, Boletzky S (1996) Cephalopod populations: definition and dynamics. Phil Trans R Soc Lond B 351 (No. 1343 The role of cephalopods in the world's oceans):985–1002

Brown C, Garwood MP, Williamson JE (2012) It pays to cheat: tactical deception in a cephalopod social signalling system. Biol Lett. doi:10.1098/rsbl.2012.0435

Budelmann BU (1994) Cephalopod sense organs, nerves and the brain: adaptation for high performance and life style. Mar Freshw Beh Physiol 25:13–33

Budelmann BU, Bleckmann H (1988) A lateral line analogue in cephalopods: water waves generate microphonic potentials in the epidermal head lines of *Sepia* and *Lolloguncula*. J Comp Physiol A 164:1–5

Bush SL, Hoving HJT, Hufford CL et al (2012) Brooding and sperm storage by the deep-sea squid *Bathyteuthis berryi* (Cephalopods: Decapoda: Decapodiformes). J Mar Biol Assn. UK 92:1629–1636

Byrne RA, Griebel U, Wood JB et al (2003) Squid say it with skin: a graphic model for skin displays in Caribbean reef squid (*Sepioteuthis sepioidea*). Berliner Palaobiol Abh 3:29–35

Chen X, Liu B, Chen Y (2008) A review of the development of Chinese distant-water squd jigging fisheries. Fish Res 89:211–221

Cheng MW, Caldwell RL (2000) Sex identification and mating in the blue-ringed octopus, *Hapalochlaena lunulata*. Anim Behav 60:27–33

Cigliano JA (1993) Dominance and den use in *Octopus bimaculoides*. Anim Behav 46:677–684

Cole PD, Adamo SA (2005) Cuttlefish (*Sepia officinalis*: Cephalopoda) hunting behavior and associative learning. Anim Cogn 8:27–30

Corner BD, Moore HT (1980) Field observations on the reproductive behavior of *Sepia latimanus*. Micronesia 16:235–260

Cortez T, Castro BG, Guerra A (1998) Drilling behaviour of *Octopus mimus* Gould. J Exp Mar Biol Ecol 224:193–203

Cosgrove JA (2002) An *In Situ* observation of webover hunting by *Enteroctopus dofleini* (Wülker, 1910). Can Field-Nat 117:117–118

Cummins SF, Boal JG, Buresch KC et al (2011) Extreme aggression in male squid induced by a β-MSP-like pheromone. Curr Biol 21(4):322–327

Darmaillacq A-S, Dickel L, Chichery M-P et al (2004) Rapid taste aversion learning in adult cuttlefish, *Sepia officinalis*. Anim Behav 68(6):1291–1298

Darmaillacq A-S, Lesimple C, Dickel L (2008) Embryonic visual learning in the cuttlefish *Sepia officinalis*. Anim Behav 76:131–134

Di Cristo C, Minnen JV, Di Cosmo A (2005) The presence of APGWamide in *Octopus vulgaris*: a possible role in the reproductive behavior. Peptides 26(1):53–62

Dickel L, Boal J, Budelmann B (2000) The effect of early experience on learning and memory in cuttlefish. Dev Psychobiol 36(2):101–110

DiMarco FP, Hanlon RT (1997) Agonistic behavior in the squid *Loligo plei* (Loliginidae, Teuthoidea): fighting tactics and the effects of size and resource value. Ethology 103:89–108

Dodge R, Scheel D (1999) Remains of the prey—recognizing the midden piles of *Octopus dofleini* (Wülker). Veliger 42(3):260–266

Domingues PM, Lopez N, Munoz JA et al (2007) Effects of a dry pelleted diet on growth and survival of the Yucatan octopus, *Octopus maya*. Aqua Nutr 13:273–280

Domingues P, Ferreira A, Marquez L et al (2008) Growth, absorption and assimilation efficiency by mature cuttlefish (*Sepia officinalis*) fed with alternative and artificial diets. Aqua Int 16(3):215–229. doi:10.1007/s10499-007-9139-5

Drickamer LC, Vessey SH, Meikle D (1996) Animal behavior, 4th ed. Wm C Brown, Boston

Duval P, Chichery M-P, Chichery R (1984) Prey capture by the cuttlefish (*Sepia officinalis* L.): an experimental study of two strategies. Behav Proc 9:13–21

Farías A, Uriarte I, Hernández J et al (2009) How size relates to oxygen consumption, ammonia excretion, and ingestion rates in cold (*Enteroctopus megalocyathus*) and tropical (*Octopus maya*) octopus species. Mar Biol 156(8):1547–1558

Ferguson GP, Messenger JB, Budelmann BU (1994) Gravity and light influence the countershading reflexes of the cuttlefish *Sepia officinalis*. J Exp Biol 191:247–256

Forsythe JW (1984) *Octopus joubini* (Mollusca: Cephalopoda): a detailed study of growth through the full life cycle in a closed seawater system. J Zool. Lond 202:393–417

Forsythe JW, Hanlon RT (1997) Foraging and associated behavior by *Octopus cyanea* Gray, 1849, on a coral atoll, French Polynesia. J Exp Mar Biol Ecol 209:15–31

Gabe SH (1975) Reproduction in the giant octopus of the North Pacific, *Octopus dofleini martini*. Veliger 18(2):146–150

Garcia S, Domingues P, Navarro JC et al (2011) Growth, partial energy balance, mantle and digestive gland lipid composition of *Octopus vulgaris* (Cuvier 1797) fed with two artificial diets. Aqua Nutr 17(2):e174-e187. doi:10.1111/j.1365-2095.2009.00746.x

Grisley MS, Boyle PR, Key LN (1996) Eye puncture as a route of entry for saliva during predation on crabs by the octopus *Eledone cirrhosa* (Lamarck). J Exp Mar Biol Ecol 202:225–237

Guerra A (2006) Ecology of *Sepia officinalis*. Vie Milieu 56(2):97–107

Guibé M, Poirier N, Houdé O et al (2012) Food imprinting and visual generalization in embryos and newly hatched cuttlefish, *Sepia officinalis*. Anim Behav. doi:10.1016/j.anbehav.2012.04.035

Hall K, Hanlon R (2002) Principal features of the mating system of a large spawning aggregation of the giant Australian cuttlefish *Sepia apama* (Mollusca: Cephalopoda). Mar Biol 140(3):533–545

Hanlon R (1990) Maintenance, rearing, and culture of Teuthoid and Sepioid squids. In: Gilbert DL, Adelman WJ, Arnold JM (eds) Squids as experimental animals. Plenum Press, New York, pp 35–64

Hanlon R, Messenger JB (1988) Adaptive coloration in young cuttlefish (*Sepia officinalis* L.): the morphology and development of body patterns and their relation to behaviour. Phil Trans Roy Soc B 320:437–487

Hanlon R, Messenger JB (1996) Cephalopod behaviour. Cambridge University Press, New York

Hanlon R, Hixon RF, Hulet WH (1983) Survival, growth, and behavior of the Loliginid squids *Loligo plei*, *Loligo pealei*, and *Lolliguncula brevis* (Mollusca: Cephalopoda) in closed sea water systems. Biol Bull 165:637–685

Hanlon R, Ament SA, Gabr H (1999a) Behavioral aspects of sperm competition in cuttlefish, *Sepia officinalis* (Sepioidea: Cephalopoda). Mar Biol 134:719–728

Hanlon R, Forsythe JW, Joneschild DE (1999b) Crypsis, conspicuousness, mimicry and polyphenism as antipredator defenses of foraging octopuses with a method of quantifying crypsis from video tapes. Biol J Linn Soc 66:1–22

Hanlon R, Smale MJ, Sauer WHH (2002) The mating system of the squid *Loligo vulgaris* Reynaudii (Cephalopoda, Mollusca) off South Africa: fighting, guarding, sneaking, mating and egg laying behavior. Bull Mar Sci 71(1):331–345

Hartwick EB (1983) *Octopus dofleini*. In: Boyle PR (ed) Cephalopod life cycles, vol I. Species Accounts. Academic Press, London, pp 277–291

Hartwick EB, Breen PA, Tulloch L (1978) A removal experiment with *Octopus dofleini* (Wülker). Veliger 21:263–264

Hendrickson L (2004) Population biology of northern shortfin squid (*Illex illecebrosus*) in the Northwest Atlantic Ocean and initial documentation of a spawning area. ICES J Mar Sci 61(2):252–266

Huffard CL, Caldwell RL, Boneka F (2008) Mating behavior of *Abdopus aculeatus* (d'Orbigny 1834) (Cephalopoda: Octopodidae) in the wild. Mar Biol 154(2):353–362

Huffard CL, Caldwell RL, Boneka F (2010) Male-male and male-female aggression may influence mating associations in wild octopuses (*Abdopus aculeatus*). J Comp Psychol 124:38–46

Hunsicker ME, Essington TE, Aydin KY et al (2010) Predatory role of the commander squid *Berryteuthis magister* in the eastern Bering Sea: insights from stable isotopes and food habits. Mar Ecol Prog Ser 415:91–108

Hurley AC (1978) School structure of the squid *Loligo opalescens*. Fish Bull 76:433–442

Ibáñez CM, Keyl F (2010) Cannibalism in cephalopods. Rev Fish Biol Fish 20(1):123–136

Ivanovich ML, Brunetti NE (1994) Food and feeding of *Illex argentinus*. Antarct Sci 6(2):185–193

Ivey GL (2007) Acoustic telemetry of the short-term movements of *Octopus cyanea* Gray, 1849 in Kaneohe Bay, Hawai'i. M.S. Thesis, University of Hawai'i, Manoa

Jantzen TM, Havenhand JN (2003a) Reproductive behavior in the squid *Sepioteuthis australis* from south Australia: interactions on the spawning ground. Biol Bull 204:305–317

Jantzen TM, Havenhand JN (2003b) Reproductive behavior in the squid *Sepioteuthis australis* from south Australia: ethogram of reproductive body patterns. Biol Bull 204:290–304

Johnson MW (1942) Some observations on the feeding habits of the octopus. Science 95(2471):478–479

Katsanevakis S, Verriopoulos G (2006a) Modelling the effect of temperature on hatching and settlement patterns of meroplanktonic organisms: the case of the octopus. Sci Mar 70(4):699–708

Katsanevakis S, Verriopoulos G (2006b) Seasonal population dynamics of *Octopus vulgaris* in the eastern Mediterranean. ICES J Mar Sci 63:151–160

King AJ, Adamo SA, Hanlon RT (2003) Squid egg mops provide sensory cues for increased agonistic behaviour between male squid. Anim Behav 66(1):49–58

Koueta N, Alorend E, Noel B et al (2006) Earlier acceptance of frozen prey by juvenile cuttlefish *Sepia officinalis* in experimental rearing: effect of previous enriched natural diet. Vie Milieu 56(2):147–152

Langridge KA, Broom M, Osorio D (2007) Selective signaling by cuttlefish to predators. Curr Biol 17: R1044–R1045

Lee P (1994) Nutrition of cephalopods: fueling the system. Mar Freshw Behav Physiol 25(1–3):35–51

Lee PG, Forsythe JW, DiMarco P et al (1991) Initial palatability and growth trials on pelleted diets for cephalopods. Bull Mar Sci 49(1–2):362–372

Leite TS, Haimovici M, Mather J (2009) *Octopus insularis* (Octopodidae), evidences of a specialized predator and a time-minimizing hunter. Mar Biol 156(11):2355–2367

Leite TS, Batista AT, Lima FD et al (in sub) Geographic variability of Octopus insularis diet: from oceanic island to continental populations. Hydrobiologia

Lorrain A, Arguelles J, Alegre A et al (2011) Sequential isotopic signature along gladius highlights contrasted individual foraging strategies of jumbo squid (*Dosidicus gigas*). PloS ONE 6(7):e22194

Lutz RA, Voight JR (1994) Close encounter in the deep. Nature 371:563

Mather JA (1978) Mating behavior of *Octopus joubini* Robson. Veliger 21(2):265–267

Mather JA (1980) Social organization and use of space by *Octopus joubini* in a semi-natural situation. Bull Mar Sci 30:848–857

Mather JA (1986) Sand-digging in *Sepia officinalis*: assessment of a cephalopod mollusc's "fixed" behavior pattern. J Comp Psychol 100:315–320

Mather JA (1988) Daytime activity of juvenile *Octopus vulgaris* in Bermuda. Malacologia 29(1):69–76

Mather JA (1991a) Foraging, feeding and prey remains in middens of juvenile *Octopus vulgaris* (Mollusca: Cephalopoda). J Zool. Lond 224:27–39

Mather JA (1991b) Navigation by spatial memory and use of visual landmarks in octopuses. J Comp Physiol A 168:491–497

Mather JA (1994) "Home" choice and modification by juvenile *Octopus vulgaris* (Mollusca: Cephalopoda): specialized intelligence and tool use? J Zool. Lond 233:359–368

Mather JA (2004) Cephalopod skin displays: from concealment to communication. In: Oller DK, Griebel U (eds) Evolution of communication systems. MIT Press, Cambridge, pp 193–214

Mather JA (2010) Vigilance and antipredator responses of Caribbean reef squid. Mar Freshw Behav Physiol 43:357–370

Mather JA (2011) Why is *Octopus cyanea* Gray in Hawaii specializing in crabs as prey? Vie Milieu 61:181–184
Mather JA (in press) Enrichment and cephalopods. Shape Enrich
Mather JA, O'Dor RK (1984) Spatial organization of schools of the squid *Illex illecebrosus*. Mar Behav Physiol 10:259–271
Mather JA, O'Dor RK (1991) Foraging strategies and predation risk shape the natural history of the juvenile *Octopus vulgaris*. Bull Mar Sci 49:256–269
Mather JA, Resler S, Cosgrove J (1985) Activity and movement patterns of *Octopus dofleini*. Mar Behav Physiol 11:301–314
Mather JA, Anderson RC, Wood JB (2010a) Octopus: the ocean's intelligent invertebrate. Timber Press, Portland
Mather JA, Griebel U, Byrne RA (2010b) Squid dances: an ethogram of postures and actions of *Sepioteuthis sepioidea* squid with a muscular hydrostat system. Mar Freshw Behav Physiol 43:45–61
Mather JA, Leite TS, Batista AT (2012) Individual prey choices of octopuses: why are they generalist or specialist? Curr Zool 58(4):596–602
McQuaid CD (1994) Feeding behaviour and selection of bivalve prey by *Octopus vulgaris* Cuvier. J Exp Mar Biol Ecol 177:187–202
Meisel DA, Byrne RA, Kuba MJ et al (2006) Comparing the activity patterns of two Mediterranean octopus species. J Comp Psychol 120:191–197
Messenger JB (1968) The visual attack of the cuttlefish, *Sepia officinalis*. Anim Behav 16:342–357
Messenger JB (1973) Learning in the cuttlefish, *Sepia*. Anim Behav 21(4):801–826
Messenger JB (2001) Cephalopod chromatophores: neurobiology and natural history. Biol Rev 76:473–528
Moltschaniwskyj NA, Hall K, Lipinski MR et al (2007) Ethical and welfare considerations when using cephalopods as experimental animals. Rev Fish Biol Fish 17:455–476
Moynihan MH, Rodaniche AF (1982) The behaviour and natural history of the Caribbean reef squid *Sepioteuthis sepioidea* with a consideration of social, signal and defensive patterns for difficult and dangerous environments. Adv Ethol 125:1–150
Neill SRStJ, Cullen JM (1974) Experiments on whether schooling by their prey affects the hunting behavior of cephalopods and fish predators. J Zool Lond 172:549–569
Nesis KN (1983) *Dosidicus gigas*. In: Boyle PR (ed) Cephalopod Life Cycles: vol. I Species Accounts. Academic Press, London, pp 215–231
Nightingail A (2012) Who's for lunch? A gull-eating octopus in Victoria, BC. http://www.birdfellow.com/journal/archive/2012/4. Accessed 21 Sept 2012
Nigmatullin CM, Nesis KN, Arkhipkin AI (2001) A review of the biology of the jumbo squid *Dosidicus gigas* (Cephalopoda: Ommastrephidae). Fish Res 54:9–19
Nixon M, Maconnachie E (1988) Drilling by *Octopus vulgaris* (Mollusca: Cephalopoda) in the Mediterranean. J Zool. Lond 216:687–716
Nixon M, Mangold K (1996) The early life of *Octopus vulgaris* (Cephalopoda: Octopodidae) in the plankton and at settlement: a change in lifestyle. J Zool. Lond 239:301–327
Norman M, Finn J, Tregenza T (1999) Female impersonation as an alternative reproductive strategy in giant cuttlefish. Proc R Soc Lond B 266:1347–1349
O'Dor RK, Webber DM (1986) The constraints on cephalopods: why squid aren't fish. Can J Zool 64:1591–1605
O'Dor RT, Dawe EG (1998) *Illex illecebrosus*. In: Rodhouse PG, Dawe EG, O'Dor RK (eds) Squid recruitment dynamics: the genus *Illex* as a model, the commercial *Illex* species and influence on variability. FAO Fish Tech Pap 376, Rome, pp 77–104
Onthank KL, Cowles DL (2011) Prey selection in *Octopus rubescens*: possible roles of energy budgeting and prey nutritional composition. Mar Biol 158(12):2795–2804
Ortiz N (2006) First description of eggs, hatchlings and hatchling behaviour of *Enteroctopus megalocyathus* (Cephalopoda: Octopodidae). J Plankton Res 28(10):881–890
Packard A (1972) Cephalopods and fish: the limits of convergence. Biol Rev 47:241–307
Peters A, Rehling M, Anderson RC (2005) Mission possible: enriched cephalopods. AZA Comm September 2005:21–23

Poirier R, Chichery R, Dickel L (2004) Effects of rearing conditions on sand digging efficiency in juvenile cuttlefish. Beh Proc 67:273–279

Poirier R, Chichery R, Dickel L (2005) Early experience and postembryonic maturation of body patterns in cuttlefish (*Sepia officinalis*). J Comp Psychol 119:230–237

Pulliam HR (1974) On the theory of optimal diets. Am Nat 108:59–74

Rigby RP, Sakurai Y (2004) Temperature and feeding related to growth efficiency of immature octopuses *Enteroctopus dofleini*. Suisanzoshoku 52(1):29–36

Rocha F, Guerra A, González AF (2001) A review of reproductive strategies in cephalopods. Biol Rev 76:291–304

Rodhouse PG (2001) Managing and forecasting squid fisheries in variable environments. Fish Res 54:3–8

Rosas C, Cuzon G, Pascual C et al (2007) Energy balance of *Octopus maya* fed crab or an artificial diet. Mar Biol 152(2):371–381. doi:10.1007/s00227-007-0692-2

Rosas C, Tut J, Baeza J et al (2008) Effect of type of binder on growth, digestibility, and energetic balance of *Octopus maya*. Aquaculture 275(1–4):291–297. doi:10.1016/j.aquaculture.2008.01.015

Rosas C, Sánchez A, Pascual C et al (2011) Effects of two dietary protein levels on energy balance and digestive capacity of *Octopus maya*. Aqua Int 19(1):165–180. doi:10.1007/s10499-010-9350-7

Sauer WHH, Lipinski MR, Augustyn CJ (2000) Tag recapture studies of the chokka squid *Loligo vulgaris reynaudii* d'Orbigny, 1845 on inshore spawning grounds on the south-east coast of South Africa. Fish Res 45:283–289

Sazima I, de Almeida LB (2006) The bird kraken: octopus preys on a sea bird at an oceanic island in the tropical West Atlantic. Mar Biodiv Rec 12/2007:1–3. doi:10.1017/S1755267206005458

Scheel D, Anderson RC (2012) Variability in the diet specialization of *Enteroctopus dofleini* in the eastern Pacific examined from midden contents. Am Malacol Bull 30(2):1–13

Scheel D, Bisson L (2012) Movement patterns of giant Pacific octopuses, *Enteroctopus dofleini* (Wülker, 1910). J Exp Mar Biol Ecol 416–417:21–31

Scheel D, Lauster A, Vincent TLS (2007) Habitat ecology of *Enteroctopus dofleini* from middens and live prey surveys in Prince William Sound, AK. In: Landman NH et al (eds) Cephalopods present and past: new insights and fresh perspectives. Springer, Dordrecht, pp 434–458

Semmens JM, Pecl GT, Gillanders BM et al (2007) Approaches to resolving cephalopod movement and migration patterns. Rev Fish Biol Fish. doi:10.1007/s11160-007-9048-8

Shaw PW, Sauer WHH (2004) Multiple paternity and complex fertilisation dynamics in the squid *Loligo vulgaris reynaudii*. Mar Ecol Prog Ser 270:173–179

Sih A (1984) Optimal behavior and density-dependent predation. Am Nat 123(3):314–326

Smith NJ (2012) The effects of seabed slope and other environmental variables on the movement patterns of the giant Pacific octopus (*Enteroctopus dofleini*). Thesis, Alaska Pacific University, Anchorage

Staudinger MD, Hanlon RT, Juanes F (2011) Primary and secondary defences of squid to cruising and ambush fish predators: variable tactics and their survival value. Anim Behav 81:585–594

Steer MA, Semmens JM (2003) Pulling or drilling, does size or species matter? An experimental study of prey handling in *Octopus dierythraeus* (Norman, 1992). J Exp Mar Biol Ecol 290:165–178

Stephens D, Krebs JR (1986) Foraging theory. Monographs in behavior and ecology. Princeton University Press, Princeton

Stowasser G, Pierce GJ, Moffat CF et al (2006) Experimental study on the effect of diet on fatty acid and stable isotope profiles of the squid *Lolliguncula brevis*. J Exp Mar Biol Ecol 333(1):97–114

Sugimoto C, Ikeda Y (2012) Ontogeny of schooling behavior in the oval squid *Sepioteuthis lessoniana*. Fish Sci doi:1007/s12562-011-0464-2

Sykes AV, Domingues P, Correia M et al (2006) Cuttlefish culture—state of the art and future trends. Vie Milieu 56(2):129–137

Tricarico E, Borelli L, Gherardi F, Fiorito G (2011) I know my neighbor: individual recognition in *Octopus vulgaris*. PLoS ONE 6:e18710

Uriarte I, Iglesias J, Domingues P et al (2011) Current status and bottle neck of octopod aquaculture: the case of American species. J World Aquac Soc 42(6):735–752

Valverde JC, Hernández MD, Aguado-Giménez F et al (2008) Growth, feed efficiency and condition of common octopus (*Octopus vulgaris*) fed on two formulated moist diets. Aquaculture 275(1–4):266–273. doi:10.1016/j.aquaculture.2008.01.012

Vaz-Pires P, Seixas P, Barbosa A (2004) Aquaculture potential of the common octopus (*Octopus vulgaris* Cuvier, 1797): a review. Aquaculture 238(1–4):221–238

Villanueva R, Nozais C, Boletzky S (1995) The planktonic life of octopuses. Nature 377:107

Villanueva R, Nozais C, Boletzky S (1996) Swimming behaviour and food searching in planktonic *Octopus vulgaris* Cuvier from hatching to settlement. J Exp Mar Biol Ecol 208:169–184

Vincent TLS, Scheel D, Brown JS et al (1996) Trade-offs and coexistence in consumer-resource models: it all depends on what and where you eat. Am Nat 148(6):1038–1058

Vincent TLS, Scheel D, Hough KR (1998) Some aspects of diet and foraging behavior of *Octopus dofleini* (Wülker, 1910) in its northernmost range. Mar Ecol 19(1):13–29

Voight JR (1991a) Enlarged suckers as an indicator of male maturity in *Octopus*. Bull Mar Sci 49(1–2):98–106

Voight JR (1991b) Ligula length and courtship in *Octopus digueti*: a potential mechanism of mate choice. Evolution 45(7):1726–1730

Voight JR, Grehan AJ (2000) Egg brooding by deep-sea octopuses in the North Pacific Ocean. Biol Bull 198:94–100

Warnke K (1994) Some aspects of social interaction during feeding in *Sepia officinalis* (Mollusca: Cephalopoda) hatched and reared in the laboratory. Vie Milieu 44:125–131

Webber DM, Aitken J, O'Dor RK (2000) Costs of locomotion and vertic dynamics of cephalopods and fish. Physiol Biochem Zool 73:651–662

Wells MJ (1978) Octopus: the physiology and behaviour of an advanced invertebrate. Chapman & Hall, London

Wodinsky J (1969) Penetration of the shell and feeding on gastropods by *Octopus*. Am Zool 9(3):997–1010

Wood JB, Anderson RC (2004) Interspecific evaluation of octopus escape behavior. J Appl Anim Welf Sci 7:95–106

Wood JB, O'Dor RK (2000) Do larger cephalopods live longer? Effects of temperature and phylogeny on interspecific comparisons of age and size at maturity. Mar Biol 136:91–99

Wood JB, Pennoyer KE, Derby C (2008) Ink as a conspecific alarm cue in the Caribbean reef squid, *Sepioteuthis sepioidea*. J Exp Mar Biol Ecol 376:11–16

Wood JB, Maynard AE, Lawlor AG et al (2010) Caribbean reef squid, *Sepioteuthis sepioidea*, use ink as a defense against predatory French grunts, *Haemulon flavolineatum*. J Exp Mar Biol Ecol 388:20–27

Yarnall JL (1969) Aspects of the behaviour of *Octopus cyanea* Gray. Anim Behav 17:747–754

Young RE, Harman RF (1988) "Larva", "Paralarva" and "Subadult" in cephalopod terminology. Malacologia 29(1):201–207

Chapter 3
Fisheries Production and Market Demand

Graham J. Pierce and Julio Portela

Abstract Since 1983, when Caddy was able to state that '… except in a few ocean regions, they (cephalopods) are not subject to exploitation', the decline in many finfish stocks has led to increased attention on other groups such as cephalopods, the increasing economic importance of which is evidenced by the rapid rise in their global landings over recent decades. World cephalopod landings (capture fisheries) rose from 500,000 t in 1950 to a peak of more than 4 million t in 2007, with landings increasing in most regions with the exception of the Northeast Atlantic. New regions such as the Southwest Atlantic and the Southeast Pacific subsequently became important cephalopod fishery areas, supplying new abundant species to world markets (notably *Illex argentinus* and *Dosidicus gigas*). In Europe, the potential expansion of cephalopod fishing will require a new focus on monitoring, assessment and management of these fisheries, taking into account the differences in life history between cephalopods and fish. The increasing demand for cephalopods on the international markets has stimulated research on rearing and 'ongrowing' of cephalopods (mainly common octopus) as an alternative to reliance on fisheries.

Keywords Cephalopod fishing · Japanese flying squid · Jumbo squid · Argentine shortfin squid · Markets · Fishery biology

---

G. J. Pierce (✉)
University of Aberdeen, Main Street,
Newburgh, Aberdeenshire, AB41 6AA, UK
e-mail: g.j.pierce@abdn.ac.uk

CESAM & Departamento de Biologia, Universidade de Aveiro,
3810–193, Aveiro, Portugal
e-mail: g.j.pierce@ua.pt

J. Portela
Centro Oceanográfico de Vigo, Instituto Español de Oceanografía (IEO),
Subida a Radio Faro, 50-52, 36390 Vigo, Spain
e-mail: julio.portela@vi.ieo.es

J. Iglesias et al. (eds.), *Cephalopod Culture,* DOI 10.1007/978-94-017-8648-5_3,

3.1 Introduction

Global marine fisheries are widely perceived to be in crisis. World capture production seems to have peaked, there is concern that we are 'fishing down the food web', many fish stocks are evidently overexploited, it has been suggested that capture fisheries will be extinct by the mid-twenty-first century, and even apparent success stories like lobster fisheries in the US Gulf of Maine have a less palatable flip side of ecological damage and vulnerability to disease outbreaks (Branch et al. 2010; Worm et al. 2009; Hilborn 2012).

As many finfish stocks decline, increased attention has been focused on other groups such as cephalopods (Caddy 1983). The cephalopods comprise the squid, octopuses, cuttlefish (grouped together as the coleoids) and nautiloids. There are about 800 living species although cephalopod taxonomy continues to be in a state of flux as molecular genetics challenges traditional views of relationships between species (Allcock 2010).

In 1983, the Food and Agriculture Organization (FAO) of the United Nations issued Fisheries Technical Paper N° 231 (Caddy 1983), in which it was stated that '...except in a few ocean regions, they (cephalopods) are not subject to exploitation. Most of the large industrial cephalopod fisheries are concentrated in the Northwest and central Pacific, the northwest African coasts, the Mediterranean and the Northwest Atlantic.' It was also said that 'in the search for resources that can locally support a high level of exploitation in the near future, cephalopods must occupy a leading place'. This situation has unquestionably changed over the past 30 years.

Cephalopods are now highly valuable commercial fishery resources with annual world catches reaching 4.3 million t in 2007 but decreasing to 3.6 million t in 2010, according to FAO statistics (FAO 2011), having increased steadily from around 600,000 t in 1950 (Jereb and Roper 2010). Caddy and Rodhouse (1998) suggested that this may be at least partly due to ecological replacement, with opportunistic cephalopod species occupying niches vacated by overfished finfish stocks. However, Nigmatullin (2010) ascribed the increase mainly to the expansion of fishing grounds, targeting new species and increasing fishing effort. Again as shown by FAO data, since the 1980s the expansion has slowed down as all the main inshore commercial stocks of cephalopods were exploited, and the past few years have shown a decrease in total cephalopod landings. Indeed, there is evidence that many coastal cephalopod stocks may have been overfished (notably in Asia: see Funge-Smith et al. 2012).

Besides their importance for human consumption, cephalopods are useful research models not only in medical and biological research, due to their nervous system and sense organs (Lee 1994; Koueta and Boucaud-Camou 1999), but also in physiology, neuroscience, nutritional biochemistry, ageing, molecular biology and immunology (Oestmann et al. 1997; Domingues et al. 2001).

Cephalopods are distributed throughout the world's oceans and it is evident that we are not yet exploiting cephalopods in many areas where they occur, especially far from land (Fries 2010). Nigmatullin (2004, 2010) proposed that significant further growth of cephalopod landings could be achieved through exploitation of oceanic resources of ommastrephid squid. Clarke (1987, 1996) estimated that sperm whales alone could be consuming almost two orders of magnitude more cephalopods than are presently taken by fisheries, although much of this amount would be oceanic species, many inaccessible due to them living far from the coast and at depth and being unpalatable to humans due to their high ammonia content.

In the European Union (EU), despite their importance in the southern European diet, cephalopods have long been regarded as minor resource species. Under the EU's Common Fisheries Policy, it remains the case that there is little assessment or regulation of cephalopod fisheries (ICES 2012).

3.2 Why are Cephalopods not Fish but Good Fishery Resources?

While researchers working on cephalopods have tended to emphasise the differences between squid and fish, fishery biologists have pointed to parallels with, for example, short-lived pelagic fish (e.g. Pauly 1998).

Firstly, to state the obvious, there are important differences between the various groups of cephalopods. Although most species (excluding cuttlefish) have planktonic paralarvae, the juveniles and adults of octopus, cuttlefish and myopsid squid are all generally benthic or demersal, and mature females lay eggs attached to the seabed or other stationary structures (including fixed fishing gears). Octopuses are more sedentary as adults than the other groups; both cuttlefish and myopsid squids normally undertake ontogenetic migrations. The oegopsid squids extend into the pelagic zone as juveniles and adults and may undertake long migrations to feeding and spawning grounds. Their eggs are embedded in gelatinous, free-floating egg masses.

Many of the shared characteristics of all coleoid cephalopods relate to being 'r-selected' or pioneer species, selected for their short life cycles with fast growth and early breeding. Also consistent with the view that cephalopods are 'r-selected' is the high metabolic rate, which helps explain why both individual physiology (metabolism, growth) and population growth parameters (juvenile survival, recruitment, migration patterns) appear to be very sensitive to environmental conditions. This has been shown by experimental studies on juveniles which suggest that temperature conditions experienced around hatching can have profound (and sometimes unexpected) consequences for growth rate, age at maturity and final adult size (e.g. Forsythe 1993; Forsythe et al. 2001; Wangvoralak 2011) and are backed up by field observations (e.g. Jackson and Domeier 2003). Studies on larger squids such as the jumbo squid *Dosidicus gigas* suggest that high temperatures can be also problematic because they lead to elevated metabolic rates in water with low oxygen content, thus effectively excluding these squids from otherwise suitable habitat (Rosa and Seibel 2008).

However, cephalopods also show some life history characteristics that may be thought of as 'k-selected' and are normally associated with 'higher' animals such as large fish, birds and mammals: many cephalopods, including familiar coastal species of squid, cuttlefish and octopus lay relatively few eggs (compared to, say, a cod: a few thousands as opposed to hundreds of thousands). Furthermore, they show complex behaviour patterns (leading to suggestions of high intelligence, especially in octopus) and mating strategies (e.g. in loliginid squids, large males guard the females on the spawning grounds while small 'sneakers' attempt to mate with the females when the larger animals are distracted—and genetic evidence suggests they often succeed; Hanlon and Messenger 1998; Shaw and Sauer 2004). Eggs from a single female *Loligo forbesii* may be fertilized by two or more different males (Shaw and Boyle 1997). Signal transmission in nerves is rapid (hence, squid giant axons are used in medical research) and the squid eye appears quite similar to the vertebrate eye in terms of general morphology. Curiously, many cephalopod species engage in highly colourful visual displays even though most cephalopods are apparently colour blind (Hanlon and Messenger 1998; Wood and Jackson 2004). The answer to this apparent contradiction may lie in the possession of high-resolution polarisation vision, as found in cuttlefish by Temple et al. (2012). Octopuses even show parental care, with the females of many species protecting their eggs (Rocha et al. 2001).

A key aspect of their ecology is the trophic role of cephalopods. Several studies on ecosystem structure have highlighted cephalopods as keystone species (e.g. Gasalla et al. 2010). Cephalopods are both active predators and important sources of prey for species ranging from fish to seabirds and marine mammals. Their fast metabolism and rapid growth result in a high ratio of production to biomass so that the amount of energy flowing through the cephalopod component of marine ecosystems is high relative to the biomass present at any one time. It should be noted, however, that energy flow is likely to show seasonal peaks, especially at higher latitudes, reflecting the seasonality of cephalopod life cycles. Ommastrephid squids such as *Illex argentinus* are major nutrient vectors and play a key role as transient 'biological pumps' linking spatially distinct marine ecosystems (Arkhipkin, 2013). Cephalopods are a major prey source for commercially important fishes (e.g. hakes, tunas and salmon), marine mammals and seabirds in ecosystems worldwide (Hunsicker et al. 2010). Two species of hake in the Benguela system are estimated to take between 565,000 and 988,000 t of cephalopods annually (Punt et al. 1992; Smale 1996). Sperm whales may eat 213 to 320 million t of cephalopods per year (Clarke 1996).

What are the implications of these life cycle and ecological characteristics for fisheries? The rapid growth rate and high productivity, coupled with relative ease of capture, good nutritional quality and acceptable taste (at least to the southern European, southern American and Asian palates), make them attractive fishery resources. Experience in many fishing grounds seems to suggest that they are relatively resilient stocks. Cephalopods contribute to fisheries in many marine ecosystems, both directly as a commodity (harvest and sale of cephalopods) and through providing an ecological support service (the portion of the landings and landed value of other species that rely on cephalopods for their production; Hunsicker et al. 2010).

However, the flip side of the 'live fast die young' life strategy is that most fished species live for only 1 year and have little-overlapping generations, so if fisheries were to fish out a generation of recruits, there would be almost no reserve of old adults to repopulate the stock. The fact that this scenario remains largely hypothetical is probably at least partly a consequence of the plasticity of the life cycle (Pecl et al. 2004). Although breeding of temperate species tends to be seasonal, it is also hugely variable and animals in breeding condition can present at most times of the year.

Their extreme environmental sensitivity also makes cephalopods rather unpredictable and unreliable fishery resources. Although the extent varies, most exploited species (and especially oceanic squids) show hard-to-predict year-to-year fluctuations from high to low abundance. It seems likely that this has discouraged the development of directed fisheries in some areas; anecdotal evidence for this exists in Scotland (Smith 2011).

3.3 What is the Status of Cephalopod Capture Fisheries in Different Parts of the World?

Cephalopod fisheries fall into three general categories. Firstly, there is large-scale directed fishing, often using jigging machines, as in the major ommastrephid (short-finned) squid fisheries mentioned below, or, sometimes, trawling. Secondly, a substantial proportion of landings of loliginid (long-finned) squid arise as by-catch, typically from demersal trawling. Finally, a wide range of small-scale directed fisheries exist, for squid, cuttlefish and octopus, using a range of artisanal gears (see Sect. 3.4).

At a global level, fishery statistics are available from FAO. Various caveats apply to these data: they refer to landings and ignore discards, they often group several species together into broad categories, they may be biased by inaccurate recording of legal landings and by illegal, unreported and unregulated (IUU) fishing, and some countries are (at best) slow in providing data. In the North Atlantic, landings data are also compiled by the International Council for the Exploration of the Sea (ICES), potentially offering greater temporal, spatial and taxonomic resolution than is available from FAO (although not consistently so), but still far from being a perfect data source. In various parts of the world, including European Atlantic coastal waters, abundance indices are also available from trawling surveys. The FAO divides the world into 18 fishery regions, of which only the three Antarctic regions report few or no landings of cephalopods. Between 1950 and 2007, world cephalopod landings rose from around 500,000 t annually to a peak of more than 4 million t. The most recent annual total (for 2010) is around 3.5 million t annually, an apparent decrease which is evident in trends from several regions (FAO 2011; see Fig. 3.1).

Two previous review papers (Caddy and Rodhouse 1998; Hunsicker et al. 2010) examined the general state of world cephalopod fisheries. Caddy and Rodhouse (1998) noted how cephalopod landings were increasing in most regions even though finfish landings were no longer increasing and suggested that overfishing on

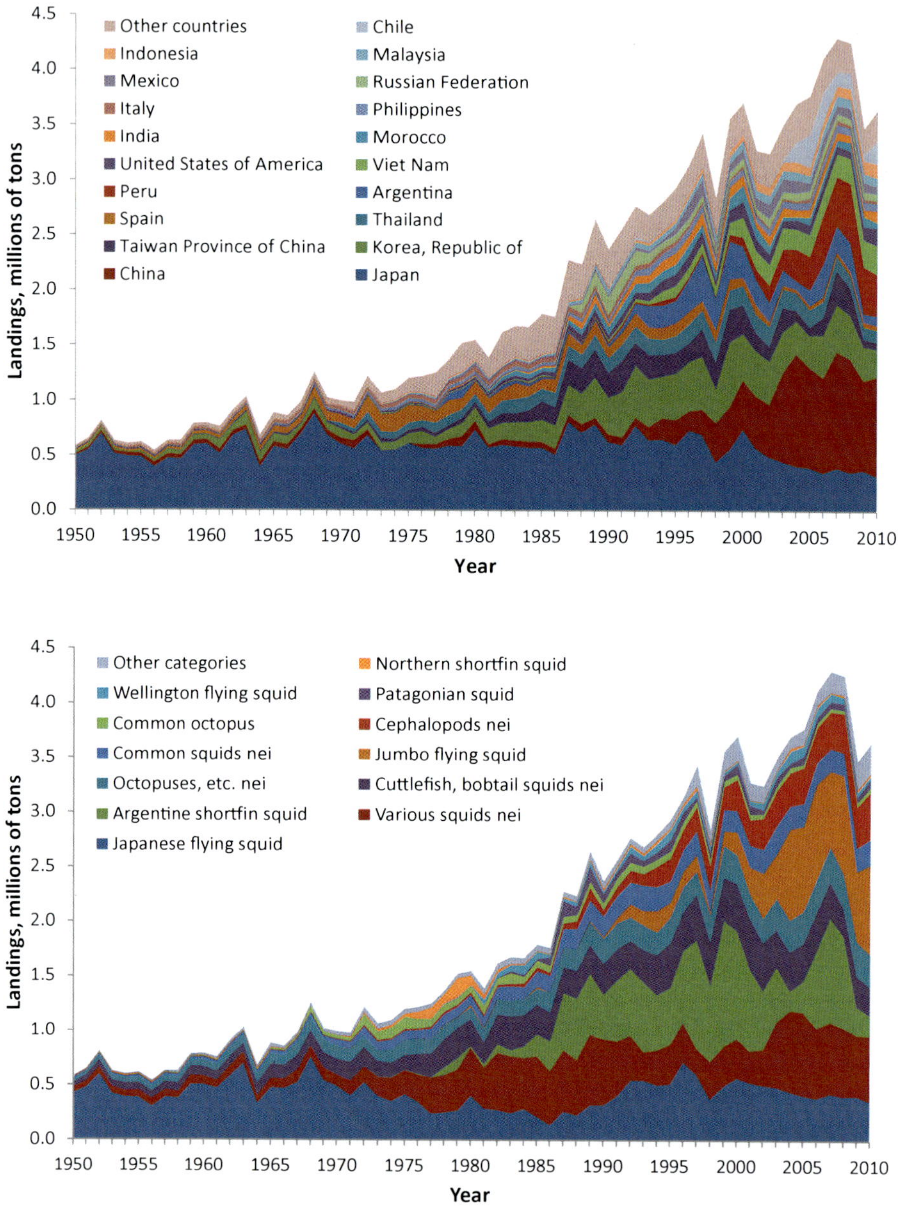

Fig. 3.1 World cephalopod landings in 1950–2010 (FAO data), by country and by commercial category

finfish was favouring cephalopod expansion. The rise in cephalopod catches was seen mainly in ommastrephid squid. The only studied region where cephalopod landings had not increased significantly over the past 25 years was the Northeast

Atlantic. In support of their general hypothesis, the authors noted the shorter life cycle of cephalopods, their more rapid turnover and lower standing stocks than seen in longer-lived finfish. They proposed that, under high fishing pressure, groundfish are outcompeted by cephalopods as they have less opportunity for spawning and replacement.

Hunsicker et al. (2010) examined the relative importance of cephalopods in various large marine ecosystems (LMEs) around the world during 1990–2004, in terms of their direct contribution to landings (and landed value) and their estimated indirect contribution as food of other landed species. They showed that the relative (direct) contribution of cephalopods to landings varied widely between areas, being the highest on the Patagonian Shelf (around 40% of landings) but low in LMEs in and adjacent to Europe, such as the Canary Current (8%), Celtic-Biscay Shelf (around 2.5%) and North Sea (< 1%). In terms of absolute amounts of cephalopods caught, there was a striking increase from 1960–1970 to 1990–2004. During the former period, only the Sea of Japan supported a large cephalopod fishery (basically that for *Todarodes pacificus*) and cephalopod catches on the Patagonian Shelf were minimal. By the latter period, the Patagonian Shelf fishery had become the most important cephalopod fishery in the world and cephalopod fisheries had expanded in most regions. As will be seen from the summary of FAO data below, this picture is already out of date, with the rise of China as a major player in squid fishing and the expansion of cephalopod fishing along the Pacific coastline of the Americas supported by a marked increase in jumbo squid *Dosidicus gigas* abundance. Hunsicker et al. (2010) expressed concern that the rapid expansion of cephalopod fishing over the past three decades may be unsustainable.

By far, the most important fishery region for cephalopods since 1950 is the *Northwest Pacific,* which accounts for 43.8% of all recorded cephalopod landings across the world during 1950–2010 (to put this in perspective, cephalopod landings from the Northeast Atlantic over the same period represent 2% of the global total). Cephalopod catches of almost 1.3 million t were landed from this region in 2010 (down from a peak of 1.47 million t in 2004). There has been a general trend of increased landings during 1990–2010 although landings have been more stable since 2000. Until 2002, Japan was the most important country for cephalopod landings, and the Japanese flying squid (*T. pacificus,* maximum mantle length (ML) = 50 cm) was the most important of cephalopods in the landings. The *Todarodes* fishery suffered a decrease in 1985 and, although it recovered by the mid-1990s, landings have again been in decline since 2000. Since 2002, squid (species not recorded) landed by China have become the most important category. Although the majority of landings from this area in recent years were not identified to species in the FAO statistics, two other squids, the neon flying squid (*Ommastrephes bartramii*) and schoolmaster gonate squid (*Berryteuthis magister*), are identified as making up a small proportion of catches. In the 1990s, the main squids taken by China appear to have been *T. pacificus* and *Uroteuthis* (formerly *Loligo*) *edulis* (Ling and Zheng 2000).

The region showing the second highest peak in landings since 1950 (1.2 million t in 1999) is the *Southwest Atlantic.* Catches in this area started to become significant in the late 1970s and climbed relatively steadily until 1999. Catches fell to just more than 200,000 t in 2004, and although they peaked at around 1 million t in 2007, they have subsequently crashed again. Although 38 different countries have officially recorded landings of cephalopods from this area, Argentina and Falkland Islands remain the main players, and catches are dominated by the Argentine shortfin squid *I. argentinus.* The only other significant catch identified to species is the Patagonian squid *Doryteuthis gahi.* Two further squid species are identified in catches, albeit in small amounts: sevenstar flying squid (*Martialia hyadesi*) and greater hooked squid (*Onykia ingens*).

The third most important fishery region for cephalopods is the *Southeast Pacific,* although cephalopod fishing was essentially nonexistent in this area until the 1990s. Landings peaked at around 820,000 t in 2006. Slightly more than half of all landings from this area go into Peru and a further 20 % are taken by Chile. Almost 95 % of landings are of the jumbo squid *Dosidicus gigas,* with Patagonian squid (2.7 %) coming a distant second.

The fourth most important cephalopod fishery region is the *western central Pacific,* which has shown a steady increase in landings since the late 1960s, reaching almost 550,000 t in 2010, with most catches being taken by Thailand and Vietnam. Although only a small proportion of the landings is identified to species, common squids, presumably Loliginidae, made up 43 % of total landings since 1950.

Landings from the *eastern central Atlantic* region increased in the mid-1960s, subsequently fluctuating between around 150,000 and 250,000 t, with highest landings seen in 1974 and 1999. However, since this latter peak, landings have fallen to around 100,000 t annually, mainly due to reduction of effort by foreign fleets. In the 1950s, almost all landings were taken by Japan, but Spain became the most important fishing nation during the 1970s, in turn being replaced by Morocco. In terms of categories of cephalopods landed, octopus is the most important, and fluctuations in octopus landings essentially drive the overall trend.

In recent years, landings from the *Northeast Atlantic* have fluctuated between 40,000 and 60,000 t. Between 1950 and 1980, Spain was the main player, but since then landings by France have increased to reach similar levels, coinciding with the increased importance of cuttlefish. There was a marked dip in landings between the mid-1970s and the mid-1980s, a trend which would be even more marked if the short-lived *T. sagittatus* fishery in Norway had not substantially boosted total landings during 1980–1985. A more detailed interpretation of the landings data is difficult since even now, when some European countries are distinguishing all the different squid and octopus species landed, a large proportion of landings is reported to the FAO under the general octopus, cuttlefish and squid categories.

Landings from the *Mediterranean* generally increased up to a peak in 1988 of slightly more than 80,000 t. There were three periods in which landings rose sharply, the mid-1960s, the late 1980s and the mid-2000s, and all were followed by marked declines in landings. Italy dominates landings from this area and until the mid-1990s was taking more than 50 % of all cephalopod landings from the area. Octopuses and cuttlefishes such as *Eledone cirrhosa, E. moschata, Octopus vulgaris* and *Sepia officinalis* have been the main categories landed, with octopus landings

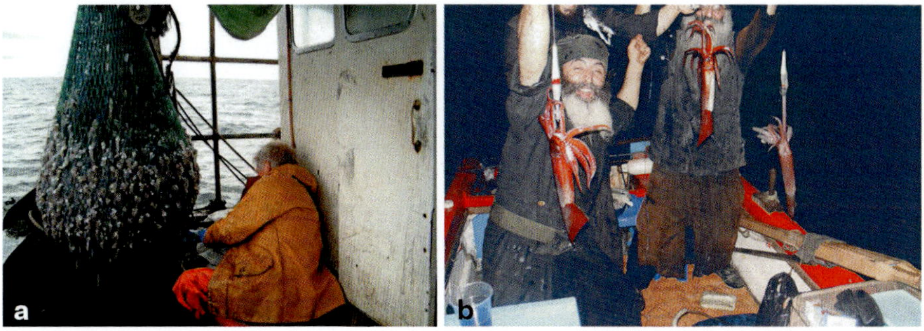

Fig. 3.2 Small-scale cephalopod fishing. **a** Directed (small-scale) trawling for *Loligo forbesii* in the Moray Firth, Scotland (UK) (Photo: Michael Wiseman). **b** Jigging for squid (*Todarodes sagittatus*) in Greece. (Photo: Jennifer Smith)

driving the general trend of an increase to 1988 and subsequent decline, but changes in cuttlefish landings contributed more to the above-mentioned sharp increases.

3.4 Small-Scale Fisheries

On a global scale, official statistics on cephalopod landings are dominated by the large-scale trawl and jig fisheries for key species such as *T. pacificus, I. argentinus, Ommastrephes bartramii* and *Dosidicus gigas*. However, an important facet of cephalopod fishing in many parts of the world, including Europe, is the high importance of small-scale artisanal fisheries targeting coastal octopus, cuttlefish and squid using a variety of traditional gears like hand jigs, pots, trammel nets and traps. In addition, squid are targeted by small trawlers, for example, in Scotland (UK; Fig. 3.2): see Pierce et al. (2010) for a recent review of European cephalopod fisheries. In Europe (and very likely in most parts of the world), information on the numbers of boats and fishers involved in small-scale fisheries and amounts landed has, historically, been poorly documented. Guerra et al. (2000) estimated that as much as half of all cephalopod landings in Galicia (Northwest Spain) did not find their way into official statistics. Socially, economically and in terms of protein supplies, small-scale fishing makes an enormous contribution to the wellbeing of coastal communities.

3.5 Stock Assessment and Fishery Management

There is no routine assessment of European cephalopod stocks although there have been several one-off assessments of particular species in particular areas under the auspices of research and data collection projects (e.g. Dunn 1999; Payne et al. 2006; Young et al. 2004, 2006a, b, etc.) and some preliminary assessments by the ICES Working Group on Cephalopod Life History and Fisheries (ICES 2008, 2009, 2010, 2011, 2012). Dunn (1999) pointed to the value of simple techniques such as catch curves while several authors have also utilised the Gomez-Muñoz model to estimate fishing effort for small-scale cephalopod fisheries (e.g. Rocha et al. 2006; Young et al. 2006b).

Traditional age-based methods are generally considered inappropriate for such short-lived species and even if the approach could be adapted by changing the timescale to follow cohorts through the course of a year, age determination is probably too time consuming to be carried out routinely for stock assessment (Ceriola and Milone 2007; Arkhipkin and Shcherbich 2012). Unlike the case in many fish, the high variability in cephalopod growth rates means that the use of age-length keys is impractical.

Historically, production models (Roa-Ureta and Arkhipkin 2007), pre-recruit/recruit surveys and depletion models have all been used with some success (see Pierce and Guerra 1994 for a review). However, the theoretical basis for using production models is of questionable relevance to cephalopods: Environmental carrying capacity is unlikely to be fixed due to the sensitivity of cephalopod population dynamics to variation in environmental conditions and it is doubtful that clear stock–recruitment relationships exist. Depletion methods work best in well-managed directed fisheries such as the Falkland Islands squid fisheries (see, e.g. Beddington et al. 1990; Rosenberg et al. 1990; Basson et al. 1996; McAllister et al. 2004; Roa-Ureta and Arkhipkin 2007), where the licence system required on-board monitoring, allowing real-time assessment, with the possibility of fishery closure if escapement was estimated to have fallen below a target value. In addition, the life cycle of the exploited species includes a well-defined recruitment period, representing the start of the assessment period.

The absence of management measures for most European cephalopod fisheries (apart from some restrictions on gear used and minimum landing size) reflects the low importance attached historically to cephalopod fishing, the lack of attention paid to nontraditional species and small-scale fishing, the impracticality of imposing catch quotas in by-catch fisheries and, possibly, the perception that cephalopods are resilient to overfishing.

However, increased interest in targeting cephalopods in Europe means that even small-scale coastal fisheries can no longer be assumed to be sustainable and, as noted above, lessons from elsewhere in the world suggest that coastal cephalopod species can be overexploited. Contingency plans are also needed to manage the expected high level of interannual variation in landings. In addition, even if catch quotas are unlikely to be useful, some safeguards need to be put in place even for the difficult-to-manage by-catch fisheries. The options include:

- Protection of spawning areas. This is fundamental for coastal demersal and benthic species which attach their eggs to the substrate. Experience in Portugal shows that fixed gear such as trammel nets and traps can be an attractive site for the attachment of loliginid squid eggs (A. Moreno, pers. comm.) and similar issues arise for cuttlefish and octopus. Loss of eggs attached to gear could represent a serious problem for several stocks. In the Saharan Bank octopus fishery, it has been suggested that coastal trap fishing impacts adversely on offshore trawl fishing by removing mature females and eggs (Faraj and Bez 2007).

- Allowing young animals to reach market size. Experience in the Moray Firth squid fishery in Scotland suggests that heavy exploitation too early in the season will yield squid that are too small to market ('powder') and adversely affect catches later in the season (Young et al. 2006a).
- Control of fishing effort in coastal areas, especially by larger vessels, and diversification of target species so that alternatives are available in years of poor recruitment.
- Use of selective gears to minimise by-catch.
- International agreements to protect high seas stocks such as the *I. argentinus* stock in the Southwest Atlantic.
- Development of cephalopod fishery certification schemes.

3.6 Certification Schemes

In recent years, there has been a proliferation of voluntary ecolabelling programmes for various products and sectors. Product certification and ecolabelling are tools that can be used to support fisheries management in order to prevent overfishing and generally improve sustainability, e.g. to prevent, deter and eliminate IUU fishing (Wessells et al. 2001). Product certification is most commonly applied in fisheries where there are particular monitoring and enforcement problems (e.g. in regulating access to a fishery). It has gained heightened importance with the adoption by the FAO Council in 2001 of the International Plan of Action (IPOA) to prevent, deter and eliminate IUU fishing. Ensuring the chain of custody of the products, from harvest to importation into final market, is critical to the effectiveness of a product certification scheme.

One of the best-known independent certification bodies for seafood products (among others like the Aquaculture Stewardship Council (ASC) and the Monterey Bay Aquarium/Seafood Watch) is the Marine Stewardship Council (MSC), which aims to promote sustainable and responsible fisheries and fishing practices worldwide. In the preamble to the Principles and Criteria of the MSC, a sustainable fishery is defined, for purposes of MSC certification, as one which is conducted fulfilling a series of criteria:

1. It can be continued indefinitely at a reasonable level.
2. It maintains, and seeks to maximise, ecological health and abundance.
3. It maintains the diversity, structure and function of the eco-system on which it depends as well as the quality of its habitat, minimising the adverse effects it causes.
4. It is managed and operated in a responsible manner, in conformity with local, national and international laws and regulations.
5. It maintains present and future economic and social options and benefits.
6. It is conducted in a socially and economically fair and responsible manner.

Labelling is also a means by which producers provide information to consumers, so that consumers may make an informed decision. All ecolabelling schemes share the common assumption that consumer product choices are not just motivated by price and mandatory product information. Consumers can be moved by other product attributes that can relate to environmental and ecological objectives as well as economic and social objectives (e.g. fair trade, support to small farmers, discouragement of child labour). Third-party certification (i.e. by a body that is not in any way involved in the production, marketing or consumption of the goods in question) can help ensure that the producer provides the consumer with truthful information (Wessells et al. 2001).

3.7 Markets and Trade: To What Extent Could Aquaculture Contribute to Supply the Markets?

Fish and fishery products are among the most widely traded natural resource-based goods. More than 40% of global fish production enters international trade. For many developing countries, foreign exchange revenues from fish exports make a major contribution to their balance of payments and are thus of strategic macro-economic importance. On the other hand, for the three major global fish importers, namely, Japan, the EU and the USA, processing, wholesaling and retailing of imported fish are of considerable economic significance, in addition to satisfying consumer demand not met by domestic production (Cochrane and Willmann 2000).

According to FAO trade data, over the period 1976–2009, only three European countries have been net exporters of cephalopods: Poland, UK and Ireland. The biggest exporters of cephalopods and cephalopod products in Europe were Poland and the UK. Poland exported more than 400,000 t of cephalopods, presumably mainly squid caught in the Southwest Atlantic, which made up the bulk of Polish landings up until 1996 (Falkland Islands Government 2005). The UK exported more than 160,000 t of cephalopods in the same period. Around two-thirds of UK landings arose from the Northeast Atlantic, most of the remainder from the Southwest Atlantic. The great majority of European countries were and probably still are net importers of cephalopods. The biggest importers of cephalopods in Europe during 1976–2009 were Spain and Italy, with total imports of more than 5 million and more than 4 million t, respectively. Italy was the biggest net importer as it had relatively low exports. The third biggest importer of cephalopods was France, whose total imports over this period were around 675,000 t.

On a world scale, over the period covered by the FAO data (1976–2009), only Japan imported more cephalopods than Italy (almost 7 million t over this period) and it had the largest total net import–export deficit of any country in the world. In total, 12 countries exported more than 1 million t of cephalopods over that period, headed by Spain (although it was a net importer) but also including Thailand, Morocco, Argentina, China, Taiwan, Republic of Korea, India, Peru, Vietnam, New Zealand and USA. Several of these countries also had total net exports of more than 1 million t, led by Morocco (2.2 million t) but also including Argentina, Thailand, Taiwan, Peru, India, Vietnam and New Zealand. In terms of total trade in cephalopods and cephalopod products (i.e. the sum of imports and exports), world leaders were Spain and Japan (more than 7 million t each), followed by China and Italy (more than 4 million t) and Republic of Korea and Thailand (more than 2 million t).

As cephalopod catches increased in the 1980s, international trade also increased—more than doubling during the decade to around 500,000 t in 1989 (Josupeit 1992; Shaw 1994). Shaw (1994) noted that both Spain and Portugal were net importers of squid, while the UK exported more squid than it imported. While the variability in landings within and between years was potentially a problem for processors, it appeared that the high market demand would favour increased squid catching.

In the past 3 years of the FAO trade data set (2007–2009), the leading importers were China, Spain, Japan and Italy. These four countries also had the highest negative trade balance and (along with the Republic of Korea) were among the top five traders of cephalopods. In this period, no European countries had a substantial positive trade balance in cephalopods and cephalopod products. Belgium (total net exports of 1,714 t) was placed highest in Europe but ranked 30th in the world in terms of net exports. Denmark, Ireland and Norway were the only other net exporters in Europe at the time.

Cephalopod trade statistics are also compiled by Globefish. Their most recent issue (Globefish 2012) makes reference to the adverse effects of the world economic situation and limited availability of octopus in 2011. Notable highlights for the first half of 2012 included increased imports of octopus into Japan, increased octopus catching as a ban on factory and freezer vessels in Mauritania was lifted and increased supplies of squid from Argentina. Imports of squid into the USA are increasing, reflecting increased consumer appetite for squid.

Imports of squid and cuttlefish from Chile, Peru, the Falkland Islands and Argentina are currently expanding along with demand for these species from Asian countries. Additionally, the high level of net imports of cephalopods into Spain and Italy and the fact that most European countries are net importers of cephalopods suggest there is a niche for cephalopod aquaculture. Now, the question is to what extent aquaculture could contribute to supply these markets. At the World Congress on Cephalopods, held in Vigo (Spain) in October 2012, it was said that cephalopods (squid, cuttlefish and also octopus) comprise a group for which supplies cannot be increased through aquaculture, at least not yet (Ferdouse 2012), but this viewpoint reflects the fact that cephalopod culture is a relatively recent development and is not yet widely perceived as commercially viable.

3.8 The Future

Examination of FAO cephalopod landings statistics suggests that the boom years of cephalopod fisheries may already be over, although the possibility remains that greater use can be made of oceanic squids, especially if technological developments permit processing of the ammonia-rich flesh of mid-water species belonging to families such as the Histioteuthidae, Cranchiidae and Chiroteuthidae (Voight et al.

1995; Siebel et al. 2005) into palatable form. In addition, global climate change seems certainly to impact species which are notoriously sensitive to environmental conditions. As more such fisheries become fully exploited or overexploited, alternatives will be needed.

As is the case for Gulf of Maine lobster fisheries, some fishing practices may favour proliferation of particular species, effectively moving from hunter–gatherer to farming. However, whether this is a desirable practice is questionable. Ongrowing in sea cages is already established for some species of octopus (see, e.g. Chaps. 9 and 24 for *Octopus vulgaris*) and further development of cephalopod aquaculture represents one of the most promising future options (Iglesias et al. 2007).

3.9 Conclusions

Since 1950, as many finfish stocks have declined, cephalopod fishery resources have come to occupy a leading place in the search for new resources that could support a high level of exploitation, reflecting their high abundance in some areas, as well as their palatability and nutritional qualities.

Cephalopods play an important role as research models in medical and biological research and are keystone species in marine ecosystems, being both active predators and important sources of prey for other species ranging from fish to seabirds and marine mammals. They are found in all the oceans and seas from coastal waters to the high seas, and from the surface to depths of more than 5,000 m.

World cephalopod landings rose from 500,000 t in 1950 to a peak of more than 4 million t in 2007. Over this period, the Northwest Pacific has been the most important fishery region for cephalopods, the Japanese flying squid (*T. pacificus*) being the most important landed species, mainly caught by the Japanese domestic fleet. In the early 1970s, around 40 % of the world catch of cephalopods came from around Japan. Since this time, the Southwest Atlantic and the Southeast Pacific have assumed high importance for catches of abundant marketable species (notably *I. argentinus* and *Dosidicus gigas*), since the mid-1980s and 1990, respectively. On a global scale, cephalopod landings are dominated by the large-scale trawl and jig fisheries, but small-scale artisanal fisheries using a variety of traditional gears play an important role in cephalopod fisheries worldwide.

Exports of fishery products including cephalopods represent a major income for many developing countries, while processing, wholesaling and retailing of imported seafood are of considerable economic significance for the three major global seafood importers (Japan, the EU and the USA), in addition to satisfying consumer demand. FAO data on cephalopod landings suggest that the boom years of cephalopod fisheries may already be over, reflecting overexploitation in some areas while, in the future, climate change may further threaten established cephalopod fisheries. Consequently, the development of culture techniques for cephalopods represents an important alternative.

Acknowledgments We thank the three anonymous referees for their useful comments on the submitted version of this chapter. The fishery statistics reported here derive from the FishStatJ databases made available by the Food and Agriculture Organization (FAO) of the United Nations (FAO Fisheries and Aquaculture Department, Statistics and Information Service. FishStatJ: Universal software for fishery statistical time series. Copyright 2011).

References

Allcock L (2010) Taxonomy. Chapter 2. In: Pierce GJ, Allcock L, Bruno I, Bustamante P, González Á, Guerra Á, Jereb P et al (eds). Cephalopod biology and fisheries in Europe. ICES Cooperative Research Report No. 303. International Council for the Exploration of the Sea, Copenhagen, pp. 6–8

Arkhipkin AI (2013) Squid as nutrient vectors linking Southwest Atlantic oceanic ecosystems. Deep Sea Res Pt II (95:7–20). http://dx.doi.org/10.1016/j.dsr2.2012.07.003

Arkhipkin AI, Shcherbich ZA (2012) Thirty years' progress in age determination of squid using statoliths. J Mar Biol Assoc UK 92:1389–1398. doi:10.1017/S0025315411001585

Basson M, Beddington JR, Crombie JA, Holden SJ, Purchase LV, Tingley GA (1996) Assessment and management techniques for migratory annual squid stocks: the *Illex argentinus* fishery in the Southwest Atlantic as an example. Fish Res 28:3–27

Beddington JR, Rosenberg AA, Crombie JA, Kirkwood GP (1990) Stock assessment and the provision of management advice for the short fin squid fishery in Falkland Islands waters. Fish Res 8:351–365

Branch TA, Watson R, Fulton EA, Jennings S, McGilliard CR, Pablico GT, Ricard D, Tracey SR (2010) The trophic fingerprint of marine fisheries. Nature 468:431–435

Caddy JF (1983) Advances in assessment of world cephalopod resources. Progrès réalisés dans l'évaluation des ressources mondiales de céphalopodes. Progresos realizados en la evaluación mundial de cefalópodos. FAO Fisheries Technical Paper. FAO Document technique sur le pêches. FAO Documento Técnico de Pesca. No.231. Rome

Caddy JF, Rodhouse PG (1998) Cephalopod and groundfish landings: evidence for ecological change in global fisheries? Rev Fish Biol Fisher 8:431–444

Ceriola L, Milone N (2007) Cephalopods age determination by statolith reading: a technical manual. Scientific Cooperation to Support Responsible Fisheries in the Adriatic Sea. GCP/RER/010/ITA/TD-22. AdriaMed Technical Documents 22:78

Clarke MR (1987) Cephalopod biomass—estimation from predation. In: Boyle PR (ed) Cephalopod life cycles, vol. 2: comparative reviews. Academic Press, London, pp 221–238

Clarke MR (1996) Cephalopods as prey. III. Cetaceans. Philos T R Soc Lond B 351:1053–1065

Cochrane K, Willmann R (2000) Eco-labelling in fisheries management. In: Nordquist MH, Moore JN (eds) Current fisheries issues and the Food and Agriculture Organization of the United Nations. Martinus Nijhoff Publishers, The Hague, pp 583–615

Domingues PM, Sykes A, Andrade JP (2001) The use of *Artemia* or mysids as food for hatchlings of the cuttlefish *Sepia officinalis*. Effects on growth and survival throughout the life cycle. Aquac Int 9:319–331

Dunn MR (1999) Aspects of the stock dynamics and exploitation of cuttlefish, *Sepia officinalis* (Linnaeus, 1758), in the English Channel. Fish Res 40:277–293

Falkland Islands Government (2005) Fisheries Department Fisheries Statistics, vol 9, 2004. FIG Fisheries Department, Stanley

FAO (2001) International Plan of Action to prevent, deter and eliminate illegal, unreported and unregulated fishing. FAO, Rome

FAO (2011) FAO Fisheries and Aquaculture Department, Statistics and Information Service. FishStatJ: Universal software for fishery statistical time series. FAO, Rome

Faraj A, Bez N (2007) Spatial considerations for the Dakhla stock of *Octopus vulgaris*: indicators, patterns, and fisheries interactions. ICES J Mar Sci 64(9):1820–1828

Ferdouse F (2012) Cephalopod markets in Asia. World Congress on Cephalopods, Vigo, Spain, 1 October 2012

Forsythe JW (1993) A working hypothesis of how seasonal temperature change may impact the field growth of young cephalopods. In: Okutani T, O'Dor RK, Kubodera T (eds) Recent advances in cephalopod fisheries biology. Tokai University Press, Tokyo, pp 133–143

Forsythe JW, Walsh LS, Turk PE, Lee PG (2001) Impact of temperature on juvenile growth and age at first egg-laying of the Pacific reef squid *Sepioteuthis lessoniana* reared in captivity. Mar Biol 138:103–112

Fries A (2010) A gap analysis of the distribution of cephalopod species worldwide with a focus on commercially important species. Masters Thesis. Duke University

Funge-Smith S, Briggs M, Miao W (2012) Regional overview of fisheries and aquaculture in Asia and the Pacific 2012. Asia-Pacific Fishery Commission, FAO Regional Office for Asia and the Pacific. RAP Publication 2012/26. pp. 139. http://www.apfic.org/downloads/finish/11-apfic-regional-overviews/9-2012-apfic-regional-overview-of-fisheries-and-aquaculture-in-asia-and-the-pacific.html. Accessed 12 April 2013

Gasalla MA, Rodrigues AR, Postuma FA (2010) The trophic role of the squid *Loligo plei* as a keystone species in the South Brazil Bight ecosystem. ICES J Mar Sci 67:1413–1424

Globefish (2012) Globefish highlights issue 3/2012, 15/08/12

Guerra A, Gracia J, González A, Rocha F (eds) (2000) Development of software to estimate unreported and misreported catch and effort data and to apply fishery management models. Final Report to the Commission of the European Communities Directorate General for Fisheries on Study No 97/0107, Instituto de Investigaciones Marinas, Vigo

Hanlon RT, Messenger JB (1998) Cephalopod behaviour. Cambridge University Press, Cambridge

Hilborn R (2012) The evolution of quantitative marine fisheries management 1985–2010. Nat Resour Model 25:122–144

Hunsicker ME, Essington TE, Watson R, Sumaila UR (2010) The contribution of cephalopods to global marine fisheries: can we have our squid and eat them too? Fish Fish 11:421–438

ICES (2008) Report of the Working Group on Cephalopod Fisheries and Life History (WGCEPH). ICES Document CM 2008/LRC:14. International Council for the Exploration of the Sea, Copenhagen

ICES (2009) Report of the Working Group on Cephalopod Life History and Fisheries (WGCEPH). ICES Document CM 2009/LRC:06. International Council for the Exploration of the Sea, Copenhagen

ICES (2010) Report of the Working Group on Cephalopod Fisheries and Life History (WGCEPH), 9–11 March 2010, Sukarrieta, Spain. ICES CM 2010/SSGEF:09. International Council for the Exploration of the Sea, Copenhagen

ICES (2011) Report of the Working Group on Cephalopod Fisheries and Life History (WGCEPH), 28 February–03 March 2011, Lisbon, Portugal. ICES CM 2011/SSGEF:03. International Council for the Exploration of the Sea, Copenhagen

ICES (2012) Report of the Working Group on Cephalopod Fisheries and Life History (WGCEPH), 27–30 March 2012, Cadiz, Spain. ICES CM 2012/SSGEF:04. International Council for the Exploration of the Sea, Copenhagen

Iglesias J, Sánchez FJ, Bersano JGF, Carrasco JF, Dhont J, Fuentes L, Linares L, Muñoz JL, Okumura S, Roo J, van der Meeren T, Vidal EAG, Villanueva R (2007) Rearing of *Octopus vulgaris* paralarvae: Present status, bottlenecks and trends. Aquaculture 266:1–15

Jackson GD, Domeier ML (2003) The effects of an extraordinary El Niño/ La Niña event on the size and growth of the squid *Loligo opalescens* off Southern California. Mar Biol 142:925–935

Jereb P, Roper CFE (eds) (2010) Cephalopods of the world. An annotated and illustrated catalogue of cephalopod species known to date. Vol 2. Myopsid and Oegopsid Squids. FAO Species Catalogue for Fishery Purposes. No. 4, Vol. 2. FAO, Rome

Josupeit H (1992) World supply and markets (English). In: Squid 91 Madrid. Proceedings; World Cephalopod Conference, 2, Madrid (Spain), 11–13 Nov 1991/ INFOFISH, Kuala Lumpur (Malaysia); Agra Europe Ltd., London, p 155–162

Koueta N, Boucaud-Camou E (1999) Food intake and growth in reared early juvenile cuttlefish *Sepia officinalis* L. (Mollusca: Cephalopoda). J Exp Mar Biol Ecol 240:93–109

Lee PG (1994) Nutrition of cephalopods: fuelling the system. Mar Freshw Behav Phy 25:35–51

Ling J, Zheng Y (2000) Stock assessment of Cephalopoda in East China Sea and Yellow Sea. Marine Fisheries Vol. 22, pp. 60–62 (2000). (in Chinese with English abstract)

McAllister MK, Hill SL, Agnew DJ, Kirkwood GP, Beddington JR (2004) A Bayesian hierarchical formulation of the De Lury stock assessment model for abundance estimation of Falkland Islands' squid. Can J Fish Aquat Sci 61:1048–1059

Nigmatullin ChM (2004) Estimation of biomass, production and fishery potential of ommastrephid squids in the World Ocean and problems of their fishery forecasting. ICES CM 2004/ CC: 06. 2004. International Council for the Exploration of the Sea, Copenhagen

Nigmatullin ChM (2010) The world cephalopod fishery history, current state and prospects: a brief overview. In: Abstracts volume of the 5th International Symposium on Pacific Squids 2010 (October 13–15 2010, La Paz, Baja California Sur. Mexico). CIBNOR, La Paz, p 53

Oestmann DJ, Scimeca JW, Forsythe JW, Hanlon RT, Lee PG (1997) Special considerations for keeping cephalopods in laboratory facilities. Contemp Top Lab Anim 36:89–93

Pauly D (1998) Why squids, though not fish, may be better understood by pretending they are. In: Payne AIL, Lipinski MR, Clarke MR, Roeleveld MAC (eds). Cephalopod biodiversity, Ecology and Evolution. S Afr J Mar Sci 20:47–58

Payne AG, Agnew DJ, Pierce GJ (eds) (2006) Trends and assessment of cephalopod fisheries—Proceedings of the CEPHSTOCK cephalopod assessment workshop. Fish Res 78:1–106

Pecl GT, Moltschaniwskyj NA, Tracey SR, Jordan AR (2004) Inter-annual plasticity of squid life history and population structure: ecological and management implications. Oecologia 139:515–524

Pierce GJ, Guerra A (1994) Stock assessment methods used for cephalopod fisheries. Fish Res 21:255–285

Pierce GJ, Allcock L, Bruno I, Bustamante P, González Á, Guerra Á, Jereb P, Lefkaditou E, Malham S, Moreno A, Pereira J, Piatkowski U, Rasero M, Sánchez P, Santos MB, Santurtún M, Seixas S, Sobrino I, Villanueva R (eds) (2010) Cephalopod biology and fisheries in Europe. ICES Cooperative Research Report No. 303. International Council for the Exploration of the Sea, Copenhagen

Punt AE, Leslie RW, Du Plessis SE (1992) Estimation of the annual consumption of food by Cape hake *Merluccius capensis* and *M. paradoxus* off the South African west coast. S Afr J Mar Sci 12:611–634

Roa-Ureta R, Arkhipkin AI (2007) Short-term stock assessment of *Loligo gahi* at the Falkland Islands: sequential use of stochastic biomass projection and stock depletion models. ICES J Mar Sci 64:3–17

Rocha F, Guerra Á, González Á (2001) A review of reproductive strategies in cephalopods. Biol Rev 76:291–304

Rocha F, Otero J, Outeiral R, González AF, Gracia J, Guerra A (2006) Modelling small-scale coastal fisheries of Galicia (NW Spain) based on data obtained from fisheries: the case of *Sepia officinalis*. Sci Mar 70:593–601

Rosa R, Seibel BA (2008) Synergistic effects of climate-related variables suggest future physiological impairment in a top oceanic predator. Proc Natl Acad Sci U.S.A. 52:20776–20780

Rosenberg AA, Kirkwood GP, Crombie JA, Beddington JR (1990) The assessment of stocks of annual squid species. Fish Res 8:335–350

Shaw R (1994) An economic description and assessment of the squid catching industry in Great Britain, Portugal and Spain. Fish Res 21:287–303

Shaw PW, Boyle PR (1997) Multiple paternity within the brood of single females of *Loligo forbesi* (Cephalopoda: Loliginidae), demonstrated with microsatellite DNA markers. Mar Ecol-Prog Ser 160:279–282

Shaw PW, Sauer WHH (2004) Evidence for multiple paternity and complex fertilisation dynamics in the squid *Loligo vulgaris reynaudii*. Mar Ecol-Prog Ser 270:173–179

Siebel BA, Goffredi SK, Thuesen EV, Childress JJ, Robison BH (2005) Ammonium content and buoyancy in midwater cephalopods. J Exp Mar Biol Ecol 313:375–387

Smale MJ (1996) Cephalopods as prey. IV. Fishes. Philos T R Soc Lond B 351:1067–1081

Smith JM (2011) Growth investment and distribution of the squid *Loligo forbesii* (Cephalopoda: Loliginidae) in Northeast Atlantic Waters. PhD thesis, University of Aberdeen

Temple SE, Pignatelli V, Cook T, How MJ, Chiou T-H, Roberts NW, Marshall NJ (2012) High-resolution polarisation vision in a cuttlefish. Curr Biol 22:R121–R122

Voight JR, Pörtner HO, O'Dor RK (1995) A review of ammonia-mediated buoyancy in squids (cephalopoda: Teuthoidea). Mar Freshw Behav Phy 25(1–3)

Wangvoralak S (2011) Life history and ecological importance of the veined squid *Loligo forbesii* in Scottish waters. PhD thesis, University of Aberdeen

Wessells CR, Cochrane K, Deere C, Wallis P, Willmann R (2001) Product certification and ecolabelling for fisheries sustainability. FAO Fisheries Technical Paper. No. 422. FAO, Rome

Wood J, Jackson K (2004) Introduction to: Cephalopod Vision http://www.thecephalopodpage.org/cephschool/CephalopodVision.pdf. Accessed 12 April 2013

Worm B, Hilborn R, Baum JK, Branch TA, Collie JS, Costello C, Fogarty MJ, Fulton EA, Hutchings JA, Jennings S, Jensen OP, Lotze HK, Mace PM, McClanahan TR, Palumbi SR, Parma AM, Rikard D, Rosenberg AA, Zeller D, Minto C (2009) Rebuilding global fisheries. Science 325:578–585

Young IAG, Pierce GJ, Daly HI, Santos MB, Key LN, Bailey N, Robin J-P, Bishop AJ, Stowasser G, Nyegaard M, Cho SK, Rasero M, Pereira JMF (2004) Application of depletion methods to estimate stock size in the squid *Loligo forbesi* in Scottish waters (UK). Fish Res 69:211–227

Young IAG, Pierce GJ, Stowasser G, Santos MB, Wang J, Boyle PR, Shaw PW, Bailey N, Tuck I, Collins MA (2006a) The Moray Firth directed squid fishery. Fish Res 78:39–43

Young IAG, Pierce GJ, Murphy J, Daly HI, Bailey N (2006b) Application of the Gómez-Muñoz model to estimate catch and effort in squid fisheries in Scotland. Fish Res 78:26–38

Chapter 4
Historical Review of Cephalopods Culture

António V. Sykes, Noussithé Koueta and Carlos Rosas

Abstract This chapter reviews the history of cephalopod culture since the 1960s until 2000, compiling the most important contributions in each decade and identifying key research laboratories and researchers. The literature found is vast in cephalopod species, methodologies and technology. Hence, this chapter focuses mainly on those species with an established aquaculture potential. It includes a description of the evolving seawater systems, broodstock acclimatisation to captivity and rearing/ culture methodologies of different life stages.

Keywords Cephalopod culture history · Cuttlefish · Model species · Octopus · Squid

4.1 Cephalopods as Model Species for Science—the Need for Aquaculture

Cephalopods have been used as experimental models in neurobiology since the beginning of the twentieth century (Gariaeff 1906; van der Sprenkel 1929; Arvanitaki et al. 1936a, b; Young 1938). Species of this class gained momentum as models for science in 1963, when Alan L. Hodgkin and Andrew F. Huxley won the Nobel Prize

A. V. Sykes (✉)
CCMAR-CIMAR L.A., Centro de Ciências do Mar do Algarve,
Universidade do Algarve, Campus de Gambelas,
8005-139, Faro, Portugal
e-mail: asykes@ualg.pt

N. Koueta
Laboratoire de Biologie des Mollusques Marins et des Ecosystèmes Associés:
BioMEA, Université de Caen, 14032, Caen Cedex, France
e-mail: noussithe.koueta@unicaen.fr

C. Rosas
Unidad Multidisciplinaria de Docencia e Investigación,
Facultad de Ciencias—UNAM, Puerto de Abrigo s/n Sisal.
Opio. Hunucma, Yucatan, Mexico
e-mail: crv@ciencias.unam.mx

J. Iglesias et al. (eds.), *Cephalopod Culture,* DOI 10.1007/978-94-017-8648-5_4,

in Physiology and Medicine for their work on the behaviour of nerve impulses, which used the giant axon of the Atlantic squid *Doryteuthis pealeii*. In addition, two other Nobel Laureates worked with squid: the American, G. Wald honoured in 1967 for his research on chemical and physiological processes in the eye and Bernard Katz of Great Britain, whose discoveries concerning the role played by chemicals in nerve impulses won him the prize in 1970. From that point on, cephalopods became the most relevant neurological model, with no peer in the animal kingdom (Young 1985), and cephalopod culture was more than ever a necessity that needed to be met. However, this was easier said than done and there was a need for studies regarding the knowledge of species biology and the development of technology that would allow for cephalopod culture.

Besides the interest on the cephalopod culture as biological models in neuroscience (Williamson and Chrachri 2004; Sio 2011), they also have been consistently used as models for research to understand processes of their biology (Wood and O'Dor 2000; Koueta et al. 2006), mechatronics (Laschi et al. 2012), behaviour (Wells 1962; Hanlon and Messenger 1996; Tricarico et al. 2011), evolution (Budelmann 1995; Strugnell et al. 2011) and climate changes (Pörtner and Farrell 2008; Melzner et al. 2009). Moreover, they represent a quality food source for human consumption (Vaz-Pires et al. 2004; Sykes et al. 2009; Uriarte et al. 2011). Other uses, but not all, are presented by Lee et al. (1994). In this sense, the present chapter provides a historical perspective of cephalopod culture in the world as informative and detailed as possible, but, inevitably, there will be gaps and inaccuracies that escaped our attention. The information is given considering a timescale on the advances on cephalopod culture since the 1960s up to 2000. Detail is given to the most researched species, while citations of the existing literature and basic description of its content are made for the remaining ones. To avoid any misunderstanding, the terminology of maintenance, rearing and culture used throughout the chapter is that proposed by Boletzky and Hanlon (1983).

4.2 1960s: The First Dedicated Studies

There are previous reports on maintaining and rearing of cephalopods of early years but the first pioneer studies on the experimental culture of cephalopods are dated from the 1960s, and many of them were performed by Korean and Japanese researchers on species of the Sepiidae family. Ohshima and Choe (1961) described the rearing success in four cephalopod species (*Sepioteuthis lessoniana, Sepia esculenta, S. lycidas* (as *S. subaculeata*) and *Sepiella inermis* (as *S. maindroni*)) with paralarvae/hatchlings being fed live mysid shrimp *Neomysis japonica*. Choe and Ohshima (1963) provided more details on the rearing of those species and of *Euprymna berryi,* namely information regarding the care of eggs, time of first feeding, food items and survival. On the other hand, Itami et al. (1963) succeeded in rearing *Octopus vulgaris* paralarvae up to the benthic juvenile stage in 33–40 days on a diet of zoeas of *Palaemon serrifer* at Hyogo Prefectural Fisheries

Experimental Station (Akashi, Japan). After settling, the octopus juveniles were fed with small pieces of ovaries, testes and hepatic glands of the crab *Charybdis japonica,* and, then after, with small shrimps or young crabs (*Gaetice depressus*). This work highlighted high mortalities, which were considered to be due to lack of proper food in quantity. Finally, Choe (1966) provided more data on *S. esculenta, S. lycidas* (*as S. subaculeata*), *S. inermis* (*as S. maindroni*), *S. lessoniana* and *E. berryi* eggs, respective rearing setup, food and growth under captive conditions. On the other hand, the culture of cephalopods for neurological studies produced some research papers that described some techniques to maintain octopuses in captivity. For instance, Coates et al. (1965) developed an automatic food dispenser for *O. vulgaris* which was useful for maintaining these animals in captivity. This apparatus was designed to make observations on the amount, time and frequency of octopus feeding.

4.3 1970s: Cephalopod Biology Studies and Maintenance Gain Momentum

In this decade, research generating further development of knowledge regarding biology and aquaculture technology of cephalopods progressed with several publications from researchers from France (Laboratoire Arago, Banyuls-sur-Mer; and Laboratoire de Biologie et d'Ecologie Marines, Université de Lille), Spain (Instituto de Investigaciones Pesqueras, Laboratorio de Cádiz and Vigo), USA (Marine Biological Laboratory, Woods Hole; and University of Miami), among others. The basic maintenance setups were developed by researchers who wanted to know more about cephalopod biology and physiology and needed seawater systems that allowed keeping animals in captivity.

For instance, the doctoral thesis of Richard (1971) dedicated a whole chapter to the culture of the European cuttlefish (*S. officinalis*). Different temperatures and their relation with the duration of embryonic development (e.g. 15 °C—87 days against 21.4 °C—31 days), female sizes and hatching sizes were provided. In addition, conditions of the seawater systems to permit cuttlefish culture were presented, such as data on open systems, waters flows (0.5 L min^{-1} in 100 L tank), water heating (to insure the minimum specific temperatures of 9–10 °C), the use of sand bottom (better acclimatisation to captive conditions), the use of amphipods as food for hatchlings and shrimps and crabs for juveniles or adults and density (D = cuttlefish 'surface'/tank bottom surface = 1/10).

LaRoe (1971) described how to culture and maintain the squids *S. sepioidea* and *D. plei* throughout the life cycle in captivity. In this study, information regarding the most important factors, such as food quantities (daily rates of 30–60 BW (body weight) day^{-1}), food conversion efficiency (10–20 %), minimum salinity (27 psu), minimum and maximum culture temperature (17.5–33 °C), oxygen concentrations and other information were provided. A different method for maintaining *Loligo vulgaris* was presented by Neill (1971), which was based on the use of collapsible plastic home swimming pools converted to ring-shaped arenas by using a suspended plastic opaque wall and continuous lighting. According to this author, squids should be kept in groups, fed live prawns and the tanks should be part of an open seawater system. Boletzky et al. (1971) described the laboratory rearing of Sepiolinae in small volumes (5–50 L) in a closed seawater system. Cuttlefish were fed on live mysids (*Leptomysis mediterranea*) and shrimp (*Leander serratus*). A description of their behaviour in captivity and feeding was provided, as well as indications on growth, maturation and spawning in captivity that were reported for the first time.

Two similar types of works were published by Boletzky (1974b, c). In these articles, Boletzky described the methods used to rear cephalopods in the laboratory for several orders, such as Sepioidea, Teuthoidea and Octopoda. It reviewed aspects such as water supply (open and semiopen seawater systems), water quality through filtering, reproduction behaviour, egg collection either from nature or from captive spawners, embryonic development periods according to given temperatures and rearing of paralarvae or hatchlings (food, temperature, light, enriched environments, feeding rates and densities). In that same year, Boletzky (1974d) also published a review on the biology of embryonic and post-embryonic development of cephalopods with the aim of defining whether cephalopod hatchlings should be called larvae, and a study on the effects of malnutrition on the development of *S. officinalis* cuttlebone (Boletzky 1974a). At the same time, Arnold et al. (1974) published a guide for the laboratory use of the squid *D. pealeii,* which included information regarding collection, maintenance, reproductive behaviour, embryonic development, the visual system, plus protocols for the use of squid for electrophysiology, among others. Summers and McMahon (1974) and Summers et al. (1974) described capture methods and seawater systems for maintenance of the same species. In addition, Overath and Boletzky (1974) described basic conditions to maintain eggs of the blue-ringed octopus *Hapalochlaena lunulata* through the embryonic development.

Boletzky (1975) described how to rear *Eledone moschata* from hatching to adulthood in aquaria, by feeding a diet of shrimp and crabs. He verified that, in this species, males mature later than females. In that same year, Richard (1975) published a very complete guide on cuttlefish culture, resulting from 10 years of research. Due to the location of the research laboratory (Station Marine de Wimereux, France), the cuttlefish culture setup used open seawater systems, water heaters (to maintain water temperature above the minimum reported for the species), sand in the bottom of the tanks and dim illumination. This author also reported values for optimal density (1/10 of the animals' surface to the tank surface) and the use of live prey items with half the size of cuttlefish (which should not be provided in excess).

O'Dor et al. (1977) described methods for collection and maintenance of *Illex illecebrosus* in a 15 m diameter tank and gave information on reproduction and life cycle in captivity. Guerra (1978) described a basic seawater setup to maintain *O. vulgaris* individuals in the laboratory. In this latter study, not only were a wide variety of live and dead prey items tested as food for octopuses but also its behaviour towards the acceptance of food was recorded. If fed sufficient quantities, octopuses would prefer crustaceans to fish.

Van Heukelem (1977) introduced *O. maya* as a better laboratory animal than other cephalopods due to its larger eggs and easier rearing and maintenance. In this work, he described how to collect, ship and incubate eggs. He also described rearing conditions such as temperature (21.8–26.7 °C), the use of egg incubators, tank volumes (20–120 L) and its maintenance and the use of several food items according to age (live brine shrimp, crabs, amphipods and isopods during the first month and frozen afterwards).

Together with Richard, Pascual (1978) was one of the first researchers who cultured *S. officinalis* for more than one generation, collecting, in this way, valuable information regarding growth, feeding and different water parameters in inbred populations. This study was one of the first to be performed aiming ultimately at mass production, for human consumption, of a given cephalopod species. Information on how to assemble an open seawater system with water temperature regulation and security alarms, plus switches, was given. In addition, the types of tanks were described considering their use for rearing all life stages of cuttlefish, with the exception of the egg stage (which was performed in a closed system). Data on embryonic development periods for the species at different culture temperatures (30 days for 22.5 °C and 54 days for 17 °C) as well as thermal tolerance of eggs were reported. This was one of the first studies reporting the removal of enriched environments in the tanks to avoid problems with food remains leading to deterioration, which were reported later by other authors for this and other cephalopod species. Several prey types were tested for the hatchling stage, such as mysid species (*Diamysis bahirensis* and *Mesodopsis slabberi*), grass shrimp (*Palaemonetes varians*) and *Artemia* spp., *P. varians* being the prey generating the best results in survival. This study also reported urethane (0.3–0.5 %) as an anaesthetic for the manipulation of cuttlefish. Under these conditions, individuals of 150–200 g were obtained in 4 months, displaying a food conversion efficiency of 40–45 % (from 2.2 to 2.5 kg of food to attain 1.0 kg of cuttlefish).

Boletzky (1979a) used 50 L round covered tanks (PVC or plexiglass), under continuous light (fluorescent white), in an open seawater system to test the acceptance of different prey types and prey quantities. Mysids, artemia and amphipods were tested for hatchlings, with a clear preference for mysids, palaemonid and crangonid prawns. Very small crabs and small fish were assessed for cuttlefish juveniles and larger fish, prawns and crabs for subadults and adults. The same author (Boletzky 1979b) also reported extremely high mortality rates in *L. vulgaris* paralarvae when fed mysids under continuous light conditions.

Detailed information on the reproduction of *I. illecebrosus* in captivity, performed in the Aquatron pool (system described by O'Dor et al. (1977)), was reported by Durward et al. (1979); while Hulet et al. (1979) described fin damage of several squid species (*D. pealeii, D. plei, Lolliguncula brevis* and *D. opalescens*) during capture, transport and maintenance or rearing in the 1,000 L circular tanks and 10,000 L raceway tanks of a closed seawater system at the National Resource Center for Cephalopods (NRCC).

4.4 1980s: The First Studies on Mass Culture

It was during this decade that methods for large-scale culture were investigated and cephalopod culture technology reported rapidly increasing numbers. This boom correlates with the establishment of the NRCC in 1975, which was a scientific programme supporting the mission of the Marine Biomedical Institute (MBI) of the University of Texas Medical Branch. The NRCC was located in Galveston (50 miles south of Houston), with access to the Gulf of Mexico, and it established a 30-year history of providing the biomedical research community with squid, cuttlefish and other cephalopods for research. It had the capacity to maintain 1,000–2,500 adult cuttlefish, *S. officinalis* and *S. pharaonis* (122,000 L of water in five systems) per year, 800 squid, *S. lessoniana* (98,000 L of water) per year, 10–40 *Nautilus pompilius* (3,000 L of water) per year. In addition, this facility had the capacity to maintain field-caught bay squid, *L. brevis* (1,000–2,000 adults year^{-1}), the sepiolids *E. scolopes* and *E. tasmanica* (200 adults year^{-1}), *O. bimaculoides* (300 year^{-1}) in relatively small 500 L tanks. Giant Pacific octopuses were held in 3,500 L chilled recirculating seawater display tanks. During this time, the NRCC was the only facility able to culture squids from egg to adulthood for multiple generations. Also, they were able to provide cuttlefish and octopuses, as well as cells, tissues and organ systems to researchers without causing the trauma of mass harvesting, thus providing excellent research animals while preserving the natural populations of cephalopods in the oceans. A great contribution of the NRCC was performing numerous experiments on the development of culture technology with more than one cephalopod species, making results applicable in a general sense to most cephalopods.

Boletzky and Hanlon (1983) reviewed the laboratory maintenance, rearing and culture of several cephalopod species. In this article, the authors provided a standardization of terminology to be used while having cephalopods in captivity: **maintenance**—holding juvenile or adult cephalopods with no intention of growing them; **rearing**—growing cephalopods through several life stages of a single generation; and **culture**—growing a cephalopod from hatching until hatchlings of a consecutive generation are obtained. The different sections of this work addressed technological limitations faced by those having a cephalopod in captivity as well as species-specific biological limitations to culture. In a general way, cephalopods need to be handled carefully to avoid skin damage, require enough space for their benthonic or nektonic mode of life, high-quality water and sufficient live food. Hanlon and Hixon (1983) wrote a chapter on laboratory maintenance and culture of octopuses and loliginid squids in *Culture of Marine Invertebrates*. In this work, the technology described for *O. joubini* was set up as an example for the culture of other octopus species, such as *O. briareus* (culture methods later described by Robaina (1983) and Hanlon and Wolterding (1989)), *O. maya* and *O. bimaculoides*. The methods were also described to have a possible application to *O. vulgaris* paralarvae. Information on rearing methodologies for these and other species was given and included details on seawater systems for several octopus and loliginid species. Two years later, Hanlon and Forsythe (1985) published the results of large-scale culture of five species of octopuses that included the four octopus species already mentioned above and

Paroctopus digueti, and in which *O. bimaculoides* was pinpointed as the best species for culture. A full description of the seawater systems was given as well as its maintenance and operability logistics. Closed seawater systems were suggested as a way of preventing high variability in salinity conditions. However, caution with the use of copper in these systems was recommended due to its toxicity to cephalopods. Details on stocking density (300–700 hatchlings m^{-2}), prey size (1/3 to 2× octopus mantle length), prey type (mysids and palaemonid shrimps) and alternative foods were presented. Moreover, the effects of temperature on growth, reproduction and duration of life cycle were assessed by Forsythe and Hanlon (1988).

Regarding squids, Yang et al. (1980) were the first to successfully rear *D. pealeii* hatchlings. A full description of the conditions (seawater systems and food) that allowed the rearing of *D. opalescens* to the juvenile stage were published by Yang et al. (1983), experiments of adaptation of *D. plei, D. pealeii* and *Lolliguncula brevis* to captive conditions by Hanlon et al. (1983) and the culture in captivity was established 3 years later by Yang et al. (1986). In that same year, Turk et al. (1986) reared *L. vulgaris* from hatching with some success. In a series of trials, they collected information regarding food type, feeding densities, growth and survival. Culture conditions for *L. forbesii* were published by Hanlon et al. (1989). All the gathered knowledge resulted in a review on how to design closed seawater systems and culture loliginid squids by Yang et al. (1989). According to these authors, squid hatchlings should be reared in 3,000 L circular tanks, with low light intensities (4–53 Lx) and fed on mysid shrimps. As for juveniles, a setup of a 10,000 L raceway tank should be used with constant illumination not exceeding 88 Lx. Further details and schemes are presented in the work.

All the previous research lead to the reviews on alternative diets for maintaining and rearing cephalopods in captivity by Toll and Strain (1988) and DeRusha et al. (1989), where the suitability of several freshwater and terrestrial food organisms (crayfish, fishes, clams, snails, earthworms, insect larvae and salamanders) as well as live and frozen diets (marine shrimps, marine worms, marine crabs, marine fish, adult brine shrimp) were assessed for different cephalopod species (*O. bimaculoides, O. maya, P. digueti, O. joubini* and *S. officinalis*) and life stages. Crayfish was the most accepted diet and the one providing the best results in growth among the freshwater and terrestrial food items tried. On the other hand, frozen diets provided similar growth compared to live diets.

During those years, the NRCC researchers also published a vast group of publications regarding cephalopod pathologies and healing methods that are revised in Chap. 6—Welfare and diseases under culture conditions.

In 1983, the first volume (Species Accounts) of the book *Cephalopod Life Cycles* was published by Academic Press and edited by Peter Boyle. Four years later, the second volume (Comparative Reviews) was published including a full chapter dedicated to reviewing cephalopod culture (Mariculture by Hanlon (1987)) and another one to cephalopod diets (Nixon 1987).

During this decade, several papers (Boucaud-Camou 1981; Boucaud-Camou and Yim 1981; Boucaud-Camou 1982; Boucaud-Camou et al. 1985) and book chapters (Boucaud-Camou and Boucher-Rodoni 1983; Boucher-Rodoni et al. 1987) were published on feeding and digestion of cephalopods, with emphasis on *S. officinalis*. In Europe, and still during the 1980s, some Italian researchers made a series of experiments (Sequi 1980; Palmegiano and Sequi 1981; D'Apote and Palmegiano 1982; Palmegiano and D'Apote 1983; Palmegiano and Sequi 1984; Sequi and Palmegiano 1984, 1985) with the objective of understanding the feasibility of cuttlefish culture in earthen ponds. Further details on these studies are given in Chap. 11—European Cuttlefish, *S. officinalis*. In fact, at the end of this decade, this species and *O. vulgaris* were identified as the most probable species to be mass cultured in Europe by some authors (Boucaud-Camou 1989; Coelho et al. 1989).

4.5 1990–2000: The Optimization of Culture Protocols

The potential for commercial aquaculture of cephalopods was engaging in its first steps but it was already acknowledged in several articles (Boucaud-Camou 1990; Bernardino 2000) and books (Barnabé 1996) regarding such activity. After establishing basic culture methods for several species in the 1980s, protocols for new species, reliable seawater systems and optimization of protocols were published during this decade.

The main optimizations (applicable to all closed and semiclosed seawater systems) were the use of biological denitrification (Whitson et al. 1993; Lee et al. 2000), which was controlled by computer hardware and software (Lee 1994, 1995), and the use of airlifts to substitute pump-driven closed recirculating marine culture systems (Hanlon et al. 1991).

Updated information on maintenance, rearing and culture of teuthoid and sepioid squids was published by Hanlon (1990) that included culture protocols for given species (*D. pealeii, D. opalescens, L. forbesii, S. lessoniana* and *S. officinalis*), critical information on how to choose the type of seawater system and its design (location, tank design, control of water physicochemical conditions) and methods for capture and transport of eggs, juveniles and adults (when these are not available near the research facilities). This comprised considerations on culture protocols and seawater systems regarding anatomical and behavioural traits of these species (e.g. delicate skin and excellent vision) and their locomotor habits (e.g. jet-propulsion escape and the coping with painted tank walls and bottom substrate). While loliginid squid were characterized as especially sensitive to captive conditions, *S. officinalis* was presented as a rather hardy species, able to tolerate relatively high amounts of nitrogenous compounds (1.0 mg L^{-1} of ammonia and nitrite and more than 125 mg L^{-1} of nitrate) and lower salinity (as low as 20 psu).

Boyle (1991) published the *Universities Federation for Animal Welfare (UFAW)* handbook on the care and management of cephalopods in the laboratory, which covered almost every single aspect regarding keeping a culture of a cephalopod for science purposes. Among those included, methods for capture, handling and transport, water quality, space requirements and housing and substrate had applicability for large-scale application.

However, regarding *S. officinalis,* Forsythe et al. (1991) and Forsythe et al. (1994) provided information on large-scale culture (in a surface area of 38 m^2 and a total volume of 18,000 L) of several consecutive captive generations in closed seawater systems (described in detail in Hanlon and Forsythe (1985) and Yang et al. (1989) for other cephalopod species). Information herein demonstrated that the species may be cultured in most tank configurations (round, square or rectangular), the tank bottom area being the most important dimension in high-density culture leading to the use of shallow tanks (5 cm for hatchlings to 30–40 cm for juvenile/adults). This setup allowed the rearing of 3,000 juveniles (up to 4 cm mantle length—ML; 250–300 hatchlings m^{-2}) or 200–400 10 cm ML juveniles (≈ 20 cuttlefish m^{-2}) or 75–100 adults with ≈ 20 cm ML (2 cuttlefish m^{-2}) in a tank with ≈ 10,000 L. These articles also provided methods for broodstock management related to the reproductive biology of the species. Tank sex ratios of 1♂:3♀ instead of 1♂:1♀ were suggested as a way to diminish the aggressive sexual behaviour displayed by *Sepia* adults. While fecundity in captivity was reported to be at similar levels to those reported for wild animals, fertility was less than 50 % after the first generation and null after only seven generations. Whether this was due to inbreeding or to the lack of symbiotic bacterial populations remains to be determined. Culture cycles were temperature related, being shorter at higher temperatures, with mean individual weights of 2.1 kg being achieved in only 14 months.

Other improvements to the culture methodologies of European cuttlefish were published. For instance, Hanley et al. (1998) described the use of a 34,500 L semiclosed system at the Marine Biological Laboratory (Woods Hole, USA.), the water being maintained between 18 and 20 °C (through the use of heating or chilling technology), but most importantly at lower nitrogenous compound concentrations. The use of UV filtration instead of ozone was suggested, due to eventual equipment malfunction, and the use of large-grain particles instead of oyster shells as substrate presented the advantage of reducing bacterial settlement and fouling and diseases resulting from skin damage. A maintenance protocol and measurements of containment were also presented in this work. The use of black plastic curtains in the tank was suggested by Hanley et al. (1999) as a way to reduce pathologies associated with wall banging. In addition, Koueta and Boucaud-Camou (1999) described the use of a semiclosed seawater system (80 % water renewal day^{-1}) for the rearing of hatchlings, which was similar in technology to what was previously reported by NRCC researchers.

The European cuttlefish was not the only one being successfully cultured for more than one generation and attaining a real culture potential derived from research at the NRCC (Lee et al. 1998). In fact, Hanlon et al. (1991) reported the culture of four generations (up to F_3) of the Indo-Pacific squid *S. lessoniana*. Not only was this species important at that time due to similar neuroanatomical features to *Loligo* spp. but it also displayed high market prices in Japan. After being reared in circular tanks (75–125 cm water column) at hatching (Lee et al. 1998), the squid attained 1–2 kg in just 6 months, with high survival, and tolerated densities of 2.5 kg m^{-3} (3 squid m^{-3}) in raceway recirculating culture (Lee et al. 1994). Again, according to the latter two publications, the use of dark tank walls was recommended, water quality was suggested to be maintained at the highest standards (lowest nitrogenous compounds concentrations as possible) by using protein skimmers to remove ink and addition of sodium bicarbonate to buffer pH. *S. lessoniana* as well as *Sepiella inermis* and *S. pharaonis* were being cultured and studied in Thailand (at the Rayong Coastal Aquaculture Station; Nabhitabhata 1996, 1997, Nabhitabhata and Nilaphat 1999, 2000) with the purpose of mass culture and restocking related to local stocks depletion (Nabhitabhata 1995). More details on the existing technology associated with the culture of *S. lessoniana* and *S. inermis* are given in Chaps. 17 and 13, respectively.

Villanueva (1995) was able to rear *O. vulgaris* paralarvae to settlement, following a previous study (Villanueva 1994) on the suitability of decapod crab zoea (*Pagurus prideaux, Dardanus arrosor, Liocarcinus depurator, Artemia* nauplii and adult, mysidacean shrimp) as first food for both *O. vulgaris* and *L. vulgaris* paralarvae. The latter trials were performed in cylindrical plastic tanks, with very small volumes (13–33 L) and low water flows (35–90 L h^{-1}). Rearing temperatures for *L. vulgaris* were ≈ 11 °C and 19 °C, while temperatures of ≈ 21 °C were used on *O. vulgaris*. The culture of the common octopus had particular interest for Mediterranean countries from the mid-1990s onward. In this sense, several Spanish research centres were dedicated to this task and some companies in Galicia (northeast region of Spain) dedicated to the fattening of *O. vulgaris* subadults (García et al. 2004). Details on these endeavours may be found in Chap. 9. On the other hand, Cagnetta (2000) and Iglesias et al. (2000) revised some critical points and rearing strategies for the species. These included capture and transport, acclimatisation, reproduction, feeding, the use of shelters, competition and aggressiveness and mortality of animals that could be fattened onshore and inshore through the use of tanks in open or closed seawater systems or in cages. The last report from 2000 was published by Cagnetta and Sublimi (2000) and concerns the productive performance of the common octopus fed different monodiets (*Carcinus mediterraneus, I. coindetii, Mugil cephalus* and *Sardina pilchardus*), with crab attaining the best results. Chapters 23 and 24 detail the seawater systems and protocols developed at that time for the different life stages.

During this time, interest in octopus culture had risen. However, the existing technology for cephalopods' fattening was considered a high-risk activity with low profits, due to the dependence on subadults collected from the wild and high prices of natural prey used in feeding. Therefore, it was generally assumed by many researchers that an adequate nutritional programme was necessary to develop a profitable commercial diet. Hence, research on the use of artificial diets for cephalopods started in this decade. Bernardino Castro and Ángel Guerra were pioneers on trophic studies that were used to design semi-purified diets for *S. officinalis* (Castro and Guerra 1990). In that study, they found that the natural diet of the European cuttlefish was mainly crustaceans (80.8 %) and fish (15.2 %). The remaining stomach content was represented by polychaetes (0.7 %) and unidentified materials (3.3 %). From that point on, several studies were performed on the use of different raw materials. Castro (1991) tested the acceptance of 2.5 cm cylindrical pellets weighing ≈ 1 g (20 % dry prawn powder, 4 % alginate and 76 % water) by *S. officinalis* juveniles with success. This led to a series of experiments on palatability and growth trials on pelleted diets for *O. bimaculoides* and *S. officinalis* juveniles. The following diets were tested by Lee et al. (1991): live and frozen shrimp, live fish and fish fillets, surimi, raw chicken meat, pureed shrimp and chicken, turkey hot dogs and pellets of penaeid shrimp, mysid shrimp and chicken. None of the species ingested surimi and both preferred live food. However, octopuses were faster in accepting the non-natural diets. Despite the acceptance and ingestion of the remaining diets, this trial was characterized by cannibalism after 30–40 days in group-reared animals, which suggested nutritional deficiencies or imbalances of the non-natural diets. The attractiveness of several chemical and crude extracts were tested in *O. maya* (Lee 1992), which demonstrated chemotaxis to proline, ATP and crab extract. In a subsequent study, Castro et al. (1993) tested the effects of fish surimi and pelleted diets on the species growth, survival and feeding rates during 45 days. Despite acceptance and ingestion of surimi in the first 30 days, survival and growth was poor with either diet (67.5% and 22.5%; 0.33% and 0.54 BW day^{-1}, respectively). Due to the results of acceptance of this study, Castro and Lee (1994) tested different surimi formulations (fish myofibrillar protein concentrate) for *S. officinalis,* varying its content with the use or not of egg albumin, casein, whole egg, menhaden oil, cholesterol and lecithin. Once again, growth was feeble with any diet, despite being accepted and ingested.

As a result of those studies, one of the main bottlenecks in cephalopod culture was identified to be nutrition and physiology. The efforts performed on solving this lack of specific knowledge are revised in the following Chap. 5—Nutrition as a key factor for cephalopod aquaculture.

Cephalopods being senescent by nature and some species hard to keep and breed in captivity, led Japanese researchers to develop the in vitro fertilization of squid, as a way to obtain hatchlings of oceanic squid (Arnold and O'Dor 1990). This technique was used mainly with ommastrephid squid species with commercial interest as *Todarodes pacificus* (Ikeda et al 1993; Ikeda and Shimazaki 1995; Sakurai et al. 1996; Watanabe et al. 1996), *I. argentinus* (Sakai and Brunetti 1997; Sakai et al. 1998), *Ommastrephes bartramii* and *Sthenoteuthis oualaniensis* (Sakurai et al. 1995). This technique has been updated and the literature reviewed recently by Villanueva et al. (2012).

4.6 Conclusions

The introduction of a new species in aquaculture requires a series of preliminary studies related to the biology, ecology and physiology of the species. Despite the increased number of researchers directly or indirectly dedicated to this task, and the increased output of information over the past 20–30 years, the main constraints in developing feasible cephalopod culture technology are considerable. This has primarily to do with the fact that cephalopods are very specialized in their physiology and are the most intelligent marine invertebrates. Up till now, the available technology, its logistics and operational costs only allow cephalopods to be cultured for their use as animal models in the laboratory. Nonetheless, there are some species which might be on the verge for mass commercial culture in the coming 5–10 years. This information is presented in detail for a given species in the next chapters by researchers who have been trying to solve these limitations since 2000.

References

Arnold JM, Summers WC, Gilbert DL et al (1974) A guide to laboratory use of the squid *Loligo pealei*. Marine Biological Laboratory, Woods Hole

Arnold JM, O'Dor RK (1990) In vitro fertilization and embryonic development of oceanic squid. J Ceph Biol 1:21–36

Arvanitaki A, Fessard A, Kruta V (1936a) Analysis of the potential action of visceral nerves in the *Sepia officinalis*. Comptes Rendus Séances Soc Biol Fil 122:1204–1206

Arvanitaki A, Fessard A, Kruta V (1936b) The repetitive method of electrical response in the visceral and stellate nerves in the *Sepia officinalis*. Comptes Rendus Séances Soc Biol Fil 122:1203–1204

Barnabé G (1996) Bases Biologiques et Écologiques de l'Aquaculture Editorial Acribia. Spain, Zaragoza

Bernardino FNV (2000) Review of aquaculture development in Portugal. J Appl Ichthyol 16:196–199

Boletzky SV (1974a) Effets de la sous-nutrition prolongée sur le développement de la coquille de *Sepia officinalis* L. (Mollusca, Cephalopoda). Bull Soc Zool Fr -Evol Zool 99:667–673

Boletzky SV (1974b) Elevage de céphalopodes en aquarium: acquis recents. BullSoc Zool Fr 114:57–66

Boletzky SV (1974c) Élevage de céphalopodes en aquarium. Vie Milieu 24:309–340

Boletzky SV (1974d) The "larvae" of cephalopod: a review. Thalassia Jugoslavica 10:45–76

Boletzky SV (1975) Le développement d'*Eledone moschata* (Mollusca, Cephalopoda) élevée au laboratoire. Soc Zool Fr 100:361

Boletzky SV (1979a) Growth and life-span of *Sepia officinalis* under artificial condition (Mollusca, Cephalopoda). Rapport Comm Internatio Mer Méditerranée 25-26:159–168

Boletzky SV (1979b) Observations on the early post-embryonic development of *Loligo vulgaris* (Mollusca, Cephalopoda). Rapport Comm Internatio Mer Méditerranée 25–26:155–158

Boletzky SV, Hanlon RT (1983) A review of the laboratory maintenance, rearing and culture of cephalopod molluscs. Memo Natl Mus Victoria 44:147–187

Boletzky SV, Boletzky MV, Frösch D et al (1971) Laboratory rearing of Sepiolinae (mollusca: cephalopoda). Mar Biol 8:82–87

Boucaud-Camou E (1981) La glande digestive, organe principal de la digestion dans *Sepia officinalis* L (Mollusque, Céphalopode). Bull Soc Zool Fr - Evol Zool 106:375–376

Boucaud-Camou E (1982) Localization of some hydrolytic enzymes in digestive organs of juvenile *Sepia officinalis* L (Mollusca, Cephalopoda). Malacol 22:685–690

Boucaud-Camou E (1989) L'aquaculture des cephalopodes: evaluation et perspectives. Haliotis 19:201–214

Boucaud-Camou E (1990) La seiche, un animal d'avenir. Peche Maritime 69:321–329

Boucaud-Camou E, Boucher-Rodoni R (1983) Feeding and digestion in cephalopods. The mollusca–physiology, Part 2. Academic Press, pp 149–187

Boucaud-Camou E, Yim M (1981) Maturation post-embryonnaire de la glande digestive de *Sepia officinalis* L (Mollusque, Céphalopode). Bull Soc Zool Fr - Evol Zool 106:377–377

Boucaud-Camou E, Yim M, Tresgot A (1985) Feeding and digestion of young *Sepia officinalis* L. (Mollusca: Cephalopoda) during post-hatching development. Vie Milieu 35:263–266

Boucher-Rodoni R, Boucaud-Camou E, Mangold K (1987) Feeding and digestion. In: Boyle PR (ed) Cephalopod Life Cycles. Academic Press London, p 85–108

Boyle PR (1991) The UFAW Handbook on the care and management of cephalopods in the laboratory. Universities Federation for Animal Welfare, Herts, pp 58

Budelmann BU (1995) The cephalopod nervous system: what evolutions has made of the molluscan design. In: Breidbach O, Kutsuch W (eds) The nervous system of invertebrates. An evolutionary and comparative approach. Birkhauser Verlag, Basel, p 115–138

Cagnetta P (2000) Preliminary observations on the productive responses of the common octopus (*Octopus vulgaris* C.) reared free or in individual nets. CIHEAM–Options Mediterraneennes 47:323–329

Cagnetta P, Sublimi A (2000) Productive performance of the common octopus (*Octopus vulgaris* C.) when fed on a monodiet. CIHEAM–Options Mediterraneennes 47:331–336

Castro BG (1991) Can *Sepia officinalis* L be reared on artificial food. Mar Behav Physiol 19:35–38

Castro BG, Guerra A (1990) The diet of *Sepia officinalis* (Linnaeus, 1758) and *Sepia elegans* (D'Orbigny, 1835) (Cephalopoda, Sepioidea) from the Ria de Vigo (NW Spain). Sci Mar 54:375–388

Castro BG, Dimarco FP, Derusha RH et al (1993) The effects of surimi and pelleted diets on the laboratory survival, growth, and feeding rate of the cuttlefish *Sepia officialis* L. J Exp Mar Biol Ecol 170:241–252

Castro BG, Lee PG (1994) The effects of semi-purified diets on growth and condition of *Sepia officinalis* L (Mollusca, Cephalopoda). Comp Biochem PhysiolA 109:1007–1016

Choe S (1966) On the eggs, rearing, habits of the fry, and growth of some cephalopoda. Bull Mar Sci 16:330–348

Choe S, Ohshima Y (1963) Rearing of cuttlefish and squids. Nature 197:307

Coates HJ, Hussey RE, Nixon M (1965) An automatic food dispenser for *Octopus vulgaris*. J Physiol 183:51–52

Coelho ML, Modesto T, Soares F (1989) *Sepia officinalis*: Potential for aquaculture and restocking. In: Boucaud-Camou E (ed) The Cuttlefish. Université de Caen. Caen, p 348–349

D'Apote MP, Palmegiano GB (1982) Inseminazione della Laguna di Lesina con uova di *Sepia officinalis* L. Studio dell'effetto di parametri ambientali sulla schiusa: temperatura e salinitá. riv. Italia Piscicultura IHIOP 3:112–119

DeRusha RH, Forsythe JW, Dimarco FP et al (1989) Alternative diets for maintaining and rearing cephalopods in captivity. Lab Anim Sci 39:306–312

Durward RD, Vessey E, O'Dor R (1979) Reproduction in the squid, *Illex illecebrosus*: First observation in captivity and implications for the life cycle. Special Meeting of STACRES. Sel Pap ICNAF, (no.6):7–13

Forsythe JW, Hanlon RT (1988) Effect of temperature on laboratory growth, reproduction and life span of *Octopus bimaculoides*. Mar Biol 98:369–379

Forsythe JW, Hanlon RT, Derusha R (1991) Pilot-Scale culture of *Sepia* in biomedical research. In: Boucaud-Camou E (ed) The Cuttlefish. Centre de Publications de l'Université de Caen, Caen, p 313–323

Forsythe JW, Derusha RH, Hanlon RT (1994) Growth, reproduction and life-span of *Sepia officinalis* (Cephalopoda, Mollusca) cultured through 7 consecutive generations. J Zool 233:175–192

García GJ, Gonzáles LMR, García GB (2004) Cost analysis of octopus ongrowing installation in Galicia. Span J Agric Res 2:531–537

Gariaeff W (1906) Le système nerveux des céphalopodes. La structure fibreuse des cellules ganglionnaires de *Octopus vulgaris*. Comptes Rendus Séances Soc Biol Fil 61:201–202

Guerra A (1978) Sobre la alimentación y el comportamiento alimentario de *Octopus vulgaris*. Inv Pesq 42:351–364

Hanley JS, Shashar N, Smolowitz R et al (1998) Modified laboratory culture techniques for the European cuttlefish *Sepia officinalis*. Biol Bull 195:223–225

Hanley JS, Shashar N, Smolowitz R et al (1999) Soft-sided tanks improve long-term health of cultured cuttlefish. Biol Bull 197:237–238

Hanlon RT (1987) Mariculture. Cephalopod life cycles. Academic Press Inc., London

Hanlon RT (1990) Maintenance, rearing and culture of teuthoid and sepioid squids. In: Guilbert DL, Adelman WJ Jr, Arnold JM (eds) Squid as experimental animals. Plenum Press, New York, p 35–61

Hanlon RT, Forsythe JW (1985) Advances in the laboratory culture of octopuses for biomedical research. Lab Anim Sci 35:33–40

Hanlon RT, Hixon RF (1983) Laboratory maintenance and culture of octopuses and loliginid squids. In: Berg CJ Jr (ed) Culture of Marine Invertebrates. Hutchinson Ross Publishing Company, Stroudsburg, p 44–61

Hanlon RT, Wolterding MR (1989) Behavior, body patterning, growth and life history of *Octopus briareus* cultures in the laboratory. Am Malac Bull 7:21–45

Hanlon RT, Hixon RF, Hulet WH (1983) Survival, growth, and behaviour of the loliginid squids *Loligo plei, Loligo pelaei*, and *Lolliguncula brevis* (Mollusca: Cephalopoda) in closed seawater systems. Biol Bull 165:637–685

Hanlon RT, Yang WT, Turk PE et al (1989) Laboratory culture and estimated life span of the Eastern Atlantic squid, *Loligo forbesi* Steenstrup, 1856 (Mollusca: Cephalopoda). Aquacult Fish Manag 20:15–34

Hanlon RT, Messenger JB (1996) Cephalopod Behaviour. Cambridge University Press, New York

Hanlon RT, Turk PE, Lee PG (1991) Squid and cuttlefish mariculture: an updated perspective. J Ceph Biol 2:31–40

Hulet WH, Villoch MR, Hixon RF et al (1979) Fin damage in captured and reared squids. Lab Anim Sci 29:528–533

Iglesias J, Sánchez FJ, Otero JJ et al (2000) Culture of octopus (*Octopus vulgaris*, Cuvier): Present knowledge, problems and perspectives. Cah Options Méditerran 47:313–321

Ikeda Y, Sakurai Y, Shimazaki K (1993) Fertilizing capacity of squid (*Todarodes pacificus*) spermatozoa collected from various sperm storage sites, with special reference to the role of gelatinous substance from oviducal gland in fertilization and embryonic development. Invertebr Reproduction Develop 23:39–44

Ikeda Y, Shimazaki K (1995) Does nidamental gland jelly induce the formation of perivitelline space at fertilization in the squid *Todarodes pacificus*? J Mar Biol Ass UK 75:495–497

Itami K, Izawa Y, Maeda S et al (1963) Notes on the laboratory culture of the octopus larvae. Bull Japan Soc Sci Fish 29:514–527

Koueta N, Boucaud-Camou E (1999) Food intake and growth in reared early juvenile cuttlefish *Sepia officinalis* L (Mollusca Cephalopoda). J Exp Mar Biol Ecol 240:93–109

Koueta N, Andrade JP, Von Boletzky S (2006) The cuttlefish *Sepia officinalis* a working model in cephalopod research. Vie Milieu 56:67–67

Laroe ET (1971) The culture and maintenance of the loliginid squids *Sepioteuthis sepioidea* and *Doryteuthis plei*. Mar Biol 9:9–25

Laschi C, Cianchetti M, Mazzolai B et al (2012) Soft robotic arm inspired by the octopus. Adv Robotics 26:709–727

Lee PG (1992) Chemotaxis by *Octopus maya* Voss et Solis in a Y-maze. J Exp Mar Biol Ecol 156:53–67

Lee PG (1994) Computer automation and intelligent control for aquaculture. Bull Natl Res Inst Aquacult Suppl I:105–110

Lee PG (1995) A review of automated control systems for aquaculture and design criteria for their implementation. Aquacult Eng 14:205–227

Lee PG, Forsythe JW, DiMarco FP et al (1991) Initial palatability and growth trials on pelleted diets for cephalopods. Bull Mar Sci 49:362–372

Lee PG, Turk PE, Yang WT et al (1994) Biological characteristics and biomedical applications of the squid *Sepioteuthis lessoniana* cultured through multiple generations. Biol Bull 186:328–341

Lee PG, Turk PE, Forsythe JW et al (1998) Cephalopod culture: physiological, behavioral and environmental requirements. Suisanzoshoku 46:417–422

Lee PG, Lea RN, Dohmann E et al (2000) Denitrification in aquaculture systems: an example of a fuzzy logic control problem. Aquacult Eng 23:37–59

Melzner F, Gutowska MA, Langenbuch M et al (2009) Physiological basis for high CO_2 tolerance in marine ectothermic animals: pre-adaptation through lifestyle and ontogeny? Biogeosciences 6:2313–2331

Nabhitabhata J (1995) Mass culture of cephalopods in Thailand. World Aquaculture 26:25–29

Nabhitabhata J (1996) Life cycle of cultured big fin squid *Sepioteuthis lessoniana* Lesson. Phuket Mar Biol Center Special Pub 16:83–95

Nabhitabhata J (1997) Life cycle of three cultured generations of spineless cuttlefish, *Sepiella inermis* (Férussac & D'Orbigny, 1848). Phuket Mar Biol Center Special Pub 17:289–298

Nabhitabhata J, Nilaphat P (1999) Life cycle of cultured pharaoh cuttlefish, *Sepia pharaonis* Ehrenberg, 1831. Phuket Mar Biol Center Special Pub 19:25–40

Nabhitabhata J, Nilaphat P (2000) Behaviour of juvenile cephalopods: preference for texture and brightness of substrata. Phuket Mar Biol Center Special Pub 21:103–112

Neill SSJ (1971) Notes on squid and cuttlefish: Keeping, handling and colour-patterns. Pubbl Staz Zool Napoli 39:64–69

Nixon M (1987) Cephalopod diets. In: Boyle PR (ed) Cephalopod life cycles. Academic Press, London, p 201–219

O'Dor RK, Durward RD, Balch N (1977) Maintenance and maturation of squid (*Illex illecebrosus*) in a 15 m circular pool. Biol Bull 153:322–335

Ohshima Y, Choe S (1961) On the rearing of young cuttlefish and squid. Bull Japan Soc Sci Fish 27:979–987

Overath H, Boletzky MV (1974) Laboratory observations on spawning and embryonic development of a blue-ringed octopus. Mar Biol 27:333–337

Palmegiano GB, Sequi R (1981) On the possibility of the introduction of cuttlefish (*Sepia officinalis* L.) in a management strategy for the coastal lagoons. Symposium International sur les Lagunes Cotieres, France, p 1–9

Palmegiano GB, D'apote MP (1983) Combined effects of temperature and salinity on cuttlefish (*Sepia officinalis* L.) hatching. Aquaculture 35:259–264

Palmegiano GB, Sequi R (1984) Allevamento estensivo di *Sepia officinalis* L. Agricoltura e Ricerca 41:49–56

Pascual E (1978) Crecimiento y alimentacíon de tres generaciones de *Sepia officinalis* en cultivo. Inv Pesq 42:421–442

Pörtner HO, Farrell AP (2008) Physiology and climate change. Science 322:690–692

Richard A (1971) Contribuition à l'étude expérimentale de la croissance et de la maturation sexuelle de *Sepia officinalis* L. (Mollusque, Céphalopode). Université de Lille, Lille. France, pp 264

Richard A (1975) L'élevage de la seiche (*Sepia officinalis* L., Mollusque Céphalopode). 10th European Symposium on Marine Biology. Ostend, Belgium 1:359–380

Robaina GO (1983) Sobre el cultivo y el mantenimiento de Octopus briareus Robson 1929 (Cephalopoda: Octopoda). Inf Técn Inst Inv Pesq 106:3–20

Sakai M, Brunetti NE (1997) Preliminary experiments on artificial insemination of the argentine shortfin squid *Illex argentinus*. Fish Sci 63:664–667

Sakai M, Brunetti NE, Elena B, Sakurai Y (1998) Embryonic development and hatchlings of *Illex argentinus* derived from artificial fertilization. South African J Mar Sci 20:255–265

Sakurai Y, Young RE, Hirota J, Mangold K, Vecchione M, Clarke MR, Bower J (1995) Artificial fertilization and development through hatching in the oceanic squids *Ommastrephes bartramii* and *Sthenoteuthis oualaniensis* (Cephalopoda: Ommastrephidae). Veliger 38:185–191

Sakurai Y, Bower JR, Nakamura Y, Yamamoto S, Watanabe K (1996) Effect of temperature on development and survival of *Todarodes pacificus* embryos and paralarvae. Am Malacol Bull 13:89–95

Sequi R (1980) Some experiences on rearing of *Sepia officinalis* L. (Mollusca Cephalopoda). Mem Biol Marina e Oceanogr Suppl X:299–303

Sequi R, Palmegiano GB (1984) Some alternatives for the use of cuttlefish (*Sepia officinalis* L.) resources in coastal lagoons. GFCM Studies and Reviews—Management of Coastal Lagoons Fisheries. Rome, p 607–614

Sequi R, Palmegiano GB (1985) Aggiornamenti sul sistema di raccolta di uova di *Sepia officinalis* L. Quademi IstIdrobiol Acquacolt Brunelli 5-6:57–63

Sio FD (2011) Leviathan and the soft animal: medical humanism and the invertebrate models for higher nervous functions, 1950s-90s. Med Hist 55:369–374

Strugnell JM, Cherel Y, Cooke IR et al (2011) The Southern Ocean: source and sink? Deep Sea Res Part 2 Top Stud Oceanogr 58:9–9

Summers WC, McMahon JJ (1974) Studies on the maintenance of adult squid (*Loligo pealei*) L. Factorial survey. Biol Bull 146:279–290

Summers WC, McMahon JJ, Ruppert GNPA (1974) Studies on the maintenance of adult squid (*Loligo pealei*) II. Empirical extensions. Biol Bull 146:291–301

Sykes AV, Oliveira AR, Domingues PM et al (2009) Assessment of European cuttlefish (*Sepia officinalis*, L.) nutritional value and freshness under ice storage using a developed Quality Index Method (QIM) and biochemical methods. Food Sci Tech 42:424–432

Toll RB, Strain CH (1988) Freshwater and terrestrial food organisms as an alternative diet for laboratory culture of cephalopods. Malacol 29:195–200

Tricarico E, Borrelli L, Gherardi F et al (2011) I know my neighbour: Individual recognition in *Octopus vulgaris*. Plos One 6:e18710

Turk PE, Hanlon RT, Bradford LA et al (1986) Aspects of feeding, growth and survival of the european squid *Loligo vulgaris* Lamarck, 1799, reared through the early growth stages. Vie Milieu 36:9–13

Uriarte I, Iglesias J, Domingues P et al (2011) Current status and bottle neck of octopod aquaculture: the case of American species. J World Aquacult Soc 42:735–752

Van Der Sprenkel HB (1929) Nerve-endings to the muscles of the arm of *Sepia officinalis*. Proceedings of the Koninklijke Akademie Van Wetenschappen Te Amsterdam 32:151–155

Van Heukelem WF (1977) Laboratory maintenance, breeding, rearing, and biomedical research potential of the Yucatan octopus (*Octopus maya*). Lab Anim Sci 27:852–859

Vaz-Pires P, Seixas P, Barbosa A (2004) Aquaculture potential of the common octopus (*Octopus vulgaris* Cuvier 1797): a review. Aquaculture 238:221–238

Villanueva R (1994) Decapod crab zoeae as food for rearing cephalopod paralarvae. Aquaculture 128:143–152

Villanueva R (1995) Experimental rearing and growth of planktonic *Octopus vulgaris* from hatching to settlement. Can J Fish Aquatic Sci 52:2639–2650

Villanueva R, Staaf DJ, Argüelles J, Bozzano A, Camarillo-Coop S, Nigmatullin CM, Petroni G, Quintana D, Sakai M, Sakurai Y, Salinas-Zavala CA, De Silva-Dávila R, Tafur R, Yamashiro C, Vidal EAG (2012) A laboratory guide to in vitro fertilization of oceanic squids. Aquaculture 342-343:125–133

Watanabe K, Sakurai Y, Segawa S, Okutani T (1996) Development of the ommastrephid squid *Todarodes pacificus*, from fertilized egg to rhynchoteuthion paralarva. Am Malacol Bull 13:73–88

Wells MJ (1962) Brain and behaviour in cephalopods. Heinemann

Whitson J, Turk P, Lee P (1993) Biological denitrification in a closed recirculating marine culture system. In: Wang J-K (ed) Techniques for modern aquaculture. American society of agricultural engineers. Washington, pp 458–466

Williamson R, Chrachri A (2004) Cephalopod neural networks. Neurosignals 13:87–98

Wood JB, O'dor RK (2000) Do larger cephalopods live longer? Effects of temperature and phylogeny on interspecific comparisons of age and size at maturity. Mar Biol 136:91–99

Yang WT, Hanlon RT, Hixon RF et al (1980) First sucess in rearing hatchlings of the squid *Loligo pealei* Lesueur 1821. Malacol Rev 13:79–80

Yang WT, Hanlon RT, Krejci ME et al (1983) Laboratory rearing of *Loligo opalescens*, the market squid of California. Aquaculture 31:77–88

Yang WT, Hixon RF, Turk PE et al (1986) Growth, behaviour, and sexual maturation of the market squid, *Loligo opalescens*, cultured through the life cycle. Fish Bull 84:771–799
Yang WT, Hanlon RT, Lee PG et al (1989) Design and function of closed seawater systems for culturing Loliginid squids. Aquacult Eng 8:47–65
Young JZ (1938) The functioning of the giant nerve fibres of the squid. J Exp Biol 15:170–185
Young JZ (1985) Cephalopods and Neuroscience. Biol Bull 168:153–158

Chapter 5
Nutrition as a Key Factor for Cephalopod Aquaculture

Juan Carlos Navarro, Óscar Monroig and António V. Sykes

Abstract Cephalopods are fast-growing animals, active swimmers and top predators, which require substantial amounts of food. As such, they show high metabolic rates dependent on a carnivorous diet, thus hypothetically linked to a predominant amino acid metabolism. Their body composition is mainly constituted by high levels of total protein, and their lipids, although quantitatively low, reveal the presence of substantial amounts of long-chain polyunsaturated fatty acids. All in all, little is known about their nutritional requirements, especially during the early stages, very prone to high mortalities under culture. This chapter is a brief account of key information concerning relevant points linked to the nutritional requirements that cephalopods have for proteins, lipids, carotenoids, carbohydrates, minerals and vitamins. Moreover, some considerations on populational metabolism are also presented.

Keywords Amino acids · Carbohydrates · Fatty acids · Lipids · Nutrition · *Octopus vulgaris* · Populational metabolism · Proteins · *Sepia officinalis*

5.1 Introduction

Cephalopods potential for aquaculture production was acknowledged by several researchers in the last decades of the twentieth century (e.g. Hanlon 1987; Boucaud-Camou 1989, 1990; Hanlon et al. 1991; Barnabé 1996). Nonetheless, several

J. C. Navarro (✉) · Ó. Monroig
Instituto de Acuicultura Torre de la Sal (IATS-CSIC), 12595 Ribera de Cabanes, Castellón, Spain
e-mail: jcnavarro@iats.csic.es

Ó. Monroig
Institute of Aquaculture, School of Natural Sciences, University of Stirling, Stirling FK9 4LA, Scotland, UK
e-mail: oscar.monroig@stir.ac.uk

A. V. Sykes
CCMAR-CIMAR L.A., Centro de Ciências do Mar do Algarve, Universidade do Algarve, Campus de Gambelas, 8005-139, Faro, Portugal.
e-mail: asykes@ualg.pt

J. Iglesias et al. (eds.), *Cephalopod Culture*, DOI 10.1007/978-94-017-8648-5_5,

bottlenecks remain and are impeding the transition of technology from pilot to full-scale, namely the existing knowledge on cephalopod nutrition. Nutrition is a key factor for proper growth and survival under captive conditions and mass culture. This chapter picks up from the last review on this subject, made by Lee (1994). Firstly, it presents a nutritional approach, which is mostly based on the biochemical composition of both cephalopods and preys, and finalizes with a metabolic hypothesis, which considers other variables such as enzymes, geographical adaptation and stress. Due to the low amount of existing information for most cephalopod species, this chapter mainly focuses on nutritional studies conducted in *Sepia officinalis* and *Octopus vulgaris*, arguably the cephalopod species of greatest commercial interests.

5.2 Proteins

Proteins are the most abundant macronutrient in cephalopods (Zlatanos et al. 2006) and, as stated by Lee (1994), large protein and amino acid contents in the diet of cephalopods are required for sustaining growth and fulfilling energy demands. According to Lee (1994), cephalopods efficiently absorb, digest and utilize dietary proteins that are further used for locomotion, structural support, energy source, oxygen transport and osmoregulation. Cephalopods display high rates of protein synthesis and retention, and low rates of protein degradation (Houlihan et al. 1990; Moltschaniwskyj and Carter 2010). In order to evaluate the dietary requirements of the common octopus *O. vulgaris* during the fast-growing stages, the total and free amino acid (FAA) composition of paralarvae and juveniles was determined by Villanueva et al. (2004). Similar to the amino acid content found in the mantle of juveniles of *O. vulgaris*, *S. officinalis* and *Loligo vulgaris* by Zlatanos et al. (2006), these authors found that glutamate and aspartate were the most abundant nonessential amino acids (NEAAs) in *O. vulgaris* paralarvae, with lysine, leucine and arginine accounting for nearly half of the essential amino acids (EAAs). Interestingly, arginine was the most abundant FAA in paralarvae, possibly indicating its further use for octopine formation produced during an anaerobic work (Baldwin et al. 1976; Hochachka et al. 1976, 1983; Storey and Storey 1979; Storey et al. 1979; Hochachka and Fields 1982) or an active metabolism for energy production and biosynthesis of other amino acids. On the other hand, cephalopods might use proline during oxidative metabolism, either as energy source or as means for augmenting the Krebs cycle (Hochachka and Fields 1982). Both arginine and proline are potentially interconvertible through glutamate and ornithine (Mommsen et al. 1982).

In a general way, there is already information on the amino acids of *O. vulgaris* prey species, such as for *Artemia* spp. (Aragão et al. 2004) and crustaceans such as the spider crab (Andrés et al. 2010), as well as for raw materials used for prepared diets (Valverde et al. 2013). Cephalopod's growth is primarily an increase in body muscle mass by protein synthesis and accretion, and individuals display very high growth rates (especially at the paralarvae and hatchling stages), which means that they have a high dietary requirement for amino acids. In addition, these high growth rates are most probably due to highly efficient ingestion, digestion (Boucher-Rodoni et al. 1987) and assimilation of protein (Domingues et al. 2005), which have to be supplied by a diet with balanced levels of amino acids, despite the capacity of some cephalopods to perform integumental amino acid uptake from seawater (de Eguileor et al. 2000; Villanueva et al. 2004). Recent results point to the possibility that cuttlefish might not be able to use protein that has been denaturated (Domingues et al. 2009) and to the favour on amino acid use through the pyruvate and tricarboxylic acid pathways in detriment of the ketogenic pathway in starving *O. vulgaris* (García-Garrido et al. 2012). Given its importance for cephalopods metabolism (Lee 1994), there is a need for characterization of the amino acids pool (these should be predominantly used as metabolic fuel, but they should also be utilized for body protein synthesis) at different life stages of different species and preys used for cephalopod rearing. The latter assumes greater importance since marine fish larvae seem to have a lower capacity to digest and absorb complex proteins than juvenile fish (Conceição et al. 2010), and display high amino acid requirements for protein deposition, turnover and catabolism to attain rapid growth (Rønnestad et al. 2003).

In a similar way to demersal fish eggs (Rønnestad et al. 1999), the FAA and protein amino acid pools of *S. officinalis* wild and culture eggs are largely dominated by taurine (more than 50%; Sykes et al. unpublished data). Taurine is an amino acid analogue that is not incorporated into protein, but well known for its multiple roles that include inotropy of high and low calcium, modulator of neuron excitability, resistance to anoxia and hypoxia, bile salt synthesis and simulation of glycolysis and glycogenesis (Huxtable 1992). Taurine is also abundant during planktonic stages of the common octopus where it might play a role in osmoregulation (Villanueva et al. 2004). Nonetheless, its effect on metabolism and growth performance in a fish like gilthead seabream was only related to the increasing methionine availability for several important physiological purposes (Pinto et al. 2013).

5.3 Lipids

Little is known about the lipid requirements of cephalopods, apart from the general information drawn from the analysis of their body composition and assumptions made from their feeding habits. It is then difficult to separate early stages from adult requirements, or establish specific differences. Efforts devoted to the culture of certain species like *O. vulgaris* and *S. officinalis* have provided some information.

Very low levels of total lipids (TL) are present in the mantle of adults (Boucaud-Camou 1990; Sykes et al. 2009b) and hatchlings (Navarro and Villanueva 2000, 2003). This fact, combined with their poor capacity for mitochondrial lipid oxidation (O'Dor et al. 1984; Hochachka 1994), has somehow put aside research on cephalopod lipid nutrition until recently, when research on the causes of the massive mortalities encountered during the culture of early stages of merobenthic species has brought back the protagonism of these essential components on cephalopod nutrition, perhaps with an overemphasis on quantitative rather than qualitative points of view, given the poor essential lipid composition of the live preys commonly used in aquaculture. Besides, other particular aspects, like the lipid-rich nervous system of hatchlings of *O. vulgaris* paralarvae representing approximately one quarter of the animal's fresh weight (Packard and Albergoni 1970), suggest the importance of lipids for suitable growth during planktonic life. Navarro and Villanueva (2000, 2003) made the first approach towards the study of lipid requirements in early stages of cephalopods to conclude that a nutritional imbalance in the lipid and fatty acid (FA) profile of the artificial feeding protocol may be responsible for the high mortalities encountered. In particular, *O. vulgaris* should require feeding on low-lipid preys, rich in polar lipids (PL), long-chain polyunsaturated fatty acids (PUFA) and possibly cholesterol (Navarro and Villanueva 2000, 2003; Okumura et al. 2005; Seixas et al. 2008). This closely resembles the composition of a 'natural' diet based on crustacean larvae and other marine planktonic organisms like copepods, but is far from the typical composition of the enriched *Artemia* spp. in any of its forms. The picture is even more complicated after evidences pointing at the paralarvae as specialist predators, contrary to the general concept by which they had been often regarded as generalist predators (Roura et al. 2012).

On the other hand, cuttlefish are unable to store lipids in the digestive gland (Fluckiger et al. 2008) and require high levels of phosphatidylcholine, phosphatidylethanolamine (PE) and cholesterol in their diets (Almansa et al. 2006). This conclusion was drawn by the latter authors based on data of whole animal (Navarro and Villanueva 2000), cuttlefish mantle (Sinanoglou and Miniadis-Meimaroglou 1998, 2000) and also of prey lipid content (Domingues et al. 2003; Domingues et al. 2004). It is, however, interesting to verify that these same lipid classes have the highest levels throughout cuttlefish wild and culture egg embryonic development and that, although there is a difference in the amount of TL of eggs from different geographical locations (Sykes et al. 2009a), there is maintenance on the amount of TL until hatching (Bouchaud and Galois 1990). According to Bouchaud and Galois (1990), the egg-yolk lipids correspond to 14% dry weight in eggs and 15% in hatchlings (mainly phospholipids that may be used for energetic purposes).

However, in the case of octopus, up to now, there is no paralarval food from the aquaculture artificial food chain that can compare with the lipid composition of natural live food, and every effort has to be made to try to increase the essential long-chain PUFA and PL content of live preys. In fact, it is not only the bulk provision of essential lipids that is important but also the adequate lipid form (Guinot et al. 2013). This scenario may be more important for early stages, and even paramount in some animal groups like cephalopods, whose FA composition is essentially constituted by palmitic acid (16:0), stearic acid (18:0), docosahexaenoic acid (22:6n-3, DHA) and eicosapentaenoic acid (20:5n-3, EPA), the latter two being essential long-chain highly unsaturated fatty acids (HUFA) for marine organisms (Sinanoglou and Miniadis-Meimaroglou 1998; Navarro and Villanueva 2000; Passi et al. 2002; Ozyurt et al. 2006; Zlatanos et al. 2006). However, the most recent results from Seixas et al. (2010) do not completely exclude the importance of DHA for the successful rearing of *O. vulgaris* paralarvae but seem to point that maybe other n-3 HUFA would be more vital. Very interesting results from Quintana (2009) have shown that EPA is particularly important in paralarvae lipids, with 1:1 EPA:DHA proportions in the PE, as opposed to the 1:2 proportion generally reported for marine fish. Likewise, EPA is higher in shrimp than fish diets tested as cuttlefish food (Domingues et al. 2003) and its importance for normal growth was proposed by Almansa et al. (2006). The latter authors also indicated that cuttlefish metabolism of both EPA and DHA might be somehow different from fish, where in general the reduction of EPA and DHA sources implies a reduction of the level of these FA in several tissues (Sargent et al. 1995). In addition, Domingues et al. (2003) indicated a 2:1 EPA:DHA proportion in the prey supplied to hatchlings and, in the Ferreira et al. (2010) work, cuttlefish juveniles attaining the best growth and survival rates were fed shrimp with a similar EPA:DHA ratio and PUFA content. On the other hand, Koueta et al. (2002) and Perrin et al. (2004) have suggested the importance of these FAs and PUFA in cuttlefish hatchling survival when facing a stressful situation (either lower water quality or improper prey size, respectively). This suggests the importance of PUFA in cephalopod nutrition and may suggest, for example, that in addition to DHA, EPA may play an important role in the brain and visual system of cephalopods.

A closer look at the available results on the FA profile of *O. vulgaris* paralarvae (Navarro and Villanueva 2000; Miliou et al. 2006) reveals that, apart from the high levels of DHA, 16:0 and EPA, arachidonic acid (20:4n-6, ARA) is one of the most abundant FA, with surprisingly high values of 18:2n-6 and other n-6 FA being found in marine species. From these and other findings in other marine molluscs (Uki et al. 1986; Dunstan et al. 1996; Durazo-Beltran et al. 2003), it has been suggested that 20:4n-6 might not be essential, since the ability for enzymatic bioconversion from adequate precursors to 20:4n-6 could be present in this species. On the other hand, 18:3n-3 is not present in octopus paralarval tissues (Navarro and Villanueva 2000, 2003; Miliou et al. 2006) although this FA is massively included through the *Artemia* spp. feeding. Moreover, 18:3n-3 may compete with 24:5n-3 for the Δ6 desaturase in the metabolic route leading to DHA production. Therefore, it might be possible that a diet rich in 18:3n-3 could be affecting the proper production of n-3 essential FA (EFA) of paralarvae. An opposite hypothesis is that the C18 desaturation–elongation pathways of 18:3n-3 and 18:2n-6 may be active in *O. vulgaris* so as to produce physiologically essential n-3 and n-6 FA. To explore these and other aspects of lipid metabolism research on dynamic aspects of lipid nutrition and metabolism by characterizing enzymes involved in lipid biosynthesis in these organisms, i.e. desaturases and elongases, is an invaluable tool. Monroig et al. (2012a) have recently isolated a complementary DNA (cDNA) with high homology to fatty acyl desaturases (Fad) in adult octopus. Functional characterization of this enzyme showed that the octopus Fad exhibited Δ5-desaturation activity towards saturated and polyunsaturated fatty acyl substrates. Thus, it efficiently converted 16:0 and 18:0 to 16:1n-11 and 18:1n-13, respectively, and desaturated PUFA substrates 20:4n-3 and 20:3n-6 to 20:5n-3 (EPA) and 20:4n-6 (ARA), respectively. Although the Δ5 Fad enables common octopus to produce EPA and ARA, the low availability of its adequate substrates 20:4n-3 and 20:3n-6, either in the diet or by limited endogenous synthesis from C18 PUFA, might indicate that EPA and ARA are indeed EFA for this species. Interestingly, the octopus Δ5 Fad can also participate in the biosynthesis of non-methylene-interrupted diene (NMID) FA, PUFA that are generally uncommon in vertebrates but have been found previously in marine invertebrates, including molluscs, and now also confirmed to be present in specific tissues of common octopus.

Molecular cloning and functional characterization of a cDNA encoding a putative elongase of very long-chain fatty acids (Elovl), a critical enzyme that catalyzes the elongation of FA including PUFA, in common octopus, suggests its phylogenetic relation to Elovl5 and Elovl2, two elongases with demonstrated roles in PUFA biosynthesis in vertebrates (Monroig et al. 2012b). Functional characterization of the octopus Elovl showed the ability to elongate some C18 and C20 PUFAs, while C22 PUFA substrates remained unmodified. Interestingly, the octopus Elovl *elongates* n-6 PUFA substrates more efficiently than their homologous n-3 substrates, suggesting that n-6 PUFA may have particular biological significance in *O. vulgaris,* as mentioned above, and stressing again the essentiality of long-chain n-3 PUFA, and in particular DHA. Besides, the elongase also plays a pivotal role in the biosynthesis of NMID FA.

Similar studies carried out in other cephalopods like *S. officinalis* show a strict parallelism in terms of both qualitative FA composition (see Navarro and Villanueva 2000; Almansa et al. 2006) and biosynthetic capacity (Monroig et al., unpublished data). This, along with tissue FA composition (Monroig et al. 2012a), emphasizes the importance and essentiality of long-chain FA for cephalopods, and validates the idea of using the information to establish the guidelines of the requirements for coastal cephalopods. The recent manuscript by Valverde et al. (2012) on lipid classes from marine species and meals intended for cephalopod feeding has provided additional information on lipids.

5.4 Carotenoids

Carotenoid deposition has been described in cephalopods at the digestive gland (Fox 1966) and in accessory nidamental glands (Decleir and Richard 1972; Van den Branden et al. 1978; Van Den Branden et al. 1980) of *S. officinalis* and proposed to be designated as sepiaxanthine, but there is scarce information about its physiological role. Carotenoid astaxanthin can also be deposited in the skin of *S. officinalis* when fed on grass shrimp (*Palaemonetes varians*; Almansa et al. 2006) and this prey, used for cuttlefish hatchlings rearing, displays up to ten times more carotenoid content (Domingues et al. 2004). *O. vulgaris* paralarvae seem to be able to deposit part of the canthaxanthin present in *Artemia* spp. and to metabolize this carotenoid to astaxanthin (Rodríguez et al., Universidad de La Laguna, unpublished data). Fisher et al. (1956) reported the existence of vitamin A and, in some species, β-carotene in cephalopods. Taking into account the pro-vitamin A and antioxidant activity of carotenoids (Liñán-Cabello et al. 2002), and the importance of this vitamin in photoreception, growth and development, it would be very interesting to carry out more studies to determine the role of carotenoids in early stages of cephalopods as suggested by Villanueva et al. (2009).

5.5 Carbohydrates

Contrary to proteins and lipids, carbohydrate (CH) nutrition in the common octopus has barely been investigated. Considering their limited abundance compared to other macronutrients (normally under 1 % of total dry weight; Vlieg 1984; Kreuzer 1984), it is generally accepted that cephalopods do not have a specific requirement for dietary CH (Lee 1994). While protein and amino acids are the primary energy source for cephalopods, it has been reported that cephalopods including *O. vulgaris* are able to rapidly catabolize dietary CH to account for energy demands in explosive activities such as prey capture and fleeing from predators (Morillo-Velarde et al. 2011). Therefore, CH may significantly contribute to fuelling metabolism in *O. vulgaris* under starvation conditions and, consequently, adequate CH inclusion of diets for octopus culture should not be underestimated. The source of CH utilized in diet formulation also needs to be considered. For instance, while glucose was easily digested by *O. vulgaris* (O'Dor et al. 1984), a recent study has revealed that other sugar types such as starch present in freeze-dried pea may exhibit extremely low digestibility (Morillo-Velarde et al. 2012).

On the other hand, the existence of a CH metabolism has been suggested by Sykes et al. (2009a), due to different temperatures of specific geographical locations that may influence the egg nutritional content and metabolism in *S. officinalis*. The sepia egg yolk is composed by a water-soluble glyco-lipoprotein (Ito et al. 1962; Blanchier 1981). This glyco-lipoprotein has 20 % lipids (with 65 % phospholipid and minor or no cholesterol contents) and 12.6 % of CHs (Ito and Fujii 1962). In fact, cuttlefish eggs from Faro (Portugal) have more CHs than lipids (Sykes et al. 2012). Bouchaud (1991) studied the energetic expenditure of *S. officinalis* during embryonic development and found that eggs with more than 0.075 g displayed a similar amount of energy (1,600 J), which led to a theory on the use of yolk for growth and catabolic purposes that is inversely correlated with temperature (e.g. higher temperature implies higher catabolism). In addition, higher temperatures will imply higher oxygen uptake by the embryo (Wolf et al. 1985), which is attained by the increased water volumes of eggs (Sykes et al. 2009a).

5.6 Minerals

The elemental requirements of *O. vulgaris*, as for cephalopods in general, are poorly understood. Nevertheless, it is accepted that octopuses, as carnivorous species, meet the majority of their elemental requirements from the diet, although direct uptake from the seawater has also been shown to occur through an ion balance mechanism regulated by the digestive gland appendages (Wells and Wells 1989). A literature review on the element concentrations in a series of tissues from cephalopods was reported by Napoleão et al. (2005a). While specific (quantitative) requirements for both essential and nonessential elements have not yet been determined, some studies aiming to determine the elemental composition of *O. vulgaris* have been conducted as a first approach to establish the dietary requirements in this species (Napoleão et al. 2005a; Napoleão et al. 2005b). Thus, Villanueva and Bustamante (2006) reported the elemental composition of the mature ovary, hatchlings, eggs at different developmental stages, wild juvenile individuals and also paralarvae fed a variety of experimental diets. Generally, S, Na, K, P and Mg were determined as the most abundant elements in *O. vulgaris*. Compared to other cephalopods, hatchlings from *O. vulgaris* contained higher levels of Ag, Cu, Mn, Ni and Zn. Compared to subadults and adults of the common octopus (Seixas et al. 2005), the contents of some nonessential elements, namely Ag, Al, Ba, Cd, Hg and Pb, were lower in hatchlings and reared paralarvae, suggesting an accumulation of such elements during development. Similar accumulation of oligoelements seems to occur in cuttlefish (Lacoue-Labarthe et al. 2008a, b, 2009, 2010a, b, Lourenço et al. 2009). Certain elements with potentially pivotal roles in the octopus and cuttlefish physiology have been studied more extensively. Copper (Cu), a key element in the respiratory function of haemocyanin (Ghiretti 1966; D´Aniello et al. 1986), has been postulated to be required for octopus paralarvae, as suggested by the high levels encountered in paralarvae fed on the natural prey *Maja brachydactyla* zoeae in comparison with *Artemia* spp. nauplii (Villanueva and Bustamante 2006). Moreover, octopuses fed on crustacean-based diets contained increased Cu levels compared to fish-fed octopuses, and therefore hypothesized to partly account for a higher cannibalism rate in the latter (García-García and Cerezo-Valverde 2006). The importance of Cu is also high in *S. officinalis* (Decleir et al. 1978), especially during maturation of haemocyanin (Decleir and Richard 1970; Decleir et al. 1971; Wolf et al. 1980; Beuerlein et al. 2004), and this probably extends to other cephalopods species (Taylor and Anstiss 1999). Similar to octopus, the low content of Cu in prepared diets has been pinpointed to provoke mortality by Castro et al. (1993).

Sulphur (S) is also regarded as an essential element for the common octopus and cuttlefish that needs to be provided in the diet at high quantities to sustain formation of muscular proteins (Lee 1994; Villanueva et al. 2004), vestigial shell (Napoleão et al. 2005b) and chitinized structures such as beaks (Hunt and Nixon 1981). Other elements such as strontium (Sr) and cobalt (Co) appear to be incorporated by the common octopus directly from the seawater and by food intake in cuttlefish. It has been demonstrated that Sr is critical for adequate statolith development and consequently for normal swimming and survival of newly hatched octopus, among other cephalopods (Hanlon et al. 1989). In addition to its potential role as an integral component of vitamin B_{12}, Co has also been pointed out as important in the development of adenochrome, a pigment found in the branchial heart, and therefore with a potential role in excretion (Miyazaki et al. 2001).

The cuttlefish *S. officinalis* has a cuttlebone which is made of calcium aragonite (Hewitt 1975) and may suffer malformation resulting from malnutrition (Boletzky 1974). Since food is the primary pathway for the accumulation of trace elements (Bustamante et al. 2004), the inclusion of calcium in a prepared diet is implicit. In fact, the mineral fraction of cuttlefish accounts for 22–32% in eggs (Sykes et al. 2012) and 6% of dry weight at hatching (Villanueva et al. 2004), which can only be attained in such way.

5.7 Vitamins

Villanueva et al. (2009) reported that information on the vitamin composition of cephalopods is mainly limited to the subadult and adult forms in relation to their edible body portions (mantle and arms) or selected organs (Fisher 1956; Sidwell et al. 1978; Motoe et al. 1997; Cho et al. 2001; Pandit and Magar 1972; Passi et al. 2002; Sikorski and Kolodziejska 1986). These authors analysed the vitamin content of the early stages of cephalopods as an approach to establish their requirements in culture. Antioxidants such as tocopherols are regarded to be very important for the prevention of lipid oxidation, particularly α-tocopherol since it is degraded to protect PUFA against oxidation in fish larvae (Sargent et al. 1997). In fact, it has been reported that the α-tocopherol requirement may depend on the dietary PUFA level (Stéphan et al. 1995; Halver 2002; Brown et al. 2005). Vitamin A and E (α- and γ-tocopherols) profiles of the European cuttlefish *S. officinalis,* European squid *L. vulgaris* and common octopus *O. vulgaris* laboratory hatchlings and wild juveniles were determined. The vitamin A content in early stages of cephalopods was not much different from that observed in other marine molluscs and fish larvae. Besides, relatively high content of vitamin E was observed in the hatchlings and juveniles. These authors postulated that the high levels of vitamin E are probably associated with the high percentage of oxidation-prone PUFA that are particularly high in paralarval and juvenile cephalopods. In general terms, they concluded that the natural and artificial preys (*Artemia* spp.) of early stages of cephalopods either fulfilled their vitamin requirements directly or provided precursors (i.e. carotenoids) that could be transformed in vitamins.

5.8 Populational Metabolism Differences

Captive cephalopods have a different physiology than those from the wild and their tissues are characterized by thicker mantles, a greater proportion of mitochondria-rich tissue, muscle fibres with smaller mitochondrial cores and fewer small muscle fibres (Pecl and Moltschaniwskyj 1999). This suggests a reduced rate of new fibre generation, indicating an alteration to the cellular growth mechanisms and not simply a change in the physiological growth rate observed in several laboratory and field studies (Semmens et al. 2004). Growth registered in captive individuals is usually lower than that estimated or verified for those from nature (an example of this is presented in Chap. 11, in the part of cuttlefish culture in earthen ponds).

As stated by Sykes et al. (2006), the particular metabolism of cephalopods needs to be considered if a successful artificial diet is to be designed. The partitioning of the dietary energy intake by cephalopods in the various types of metabolism, excretion and growth was reviewed by O'Dor and Wells (1987). Bearing that in mind, the nutritional content of the diet must be good enough to sustain the existing cephalopod metabolic costs, so the animal is able to allocate an optimal distribution of surplus energy to somatic growth and, later, to reproduction (Wells and Clarke 1996). Therefore, the animal has to feed on a diet correctly balanced to its metabolic needs at a given temperature (André et al. 2009). The existing information is resumed to two theories that clash in terms of what the cephalopods use as energy substrate. The first one, by Lee (1994) and Boucher-Rodoni and Mangold (1994), considers that under normal feeding conditions, both growth and energy use the protein fraction as fuel. The second theory, by Storey and Storey (1983) and Hochachka (1994), considers that the CH fraction is used as energy source and the protein fraction is exclusively used for growth. Such contrasting theories may originate from the fact that the same species is physiologically adapted to a given geographical location, with different temperature regimes and food nutritional composition. For instance, *S. officinalis* populations from the English Channel and southern Portugal have been reported to be genetically different (Wolfram et al. 2006). In addition, this species displays a physiological plasticity (Oellermann et al. 2012), which is temperature- and food-dependent (reliant on the lifestyle of cephalopods and their low energy reserves).

Cephalopods have appropriate catabolic pathways to breakdown protein to amino acids to obtain energy (Ballantyne et al. 1981). However, most cephalopods' living strategy is to lay down protein reserves into rapid growth to convert them into gametes (O'Dor et al. 1984; Moltschaniwskyj and Carter 2013). Hypothetically, it would be a waste to partially use amino acids for energy, these being reserves only used in case of starving or at reproduction. This would point to the use of other reserves before protein and amino acids at early stages of life. Fast-growing cephalopods, when fed properly, are extremely efficient in converting food to protein, display low protein degradation and exhibit increased efficiency of retaining synthesized protein (Carter et al. 2009). On the other hand, cephalopods are said to have a limited capacity for lipid oxidation (Ballantyne et al. 1981), and its digestion becomes slow and inefficient due to the lack of emulsifiers (biliary salts) in the digestive tract (Vonk 1962). On the contrary, Moltschaniwskyj and Johnston (2006) have shown that *Euprymna tasmanica* has the ability to digest lipids (very high levels of lipase were found in the digestive gland) but these are not stored in the digestive gland, which indicates the species capacity of using lipids as a source of fuel (Swift et al. 2005). This species is known to have a very sedentary lifestyle (it does not move a lot, so there are not many mantle burst activities) in Southern Australia temperate waters. A similar ability to use lipids had already been displayed by individuals of *S. officinalis* from the English Channel populations, which metabolized lipids instead of protein or CHs when facing prolonged starvation (Castro et al. 1992). On the other hand, Lamarre et al. (2012) observed the mixed use of lipids and protein, and after 8 days the exclusive use of protein in short starvation, in cuttlefish from the Mediterranean Sea. This use of lipids by cuttlefish might be eventually identical, at the cellular level, to what is described by Finn and Dice (2006) in vertebrates.

The higher growth rates observed in cephalopods that live or are cultured in subtropical regions (with high temperatures than those observed in temperate waters) would mean that from a given temperature threshold, cephalopods would use CHs as energy, while at lower temperatures, they would preferably use other sources such as lipids and protein (depending on food availability). The enzyme content of the different life stages helps to understand the eventual existence of different metabolisms. While the enzyme content of the English Channel population of *S. officinalis* has been reported, for several years and by numerous researchers, those of Faro or Mediterranean populations have never, to the best of our knowledge, been reported. Regarding the first, Boucaud-Camou (1969, 1947) characterized amylase and protease activities in different digestive organs of juvenile cuttlefish and Yim (1978) detected amylase activity in mature cuttlefish. According to Koueta et al. (2000), this activity increases with age, not being present at hatching, which is concomitant with the maturation of the sepia digestive system (Boucaud-Camou et al. 1985). Nonetheless, the use of silage as enrichment for shrimp given as prey to cuttlefish hatchlings promoted an increase of total CH and peptides in the diet and supported 100% survival plus increased growth, when compared with natural diets, and despite the lower content in total protein (Le Bihan et al. 2006). However, these same authors reported an inhibition of amylase activity but higher proteolytic activity.

Higher temperatures promote oxygen-efficient adenosine triphosphate (ATP) production due to limitations in available oxygen (Hochachka 1994; Pörtner 2010). However, most studies on cephalopod metabolism have been performed in fasting animals, where the stressful situation of meeting the energy requirements will promote the use of protein reserves (McCue 2010). In addition, one has to consider that all animals exhibit adaptive biochemical and physiological responses to the lack of food (Wang et al. 2006). This is particularly true regarding cephalopods, most of which inhabit environments in which food availability fluctuates or encounters with appropriate food items might be rare and unpredictable at given geographical locations or seasons.

Houlihan et al. (1990) studied protein metabolism in *O. vulgaris* and concluded that the high growth rates displayed by this species rely on high rates of protein synthesis and high efficiencies of retention of synthesized protein and little protein degradation. According to Oellermann et al. (2012), the European cuttlefish has the capability of adjusting its cellular and mitochondrial energetics over short- and long-term changes of temperature and environmental conditions, which is an evolutionary adaptation of given populations, such as *S. officinalis*. In addition, temperature has a significant effect on oxygen consumption (Grigoriou and Richardson 2009) and energy metabolism of cuttlefish (Mark et al. 2008). Furthermore, temperate cuttlefish (English Channel populations) display a predominant oxidation of proline in systemic heart, while subtropical cuttlefish (Mediterranean populations) exhibit enhanced pyruvate oxidation. The latter is supported by the findings of Ballantyne et al. (1981) on octopine dehydrogenase linking amino acid (arginine) and CH (pyruvate) metabolism, which are said to occur during hypoxic conditions, burst activity or both.

In this sense, cold-adapted metabolism in cuttlefish will show a suppressed CH metabolism, favoring the use of lipids (e.g. results of Koueta et al. (2002); Perrin et al. (2004) and Koueta et al. (2006)) and proline, which are less oxygen efficient (Hochachka 1994). At lower temperatures, amino acids such as glutamate, ornithine or arginine may sustain the supply of succinate (Ballantyne et al. 1981; Hochachka et al. 1983). Cephalopods should have developed an aerobic CH–amino acid metabolism that maximizes ATP yield per unit of consumed oxygen due to the lack of an intracellular myoglobin analogue and the exclusive localization of mitochondria in interfibrillar zones (Hochachka 1994; Ballantyne 2004). CH are catabolized rapidly following ingestion, and are not used during fasting, but mobilized for locomotor activity (O'Dor et al. 1984). According to Hochachka and Fields (1982) and Storey and Storey (1978), glucose metabolism may be coupled to that of proline through glucose conversion to pyruvate, which is oxidized in the Krebs cycle, and proline simultaneously augments the cycle intermediates. These first authors still add that glutamate, proline and most probably arginine can be used directly as gluconeogenic precursors.

5.9 Conclusions

Proteins are the most abundant macronutrient in cephalopods, and large protein and amino acid contents in the diet are required for sustaining growth and eventually fulfilling energy demands. Although low in quantitative terms, lipids are essential in cephalopod nutrition, with long-chain PUFA playing a pivotal role, since the enzymatic machinery unveiled through molecular research points at their inability in the synthesis of these important nutrients. Therefore, the effect of dietary protein/lipid ratios on amino acid absorption efficiency and metabolism needs to be researched.

Cephalopods also require an adequate provision of antioxidants like pro-vitamin A and carotenoids, since the importance of vitamins in photoreception, growth and development must not be neglected. It is generally accepted that cephalopods do not have a specific requirements for dietary CH, but it has been reported that they are able to rapidly catabolize dietary CH to account for energy demands in explosive activities such as prey capture and fleeing from predators, and the existence of a CH metabolism has been established. However, the CH theory presented in this chapter needs to be proven. Finally, it is accepted that octopuses and cuttlefish, as carnivorous species, meet the majority of their elemental requirements from the diet, although direct uptake from the seawater has also been shown to occur through an ion balance mechanism regulated by the digestive gland appendages.

Methods for the determination of possible metabolic pathways have been developed for fish larvae (Conceição et al. 2003; Morais et al. 2004) and these may be adapted for the study of metabolism in cephalopods. Additionally, the use of systems biology modelling to cephalopod nutrition, similar to that published by Hormiga et al. (2010), might help on a faster progress towards our understanding of cephalopods' physiology and nutrition.

References

Almansa E, Domingues P, Sykes A et al (2006) The effects of feeding with shrimp or fish fry on growth and mantle lipid composition of juvenile and adult cuttlefish (*Sepia officinalis*). Aquaculture 256:403–413

André J, Grist EPM, Semmens JM et al (2009) Effects of temperature on energetics and the growth pattern of benthic octopuses. Mar Ecol Prog Ser 374:167–179

Andrés M, Estévez A, Simeó CG et al (2010) Annual variation in the biochemical composition of newly hatched larvae of *Maja brachydactyla* in captivity. Aquaculture 310:99–105

Aragão C, Conceição LEC, Dinis MT et al (2004) Amino acid pools of rotifers and *Artemia* under different conditions: Nutritional implications for fish larvae. Aquaculture 234:429–445

Baldwin J, Fields JHA, Hochachka PW (1976) Role of octopine dehydrogenase in energy-metabolism of cephalopods. Proc Australian Biochem Soc 9:5–5

Ballantyne JS (2004) Mitochondria: aerobic and anaerobic design—lessons from molluscs and fishes. Comp Biochem Physiol B 139:461–467

Ballantyne JS, Hochachka PW, Mommsen TP (1981) Studies on the metabolism of the migratory squid, *Loligo opalescens*—enzymes of tissues and heart-mitochondria. Mar Biol Let 2:75–85

Barnabé G (1996) Bases Biologiques et Écologiques de l'Aquaculture. Editorial Acribia SA, Zaragoza

Beuerlein K, Ruth P, Scholz FR et al (2004) Blood cells and the biosynthesis of hemocyanin in *Sepia* embryos. Micron 35:115–116

Blanchier B (1981) Etude des lipides totaux et des steroides dans la glande digestive et la gonade chez la seiche *Sepia officinalis* L. (Mollusque, Cephalopode). Université de Caen, France

Boletzky SV (1974) Effects of continuous malnutrition on development of cuttlebone in *Sepia officinalis* L (Mollusca, Cephalopoda). Bull Soc Zool France—Evol Zool 99:667–673

Bouchaud O (1991) Energy consumption of the cuttlefish *Sepia officinalis* L. (Mollusca: Cephalopoda) during embryonic development, preliminary results. Bull Mar Sci 49:333–340

Bouchaud O, Galois R (1990) Utilization of egg-yolk lipids during the embryonic development of *Sepia officinalis* L. in relation to temperature of the water. Comp Bioch Physio 97B:611–615

Boucaud-Camou E (1969) Localization of amylase and protease activities in digestive system of *Sepia officinalis* L. Cr Acad Sci D Nat 269:2564–2566

Boucaud-Camou E (1974) Localisation d'activités enzymatiques impliquées dans la digestion chez *Sepia officinalis* L. Arch Zool Exp Gén 115:5–27

Boucaud-Camou E (1989) L'aquaculture des cephalopodes: evaluation et perspectives. Haliotis 19:201–214

Boucaud-Camou E (1990) La seiche, un animal d'avenir. Peche Maritime 69:321–329

Boucaud-Camou E, Yim M, Tresgot A (1985) Feeding and digestion of young *Sepia officinalis* L. (Mollusca: Cephalopoda) during post-hatching development. Vie Milieu 35:263–266

Boucher-Rodoni R, Mangold K (1994) Ammonia production in cephalopods, physiological and evolutionary aspects. Mar Freshw Behav Physiol 25:53–60

Boucher-Rodoni R, Boucaud-Camou E, Mangold K (1987) Feeding and digestion. In: Boyle PR (ed) Cephalopod Life Cycles. Academic Press London, p 85–108

Brown MR, Battaglene SC, Morehead DT, Brock M (2005) Ontogenetic changes in amino acid and vitamins during early larval stages of striped trumpeter (*Latris lineata*). Aquaculture 248:263–274

Bustamante P, Teyssie JL, Danis B et al (2004) Uptake, transfer and distribution of silver and cobalt in tissues of the common cuttlefish *Sepia officinalis* at different stages of its life cycle. Mar Ecol—Progr Ser 269:185–195

Carter CG, Lynch KA, Moltschaniwskyj NA (2009) Protein synthesis in a solitary benthic cephalopod, the Southern dumpling squid (*Euprymna tasmanica*). Comp Biochem Physio A: Mol Int Physiol 153:185–190

Castro BG, Garrido JL, Sotelo CG (1992) Changes in composition of digestive gland and mantle muscle of the cuttlefish *Sepia officinalis* during starvation. Mar Biol 114:11–20

Castro BG, Dimarco FP, Derusha RH et al (1993) The effects of surimi and pelleted diets on the laboratory survival, growth, and feeding rate of the cuttlefish *Sepia officialis* L. J Exp Mar Biol Ecol 170:241–252

Cho SY, Joo DS, Choi HG, Nara E, Miyashita K (2001) Oxidative stability of lipids from squid tissues. Fish Sci 67:738–743

Conceição LEC, Grasdalen H, Ronnestad I (2003) Amino acid requirements of fish larvae and post-larvae: new tools and recent findings. Aquaculture 227:221–232

Conceição L, Aragão C, Ronnestad I (2010) Protein metabolism and amino acid requirements in fish larvae. In: Cruz-Suarez LE, Ricque-Marie D, Tapia-Salazar M, Nieto-López MG, Villarreal-Cavazos DA, Gamboa-Delgado J (eds) Avances en Nutrición Acuícola X—Memorias del Décimo Simposio Internacional de Nutrición Acuícola. Monterrey, México, p 250–263

D'Aniello A, Strazzullo L, D'Onofrio G, Pischetola M (1986) Electrolytes and nitrogen compounds of body fluids and tissues of *Octopus vulgaris* Lam. J Comp Physiol 156:503–509

De Eguileor ML, Grimaldi A et al (2000) Integumental amino acid uptake in a carnivorous predator mollusc (*Sepia officinalis*, Cephalopoda). Tissue Cell 32:389–398

Decleir W, Richard A (1970) A study of the blood proteins in *Sepia officinalis* L. with special reference to embryonic hemocyanin. Comp Biochem Physiol 34:203–211

Decleir W, Richard A (1972) A study of the orange-red pigment from the accessory nidamental glands of the cephalopod *Sepia officinalis* L. Biol Jb Dodonaea 40:188–197

Decleir W, Lemaire J, Richard A (1971) The differentiation of blood proteins during ontogeny in *Sepia officinalis* L. Comp Biochem Physiol 40B:923–930

Decleir W, Vlaeminck A, Geladi P et al (1978) Determination of protein-bound copper and zinc in some organs of the cuttlefish *Sepia officinalis* L. Comp Biochem Physiol 60B:347–350

Domingues P, Poirier R, Dickel L et al (2003) Effects of culture density and live prey on growth and survival of juvenile cuttlefish, *Sepia officinalis*. Aquac Int 11:225–242

Domingues P, Sykes A, Sommerfield A et al (2004) Growth and survival of cuttlefish (*Sepia officinalis*) of different ages fed crustaceans and fish. Effects of frozen and live prey. Aquaculture 229:239–254

Domingues PM, Dimarco FP, Andrade JP et al (2005) Effect of artificial diets on growth, survival and condition of adult cuttlefish, *Sepia officinalis* Linnaeus, 1758. Aquac Int 13:423–440

Domingues P, Marquez L, López N et al (2009) Effects of food thermal treatment on growth, absorption, and assimilation efficiency of juvenile cuttlefish, *Sepia officinalis*. Aquac Int 17:283–299

Dunstan GA, Baillie HJ, Barrett SM, Volkman JK (2006) Effect of diet on the lipid composition of wild and cultured abalone. Aquaculture 140:115–127

Durazo-Beltrán E, D'Abramo LR, Toro-Vazquez JF et al (2003) Effect of triacylglycerols in formulated diets on growth and fatty acid composition in tissue of green abalone (*Haliotis fulgens*). Aquaculture 224:257–270

Ferreira A, Marquez L, Almansa E et al (2010) The use of alternative diets to culture juvenile cuttlefish, *Sepia officinalis*: effects on growth and lipid composition. Aquac Nut 16:262–275

Finn PF, Dice JF (2006) Proteolytic and lipolytic responses to starvation. Nutrition 22:830–844

Fisher LR, Kon SK, Thompson SY (1956) Vitamin A and carotenoids in certain invertebrates. V Mollusca: Cephalopoda. J Mar Biol Assoc UK 35:63–80

Fluckiger M, Jackson GD, Nichols P et al (2008) An experimental study of the effect of diet on the fatty acid profiles of the European cuttlefish (*Sepia officinalis*). Mar Biol 154:363–372

Fox DL (1966) Pigmentation of Molluscs. In: Yonge CM (ed) Physiology of Mollusca. Academic Press Inc., New York, p 240–273

García-García B, Cerezo-Valverde J (2006) Optimal proportions of crabs and fish in diet for common octopus (*Octopus vulgaris*) ongrowing. Aquaculture 253:502–511

García-Garrido S, Hachero-Cruzado I, Rosas C et al (2012) Protein and amino acid composition from the mantle of juvenile *Octopus vulgaris* exposed to prolonged starvation. Aquac Res 44:1741–1751

Grigoriou P, Richardson CA (2009) Effect of body mass, temperature and food deprivation on oxygen consumption rate of common cuttlefish *Sepia officinalis*. Mar Biol 156:2473–2481

Guinot D, Monroig Ó, Navarro JC et al (2013). Enrichment of *Artemia* metanauplii in phospholipids and essential fatty acids as a diet for common octopus (*Octopus vulgaris*) paralarvae. Aquac Nut 19:837–844

Ghiretti F (1966) Molluscan hemocyanins. In: Wilbur KM, Yonge CM (eds) Physiology of Mollusca, vol II. Academic Press, London

Halver JE (2002) The vitamins In: Halver JE, Hardy RW (eds) Fish Nutrition, 3rd edn. Academic Press, San Diego, p 61–141

Hanlon RT (1987) Mariculture. In: Cephalopod Life Cycles. Academic Press Inc., London

Hanlon RT, Bidwell JP, Tait R (1989) Strontium Is required for statolith development and thus normal swimming behavior of hatchling cephalopods. J Exp Biol 141:187–195

Hanlon RT, Turk PE, Lee PG (1991) Squid and cuttlefish mariculture: an updated perspective. J Ceph Biol 2:31–40

Hewitt RA (1975) Analysis of aragonite from cuttlebone of *Sepia officinalis* L. Mar Geol 18:M1–M5

Hochachka PW (1994) Oxygen efficient design of cephalopod muscle metabolism. Mar Fresh Behav Physiol 25:61–67

Hochachka PW, Fields JHA (1982) Arginine, glutamate, and proline as substrates for oxidation and for glycogenesis in cephalopod tissues. Pac Sci 36:325–335

Hochachka PW, Hartline PH, Fields JHA (1976) Octopine as an end product of anaerobic glycolysis in the chambered nautilus. Science 195:72–74

Hochachka PW, Mommsen TP, Storey J et al (1983) The relationship between arginine and proline metabolism in cephalopods. Mar Biol Lett 4:1–21

Hormiga JA, Almansa E, Sykes AV et al (2010) Model based optimization of feeding regimens in aquaculture: Application to the improvement of *Octopus vulgaris* viability in captivity. J Biotechnol 149:209–214

Houlihan DF, Mcmillan DN, Agnisola C et al (1990) Protein synthesis and growth in *Octopus vulgaris*. Mar Biol 106:251–259

Hunt S, Nixon M (1981) A comparative study of protein composition in the chitin-protein complexes of the beak, pen, sucker disk, radula and esophageal cuticle of cephalopods. Comp Biochem Physiol B 68:535–546

Huxtable RJ (1992) Physiological actions of taurine. Physiol Rev 72:101–163

Ito Y, Fujii T (1962) Chemical composition of the egg-yolk lipoproteins. J Biochem 52:221–222

Ito Y, Fujii T, Otake M (1962) On peptide constituent of the egg-yolk lipoproteins. J Biochem 52:223–225

Koueta N, Le CA, Noel B et al (2000) Changes of digestive enzymes during growth of cultured juvenile cuttlefish *Sepia officinalis* L. (Mollusca Cephalopoda). Effect of enriched diet and ration. In: ICES (ed) ICES Annual Science Conference. Brugge, Belgium, p 17

Koueta N, Boucaud-Camou E, Noel B (2002) Effect of enriched natural diet on survival and growth of juvenile cuttlefish *Sepia officinalis* L. Aquaculture 203:293–310

Koueta N, Alorend E, Noel B et al (2006) Earlier acceptance of frozen prey by juvenile cuttlefish *Sepia officinalis* in experimental rearing: effect of previous enriched natural diet. Vie Milieu 56:147–152

Kreuzer R (1984) Cephalopods: handling, processing and products. FAO Fish Tech Pap 254:108

Lacoue-Labarthe T, Warnau A, Oberhansli F et al (2008a) Differential bioaccumulation behaviour of Ag and Cd during the early development of the cuttlefish *Sepia officinalis*. Aquat Toxicol 86:437–446

Lacoue-Labarthe T, Warnau M, Oberhansli F et al (2008b) First experiments on the maternal transfer of metals in the cuttlefish *Sepia officinalis*. Mar Pollut Bull 57:826–831

Lacoue-Labarthe T, Warnau M, Metian M et al (2009) Biokinetics of Hg and Pb accumulation in the encapsulated egg of the common cuttlefish *Sepia officinalis*: Radiotracer experiments. Sci Total Environm 407:6188–6195

Lacoue-Labarthe T, Le Bihan E, Borg D et al (2010a) Acid phosphatase and cathepsin activity in cuttlefish (*Sepia officinalis*) eggs: the effects of Ag, Cd, and Cu exposure. ICES J Mar Sci 67:1517–1523

Lacoue-Labarthe T, Warnau M, Oberhansli F et al (2010b) Contrasting accumulation biokinetics and distribution of Am-241, Co, Cs, Mn and Zn during the whole development time of the eggs of the common cuttlefish, *Sepia officinalis*. J Exp Mar Biol Ecol 382:131–138

Lamarre SG, Ditlecadet D, Mckenzie DJ et al (2012) Mechanisms of protein degradation in mantle muscle and proposed gill remodeling in starved *Sepia officinalis*. Am J Physiol-Reg I 303:R427–R437

Le Bihan E, Perrin A, Koueta N (2006) Influence of diet peptide content on survival, growth and digestive enzymes activities of juvenile cuttlefish *Sepia officinalis*. Vie Milieu 56:139–145

Lee PG (1994) Nutrition of cephalopods: fueling the system. Mar Fresh Behav Physiol 25:35–51
Liñán-Cabello MA, Paniagua-Michel J, Hopkins PM (2002) Bioactive roles of carotenoids and retinoids in crustaceans. Aquac Nut 8:299–309
Lourenço HM, Anacleto P, Afonso C et al (2009) Elemental composition of cephalopods from Portuguese continental waters. Food Chem 113:1146–1153
Mark F, Melzner F, Bock C et al (2008) Thermal effects on cephalopod energy metabolism—a case study for *Sepia officinalis*. Comp Biochem Physiol A 150:S171–S171
McCue MD (2010) Starvation physiology: reviewing the different strategies animals use to survive a common challenge. Comp Biochem Physiol A 156:1–18
Miliou H, Fintikaki M, Tzitzinakis M et al (2006) Fatty acid composition of the common octopus, *Octopus vulgaris*, in relation to rearing temperature and body weight. Aquaculture 256:311–322
Miyazaki T, Nakahara M, Ishii T et al (2001) Accumulation of cobalt in newly hatched octopus, *Octopus vulgaris*. Fish Sci 67:170–172
Moltschaniwskyj N, Johnston D (2006) Evidence that lipid can be digested by the dumpling squid *Euprymna tasmanica*, but is not stored in the digestive gland. Mar Biol 149:565–572
Moltschaniwskyj NA, Carter CG (2010) Protein synthesis, degradation, and retention: mechanisms of indeterminate growth in cephalopods. Physiol Biochem Zool: PBZ 83:997–1008
Moltschaniwskyj MA, Carter CG (2013) The adaptive response of protein turnover to the energetic demands of reproduction in a cephalopod. Physiol Biochem Zool 86:119–126
Mommsen TP, French CJ, Emmett B et al (1982) The fate of arginine and proline carbon in squid tissues. Pac Sci 3:343–348
Monroig Ó,G, Hontoria F et al (2012a) Biosynthesis of essential fatty acids in *Octopus vulgaris* (Cuvier, 1797): Molecular cloning, functional characterization and tissue distribution of a fatty acyl elongase. Aquaculture 360–361:45–53
Monroig Ó,N, Dick JR et al (2012b) Identification of a Δ5-like fatty acyl desaturase from the cephalopod *Octopus vulgaris* (Cuvier 1797) involved in the biosynthesis of essential fatty acids. Mar Biotech 14:411–422
Morais S, Conceição L, Dinis MT, Rønnestad I (2004) A method for radiolabeling *Artemia* with applications in studies of food intake, digestibility, protein and amino acid metabolism in larval fish. Aquaculture 231:469–487
Morillo-Velarde PS, Cerezo Valverde J, Serra Llinares RM et al (2011) Energetic contribution of carbohydrates during starvation in common octopus (*Octopus vulgaris*). J Mollusc Stud 77:318–320
Morillo-Velarde PS, Cerezo Valverde J, Hernández MD et al (2012) Growth and digestibility of formulated diets based on dry and freeze-dried ingredients in the common octopus (*Octopus vulgaris*). Aquaculture 368–369:139–144
Motoe K, Ooizumi T, Hayashi S, Kawasaki K (1997) Seasonal changes in the contents of proximate composition, minerals and vitamins of firefly squid. J Jpn Soc Food Sci Tech 44:133–139
Napoleão P, Pinheiro T, Reis CS (2005a) Elemental characterization of tissues of *Octopus vulgaris* along the Portuguese coast. Sci Total Environm 345:41–49
Napoleão P, Reis CS, Alves LC et al (2005b) Morphologic characterization and elemental distribution of *Octopus vulgaris* Cuvier, 1797 vestigial shell. Nuclear Instrum Meth Physic Research B 231:345–349
Navarro JC, Villanueva R (2000) Lipid and fatty acid composition of early stages of cephalopods: an approach to their lipid requirements. Aquaculture 183:161–177
Navarro JC, Villanueva R (2003) The fatty acid composition of *Octopus vulgaris* paralarvae reared with live and inert food: deviation from their natural fatty acid profile. Aquaculture 219:613–631
O'Dor RK, Mangold K, Boucher-Rodoni R et al (1984) Nutrient absorption, storage and remobilization in *Octopus vulgaris*. Marine Behavioral Physiology 11:239–258
O'Dor RK, Wells MJ (1987) Energy and nutrient flow. In: Boyle PR (ed) Cephalopod life cycles. Academic Press, London, p 109–133
Oellermann M, Portner HO, Mark FC (2012) Mitochondrial dynamics underlying thermal plasticity of cuttlefish (*Sepia officinalis*) hearts. J Exp Biol 215:2992–3000

Okumura S, Kurihara A, Iwamoto A et al (2005) Improved survival and growth in *Octopus vulgaris* paralarvae by feeding large type *Artemia* and Pacific sandeel, *Ammodytes personatus*: Improved survival and growth of common octopus paralarvae. Aquaculture 244:147–157

Ozyurt G, Duysak O, Akamca E et al (2006) Seasonal changes of fatty acids of cuttlefish *Sepia officinalis* L. (Mollusca: Cephalopoda) in the north eastern Mediterranean sea. Food Chem 95:382–385

Packard A, Albergoni V (1970) Relative growth, nucleic acid content and cell numbers of brain in *Octopus vulgaris* (Lamarck). J Exp Biol 52:539–552

Pandit AR, Magar NG (1972) Chemical composition of *Sepia orientalis* and *Loligo vulgaris*. Fish Technol 9:122–125

Passi S, Cataudella S, Di Marco P et al (2002) Fatty acid composition and antioxidant levels in muscle tissue of different Mediterranean marine species of fish and shellfish. J Agric Food Chem 50:7314–7322

Pecl GT, Moltschaniwskyj NA (1999) Somatic growth processes: how are they altered in captivity? Proc Royal Soc London 266:1133–1139

Perrin A, Le Bihan E, Koueta N (2004) Experimental study of enriched frozen diet on digestive enzymes and growth of juvenile cuttlefish *Sepia officinalis* L. (Mollusca Cephalopoda). J Exp Mar Biol Ecol 311:267–285

Pinto W, Figueira L, Santos A et al (2013) Is dietary taurine supplementation beneficial for gilthead seabream (*Sparus aurata*) larvae? Aquaculture 384–387:1–5

Pörtner H-O (2010) Oxygen- and capacity-limitation of thermal tolerance: a matrix for integrating climate-related stressor effects in marine ecosystems. J Exp Biol 213:881–893

Quintana D (2009) Valoración de los requerimientos nutricionales de reproductores de pulpo común (*Octopus vulgaris*). PhD Thesis. University of La Laguna, Spain

Rønnestad I, Thorsen A, Finn RN (1999) Fish larval nutrition: a review of recent advances in the roles of amino acids. Aquaculture 177:201–216

Rønnestad I, Tonheim SK, Fyhn HJ et al (2003) The supply of amino acids during early feeding stages of marine fish larvae: a review of recent findings. Aquaculture 227:147–164

Roura Á,G, Redd K et al (2012) Molecular prey identification in wild *Octopus vulgaris* paralarvae. Mar Biol 159:1335–1345

Sargent J, Bell MV, Bell JG et al (1995) Origins and functions of n-3 polyunsaturated fatty acids in marine organisms. In: Phospholipids: Characterization, Metabolism and Novel Biological Applications. Americal Oil Chemical Society, Champaign, p 248–259

Sargent J, McEvoy LA, Bell JG (1997) Requirements, presentation and sources of polyunsaturated fatty acids in marine fish larval feeds. Aquaculture 155:117–127

Seixas S, Bustamante P, Pierce GJ (2005) Interannual patterns of variation in concentrations of trace elements in arms of *Octopus vulgaris*. Chemosphere 59:1113–1124

Seixas P, Rey-Mendez M, Valente LMP et al (2008) Producing juvenile *Artemia* as prey for *Octopus vulgaris* paralarvae with different microalgal species of controlled biochemical composition. Aquaculture 283:83–91

Seixas P, Rey-Méndez M, Valente LMP et al (2010) High DHA content in *Artemia* is ineffective to improve *Octopus vulgaris* paralarvae rearing. Aquaculture 300:156–162

Semmens JM, Pecl GT, Villanueva R et al (2004) Understanding octopus growth: patterns, variability and physiology. Mar Freshw Res 55:367–377

Sidwell VD, Loomis AL, Foncannon PR, Buzzell DH (1978) Composition of the edible portion of raw (fresh or frozen) crustaceans, finfish, and mollusks. IV. Vitamins. Mar Fish Rev 40:1–16

Sikorski ZE, Kolodziejska I (1986) The composition and properties of squid meat. Food Chem 20:213–224

Sinanoglou VJ, Miniadis-Meimaroglou S (1998) Fatty acid of neutral and polar lipids of (edible) Mediterranean cephalopods. Food Res Int 31:467–473

Sinanoglou VJ, Miniadis-Meimaroglou S (2000) Phospholipids in Mediterranean cephalopods. Z Naturforsch C 55:245–255

Stéphan G, Guillaume J, Lamour F (1995) Lipid-peroxidation in turbot (*Scophthalmus maximus*) tissue: effect of dietary vitamin E and dietary n–6 or n–3 polyunsaturated fatty acids. Aquaculture 130:251–268

Storey KB, Storey JM (1978) Energy metabolism in the mantle muscle of the squid, *Loligo pealeii*. J Comp Physiol 123:169–175

Storey KB, Storey JM (1979) Octopine metabolism in the cuttlefish, *Sepia officinalis*—octopine production by muscle and its role as an aerobic substrate for non-muscular tissues. J Comp Physiol 131:311–319

Storey KB, Storey JM (1983) Carbohydrate metabolism on cephalopod molluscs. In: Hochachka PW (ed) The Mollusca. Academic Press Inc., New York, p 91–136

Storey KB, Storey JM, Johansen K et al (1979) Octopine metabolism in *Sepia officinalis*—effect of hypoxia and metabolite loads on the blood-levels of octopine and related-compounds. Can J Zool 57:2331–2336

Swift K, Johnston D, Moltschaniwskyj N (2005) The digestive gland of the Southern Dumpling Squid (*Euprymna tasmanica*): structure and function. J Exp Mar Biol Ecol 315:177–186

Sykes AV, Domingues PM, Correia M et al (2006) Cuttlefish culture—state of the art and future trends. Vie Milieu 56:129–137

Sykes AV, Almansa E, Lorenzo A et al (2009a) Lipid characterization of both wild and cultured eggs of cuttlefish (*Sepia officinalis* L.) throughout the embryonic development. Aquac Nut 15:38–53

Sykes AV, Oliveira AR, Domingues PM et al (2009b) Assessment of European cuttlefish (*Sepia officinalis*, L.) nutritional value and freshness under ice storage using a developed Quality Index Method (QIM) and biochemical methods. Food Sci Tech 42:424–432

Sykes AV, Pereira D, Rodríguez C et al (2012) Effects of increased tank bottom areas on cuttlefish (*Sepia officinalis*, L.) reproduction performance. Aquac Res. doi:10.1111/j.1365–2109.2012.03106.x

Taylor HH, Anstiss JM (1999) Copper and haemocyanin dynamics in aquatic invertebrates. Mar Freshw Res 50:907–931

Uki N, Kemuyama A, Watanabe T (1986) Optimum protein level in diets for abalone. Bull Jpn Soc Sci Fish 52:1005–1012

Valverde JC, Hernandez MD, Garcia-Garrido S et al (2012) Lipid classes from marine species and meals intended for cephalopod feeding. Aquac Int 20:71–89

Valverde J, Martínez-Llorens S, Vidal A et al (2013) Amino acids composition and protein quality evaluation of marine species and meals for feed formulations in cephalopods. Aquac Int 21:413–433

Van Den Branden CR, Lemaire J et al. (1978) La glande nidamentaire accessoire de *Sepia officinalis* L.: analyses biochimiques des pigments des bactéries symbiotiques. Annales Soc R Zool Belg 108:123–139

Van Den Branden C, Gillis M, Richard A (1980) Carotenoid producing bacteria in the accessory nidamental glands of *Sepia officinalis* L. Comp Biochem Physiol B 66:331–334

Villanueva R, Bustamante P (2006) Composition in essential and non-essential elements of early stages of cephalopods and dietary effects on the elemental profiles of *Octopus vulgaris* paralarvae. Aquaculture 261:225–240

Villanueva R, Riba J, Ruiz-Capillas C et al (2004) Amino acid composition of early stages of cephalopods and effect of amino acid dietary treatments on *Octopus vulgaris* paralarvae. Aquaculture 242:455–478

Villanueva R, Escudero JM, Deulofeu R et al (2009) Vitamin A and E content in early stages of cephalopods and their dietary effects in *Octopus vulgaris* paralarvae. Aquaculture 286:277–282

Vlieg P (1984) Proximate composition of New Zealand squid species. New Zealand J Sci 27:45–150

Vonk HJ (1962) Emulgators in the digestive fluids of invertebrates. Arch Int Physiol Biochimie 70:67–85

Wang T, Hung CCY, Randall DJ (2006) The comparative physiology of food deprivation: from feast to famine. Annu Rev Physiol 68:223–251

Wells MJ, Wells J (1989) Water uptake in a cephalopod and the function of the so-called pancreas. J Exp Biol 145:215–226

Wells MJ, Clarke A (1996) Energetics: the costs of living and reproducing for an individual cephalopod. Philo Trans Royal Soc London Ser B 351:1083–1104

Wolf G, Witters R, Decleir W et al (1980) Immunological evidence for hemocyanin-related proteins in mature eggs and embryos of *Sepia officinalis* L. Arch Int Physiol Bio 88:B254–B254

Wolf G, Verheyen E, Vlaeminck A et al (1985) Respiration of *Sepia officinalis* during embryonic and early juvenile life. Mar Biol 90:35–39

Wolfram K, Mark FC, John U et al (2006) Microsatellite DNA variation indicates low levels of genetic differentiation among cuttlefish (*Sepia officinalis* L.) populations in the English Channel and the Bay of Biscay. Comp Biochem Physiol D 1:375–383

Yim M (1978) Développement post-embryonnaire de la glande digestive de *Sepia officinalis* L. (Mollusque Céphalopode). Université de Caen, Caen, p 81

Zlatanos S, Laskaridis K, Feist C et al (2006) Proximate composition, fatty acid analysis and protein digestibility-corrected amino acid score of three Mediterranean cephalopods. Mol Nut Food Res 50:967–970

Chapter 6
Welfare and Diseases Under Culture Conditions

António V. Sykes and Camino Gestal

Abstract This chapter reviews the welfare and diseases that have been reported since cephalopods are maintained, reared or cultured in captivity. Although cephalopod welfare is only going to be assured in terms of the European Union (EU) legislation from January 2013, it has long been enforced in other regions or countries all over the world. Pathologies registered under captive conditions derive, most of the times, from bad welfare practices. A revision of cephalopods' immune system and the most important pathologies are presented, which are divided into viral, bacterial, fungal and parasitic pathogenic agents as well as chemical and mechanical damages. In addition, information regarding healing, antibiotics application and surgery is provided. Welfare under research and commercial culture conditions is discussed in terms of the use of anaesthesia and euthanasia agents and their assessment in terms of effectiveness. Further research on the different aspects considered is suggested.

Keywords Welfare and diseases under research and culture conditions · Cephalopods immune system · Pathological agents · Mechanical and chemical damage · Healing · Antibiotics and surgery

6.1 Welfare in Cephalopods Under Captive Conditions

6.1.1 Legislation

According to Ohl and van der Staay (2012), animal welfare is a complex concept which needs to have a scientific definition. These authors define positive welfare as

A. V. Sykes (✉)
CCMAR-CIMAR L.A., Centro de Ciências do Mar do Algarve, Universidade do Algarve, Campus de Gambelas, 8005-139, Faro, Portugal
e-mail: asykes@ualg.pt

C. Gestal
Instituto de Investigaciones Marinas (IIM), CSIC, Eduardo Cabello 6, 36208, Vigo, Spain
e-mail: cgestal@iim.csic.es

J. Iglesias et al. (eds.), *Cephalopod Culture*, DOI 10.1007/978-94-017-8648-5_6,

being dependable "on the freedom to react appropriately and adequately to" a set of different aspects, such as: "hunger, thirst or incorrect food; thermal and physical discomfort; injuries or diseases; fear and chronic stress, and thus, the freedom to display normal, species-specific behavioural patterns and adapt to changing living conditions up to a level that is perceived as positive". Despite being quite reasonable, dynamic and complete, these authors warned that this definition still does not consider positive emotions. On the other hand, welfare should be considered "as a measure of biological function, (…) not considered a static concept", but should contemplate "(…) the dynamics of the individual's interaction with its environment over time".

"Cephalopods welfare" is a recent subject in most of the European countries due to the inclusion of this class on the latest European Union (EU) animal welfare legislation ("Directive 2010/63/EU" (EU 2010)). Nonetheless, legislation regarding the safeguard of cephalopod welfare has long been enforced in non-European countries such as New Zealand, Australia (ACT 2012) and Canada (Harvey-Clark 2011). Within the EU, animal welfare legislation was only applied to cephalopods in the UK under the scope of the Animals Act 1986 updated by the Animal Welfare Bill 58 (Bennett 2005) and to octopus and squid in Norway by the Animal Welfare Act 2010.

Cephalopod researchers and technicians have been trying to promote the best welfare practices since there is a need to keep these animals for research or public display purposes. This new EU legislation, based on the enforcement of the 3 Rs (reduction, refinement and replacement) policy suggested by Russell et al. (1992) and with the ultimate goal of replacing the use of animals in experimentation, is officially endorsed by the European Science Foundation since 2001 and, for the first time, includes the class Cephalopoda. This Directive is likely to have a high impact and many practical implications for researchers that use cephalopods in their investigation, irrespective of its field of research, and for those performing research with aquaculture development purposes (Sykes et al. 2012). It covers animal welfare, experimental design, housing, husbandry and experimental protocols. Implementation of the Directive in cephalopods welfare will also require guidelines for different species. There is lack of scientific data to support the fact that cephalopods are actually able to feel pain and suffering (see a review on this discussion in Sykes et al. 2012). Nonetheless, not only do cephalopod researchers cherish their working models but they also agree that, despite no scientific evidence is known or published, the enforcement of pre-emptive measures is needed. This chapter does not intend to accomplish that mission. Actually, it just centres on the cephalopod immune system, disruption of welfare under culture conditions and the resulting pathologies.

6.1.2 "Pathology" in Cephalopods

The practice of good (positive) or bad (negative) welfare in research, maintenance, rearing or culture conditions will determine the existence of pathologies. The cause of pathologies may be divided into viral, bacterial, fungal, parasitic, chemical and mechanical agents. Despite the fact that cephalopods have a "rudimental immunological system" when compared to vertebrates, this system is quite effective. In fact, the simplicity of cephalopods as living animals and of this system has determined the scarce reporting of illnesses in captivity for this class over the years. Cephalopods have such a simple way of life that when they suffer injury, they will, depending on the importance to their survival, regenerate some structures such as arms (Feral 1978, 1979, 1988; Rohrbach and Schmidtberg 2006). This way of living fast and dying young is particularly evident in their protein turnover (Houlihan et al. 1990) and in their fast growth.

Most literature regarding the pathology of cephalopods in captivity, covering an extensive list of pathological causes, was elaborated at the end of the 1980s and in the beginning of the 1990s, when interest on the culture of these classes emerged. In captive conditions, pathological findings are usually related to problems in seawater systems' design and daily running and transportation of live animals to and from the laboratory. Since this is an area of research with a very short report list, this section includes published and unpublished data.

6.1.2.1 "Cephalopods Immune System"

Cephalopods have a well-developed circulatory system conformed by a systemic heart and two accessory hearts (branchial hearts) that function co-ordinately to distribute haemolymph through arteries and capillaries to the whole body (Fiedler and Schipp 1987, 1991; Fiedler 1992; Schipp 1987a; Versen et al. 1997; Wells and Smith 1987). Additionally, haemocyanin production and the elimination of particles are attributed to the branchial hearts (Beuerlein et al. 1998; Beuerlein and Schipp 1998; Beuerlein et al. 2002b). In molluscs, the haemocytes play a main role in the internal defence by the recognition and the elimination of foreign materials as well as shell and wound repair (Cheng 1975). In cephalopods, "haemocytes" are produced in the white body located behind the eyes in the orbital pits of the cranial cartilages (Cowden 1972). The functional morphology of the white bodies of *Sepia officinalis* and comparisons with other cephalopod species were studied by Claes (1996), who found similarities that support the hypothesis that the white blood cells are involved in haemopoiesis and reticuloendothelial functions. A single type of hemocytes has been identified classically in cephalopods. However, recently Castellanos-Martínez et al. (2013) described two types of hemocytes in the common octopus Octopus vulgaris, involved as in other molluscs, in the repair of damaged has been identified in cephalopods and it is involved in the repair of damaged tissues, nutrient transport and digestion and internal defence against nonself material (Cheng 2000; Chu 2000). Wound repair involves the movement and aggregation of haemocytes at the injured site to prevent bleeding until epithelial cells grow over the wound to complete the healing (Chu 2000). Haemocytes are capable of forming a plug which is accompanied with vasoconstriction and collagen synthesis to repair a lesion (Feral 1988).

Similar to other molluscs, cephalopods have a nonadaptive (or innate) immune system; they do not have immunoglobulins and therefore they do not have immunological memory (Fisher and Di Nuzzo 1991; Nyholm and McFall-Ngai 2004). Nonetheless, this system is quite effective to attach the pathogens they can be exposed to and to manage the communities of resident bacteria present in the gut (McFall-Ngai 2007). The cephalopod immune system works based on cellular factors. The haemocytes respond by phagocytosis, encapsulation, infiltration or cytotoxic activities to destroy or isolate pathogens (Beuerlein et al. 2002a). In addition, haemocytes are also involved in the production of oxygen and nitrogen radicals (Castellanos-Martínez et al. 2013; Ford 1992; Malham et al. 1997; Malham and Runham 1998; Nyholm et al. 2009; Rodríguez-Domínguez et al. 2006). Destruction of pathogens through phagocytosis or under haemocyte stimulation is complemented with the production of oxidative chemicals, frequently, the release of reactive oxygen intermediates (ROIs), collectively known as respiratory burst (Chu 2000). Another kind of oxidative chemical, which is part of the innate immune response, is nitric oxide (NO). NO is a highly reactive free radical gas that is not stored and readily diffuses through membranes (Jacklet 1997), so it is an effective agent against pathogens. Both oxidative chemicals have been identified in different cephalopod species in response to stress and against pathogens (Castellanos-Martínez and Gestal 2013; Castellanos-Martínez et al. 2014; Malham et al. 2002; McFall-Ngai et al. 2010).

In addition, the dissolved molecules in the serum (opsonins, agglutinins, lysozyme) also contribute to the immune response of cephalopods (Ford 1992).These humoral factors complement the cellular activity. Recently, Alpuche et al. (2010) described a new lectin of 66 kDa (OmA) found in *Octopus maya,* and a homologue to the type A haemocyanin from *O. dofleini*. Due to the specificity of this lectin to galactosamine, mannose and fucose, these authors suggested that it could work in the immune response by recognizing and agglutinating oligosaccharides from pathogens. The enzyme lysozyme is also part of the defence mechanism. It has been found in haemocytes and tissue from *Eledone cirrhosa,* showing higher activity in haemocytes of octopuses infected by *Vibrio anguillarum* when measured immediately, compared to when measured after 4 and 24 h (Malham et al. 1998).

6.1.2.2 "Pathological Agents"

Virus

Viruses are the most abundant components of aquatic microbial communities (Wommack and Colwell 2000). However, there are few records of viruses producing pathologies in cephalopods. Virus-like particles have been associated with tumours in *O. vulgaris* embedded in arm musculature and may be found in up to 8% of wild specimens (Hanlon and Forsythe 1990a; Hanlon and Forsythe 1990b; Rungger et al. 1971). In addition, virus-like particles have also been identified in the stomach epithelium of *S. officinalis,* but without any detail related to what produced the disease (Devauchelle and Vago 1971). Nonetheless, the latter authors suggested that the development and structure of these particles appeared to be similar to those of the vertebrate reovirus.

Other reports include the identification of virus-like particles in the epithelial cells of the tubules of the digestive gland of *Loligo pealei* and in the renal appendages of several octopod species (Hanlon and Forsythe 1990b).

Bacteria

Bacterial infections are usually developed on the cephalopod's mantle after injuries produced during collection or on lesions suffered in captivity (Hanlon et al. 1988), which cause the spread of the infection to other organisms sharing the tank (Hanlon et al. 1983). A number of bacterial infections of cultured cephalopods were summarized by Hanlon and Forsythe (1990a) in their review of cephalopod diseases. Vibriosis is the most significant bacterial disease encountered in cephalopods (Sherrill et al. 2000).

Hanlon et al. (1984) observed "skin ulcers" on the dorsal mantle, head and arms showing hyperplasia of the epidermis and increased mucus production in high-density rearing of *O. joubini* and *O. briareus*. Gram-negative bacteria including *V. alginolyticus, V. damsela, Pseudomonas stutzeri* and *Aeromonas caviae* were isolated from *O. joubini* ulcers, while *V. parahaemolyticus, V. damsela* and *P. stutzeri* where isolated from ulcers of *O. briareus* which resulted in chromatophore damage.

Ford et al. (1986) described the bacterial populations of normal and ulcerated mantle in *Lolliguncula brevis* and found that laboratory-maintained squids had significantly higher viable bacterial cells per cm^2 of mantle tissue than wild squids, although the latter had a slightly higher species diversity. Thus, the gram-negative bacteria *Aeromonas* sp., *Flavobacterium* sp., *Pseudomonas* sp., *Vibrio* sp., *Proteus* sp., *V. parahaemolyticus, V. alginolyticus* and *Flexibacter* as well as the gram-positive bacteria *Bacillus* sp., *Streptococcus* sp., *Staphylociccus* sp., *Micrococcus* sp. and *Planococcus* sp. were identified in wild squids. In laboratory-maintained squids, *V. alginiolyticus, Vibrio* sp., *Aeromonas* sp., *V. paraahaemolyticus, V. metschnikovii, Flavobacterium* sp., *Proteus* sp., *Bacillus* sp. and *Streptococcus* sp. were identified in normal-skin squids, and *V. alginolyticus, Vibrio* sp., *V. metschinikovii, Aeromonas* sp., *Pseudomonas* sp., *Bacillus* sp. and *Staphylococcus* sp. were identified in squids showing ulcerated skin.

Mantle and arms are not the only tissues susceptible to bacterial infection. The gills of reared *O. vulgaris* have been found to be infected by Rickettsiales-like organisms observed in the form of basophilic intracytoplasmic microcolonies within epithelial cells. The invaded cells were hypertrophic and necrosis was occasionally observed. No significant harm has been observed in the host, but under culture conditions, bacteria are able to have a detrimental effect on the host respiratory gaseous exchange (Gestal et al. 1998).

Farto et al. (2003) identified the gram-negative bacteria *V. lentus* isolated for the first time from an octopus's internal organ, in the gill heart of wild *O. vulgaris*. The specimens also showed lesions in the mantle tissue. Experimental infections in the laboratory demonstrated that *V. lentus* induced mortality in 50 % of octopuses in the first 6 h. The lesions showed a typical round pattern in the arms or head. According to these authors, low salinity (29 %) could be a stressing factor related to the differences in the immune response of the octopuses against the opportunistic pathogen *V. lentus* (Ford et al. 1986).

Hanlon and co-authors (1989) reported bacterial infection of damaged eyes in *L. forbesi* and *Sepiotheuthis lessoniana*. Infection of the vitreous humor with the gram-positive bacteria *Micrococcus* sp. provoked one swollen eye much larger than the other in occasions with cloudy-to-opaque corneal tissue.

Fungi

Reports on fungal activity are very scarce and most of them related to eggs' embryonic development. A simple procedure to avoid problems during this stage is given by Sykes et al. (2006) and in Chap. 11 of this book. Hanlon and Forsythe (1990a) also described occasional fungal activity in *O. vulgaris*.

Parasites

Most wild cephalopods host parasites including protozoans, dicyemids and metazoans. Generally, cephalopod parasites are found in skin, gills, digestive tract, digestive glands and kidneys (Gestal et al. 1999; Hochberg 1982, 1990; McLean et al. 1987; Pascual et al. 1996). Among the protozoans, one of the main epizootiological agents in both wild and cultured cephalopods is the gastrointestinal coccidian of the genus *Aggregata*. This protozoan parasite produces a severe disease in octopods and cuttlefish causing a malabsorption syndrome, diminishing nutrient absorption and reducing immune response capability (Gestal et al. 2002; Gestal et al. 2007).

Dicyemids are endosymbionts that inhabit the renal sacs of cephalopods including octopuses, cuttlefishes and loliginid squids (Furuya et al. 2004; Hochberg 1990). No damage has been reported for dicyemids. In fact, a possible contribution to eliminate ammonium from the host's urine has been proposed (Hochberg 1990). However, dicyemids could be a problem when the population rises so much that it blocks the ducts of the renal sacs conducts.

Cephalopods are secondary, third or parathenic hosts for different metazoan parasites, namely trematodes digenea, cestodes and nematodes infected through the food chain (Hochberg 1990; Pascual et al. 1996). Among them, the nematodes anisakids are one of the most abundant and frequent parasites found in cephalopods with important pathological effects (Abollo et al. 2001). Crustaceans, such as copepods and isopods, are also encountered parasitizing gills and mantle cavity of cephalopods, therefore affecting the body condition of the host (Pascual et al. 1996).

Mechanical and chemical damage

These types of damages share a very small occurrence but have been reported over the years by several authors (Forsythe et al. 1987; Hanlon et al. 1984). Despite

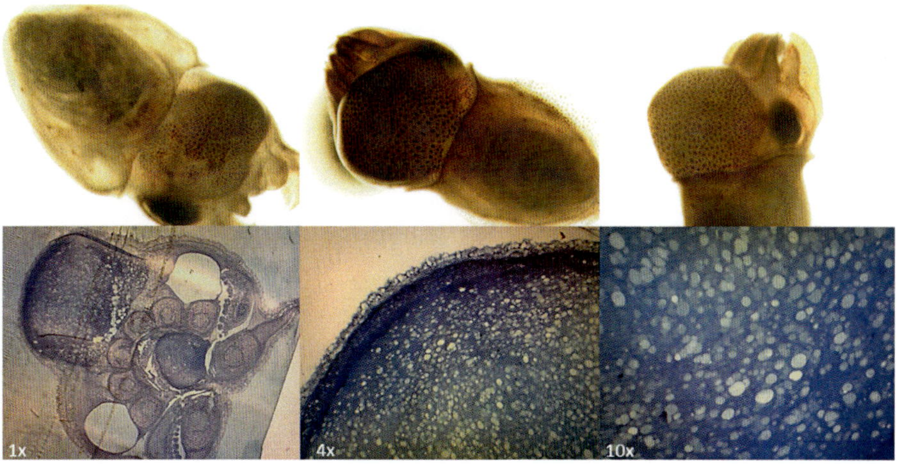

Fig. 6.1 Cuttlefish hatchlings suffering from neoplasia. Histological pictures result from 3-μm sections stained with toluidine blue O. (Photos by A. Sykes)

congenital morphology, related alterations have ever been reported; they have been occasionally observed (Fig. 6.1).

Neoplasia of tissues in the head of hatchlings (Fig. 6.1) was observed in animals resulting from eggs of a captive population cultured in an open seawater system. The reasons for this malformation are unknown and this was only observed at this time, so this pathology might be related to chemicals in the water during embryonic development.

On the other hand, cephalopods have occasionally shown injuries produced during collection and maintenance in the laboratory (Boyle 1991). In addition, cephalopods reared or cultured in captivity may develop skin lesions provoked by individual interactions under crowded conditions or inadequate seawater system designs (Forsythe et al. 1987; Hanlon et al. 1984; Hanlon et al. 1988; Hanlon and Forsythe 1990b; Oestmann et al. 1997). Hanlon and Forsythe (1990a) described four stages of progressive infection after skin abrasion (Table 6.1).

In cuttlefish, several mechanical and eventual chemical damage descriptions may be found in the literature. Hanley et al. (1998) described ulcerative dermatitis and cellulitis in the mantle of a small number of cuttlefish (Fig. 6.2a), which may be followed by septicaemia and fluid accumulation between the cuttlebone and the dermis (Fig. 6.2b). This has also been observed at Centro de Ciências do Mar do Algarve (CCMAR) but no mortality was ever registered even in animals where most of the mantle covering the cuttlebone was removed (Fig. 6.2c).

Other less common pathologies have been observed over these years, such as those related to the cuttlebone. For instance, the development of vesicles (Fig. 6.2d) and partial or full fracture of this structure has been seen in captive animals (Fig. 6.2e, f). Whether this type of abnormality is congenital or related to trauma due to fighting or banging against the tank walls remains to be determined. Several skin pathologies have been registered during reproduction over the years. For

Table 6.1 Stages of progressive infection after skin abrasion in *Octopus joubini* and *O. briareus*. (Adapted from Hanlon and Forsythe 1990b)

Stage	Anatomical manifestations	Dead animals?
One	Extensive damage to microvillus epidermis of the dorsal mantle and dysfunction of the underlying chromatophores (destruction of nerves and radial muscle)	No
Two	Destruction of epidermis and dermal chromatophores (clear or white lesions) Penetration of infection into the underlying muscle layers of the mantle	No
Three	Lesion spread to the lateral and ventral surfaces of the mantle Exposure of internal organs	Yes
Four	Lesion spread to head and arms	Yes

instance, idiopathic bulbous protrusions of one or two eyes (Hanley et al. 1998) have been seen, especially in females (Fig. 6.3a, *arrows in black*), which might promote opaque corneas and swelling of periorbital tissues. Some animals develop skin wounds resulting from a fast escape during mating, which will eventually expose the cuttlebone and interfere with buoyancy (Fig. 6.3b, c). Cuttlefish are extremely aggressive when they copulate or fight for copula, and this results in multiple skin wounds (Fig. 6.3a, *arrows in white*) or even extended biting (Fig. 6.3d), with eventual internal organ exposure. Nonetheless, reports on mortality associated with this type of trauma are inexistent. As senescence takes over, the degree of skin "wounds" increases (Anderson et al. 2002). By this time and until death, cuttlefish will mostly display an increased destruction of the mantle epidermis (Fig. 6.3f, i). Nonetheless, sometimes this effect will only be seen mostly on the animal's head (Fig. 6.3e) or on both head and mantle (Fig. 6.3g, h). The use of euthanasia as a way to avoid pain at this stage is currently under discussion and shall be included in the EU cephalopod guidelines.

The polluted environment in open, semiopen and closed seawater systems as well as in offshore, onshore or inshore rearing cages, tanks or ponds might also be detrimental to cephalopods' health. The impact of diseases increases with stress and pollutant conditions, which can suppress the immune response and allow microorganisms, such as virus, bacteria and fungi, to proliferate. Therefore, the opportunistic microorganisms will generate secondary infections to these lesions.

6.1.2.3 Healing, Antibiotics and Surgery

According to Hanlon and Forsythe (1990a), wound healing in cephalopods involves a sequence of muscular contraction, dermal cellular reaction and epidermal migration.

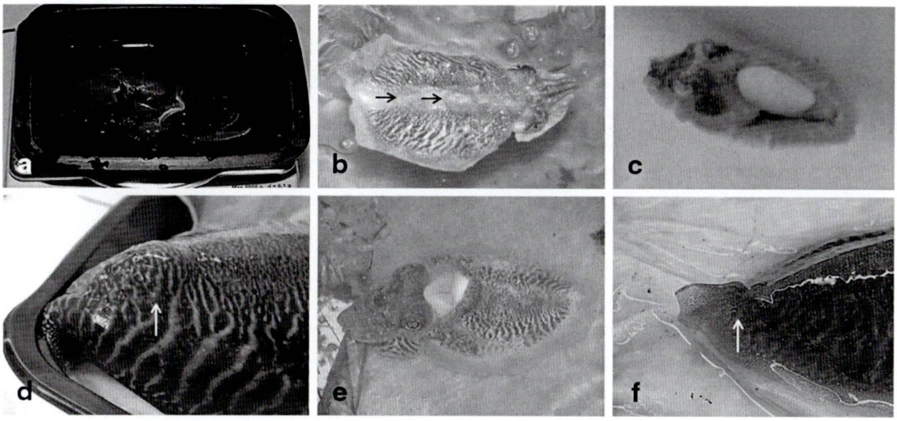

Fig. 6.2 Examples of mechanical damage in cuttlefish juveniles and adults. **a** Ulcerative dermatitis and cellulitis. **b** Septicaemia and fluid accumulation. **c** Mantle absence over the cuttlebone. **d** Vesicle in the distal tip of the mantle. **e** Full fracture of cuttlebone. **f** Partial fracture of the cuttlebone. (Photos by A. Sykes)

Fig. 6.3 Examples of mechanical damage in cuttlefish juveniles and adults. **a** Idiopathic bulbous protrusions of the eyes. **b, c** Skin wounds at the distal tip of the mantle resulting from banging against the tank walls. **d** Bite wound in male. **e** Senescent female. **f, i** Senescent animals' destruction of the mantle epidermis. **g, h** Increased destruction of the mantle and head epidermis. (Photos by A. Sykes)

Tetracycline injected in food was used to treat lesions with spreads of *Pseudomonas* sp. and *Acinetobacter anitratus* with success in *O. dofleini* by Stoskopf et al. (1987). Hanlon and Forsythe (1990a) described the use of chloramphenicol injected in food as having some success in treating infections by *V. carchariae*. The same authors described the use of chloramphenicol (40 mg kg^{-1}) and gentamycin (20 mg kg^{-1}), either by intramuscular injection or via food, to treat *S. officinalis* infected by *V. pelagicus, V. splendidus* and *P. stutzeri* related to internal infection of the blood vessels.

Forsythe et al. (1987) and Forsythe et al. (1990) described the use of several antibacterial and antiprotozoacidal agents administered by short-term baths or dips, continuous baths, intramuscular injection or by injection in food, which may be used on infected animals or eggs. In the latter work, safe concentrations for the application of these therapeutic agents are also given. In general, these treatments are only effective for cuttlefish and octopus and the preferable method of application is oral administration. There is still need for further research in this field.

Reports on surgery performed in cephalopods are scarce and may be an additional focus of infections, as described by Harms et al. (2006).

6.2 Welfare Under Research and Commercial Culture Conditions

To avoid the pathologies mentioned in the previous sections, a good welfare practice must be implemented at different stages of maintenance, rearing and culture of cephalopod species. Both Mather and Anderson (2007) and Moltschaniwskyj et al. (2007) reviewed the ethical and welfare considerations when using cephalopods as research animals. These authors presented the refinement of methods as the most important task to ensure protection of cephalopods both in the laboratory and in the field. However, by considering cephalopods as sentient animals, one cannot talk about welfare without defining stressors and stress in cephalopods. More recently, Sykes et al. (2012) reviewed the application of Directive 2010/63/EU concerning cephalopod breeding and experimentation in aquaculture research. In that work, the authors proposed a definition of "acceptable stress" for cephalopods as the biological response of the animal when subjected to or when it perceives a stressor. To be defined as "acceptable," this biological response must not promote the development of a state of illness, provoke physical pain nor extended suffering (although the latter two have not been proved yet to occur in cephalopods). This will be the borderline to the use of anaesthetics, analgesia or euthanasia in aquaculture -related procedures or protocols. These authors also reviewed the application of several anaesthesia and euthanasia agents in cephalopods.

The use of effective anaesthesia and euthanasia agents as welfare promoters in fish has been shown by several authors (such as Pottinger (2008) and Ross and Ross (1984)), but similar works in the class Cephalopoda are scarce, lack solid grounds or have never been performed (Sykes et al. 2012).

The use of anaesthetics facilitates work at the research level and is required for invasive studies, where cephalopods must be held immobile for extended periods of time. Nonetheless, anaesthetics are also used to reduce the level of stress associated with such procedures. In addition, an overdose of anaesthetics is also routinely used as an effective and humane mean of euthanizing fish, and similar procedures are already in use for cephalopods. However, the correct assessment of anaesthesia, analgesia and euthanasia agents is a key aspect to determine the best agents, and the physiology-based methods used for fish still need to be validated in cephalopods. It is also important that the study of effective agents should be performed using a welfare-based approach and this requires the development of new methods to accomplish this mission in the most ethical and humane way as possible. Studies to determine a given agent and the best dose of an anaesthetic usually only display information regarding (1) times of induction, sedation and recovery; (2) scarce behaviour description; and (3) measurements of hormonal levels (circulating concentrations of catecholamines and cortisol) and immune functions (circulating haemocytes and haemocyte phagocytotic activity). In cephalopods, not only the amount of anaesthetics tested is high and not fully proven (Sykes et al. 2012) but also the great majority of these types of studies only include items concerning (1) and, recently, (2) (e.g. Seol et al. (2007) and Estefanell et al. (2011)). According to Broom (2008), this is insufficient information to determine the viability of a given anaesthetic.

The development of new tools and the correct application of the existing ones for the assessment of anaesthetics are of utmost importance to obtain a clear and full understanding of animal reaction during the evaluation of anaesthesia and euthanasia procedures. There is increasing evidence that alternative methods (Kittilsen et al. 2009) and available technology may provide simpler and more humane methods of assessing stress responsiveness than invasive methods, such as determination of plasma cortisol, which is widely used in fish (Ellis et al. 2012). In addition, the most established methods of assessing primary and secondary stress response in fish (such as plasma cortisol and glucose (Pottinger 2008)) were never applied to cephalopods nor do they have a non-humane application (small amount of available blood and difficult access for extraction will require slaughtering), whether due to the type and location of the circulatory system of this class (Schipp 1987b) or by lack of reliability of these methods at given conditions, as suggested by Martínez-Porchas et al. (2009). Moreover, according to Ashley (2007), both behavioural and physiological measures of welfare are necessary for the correct interpretation of stress responsiveness. While there is a wealth of knowledge on the physiological consequences of many aquaculture practices in fish, it is now equally important to understand the behavioural responses to these practices and to relate them to welfare in cephalopods (Gonçalves et al. 2012). The assessment of welfare during the application of these agents should be performed using noninvasive and multiple methods that need to be certified or developed. Furthermore, the study of stress responsiveness of anaesthesia, analgesia and euthanasia agents should be accomplished through refined animal welfare methodologies (based on the 3 Rs principle) in cephalopods.

6.3 Conclusions

While the study of pathologies has been performed since cephalopods are kept in captivity, welfare is a new growing line of research that started in the twenty-first century. Both are, however, in need of more attention in the years to come, not only because of welfare legislations arising in most countries but also because they have an intrinsic importance in cephalopod culture. In fact, the existing knowledge is poor when compared with what exists in fish. In terms of welfare, while several agents for anaesthesia and euthanasia have been tested, there is not a single study on the physiology behind it (anaesthetic site of action). Therefore, we cannot be sure that when we are suggesting a given agent, this is in fact an anaesthetic. In addition, the existing methods used to assess anaesthesia and euthanasia need revision and a common methodology for assessment needs to be developed.

Regarding pathologies of cephalopods in captivity, with the development of culture technology, these animals will be subjected to different conditions than those observed in small-scale culture. Although the methods under development consider the practice of good welfare, diseases will be predictably more common in industrial culture. Several pathologies have been identified in the laboratory and these have been used as a basis to gather information regarding healing, antibiotics application and changes to culture setups. Nonetheless, further research is needed.

References

Abollo E, Gestal C, Pascual S (2001) Anisakis infestation in marine fish and cephalopods from Galician waters: an updated perspective. Parasitol Res 87:492–499

Alpuche J, Pereyra A, Mendoza-Hernández G et al (2010) Purification and partial characterization of an agglutinin from Octopus maya serum. Comp Biochem Physiol B: Biochem Mol Biol 156:1–5

Anderson RC, Wood JB, Byrne RA (2002) Octopus senescence: the beginning of the end. J Appl Anim Welf Sci 5:275–283

Ashley PJ (2007) Fish welfare: current issues in aquaculture. Appl Anim Behav Sci 104:199–235

Australian Capital Territory (2012) Animal welfare act 1992. In: Australian capital territory parliamentary counsel, p 123

Bennett O (2005) The animal welfare bill, bill no 58 of 2005–06. In: House of commons, p 90

Beuerlein K, Schipp R (1998) Cytomorphological aspects on the response of the branchial heart complex of *Sepia officinalis* L. (Cephalopoda) to xenobiotics and bacterial infection. Tissue Cell 30:662–671

Beuerlein K, Schimmelpfennig R, Westermann B et al (1998) Cytobiological studies on hemocyanin metabolism in the branchial heart complex of the common cuttlefish *Sepia officinalis* (Cephalopoda, Dibranchiata). Cell Tissue Res 292:587–595

Beuerlein K, Lohr S, Westermann B et al (2002a) Components of the cellular defense and detoxification system of the common cuttlefish *Sepia officinalis* (Mollusca, Cephalopoda). Tissue Cell 34:390–396

Beuerlein K, Ruth P, Westermann B et al (2002b) Hemocyanin and the branchial heart complex of *Sepia officinalis*: are the hemocytes involved in hemocyanin metabolism of coleoid cephalopods? Cell Tissue Res 310:373–381

Boyle PR (1991) The UFAW handbook on the care and management of the cephalopods in the laboratory. UFAW, Herts

Broom DM (2008) Welfare assessment and relevant ethical decisions: key concepts. Annu Rev Biomed Sci 10:79–90

Castellanos-Martínez S, Gestal C (2013) Pathogens and immune-response of cephalopods. J Exp Mar Biol Ecol 447:14–22

Castellanos-Martínez S, Prado-Alvarez M, Lobo-da-Cunha A et al (2014) Morphologic, cytometric and functional characterization of the common octopus (Octopus vulgaris) hemocytes. Dev Com Immunol 44:50–58

Cheng TC (1975) Functional morphology and biochemistry of molluscan phagocytes. Ann Ny Acad Sci 266:343–379

Cheng TC (2000) Cellular defense mechanisms in oysters. In: Fingerman M, Nagabhushanam R (eds) Recent advances in marine biotechnology. Science Publishers Inc., U.K., pp 43–83

Chu FL (2000) Defense mechanisms of marine bivalves. In: Fingerman M, Nagabhushanam R (eds) Recent advances in marine biotechnology. Science Publishers Inc., U.K., pp 1–42

Claes MF (1996) Functional morphology of the white bodies of the cephalopod mollusc *Sepia officinalis*. Acta Zool-Stockholm 77:173–190

Cowden RR (1972) Some cytological and cytochemical observations on the leucopoietic organs, the "white bodies" of *Octopus vulgaris*. J Inv Pathol 19:113–119

Devauchelle G, Vago C (1971) Particulles d'allure virale dans les cellules de l'estomac de la seiche, *Sepia officinalis* L. (Mollusques, Céphalopodes). C. r. hebd. Séanc. Acad Sci Paris 272:894–896

Ellis T, Yildiz H, López-Olmeda J et al (2012) Cortisol and finfish welfare. Fish Physiol Biochem 38:163–188

Estefanell J, Socorro J, Afonso JM et al (2011) Evaluation of two anaesthetic agents and the passive integrated transponder tagging system in *Octopus vulgaris* (Cuvier 1797). Aquac Res 42:399–406

European Union (2010) Directive 2010/63/EU of The European parliament and of the council of 22 September 2010 on the protection of animals used for scientific purposes. Official Journal of European Union, pp 33–79

Farto R, Armada SP, Montes M et al (2003) *Vibrio lentus* associated with diseased wild octopus (*Octopus vulgaris*). J Inv Pathol 83:149–156

Feral JP (1978) Regeneration of arms of Cuttlefish *Sepia officinalis* L (Cephalopoda, Sepioidea):1—morphological study. Cahiers De Biologie Marine 19:355–361

Feral JP (1979) Regeneration of the arms of *Sepia officinalis* L (Cephalopoda, Sepioidea): 2—histologic and cytologic study. Cahiers De Biologie Marine 20:29–42

Feral JP (1988) Wound-healing after arm amputation in *Sepia officinalis* (Cephalopoda, Sepioidea). J Inv Pathol 52:380–388

Fiedler A (1992) The role of venous filling pressure in autoregulation of the branchial hearts of *Sepia officinalis* L (Cephalopoda). Zool Jahrb Allg Zool 96:265–278

Fiedler A, Schipp R (1987) The role of the branchial heart complex in circulation of coleoid cephalopods. Experientia 43:544–553

Fiedler A, Schipp R (1991) Localization of catecholamine-containing nerve-fibers in the branchial heart and cardiac ganglion of the common cuttlefish *Sepia officinalis* L (Cephalopoda). Tissue Cell 23:813–819

Fisher WS, Di Nuzzo AR (1991) Agglutination of bacteria and erythrocytes by serum from six species of marine molluscs. J Inv Pathol 57:380–394

Ford LA (1992) Host defense mechanisms of cephalopods. Annu Rev Fish Diseases 2:25–40

Ford LA, Alexander SK, Cooper KM et al (1986) Bacterial populations of normal and ulcerated mantle tissue of the squid, *Lolliguncula brevis*. J Inv Pathol 48:13–26

Forsythe JW, Hanlon RT, Lee PG (1987) A synopsis of cephalopod pathology in captivity. In: Williams TD (ed) 18th annual conference of the international association for aquatic animal medicine. Monterey bay aquarium, Monterey, California, pp 130–135

Forsythe JW, Hanlon RT, Lee PG (1990) A formulary for treating cephalopod mollusc diseases. In: Perkins FO, Cheng TC (eds) Third international colloquium on pathology in marine aquaculture, Gloucester Point, Virginia (USA), 2–6 October 1988, Academic Press Inc., pp 51–63

Furuya H, Ota M, Kimura R et al (2004) Renal organs of cephalopods: A habitat for dicyemids and chromidinids. J Morphol 262:629–643

Gestal C, Abollo E, Pascual S (1998) Rickettsiales-like organisms in the gills of reared *Octopus vulgaris* (Mollusca, Cephalopoda). Bull Eur Assoc Fish Pathol 18:13–14

Gestal C, Belcari P, Abollo E et al (1999) Parasites of cephalopods in the northern Tyrrhenian Sea (western Mediterranean): new host records and host specificity. Scientia Marina 63:39–43

Gestal C, Guerra A, Pascual S et al (2002) On the life cycle of *Aggregata eberthi* and observations on *Aggregata octopiana* (Apicomplexa, Aggregatidae) from Galicia (NE Atlantic). Eur J Protistol 37:427–435

Gestal C, Guerra A, Pascual S (2007) *Aggregata octopiana* (Protista: Apicomplexa): a dangerous pathogen during commercial *Octopus vulgaris* ongrowing. ICES Mar Sci 64:1743–1748

Gonçalves RA, Frias PA, Aragão C et al (2012) The use of different anaesthetics as welfare promoters during short-term human manipulation of European cuttlefish (*Sepia officinalis*) juveniles. Aquaculture 370:130–135

Hanley JS, Shashar N, Smolowitz R et al (1998) Modified laboratory culture techniques for the European cuttlefish *Sepia officinalis*. Biol Bull 195:223–225

Hanlon RT, Forsythe JW (1990a) Diseases caused by microorganisms. In: Kinnie O (ed) Diseases of Mollusca: Cephalopoda. Biologische Anstalt Helgoland, Hamburg, pp 23–46

Hanlon RT, Forsythe JW (1990b) Structural abnormalities and neoplasia. In: Kinnie O (ed) Diseases of marine animals. Biologische Anstalt Helgoland, Hamburg, pp 203–204

Hanlon RT, Hixon RF, Hulet WH (1983) Survival, growth, and behaviour of the loliginid squids *Loligo plei*, *Loligo pelaei*, and *Lolliguncula brevis* (Mollusca: Cephalopoda) in closed seawater systems. Biol Bull 165:637–685

Hanlon RT, Forsythe JW, Cooper KM et al (1984) Fatal penetrating skin ulcers in laboratory-reared octopuses. J Inv Pathol 44:67–83

Hanlon RT, Forsythe JW, Lee PG (1988) External pathologies of cephalopods in captivity. In: Perkins FO, Cheng TC (eds) Third international colloquium on pathology in marine aquaculture. Academic Press Inc., Gloucester Point, Virginia (USA), 2–6 October 1988, pp 17–18

Hanlon RT, Yang WT, Turk PE et al (1989) Laboratory culture and estimated life span of the Eastern Atlantic squid, *Loligo forbesi* Steenstrup, 1856 (Mollusca: Cephalopoda). Aquacult Fish Manag 20:15–34

Harms CA, Lewbart GA, Mcalarney R et al (2006) Surgical excision of mycotic (*Cladosporium* sp.) granulomas from the mantle of a cuttlefish (*Sepia officinalis*). J Zoo Wildlife Med 37:524–530

Harvey-Clark C (2011) IACUC challenges in invertebrate research. ILAR J 52(2):213–220

Hochberg FG (1982) The "kidneys" of cephalopods: a unique habitat for parasites. Malacologia 23:121–134

Hochberg FG (1990) Diseases of Mollusca: Cephalopoda. In: Kinne O (ed) Diseases of marine animals. Biologisches Anstalt Helgoland, Hamburg, pp 47–227

Houlihan DF, Mcmillan DN, Agnisola C et al (1990) Protein synthesis and growth in *Octopus vulgaris*. Mar Biol 106:251–259

Jacklet JW (1997) Nitric oxide signaling in invertebrates. Invert Neurosci 3:1–14

Kittilsen S, Ellis T, Schjolden J et al (2009) Determining stress-responsiveness in family groups of Atlantic salmon (*Salmo salar*) using non-invasive measures. Aquaculture 298:146–152

Malham SK, Runham NW (1998) A brief review of the immunobiology of *Eledone cirrhosa*. South African J Mar Sci 20:385–391

Malham S, Runham NW, Secombes CJ (1997) Phagocytosis by haemocytes from the lesser octopus *Eledone cirrhosa*. Iberus 15:1–11

Malham SK, Runham NW, Secombes CJ (1998) Lysozyme and antiprotease activity in the lesser octopus *Eledone cirrhosa* (Lam.) (Cephalopoda). Develop Comp Immunol 22:27–37

Malham SK, Lacoste A, Gelebart F et al (2002) A first insight into stress-induced neuroendocrine and immune changes in the octopus *Eledone cirrhosa*. Aquat Living Resour 15:187–192
Martínez-Porchas M, Martínez-Córdova LR, Ramos-Enriquez R (2009) Cortisol and glucose: reliable indicators of fish stress? Pan-American J Aquatic Sci 4(2):158–178
Mather JA, Anderson RC (2007) Ethics and invertebrates: a cephalopod perspective. Diseases Aquatic Org 75:119–129
Mcfall-Ngai M (2007) Adaptive immunity: care for the community. Nature 445:153–153
Mcfall-Ngai M, Nyholm SV, Castillo MG (2010) The role of the immune system in the initiation and persistence of the *Euprymna scolopes–Vibrio fischeri* symbiosis. Semin Immunol 22:48–53
Mclean N, Hochberg FG, Shinn GL (1987) Giant protistan parasites on the gills of cephalopods (Mollusca). Diseases Aquatic Org 3:119–125
Moltschaniwskyj NA, Hall K, Marian JEAR et al (2007) Ethical and welfare considerations when using cephalopods as experimental animals. Rev Fish Biol Fish 17:455–476
Nyholm SV, Mcfall-Ngai M (2004) The winnowing: establishing the squid-vibrio symbiosis. Nat Rev Micro 2:632–642
Nyholm SV, Stewart JJ, Ruby EG et al (2009) Recognition between symbiotic *Vibrio fischeri* and the haemocytes of *Euprymna scolopes*. Environm Microbiol 11:483–493
Oestmann DJ, Scimeca JM, Forsythe J et al (1997) Special considerations for keeping cephalopods in laboratory facilities. Contemp Topics Lab Anim Sci 36:89–93
Ohl F, Van Der Staay FJ (2012) Animal welfare: at the interface between science and society. Vet J 192:13–19
Pascual S, Gestal C, Estevez JM et al (1996) Parasites in commercially-exploited cephalopods (Mollusca, Cephalopoda) in Spain: an updated perspective. Aquaculture 142:1–10
Pottinger TG (2008) The stress response in fish—mechanism, effects and measurement. In: Branson EJ (ed) Fish welfare. Blackwell, Oxford, pp 32–48
Rodríguez-Domínguez H, Soto-Búa M, Iglesias-Blanco R et al (2006) Preliminary study on the phagocytic ability of *Octopus vulgaris* Cuvier, 1797 (Mollusca: Cephalopoda) haemocytes in vitro. Aquaculture 254:563–570
Rohrbach B, Schmidtberg H (2006) Sepia arms and tentacles: Model systems for studying the regeneration of brachial appendages. Vie Milieu 56:175–190
Ross LG, Ross B (1984) Anaesthetic and sedative techniques for fish. Institute of Aquaculture—University of Stirling, Stirling
Rungger D, Rastelli M, Braendle E et al (1971) A viruslike particle associated with lesions in the muscles of *Octopus vulgaris*. J Invertebrate Pathol 17:72–80
Russell WM, Burch RI, Hume CW (1992) Principles of human experimental technique. Universities Federation for Animal Welfare (UFAW), Wheathampstead
Schipp R (1987a) General morphological and functional characteristics of the cephalopod circulatory system. An introduction. Experientia 43:474–477
Schipp R (1987b) The blood vessels of Cephalopods. A comparative morphological and functional survey. Experientia 43:525–537
Seol D-W, Lee J, Im S-Y et al (2007) Clove oil as an anaesthetic for common octopus (*Octopus minor*, Sasaki). Aquac Res 38:45–49
Sherrill J, Spelman LH, Reidel CL et al (2000) Common cuttlefish (*Sepia officinalis*) mortality at the national zoological park: implications for clinical management. J Zoo Wildlife Med 31:523–531
Stoskopf MK, Nevy S, Flynn S (1987) Treatment of ulcerative mantle disease due to *Pseudomonas* spp. In: *Octopus dofleini* and *Octopus bimaculoides* with oxytetracycline. In: Annual meeting of the international association of aquatic animal medicine, Monterey, California
Sykes AV, Domingues PM, Correia M et al (2006) Cuttlefish culture—State of the art and future trends. Vie Milieu 56:129–137
Sykes AV, Baptista FD, Gonçalves RA et al (2012) Directive 2010/63/EU on animal welfare: a review on the existing scientific knowledge and implications in cephalopod aquaculture research. Rev Aquacult 4:142–162

Versen B, Gokorsch S, Lücke J et al (1997) Auricular-ventricular interacting mechanisms in the systemic heart of the cuttlefish *Sepia Officinalis* L. (Cephalopoda). Vie Milieu 47:123–130

Wells MJ, Smith PJS (1987) The performance of the octopus circulatory system: A triumph of engineering over design. Experientia 43:487–499

Wommack KE, Colwell RR (2000) Virioplankton: viruses in aquatic ecosystems. Microbiol Mol Biol Rev 64:69–114

Chapter 7
Aquaculture to Restocking

Jaruwat Nabhitabhata and Susumu Segawa

Abstract Restocking of cephalopods originated from the awareness of the depletion of natural resources and a need for conservation from stakeholders, public and private sectors. The concept of cephalopod restocking activities is to produce the cephalopod seeds and then release them back into the habitat where the species occur. The process of seed production comprises the collection of broodstocks from the wild, incubation of egg masses, post-hatching management and releasing the seeds at selected locations. The aquaculture methodology that enhances the hatching rate of the eggs and survival rate of the hatchling must produce high yields for the success of restocking. The neritic species, particularly *Sepioteuthis lessoniana, Sepia* spp. and *Octopus vulgaris* have been one key focus for restocking due to its success in previous studies on aquaculture as well as their importance to fisheries. The success of aquaculture restocking, supported by the public sector, has been outstanding in Japan and Thailand with a long historical background. Millions of cephalopod seeds were annually released during the peak activities. Although the biological significance of restocking activities to natural stocks requires further research and evaluation, the activities are considered to have produced a significant social success.

Keywords Restocking · Seed production · Neritic species · Stakeholders

---

J. Nabhitabhata (✉)
Excellence Centre for Biodiversity of Peninsular Thailand (CBIPT), Faculty of Science, Prince of Songkla University, Hatyai, Songkhla 90112, Thailand
e-mail: jaruwat.n5@gmail.com; jaruwat.n@psu.ac.th

S. Segawa
Department of Ocean Sciences, Faculty of Marine Science, Tokyo University of Marine Science and Technology, 4-5-7 Konan, Minato-ku, Tokyo 108-8477, Japan
e-mail: segawa@kayodai.ac.jp; s-ika@h04.itscom.net

J. Iglesias et al. (eds.), *Cephalopod Culture,* DOI 10.1007/978-94-017-8648-5_7,

7.1 Introduction

Marine natural resources, including cephalopods, are declining due to the consequences of climate change in the global environment and anthropogenic activities. Cephalopods are more commercially vulnerable due to their short lifespan, compared to fish, crustaceans and other molluscs. Based on this awareness, public sectors in Japan and Thailand have initiated the restocking programme using the aquaculture methodology. The aquaculture-restocking activities in the two countries have been outstanding. Cephalopod restocking has a long history in Japan for as long as 50 years, i.e. since 1963 (JASFA 2003). The attempt began with studies on aquaculture and restocking of *Octopus vulgaris* and *Sepioteuthis lessoniana*. In Thailand, the history of cephalopod restocking is more than 20 years old (since 1990), but the attempt on developing the aquaculture technology had started 10 years earlier, in 1978 (Nabhitabhata 1978a, b). The economic status of the two countries is different; therefore, this has led to differences in the details of the two developed methodologies, although the objective is similar, since the reduction of cost of production is one of the main common concerns.

7.2 Restocking in Japan

7.2.1 Background

A total of ten species of cephalopods found in Japanese waters have been reported to be reared or cultured from eggs for the purpose of studying their basic life histories and the culture of cephalopods (Boletzky and Hanlon 1983) but none of these studies were initially aimed at restocking. Nominated species were *Sepia esculenta, Sepia lycidas* (as *Sepia subaculeata*), *Sepia latimanus, Euprymna berryi, Idiosepius paradoxus, S. lessoniana, Heterololigo bleekeri, Todarodes pacificus, O. vulgaris* and *Amphioctopus fangsiao* (as *Octopus ocellatus*) (Choe 1966; Choe and Oshima 1963; Inoha 1971; Natsukari 1970; Inoha and Sezoko 1967; Soichi 1976; Matsumoto 1976; Hamabe 1963; Itami et al. 1963; Yamauchi and Takeda 1964). Among them, *S. esculenta, S. latimanus, S. lessoniana, O. vulgaris* and *A. fangsiao* (Fig. 7.1, Table 7.1) are presently reared from eggs to their sub-adult stages to help to clarify the life history of the species and to provide basic information for aquaculture (Boletzky and Hanlon 1983).

In Japan, the production of finfish seed for restocking has been promoted by the Seto Inland Sea Fish Farming Association (SISFFA), established in 1963. This organization was, later, merged with the Japan Sea-Farming Association (JASFA) in 1978 with one of its aims being to develop the technology for restocking of aquatic resources under the Japan Fisheries Agency. SISFFA and JASFA have played a major role in the development of the technology for seed production for restocking not only for finfish (Fushimi 2001) but also for cephalopods. In 1963, SISFFA began experiments to incubate the egg masses of *O. vulgaris* and *S. lessoniana* collected from the sea for

Fig. 7.1 Localities, names and target species of the stations of the Japan Sea-Farming Association (JASFA) that produced cephalopod seeds

the release of hatchlings. In 1964, *S. lessoniana* hatchlings were reared up to 70 mm in total length (TL) in 2 months and *O. vulgaris* from eggs to the paralarval stage with 14–15 suckers in 20–25 days (JASFA 2003). In 1982, JASFA succeeded in rearing *O. vulgaris* to the stage of paralarval settlement (JASFA 2003). The targeted cephalopod species for the development of technology for restocking by JASFA were *O. vulgaris, S. lessoniana, H. bleekeri* (as *Loligo bleekeri*), *S. esculenta* and *S. latimanus.* The Fisheries Research Agency (FRA) took over the duties of JASFA in 2003.

In addition to the activities of developing the technology for restocking, artificial spawning substrata were installed in the spawning ground to enhance the stock of the target species such as for *S. lessonina* (Ueta et al. 1995; Ueta and Kitakado 1996) and Yariika (*H. bleekeri*) in the coastal area of Japanese waters. For the propagation of *O. vulgaris* to enhance the stock in fishing grounds, octopus pots for spawning were set in the octopus spawning areas in Seto Inland Sea, Japan (Itami 1975).

Table 7.1 Species, numbers of cephalopod seeds (× 10^3 individuals) produced and released for restocking in Japan by the Japan Sea-Farming Association (JASFA) during the 19 years from 1983 to 2001 and average total length (mm) in brackets; + is a number less than 10^3 individuals. (JASFA 2003)

Year	*Octopus vulgaris*	*Sepia esculenta*	*Sepia latimanus*		*Loligo bleekeri*	*Sepioteuthis lessoniana*
	produced	produced	produced	released	produced	produced
1983	1 (11)	–	–	–	–	2 (35)
1984	1 (11)	–	–	–	–	–
1985	–	–	–	–	+(6)	1 (12)
1986	+(11)	+(30)	+(30)	–	+(8)	+(23)
1987	1 (12)	–	1 (30)	–	+(6)	+(19)
1988	7 (8)	–	1 (30)	+(80)	1 (6)	1 (9)
1989	26 (7)	–	6 (30)	2 (68)	5 (6)	–
1990	17 (15)	–	6 (25)	1 (83)	1	1
1991	5 (7)	–	8 (30)	+(100)	+(12)	+
1992	33 (6)	–	37 (17)	–	–	–
1993	15 (7)	–	–	–	+(4)	–
1994	25 (7)	–	–	–	–	–
1995	17 (7)	–	–	–	–	–
1996	38 (8)	–	–	–	–	–
1997	3 (8)	–	119 (14)	94 (14)	–	–
1998	10 (7)	–	52 (14)	51 (14)	–	–
1999	13 (6)	–	20 (13)	20 (13)	–	–
2000	8 (6)	–	77 (13)	76 (13)	–	–
2001	18 (8)	–	+(30)	+(80)	–	–

7.2.2 Hatchery Seed Production and Seed Release

7.2.2.1 *Octopus vulgaris*

The common octopus, *O. vulgaris,* is an economic marine species especially in the Seto Inland Sea of Japan. Octopus pots have been introduced into the fishing ground as a spawning aid during the spawning period of *O. vulgaris* since 1931 with the aim to protect females for enhancement of the stock (Tauti and Matsumoto 1954). Itami et al. (1963) reared the planktonic hatchlings (paralarvae) of this species to its settling young stage and beyond by feeding with the zoea larvae of *Palaemon serrifer*. SISFFA also started a project in 1964 to culture *O. vulgaris* for restocking. The project produced 150,000 paralarval individuals (with 14–15 suckers over 20–25 days after hatching) in the first year, and 157 settling paralarvae (with 12–19 suckers over 31 days after hatching) in 1982 with a yield rate of 13.1 % from hatching (JASFA 2003). Except from mid-1967 to 1980, JASFA and FRA have continued this project until the present day (JASFA 2003). In addition to seed production for restocking octopus, the Kagawa Prefecture has irregularly transplanted juvenile octopus into the spawning grounds in the Seto Inland Sea every year in order to enhance the stocks: 15,000 juveniles in 1979 and 5,432 juveniles of 64 mm mean TL in 2002.

Research on providing adequate food and nutrition to grow healthy paralarvae and juvenile was also performed from the beginning of the SISFFA and JASFA projects. In 2001, JASFA achieved the production of settling paralarvae at a high survival rate through feeding with the frozen Pacific sand eel *Ammodytes personatus* and large-sized nauplii of *Artemia tibetiana* (Shiraki 1997; Okumura et al. 2005).

7.2.2.2 *Sepioteuthis lessoniana*

Choe and Oshima (1963) reared *S. lessoniana* from hatching to the age of 45 days fed on *Neomysis japonica*. Later, there were several studies to rear this species for the purpose of aquaculture (Choe 1966; Inoha and Sezoko 1967; Tsuchiya 1982; Segawa 1987). SISFFA began incubating wild-collected egg masses of this species for the release of hatchlings in 1963 and reared them to juveniles of 70 mm TL at the age of 2 months in 1964 (JASFA 2003). In 1982, JASFA restarted experiments to develop the technology for seed production to restock *S. lessoniana* in Goto Island, Nagasaki Prefecture (Fig. 7.1), because of the declining catches of the species, but this activity ended in 1992 (JASFA 2003). In the first year, eggs were collected from live females spawning in captivity and from an artificial substrate in the wild. The artificial substrate was made from tree branches immersed in the spawning ground as an artificial spawning bed. The collected eggs were then incubated in a tank and hatchlings were produced. One of the purposes of this project was to develop effective artificial spawning beds for the species. Ueta et al. (1995) developed several types of effective man-made spawning beds for *S. lessoniana*.

7.2.2.3 *Heterololigo bleekeri*

Matsumoto (1976) maintained adults of this species in tanks for 3 weeks. JASFA addressed the development of technology for spawning and hatching of this species for use in restocking between 1985 and 1993 in Miyazu which faces Wakasa Bay in the Kyoto Prefecture (JASFA 2003). In 1985, JASFA reared hatchlings fed on live zooplankton, *Artemia* and hatchlings of scorpion fish and produced 134 squid (6–7 mm in mantle length (ML), with a 0.35 % survival rate from hatchlings) at 40 days after hatching. In 1986, 81,000 eggs were collected from an artificial spawning bed experiment in the sea (JASFA 2003). This project accomplished certain results in developing adequate spawning beds and ended because of the difficulties encountered for mass culturing of this species.

7.2.2.4 *Sepia latimanus*

Inoha (1971) observed that *S. latimanus* laid eggs and attached them to corals of the genus *Millepora* in Okinawa and they were easy to rear in the aquarium. JASFA has conducted research and development of the technology for seed production of this species and restocking in Okinawa and Ishigaki Island, Okinawa, since 1987. Technologies were established for collecting eggs in the laboratory and producing hatchlings for release (Oka et al. 1989; Oka 1993). In 1989, JASFA started a lagoon nursery project in Ishigaki Island, Okinawa, and *S. latimanus* was nominated as one of the main target species for aquaculture (JASFA 2003). Since this species has large hatchlings of 11–15 mm ML competent in camouflage and foraging, restocking with these hatchlings was effective (Dan et al. 2008). Based on these results, JASFA began a large-scale project to release hatchlings for restocking of this species in 1992. To evaluate the restocking effectiveness, hatchlings were released with their cuttlebones stained with Alizarin red S and these were subsequently examined for specimens collected at fish markets and from beaches during the period from 2000 to 2003. The stained cuttlebones were mostly collected on the beaches around the release sites. Cuttlefish were caught in the bays after they were released from the autumn of the same year to the spring 2 years later. The proportion of released cuttlefish in different catches ranged from 4.5 to 18% (Dan et al. 2008).

7.2.2.5 Other Cephalopod Species

Oshima and Choe (1961) and Choe (1966) reared *S. esculenta, S. lycidas* (which they identified as *S. subaculeata*) and *Sepiella japonica* (which they identified as *Sepiella maindroni*) in the laboratory from hatchlings for up to 4 months by feeding them on *N. japonica*. Arima et al. (1963, 1964) studied the technologies for developing spawning beds, seed production and cultivation of *S. esculenta, S. lycidas and S. japonica* and grew them from eggs spawned in tanks and fed on the larvae of shrimp, obtaining detailed results on feeding rate, food conversion rate and growth. JASFA tried once to rear *S. esculenta* and produced 30,000 hatchlings in 1986 for the purpose of restocking.

Hamabe (1963) obtained hatchlings of *Todarodes pacificus* spawned in the laboratory but they did not survive. Soichi (1976) kept juveniles of this species collected from the wild for 67 days in a public aquarium by feeding with juvenile mullet. *T. pacificus* has not been raised from eggs as the diet of the paralarvae is unknown. Techniques for the artificial fertilization of ommastrephid squids were established by Sakurai et al. (1995). The embryonic developmental process from artificial fertilization to hatching of *T. pacificus* was described by Watanabe et al. (1996). There is a possibility that technologies for seed production will be developed for this species in the future, if the food organisms and feeding behaviour of the paralarvae of this species in the wild were determined.

7.3 Restocking in Thailand

7.3.1 Background

Squid trapping is an artisanal fishery activity that is very popular around the coasts of Thailand since its origin in about 1985. The trap itself is made from local materials

Fig. 7.2 A squid trap with an egg cluster hung inside covered with coconut leaves to create a protective environment, on a beach in eastern Thailand. (Photograph J Nabhitabhata)

and the fishing technique operates from the experience and understanding of squid behaviour by local fishermen (see also Chap. 17 '*Sepioteuthis lessoniana*'). The trap frame is made of locally available wood and covered by a polyethylene fishing net of 25 mm mesh size. The trap size is 50–80 × 80–120 × 50–65 cm, typically 60 × 100 × 50 cm. Coconut fronds are used to cover the trap to give it the appearance of a natural shelter. Bigfin squid egg clusters are hung inside the trap to persuade the squid to enter (Fig. 7.2). Stripped plastic shopping bags can also be used as substitutes when egg clusters are in short supply. For fishing, the trap is suspended in the water column at 2–3 m above the bottom using rock sinkers and styrofoam and bamboo floats. The fishing depth is from approximately 6 to 40 m, depending on the maximum yield. The female squid, escorted by her male mate, will enter the trap searching for appropriate substrates for spawning. The trap is left overnight or more. The fishermen come back, pull up the traps and collect the yields. The main target species of squid traps is the bigfin squid, *S. lessoniana,* 90–95 % of the catch, but *Sepia pharaonis* is a major by-catch, 5–10 % with a maximum of 30 % (Boongerd and Rachaniyom 1990; Chotiyaputta and Yamrungreung 1998).

The squid trap is likely to be an appropriate gear in view of natural resource conservation. There are two reasons to support this view. Firstly, the fishing period is about 9 months in each year and only 20 days in each of those months, with the remaining period reserved for natural stock recruitment. Secondly, most of the squids captured in traps are fully mature, since 80 % are larger than 140 mm, equal to the first maturation size (Chotiyaputta 1988). However, one disadvantage is that the female enters the trap for spawning and to find shelter for her eggs. The egg capsules are attached all over the covering net and any suspending ropes and floats. These capsules will be discarded or left for sun drying when the trap is landed for cleaning or replacing. Some capsules might be thrown back into the sea and sink to the bottom, but these have a low chance of hatching (Chindamaikul et al. 1994).

This disadvantage of squid trapping in view of conservation has not been overlooked by the biologists and a primary solution was initiated. In 1990, the Thailand Department of Fisheries (DOF) launched a research programme on cephalopod culture at the Rayong Coastal Fisheries Research and Development Center (RCRC) in Rayong Province, about 200 km from Bangkok. One of the activities in this programme was for restocking through aquaculture methods. The RCRC was in a good location where artisanal squid trapping was popular and this enhanced the possibilities for tackling the mentioned disadvantages. The first step of a possible solution was to collect the egg capsules before they were discarded. The fishermen were paid for the live egg capsules they collected at about US$ 0.33 per 1 kg. This payment was intended to be an additional source of income for the fishermen and, on the other hand, to raise their awareness of the economic value of natural resources. The second step was to endorse hatching of those eggs in a hatchery through aquaculture management. Previous studies revealed that the hatching rate of egg capsules incubated in plastic baskets was 20% higher than when left at the bottom (Chindamaikul et al. 1994). The third step was the release of the produced squid seeds back into the actual fishing grounds in order to maximise restocking efficiency. Unfortunately, this activity has declined since 2003 because there have been changes in policy of the Thailand DOF, and the cephalopod hatchery has been shut down.

7.3.2 Methods

7.3.2.1 Hatchery Seed Production

***Sepia pharaonis* Broodstock** Live cuttlefish are collected from traps and maintained in concrete tanks as broodstock. The fisherman maintains the live cuttlefish in a partition in his boat through which fresh seawater can flow. The cuttlefish can survive for more than 24 h in this chamber. The body weight of the spawners range from 400 to 1,500 g at 200–300 mm ML. At landing, the cuttlefish are transported to the hatchery in a plastic transportation tank of 500-L capacity. They are maintained in 2 m^3 concrete tanks at a male-to-female ratio of 1:2 (Fig. 7.3). The cuttlefish will mate in less than an hour after being released into the tank. Pieces of fishing net, 25 mm mesh size, with stone sinkers are put in the tanks as artificial substrates.

Spawning in the tank proceeds within that day or during the next morning. The females attach their egg capsules in clusters to the artificial substrates. Each female is able to spawn approximately 1,600 (500–3,000) egg capsules in total. (Nabhitabhata 1978b, Nabhitabhata and Nilaphat 1999). One female lays eggs in one cluster (Fig. 7.4) or more every day or every other day if they are interrupted while spawning. The spawning period can be more than 3 weeks, depending on the female size, the numbers of eggs produced and her condition. The male escorts and protects his mate from other males during spawning, but will turn to form a pair

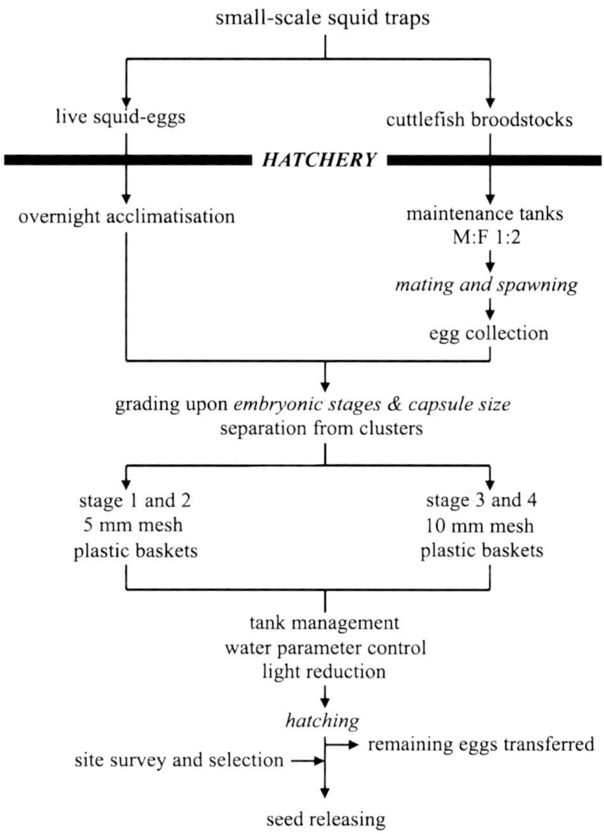

Fig. 7.3 Diagram of the cephalopod seed production methods in Thailand

with another female after her last spawning. Mating and spawning are observed in the early morning and late afternoon.

Eggs are collected daily from the broodstock tanks and transferred to be nursed in other tanks in the hatchery under controlled conditions. The capsules are collected by hand each day after the females pause their spawning. The egg clusters are also available from some squid traps.

***Sepioteuthis lessoniana* Broodstock** Live bigfin squid can be collected using a similar method to that for cuttlefish, but the collection of only egg masses is an appropriate current choice. Firstly, it is more economical and convenient because the percentage of egg capsules attached to traps is more than 95 % in quantity compared to 5 % for the cuttlefish. Secondly, most of the collected broodstock squid are not healthy enough to produce healthy eggs due to their pelagic habit and high metabolic rate.

Primary Handling of Egg Capsules The egg capsules collected from the wild occur in various sizes, hence differ in their embryonic stages, and have to be graded. Squid egg capsules are graded into four stages according to their size and shape. The first stage is the newly spawned egg capsule with a cylindrical shape and is at

Fig. 7.4 An egg cluster of the pharaoh cuttlefish, *Sepia pharaonis*. (Photograph J Nabhitabhata)

least 50 % smaller than those near to hatching. Second-stage egg capsules are similar in shape but with comparatively larger egg chambers. In the third stage, each chamber of the egg capsule is globular in shape, becoming cubic in the fourth stage with a finger-like shape and comparatively more transparent. The capsules collected and transported at the early stages can yield a better hatching rate than those collected nearer to hatching (Nabhitabhata 1978a, b).

The egg capsules require good aeration and ventilation to yield a good hatching rate. After being transported to the hatchery and being left overnight for acclimatization (Fig. 7.5), the capsules have to be separated from their clusters by cutting with scissors. Squid egg capsules are put into baskets of 5 mm mesh size for those in the first and second stages and of 10 mm mesh size for those in the third and the fourth stages (Fig. 7.6). Cuttlefish egg capsules require just one kind of plastic basket, with a 5 mm mesh size. The capsules are maintained until hatching in such baskets floating in concrete tanks of 4,000-L capacity and 1.8 m diameter (Nabhitabhata 1978a, b, 1996; Nabhitabhata and Kbinrum 1981). The egg capsules of squid and cuttlefish are always in separate tanks, due to differences in their embryonic periods, and hence the incubation periods are required.

Nursing of Egg Capsules Egg capsules are nursed under controlled conditions in tanks of similar size and water quality to the acclimatization tanks in the hatchery. An open system is employed where any changes in physical and chemical parameters are minimized. Egg capsules are visually checked every morning, and empty capsules and those with dead or abnormally developed embryos are discarded in

Fig. 7.5 Egg clusters of the bigfin squid, *Sepioteuthis lessoniana* acclimatized in a plastic basket. (Photograph J Nabhitabhata)

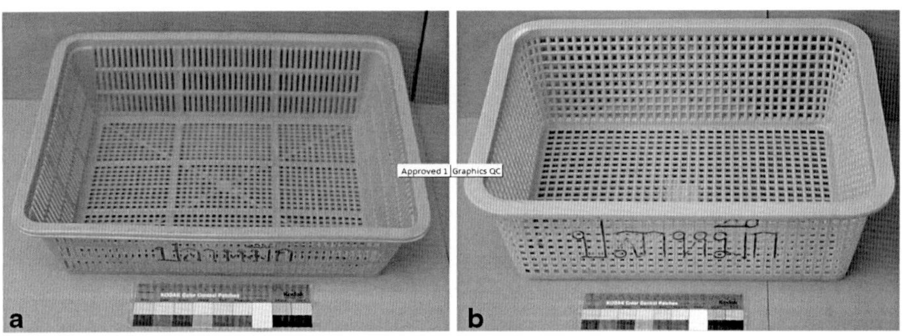

Fig. 7.6 Two types of plastic baskets used for nursing the egg capsules, 5 mm mesh (*left*) and 10 mm mesh (*right*). (Photograph J Nabhitabhata)

order to prevent microbial infection and self-decomposition, which will reduce water quality. Brief changes in temperature, salinity and mechanical stimuli, which can cause premature hatching, are avoided. In order to minimize temperature fluctuations, a water flow-through method is applied at a rate of 1 L min^{-1}. Fresh seawater is pumped through an activated carbon filter and stored in a 200-m^3 concrete tank before gravity-feeding to the hatchery (Fig. 7.7). The current generated by the water feed also assists in removing waste products by producing an upwelling outflow through drainage pipes at the centre of the tank. This system is able to maintain an average temperature of 28°C, a salinity of 30–33 practical salinity unit (psu) and a

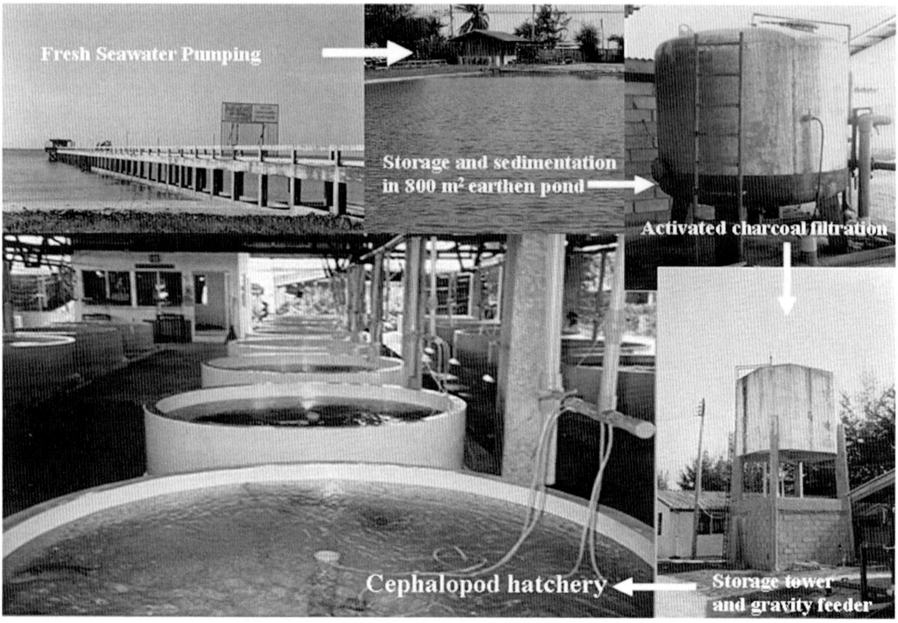

Fig. 7.7 Seawater system for the open-system cephalopod hatchery in Thailand. (Photograph J Nabhitabhata)

pH of 6.0–8.0 (Nabhitabhata et al. 1991a, 2005). The salinity range to obtain a 50 % hatching rate of eggs is 22–38 psu (Nabhitabhata et al. 1991b, 1993a, 2001).

The length of the embryonic period depends on both biotic and abiotic factors, but particularly on temperature. The average embryonic period is 14 days (9–25) for the cuttlefish *S. pharaonis* (Anil et al. 2005; Nabhitabhata 1978b, Nabhitabhata and Nilaphat 1999) and 20 days (17–23) for the squid, *S. lessoniana* (Nabhitabhata 1978a, 1995, 1996; Nabhitabhata and Kbinrum 1981) each at about 28°C. The embryonic period can be shortened at a higher temperature and extended at a lower one. The hatching period, from the hatching of the first egg to the hatching of the last eggs of the same cluster from the same spawning, is 3–10 days. Most of the hatching, i.e. more than 50 %, occurs at night on the second and third day from the first hatching. The hatching rate is more than 90 % (Nabhitabhata and Nilaphat 1999) under the above-mentioned management conditions.

Other important physical factors are turbidity or suspended solids and light. Turbidity should be maintained as low as possible (Nabhitabhata 1995) through filtration and/or prior sedimentation, particularly in open systems. Cephalopods have a high metabolic rate and their gills have a low efficiency for removing sediment. High levels of suspended solids or high turbidity are critical, since it can easily reduce or block oxygen exchange. Any unnatural or artificial lighting as well as unnatural daily light–dark periods should be avoided. The growth of algae and fungi on the surface of the egg capsule is due to excess light and this reduces the hatching rate by blocking the oxygen supply. The attachment of algae also initiates fungal

Table 7.2 Numbers of cephalopod seeds ($\times 10^4$ individuals) produced and released for restocking in Thailand during the 14 years from 1999 to 2003. (Nabhitabhata et al. 1993b, 2002, 2003, 2004, 2005; Timdee and Wongwiwatanawute 2000; Timdee et al. 2003)

Year	*Sepioteuthis lessoniana*	*Sepia pharaonis*	Others[a]	Total
1990	144.68	1.26	–	145.94
1991	166.86	2.66	–	169.52
1992	304.89	15.16	–	320.05
1993	198.38	20.21	–	218.59
1994	378.64	27.00	–	405.64
1995	86.42	53.09	–	139.51
1996	187.45	14.28	–	201.73
1997	225.21	33.26	–	258.47
1998	104.65	12.76	–	117.41
1999	147.40	44.16	–	191.56
2000	166.08	12.16	2.86	181.10
2001	173.83	32.36	6.71	213.58
2002	22.01	8.36	1.28	31.65
2003	–	10.73	–	10.73
Total	2,306.50	287.45	10.85	2,605.48
Average	177.42	20.53	3.62	186.11
SE	24.34	4.12	0.97	27.42

[a] Spineless cuttlefish, *Sepiella inermis*; bobtail squid, *Euprymna hyllebergi*; octopus, *Amphioctopus aegina*, *Amphioctopus neglectus* and *Amphioctopus rex*

growth on the capsule membrane and, later, infection to the embryo. In order to reduce light, curtaining of the hatchery with a camouflage net with a light-reducing efficiency of 80 % (Nabhitabhata et al. 2005) is the most convenient method.

Other Cephalopod Species Rearing and culture experiments associated with other species of cephalopods occasionally produce excess numbers of seeds for experimental performance, which can then be released for restocking (Table 7.2). Detailed methods are provided in other chapters of this volume, such as for the spineless cuttlefish *Sepiella inermis* (Chap. 13), the bobtail squid *Euprymna hyllebergi* (Chap. 15), the octopus *Amphioctopus aegina* (Chap. 18) and *Amphioctopus neglectus* and *Amphioctopus rex* (following the methods described for *A. aegina*).

Post-hatching Management and Seed Releasing The hatching of cephalopod eggs has unique characteristics. The period, from the hatching of the first egg to the hatching of the last eggs of the same cluster of the same spawning, extends for 3–10 days. Most hatching (more than 50 %) occurs at night on the second and third day from the first hatching. The overall hatching rate is more than 90 % (Nabhitabhata and Nilaphat 1999) under the above-mentioned management conditions. The hatchlings pass through the basket mesh into the nursing tank. When the optimum density in the tank is achieved, the egg baskets are transferred to another tank of previously prepared water of similar quality.

The cephalopod seeds are collected for release during the next day. The water in the nursing tanks is drained through the centre pipe into a 50-L tank with an immersed hand scoop for reducing the flow rate and collecting the cephalopods. With aeration from a portable air pump, the tank with seeds is transported to the selected releasing sites.

The total numbers of released seeds was 25.95 million over 14 years (1990–2003): 23.06 million for bigfin squid, 2.77 million for pharaoh cuttlefish and 1.08 million for other species (Table 7.2). The average number released was 2.14 million seeds per year. About 90 % of the seed was the bigfin squid and the remaining 10 % was mostly pharaoh cuttlefish but included other species.

7.3.2.2 In situ Seed Production

The project is being conducted by a community of squid-trapping fishermen who are aware of the need to conserve their livelihood through the conservation of natural resources. The first trial, the so-called Cephalopod Seed Bank, was conducted in June 2012, in Chumporn Province, southern Thailand, Gulf of Thailand. The activity was funded by the Thailand DOF with technical consultation by the Prince of Songkla University. The eggs were collected from traps owned by the fishermen themselves.

The idea was to incubate the collected egg capsules in field conditions. The egg capsules were placed in net bags made from fishing net of 0.5 cm mesh size (Fig. 7.8). Each bag was about 50 cm long and 30 cm in diameter. The bags were hung from a rope with a 30-cm interval at one end and kept suspended with a 100-mm plastic float at the other end. Seven strings of rope were fastened to the frame that was made from iron pipes and floated with empty polyvinyl chloride (PVC) barrels. A net cage of $300 \times 200 \times 200$ cm, 0.5 cm mesh size was also hung from this frame to serve as a fence for the egg bags. The 'set' was then covered by a piece of shutter net to reduce excess light. The sites were selected by the community. In order to reduce the cost, bamboo frames were used as a substitute for the iron pipe frame. This project has been continued up to the present time and has been extended to neighbouring fishing communities.

The advantage is that there is little cost for the hatchery operation. The fisherman will stop by the site during his homeward journey and put his collected egg capsules into the bags. The hatchlings pass through the mesh into natural waters, using the floating facility as a temporary shelter for primary acclimatization. This method can also be applied for aquaculture as an in situ nursery of eggs and hatchlings that requires little expense and is cost effective compared to a land-based hatchery.

7.4 Conclusions

Aquaculture to restock cephalopods has been performed by public sectors in both Japan and Thailand with a long history of about 50 years. The restocking activities are based on public awareness about the depletion of natural resources. Most

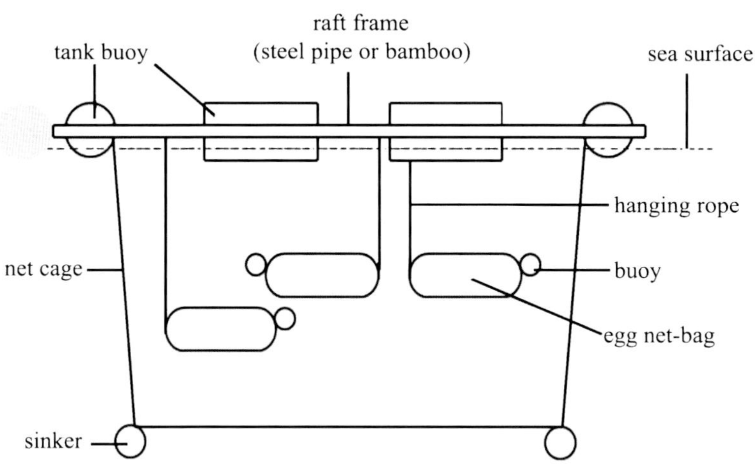

Fig. 7.8 The in situ hatchery operated by a fishing community in southern Thailand, illustrating a raft with an iron-piped frame and hanging ropes for egg net bags (*above*) and in the diagram (*below*). (Photograph courtesy of J Petchkamnerd)

cephalopod species in focus are neritic, including the sepiid cuttlefish and benthic octopus. The major species are *S. lessoniana*, *H. bleekeri*, *S. latimanus*, *S. pharaonis* and *O. vulgaris*. Broodstocks are collected from the wild and seeds are produced in the hatchery before their release. Attempts at aquaculture have yielded a certain level of success (see other chapters on corresponding species in this volume) and restocking activity has been routinely performed. The reduction of the costs of a restocking process is a primary concern and has generated methods of in situ seed production which can be applied to commercial-scale production. The scientific or biological significance of the restocking of natural populations and fishery yields requires further research and evaluation. The methods used in the studies of fishery biology, i.e. tagging and labelling and release and recapture, should be applied to such research. The restocking activities require the cooperation of local fishermen or local fishery associations who are the most important stakeholders for utilization of natural resources, and who ultimately ensure the success of these activities. However, previous results in both Japan and Thailand have succeeded in raising public awareness of the importance of cephalopod natural resources and restocking.

References

Anil MK, Andrews J, Unnikrishnan C (2005) Growth, behavior and mating of pharaoh cuttlefish (*Sepia pharaonis* Ehrenberg) in captivity. Israel J Aquacult 57:25–31

Arima S, Hiramatsu T, Segawa K, Tako N, Terada K (1963) Research on seeding production techniques and cultivation technology of the cuttlefish II. Res Rep Fukuoka Pref Buzen Fish Exp Stn 1962:27–54 (in Japanese)

Arima S, Hiramatsu T, Norimatsu T, Segawa K, Tako N (1964) Research on seeding production techniques and cultivation technology of the cuttlefish III. Res Rep Fukuoka Pref Buzen Fish Exp Stn 1963:1–56 (in Japanese)

Boletzky Sv, Hanlon RT (1983) A review of the laboratory maintenance, rearing and culture of cephalopod molluscs. Mem Natl Mus Victoria 44:147–187

Boongerd S, Rachaniyom S (1990) Squid trap fishing. Tech Pap 1/1990, Fish Tech Subdiv, Mar Fish Div, Dept Fish (in Thai)

Chindamaikul T, Nabhitabhata J, Somsueb S, Krainara T (1994) Comparison on different methods for hatching of bigfin squid, *Sepioteuthis lessoniana*. Tech Pap 7/1994, Phuket Coast Aquacult Develop Cent, Coast Aquacult Div, Dept Fish (in Thai with English abstract)

Choe S (1966) On the growth feeding rates and the efficiency of food conversion for cuttlefishes and squids. Korean J Zool 9:12–20

Choe S, Oshima Y (1963) Rearing of cuttlefishes and squids. Nature 197:307

Chotiyaputta C (1988) Reproductive biology of bigfin reef squid from squid trap. Tech Pap 5, Mar Life Hist Sect, Mar Fish Div, Dept Fish (in Thai with English abstract)

Chotiyaputta C, Yamrungreung A (1998) Trap fisheries for squid and their impact on spawning. S Afr J Mar Sci 20:285–291

Dan S, Hamasaki K, Yamashita T, Oka M, Kitada S (2008) Evaluation of the stocking efficiency of giant cuttlefish *Sepia latimanus* hatchlings through marking experiments. Nippon Sui Gakk 74:615–624

Fushimi H (2001) Production of juvenile marine finfish for stock enhancement in Japan. Aquaculture 200:33–53

Hamabe M (1963) Spawning experiments on the common squid *Ommastrephes sloani pacificus* Steenstrup in an indoor aquarium. Bull Jap Soc Sci Fish 29:930–934

Inoha M (1971) Studies on the seedling production of Kobushime *Sepia (Platisepia) latimanus* Quoy and Gaimard. Proj Rep 1968, Ryukyu (Okinawa) Fish Res Lab:136–138 (in Japanese)

Inoha M, Sezoko M (1967) The experiment on the culture of Aori-ika (white squid *Sepioteuthis lessoniana*. Proj Rep 1966, Ryukyu (Okinawa) Fish Res Lab:74–80 (in Japanese)

Itami K (1975) The Seto Inland Sea octopus fisheries mainly based on the development of resource culture techniques. Bull Hyog Pref Fish Exp Stn 15:109–118

Itami K, Izawa Y, Maeda S, Nakai K (1963) Note on the laboratory culture of the octopus larvae. Bull Jap Soc Sci Fish 29:514–520

JASFA (2003) The forty years' history of the Japan Sea-Farming Association (JASFA). Japan Sea-Farming Association, Tokyo (in Japanese)

Matsumoto G (1976) Transportation and maintenance of adult squid (*Doryteuthis bleekeri*) for physiological studies. Biol Bull 150:279–285

Nabhitabhata J (1978a) Rearing experiment on economic cephalopod -I: long-finned squid *Sepioteuthis lessoniana* Lesson. Tech Pap 1978, Rayong Brackish Fish Sect, Rayong Fish Stn, BrackishFish Div, Dept Fish (in Thai with English abstract)

Nabhitabhata J (1978b) Rearing experiment on economic cephalopod-II: cuttlefish *Sepia pharaonis* Ehrenberg. Tech Pap 1978, Brackish Fish Sect Rayong Brackish Fish Stn, Brackish Fish Div, Dept Fish (in Thai with English abstract)

Nabhitabhata J (1995) The culture of cephalopods: commercial scale attempt in Thailand. In: Nambiar KPP, Singh T (eds) Proc INFOFISH-AQUATECH' 94, Int Conf Aquacult. INFOFISH, Columbo

Nabhitabhata J (1996) Life cycle of cultured bigfin squid, *Sepioteuthis lessoniana* Lesson. Phuket Mar Biol Cent Spec Pub 16:83–95

Nabhitabhata J, Kbinrum S (1981) The culture of long-finned squid, *Sepioteuthis lessoniana* Lesson. Rayong Brackish Fish Stn, Brackish Fish Div, Dept Fish, Ann Rep 1981:74–115 (in Thai with English abstract)

Nabhitabhata J, Nilaphat P (1999) Life cycle of cultured pharaoh cuttlefish, *Sepia pharaonis* Ehrenberg, 1831. Phuket Mar Biol Cent SpecPub 19:25–40

Nabhitabhata J, Kbinrum S, Soodmee Y (1991a) Effect of pH on survival rate of cephalopods hatchlings. Tech Pap 17/1991, Rayong Coast Aquacult Stn, Coast Aquacult Div, Dept Fish (in Thai with English abstract)

Nabhitabhata J, Kbinrum S, SoodmeeY (1991b) Effect of salinity on hatching rate of *Sepioteuthis lessoniana* Lesson. Tech Pap 16/1991, Rayong Coast Aquacult Stn, Coast Aquacult Div, Dept Fish (in Thai with English abstract)

Nabhitabhata J, Phudtaraksa W, Nilaphat P, Phetrasatian G (1993a) Effect of salinity on hatchling rate of pharaoh cuttlefish, *Sepia pharaonis* Ehrenberg. Tech Pap 10/1993, Rayong Coast Aquacult Stn, Coast Aquacult Div, Dept Fish (in Thai with English abstract)

Nabhitabhata J, Sangkasem P, Phudtaraksa W, Sriakaphrom V, Chenkitkarn S, Soodmee Y, Nilaphat P, Phetrasatian G, Kongrot J, Ratanasoot S (1993b) On Mass production of cephalopod seeds for releasing as stock enhancement, 1990–1992. Tech Pap 9/1993, Rayong Coast Aquacult Stn, Coast Aquacult Div, Dept Fish (in Thai with English abstract)

Nabhitabhata J, Asawangkune P, Amornjaruchit S, Promboon P (2001) Tolerance of eggs and hatchlings of neritic cephalopods to salinity changes. Phuket Mar Biol Cent Spec Pub 25:91–99

Nabhitabhata J, Nilaphat P, Promboon P, Jaroongpattananon C, Nilaphat G, Reunreng A (2002) Broodstock collection and production of cephalopod seeds for releasing as stock enhancement, 1999–2001, with relationships to environmental factors. Contrib 25, Rayong Coast Aquacult Stn, Coast Aquacult Div, Dept Fish (in Thai)

Nabhitabhata J, Nilaphat P, Promboon P, Jaroongpattananon C, Nilaphat G, Reunreng A (2003) Broodstock collection and production of cephalopod seeds for releasing as stock enhancement, 2002, with relationships to environmental factors. Contrib 26, Rayong Coast Aquacult Stn, Coast Aquacult Div, Dept Fish (in Thai)

Nabhitabhata J, Nilaphat P, Promboon P, Reunreng A, Jaroongpattananon C, Nilaphat G (2004) Relationships of environmental factors and production of cephalopod seeds for releasing as stock enhancement, 2003. Contrib 27, Rayong Coast Aquacult Stn, Coast Aquacult Div, Dept Fish (in Thai)

Nabhitabhata J, Nilaphat P, Promboon P, Jaroongpattananon C, Nilaphat G, Reunreng A (2005) Performance of simple large-scale cephalopod culture system in Thailand. Phuket Mar Biol Cent Res Bull 66:337–350

Natsukari Y (1970) Egg-laying behavior, embryonic development and hatched larva of the pygmy cuttlefish, *Idiosepius pygmaeus paradoxus* Ortmann. Bull Fac Fish Nagasaki Univ 30:15–29

Oka M (1993) Methods to collect eggs of giant cuttlefish in an aquarium. In: Okutani T, O'Dor RK, Kubodera T (eds) Recent advances in cephalopod fisheries biology. Tokai Univ Press, Tokyo

Oka M, Tezuka N, Fushimi H (1989) Studies on collecting eggs in an aquarium and hatching of *Sepia latimanus*. Saibai Giken 18:1–14 (in Japanese)

Okumura S, Kurihara A, Iwamoto A, Takeuhi T (2005) Improved survival and growth in *Octopus vulgaris* paralarvae by feeding large type *Artemia* and Pacific sandeel, *Ammodytes personatus*. Aquaculture 244:147–157

Ohshima Y, Choe S (1961) On the rearing of young cuttlefish and squid. Bull Jap Soc Sci Fish 27:979–986

Sakurai Y, Young RE, Hirota J, Mangold K, Vecchine M, Clarke MR, Bower J (1995) Artificial fertilization and development through hatching in the oceanic Squids *Ommastrephes bartramii* and *Sthenoteuthis oualaniensis* (Cephalopoda: Ommastrephidae). Veliger 38:185–191

Segawa S (1987) Life history of the oval squid, *Sepioteuthis lessoniana* in Kominato and adjacent waters central Honshu, Japan. J Tokyo Univ Fish 74:67–105

Shiraki M (1997) Developmental program of seed production techniques for common octopus – finding the food of octopus larvae - Saibai 84:41–43 (in Japanese)

Soichi M (1976) On the growth and food quantity of young common squid, *Todarodes pacificus*, in captivity. Dosuishi 13:79–82 (in Japanese)

Tauti M, Matsumoto I (1954) On the protection of spawner of octopus (*Polypus vulgaris* Lamarck) in Hyogo Prefecture. Bull Jap Soc Sci Fish 20: 479–482

Timdee W, Wongwiwatanawute C (2000) Mass production of spineless cuttlefish (*Sepiella inermis*) seeds for releasing as stock enhancement, year 2000. Tech Paper 2000, Chonburi Coast Aquacult Stn, Coast Aquacult Div, Dept Fish (in Thai with English abstract)

Timdee W, Wongwiwatanawute C, Nabhitabhata J (2003) Broodstock collection and production of spineless cuttlefish (*Sepiella inermis*) seeds for releasing as stock enhancement, 2000–2002. Contrib 26, Chonburi Coast Aquacult Stn, Coast Aquacult Div, Dept Fish (in Thai with English abstract)

Tsuchiya M (1982) On the culture of the squid, *Sepioteuthis lessoniana* Lesson (Loliginidae), from hatching to the whole life in Iriomote Island, Okinawa. Bull Inst Oceanic Res Develop Tokai Univ Notes 4:49–70

Ueta Y, Kitakado I (1996) Experimental Studies on the form of spawning bed by oval squid, *Sepioteuthis lessoniana*. Suisanzoshoku 44:67–72

Ueta Y, Kitakado I, Segawa S, Tenjin M, Jo Y, Fukunaga M, Kangawa Y (1995) Selection of spawning ground and spawning bed in oval squid *Sepioteuthis lessonina*. Fish Eng 31:189–194

Watanabe K, Sakurai Y, Segawa S, Okutani T (1996) Development of the ommastrephid squid *Todarodes pacificus* from fertilized egg to the rhynchoteuthion paralarva. Amer Malacol Bull 13:73–88

Yamauchi K, Takeda B (1964) On the rearing of an octopod, *Octopus ocellatus* Gray-1. Aquac 12:1–9 (in Japanese)

Chapter 8
Applications, Uses and By-products from Cephalopods

Noussithé Koueta, Helene Viala and Estelle Le Bihan

Abstract The worldwide catch of cephalopod species (squid, cuttlefish and octopuses) was close to 4 million t in 2008. The utilization and the processing of cephalopods result in the generation of a large amount of by-products. Generated by-products represent 35% of the total mass caught and these include head, viscera, skin, bone, etc. These wastes are currently unvalued. Serious environmental problems can be created without appropriate management. Numerous studies have demonstrated that cephalopod by-products are suitable for human consumption, animal food and other applications with high market value. Indeed, cephalopod by-products are a source of interest for molecules such as polyunsaturated acids, chitin, collagen, etc. This chapter provides an overview of the extraordinary potential of cephalopod by-products and their applications.

Keywords Cephalopod · Processing · Underutilized by-product · Valorisation · High value molecule

8.1 Introduction

The world total catch of cephalopods (squids, cuttlefish and octopuses) continues to increase and had reached a threshold of 4 million t in 2008 (FAO 2010). The total catch of this species group suffered a decrease of 0.8 million t in 2009, before beginning to grow again in 2010 and reach 4% in the world fish trade (FAO 2012). Such a growth in the exploitation of cephalopods has resulted in a similar expansion in processing industries and the generation of large amounts of waste, with considerable disposal costs. As a result, the valorisation of by-products would allow increased company competitiveness and reduce costs. By-products of the cephalopod fishery are widely used—in food or animal feed, cosmetics and medicine. By-product valorisation is a

N. Koueta (✉)
CNRS INEE—FRE3484 biomea Biologie des Mollusques marins et des Ecosystèmes Associés, Esplanade la Paix, Université de Caen, 14032 Caen Cedex, France
e-mail: Noussithé.koueta@unicaen.fr

H. Viala · E. Le Bihan
IVAMER, Innovation Marine, 32 rue Fragonard, 14220 Thury-Harcourt, France
e-mail: contact@ivamer.fr

J. Iglesias et al. (eds.), *Cephalopod Culture*, DOI 10.1007/978-94-017-8648-5_8,

historical practice. Fish sauce fermentation is a common practice in Southeast Asia as a means of preserving and producing value-added products from underutilized fish species. Associated with the sophistication of technical fishing, the valorisation of marine by-products allows better management and exploitation of marine resources. Attitudes have evolved and now it is inconceivable to waste any of the biomass: By-products are a wealth (Le Bihan and Koueta 2010). The development of technological tools opens new possibilities for the valorisation of markets which arc aspirant and increasing. Furthermore, it is important to develop complementary approaches to valorisation. Further research is needed into the development of green processes to limit the impact on the environment (Le Bihan and Koueta 2010). The wide chemical and biological diversity observed in the marine environment makes the ocean an extraordinary source of high value-added compounds which can be used in many applications. The cephalopod-processing industry generates large amounts of solid and liquid wastes, which can cause major problems for the environment. After processing, the solid wastes from the cuttlefish-processing industry may represent 35% of the original material (Le Bihan 2006; Balti et al. 2010). Occasionally, the wastes may reach 75% of the total catch weight from squid fishery, in forms of skin, head, cuttlebone, pen, ink and viscera (Shahidi 2006; Fig. 8.1). These wastes constitute an important source of proteins, lipids and other bioactive molecules with high commercial value (Fig. 8.2).

8.2 Uses and Applications

8.2.1 Marine Oil

Marine oil can be obtained by extracting fatty acids from cephalopods, particularly using viscera or skin. Ω-3 fatty acid concentrates remain a topic of general interest for the pharmaceutical and food industries, for the production of drugs which enhance performance and for the production of nutritional supplements. The valorisation of fish by-products by recovering their oil is of great interest in the fish-processing industry because of the high quality of seafood oil and the natural sourcing of Ω-3 polyunsaturated fatty acids (PUFAs). In addition, due to the demand–supply issues, the economic value of fish oil continues to increase. By-products can be an alternative to commercial fish oil. Fish oil production is a good opportunity for adding value to fish wastes and increasing the competitiveness of the fish industry (Chow 2000). Several authors have explored the potentiality to extract oil from cephalopod by-products.

Sinanoglou and Miniadis-Meimaroglou (1999) specify that for cephalopods, the major PUFAs are eicopentaenoic acid (EPA; C20:5n-3) and docosahexaenoic acid (DHA; C22:6n-3). The percentage composition of EPA and DHA with respect to the total fatty acids was studied by several authors; for example, it was 30.2% in the cuttlefish oil, 36.2% in *Octopus vulgaris* at early stage, 48% in *Sepia officinalis* and 45.5% in *Loligo vulgaris* (Navarro and Villanueva 2000). Hwanga and Liang (2001)

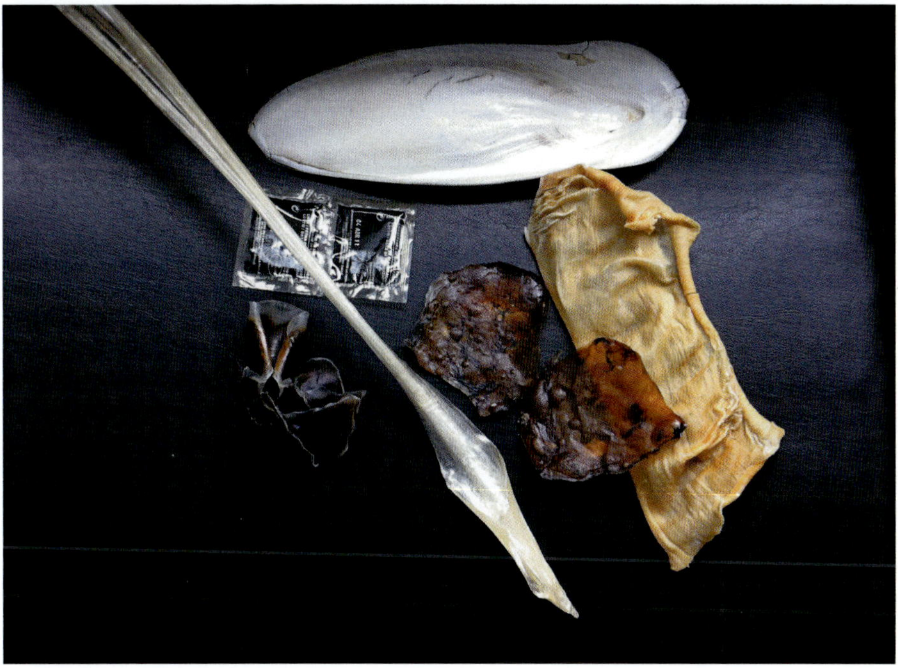

Fig. 8.1 Some cephalopod by-products: bone, ink, beak, skin. (Source: IVAMER)

studied the viscera oil composition of squid and explained that this oil contains 15–25 % of EPA and DHA. Squid contains more DHA than EPA in its lipids (Liang and Hwang 2000). The oil extracted from the viscera of cuttlefish (*Sepiella maindroni* de Rochebruns) was studied by Shen et al. (2007). Fatty acid composition, cholesterol content and volatile compound composition were analysed by these authors. The composition of fatty acids was 50 % monounsaturated fatty acids (MUFAs), followed by 31 % PUFAs and finally 19 % saturated fatty acids (Shen et al. 2007). In their study, Shen et al. (2007) revealed that cuttlefish oil from East China had an unsaturated fatty acid content of 81 %. Cuttlefish oil is also suitable as a high-energy feedstuff in aquaculture because of its fatty acid content and the strong fishy odour, which is an attractant for fish or shrimp. Even the residual portion after the extraction of PUFA and MUFA can be used for biodiesel production (Shen et al. 2007). Tavakoli and Yoshida (2006) studied squid oil hydrolysis as a mechanism for the production of Ω-3 PUFA and fatty acids for biodiesel. When using cuttlefish oil for functional ingredients, the EPA and DHA yields can be optimized. Park et al. (2011) developed a refined oil from raw cuttlefish (*Todarodes pacificus*) viscera and obtained a significantly higher composition, 14.7 % of EPA and 28.6 % of DHA; this refined product has a good stability. The cuttlefish digestive gland, which constitutes about 7–12 % of body weight, is also rich in EPA and DHA (Joseph et al. 2005). The study of Joseph et al. (2005) has shown that the use of Ω-3 PUFA, from cuttlefish digestive

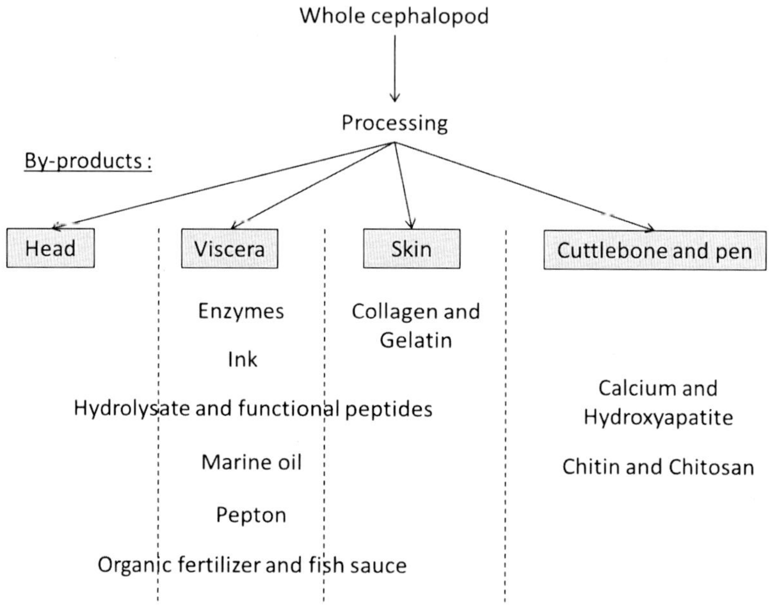

Fig. 8.2 General scheme for the utilization of cephalopod by-products. (Source: IVAMER)

gland, as a feeding ingredient has beneficial effects in rats on cellular proliferation (spleen cells and bone marrow cells) and on humoral immune functions. Sinanoglou and Miniadis-Meimaroglou (1999) have examined the mantles of three species of cephalopod molluscs from Saronicos Bay (Greece) for its fatty acids. They found that cephalopod skins can be excellent sources for PUFAs, especially Ω-3. Total lipids in the mantle of the cephalopods *Eledone moschata, S. officinalis* and *Todarodes sagittatus* constituted 2.0, 1.4 and 1.7% of wet tissue, respectively (Sinanoglou and Miniadis-Meimaroglou 1999).

8.2.2 Chitin and Chitosan

Cuttlebones and pens from cephalopods are rich in chitin. In general, squid and cuttlefish contained 3–20% of chitin (Patil and Satam 2002). Sometimes, higher quantities of chitin were observed, for example, the chitin content was about 40% of the original weight of the dried pens (Tolaimate et al. 2000). Chitin can be industrially converted into a more applicable chitosan, a structural modification of chitin often performed by alkaline hydrolysis (Wang et al. 2009b). Chitosan, produced by thermochemical alkaline deacetylation of chitin, is a biopolymer with unique properties favourable for a broad variety of industrial and biomedical applications (Tolaimate et al. 2000). Chitin has been identified in nature as the main supporting structure in a wide variety of organisms such as the exoskeleton or cuticle of different invertebrates (Cortizo et al. 2008). Chitin in cuttlebone and pen is in the β-conformation. The β-chitin is more easily solubilized and more reactive than the α-chitin (Cortizo et al. 2008). Muzzarelli and Muzzarelli (2005) describe many chitin and chitosan derivatives of major importance (complexes with hyaluronic acid, chitin ethers and esters, oxychitin, etc.) and in various forms (nanoparticles, hydrogels, films, etc.). Chitosan can be produced by chemical methods or by hydrolysis. Chitosanase hydrolysis is becoming the preferential method over chemical processes. Chitosanase hydrolysis has many advantages, in terms of the environment, cost and reproducibility (Wang et al. 2009a). Recent studies concern the optimization of bacterial culture conditions for protease and chitosanase productivity by using a sole carbon/nitrogen source from squid pen powder and deproteinisation of squid pen for β-chitin (Wang et al. 2009a, b). The process developed by these authors is a less expensive way to produce chitosanase, with the aim of obtaining a high value-added product, such as chitosan oligosaccharides, with high potential in the production of functional foods. Revathi et al. (2012) developed a process to produce and characterize chitin from *Vibrio* species and a combination of waste from two species, head waste of shrimp and cuttlebone chitin. Chitin and its derivatives, mainly chitosan, have attracted the interest of many researchers and industries in the past 30 years owing to its physical–chemical properties. Recently, Barwin et al. (2011) determined the physicochemical characterization of biopolymers chitin and chitosan, extracted from *Doryteuthis sibogae* squid pen. Moreover, the same authors (Barwin et al. 2012) characterized chitosan and sulphated chitosan from cuttlebone. The obtained yield is 21 % of chitin and 49.71 % of chitosan. These authors suggested that sulphated chitosan contribute significantly towards the observed antioxidant effect of cuttlefish food products. These polymers also display antimicrobial activity, biocompatibility, biodegradability and they also interact strongly with pesticides and metal ions in aqueous solutions (Lavall et al. 2007; Al-Sagheer et al. 2009). Thus, they display a wide range of applications in different fields such as cosmetic manufacture, medicine, agriculture, food production, pharmacy, biomedicine, the paper industry and also as absorbent materials for wastewater treatment for the uptake of metal ions from polluted water as well as for analytical applications (Lavall et al. 2007; Al-Sagheer et al. 2009). Moreover, chitosan can be used to modify the surface of nonwoven fabrics and polypropylene films to improve antimicrobial properties (Al-Sagheer et al. 2009). Composite films based on gelatine and chitosan have a potential application as preservatives in fish products (Gómez-Estaca et al. 2009). Uriarte-Montoya et al. (2010) detected a positive plastizer effect of squid collagen over a chitosan film. The blending of acid-soluble collagen from jumbo squid mantle and commercial chitosan gives the possibility of producing new material with food or biomedical applications. Various functions of chitin, such as moisture retention, adsorption and physiological activity, have been discovered and studies on the application of chitin have been performed in the textile, medicine and food fields (Yamashita et al. 2003). Chitin can suppress protein denaturation and increase the amount of unfrozen water in cells. The results revealed that, with a concentration of more than 5.0 % chitin hydrolysate, the freeze denaturation of myofibrillar protein was completely suppressed and the unfrozen water content of myofibrillar protein increased (Yamashita et al. 2003). Figueiredo et al. (2005) assayed natural waste materials containing chitin as adsorbents for textile dyestuffs. They tested cuttlefish (*S. officinalis*) bone and squid (*L. vulgaris*) pens. These authors concluded that the deproteinised squid pen had the highest adsorption capacity but showed an inferior performance to the granular activated carbon tested in the column experiments (Figueiredo et al. 2005).

8.2.3 Collagen and Gelatine

Collagen is one of the major protein components of connective tissue in multicellular animals. It is also a food constituent and important in developing the texture of edible tissues and their processed products. A number of studies have demonstrated that the collagen present in many aquatic animals is an important determinant of the texture of the processed meats obtained from them. Cephalopods are rich in collagen, at different concentrations, from 3 to 11 % in the mantle of squid like *Illex* and *Loligo* (Sikorski and Kolodziejska 1986), 18.33 % in the mantle of *Dosidicus gigas* (Torres-Arreola et al. 2008). Also, skins are rich in collagen: About 70–80 % of the squid skin dry matter is collagen (Lin and Li 2012). Two genetically distinct types of collagen have been identified in the mantle muscle and skin of the common squid *T. pacificus* (Mizuta et al. 2009), suggesting that collagens from these species may have relatively high resistance to hot-water extraction even after their denaturation. This thermal behaviour of squid collagens may contribute to the maintenance of mechanical strength or structural integration of heat-processed squid meats (Mizuta et al. 2009).

Gelatine is a thermally denatured protein obtained from collagen, generally by acidic or alkaline processes. Gelatine is a highly nutritious food and low in calories. Its use in the manufacture of low-fat products or diet products is very common. Generally, gelatine has a wide range of food, cosmetic, biomedical and pharmaceutical applications, including leather and encapsulation (Balti et al. 2011). Recently, and especially in the food industry, an increasing number of new applications have been found for gelatine in products such as emulsifiers (Aewsiri et al. 2011), foaming agents, colloid stabilizers, biodegradable film-forming materials (Hoque et al. 2011b) and microencapsulating agents, in line with the growing trend to replace synthetic agents with more natural ones. Hoque et al. (2011b) investigated the properties of blend film based on cuttlefish skin gelatine. During the processing of cuttlefish, the skin is generated as a by-product with a low market value. To increase its profitability, cuttlefish skin has recently been used for gelatine extraction (Hoque et al. 2011a; Aewsiri et al. 2011). However, gelatine yield is low, 2.6 % obtained from skin of the giant squid *D. gigas* (Gómez-Guillén et al. 2002). A higher extraction temperature can improve the yield but can negatively affect functional properties. Nagarajan et al. (2012) successfully extracted gelatine from splendid squid skin at an appropriate temperature. Cuttlefish skin gelatine, modified with tannic acid, has emulsifying activity and can improve the antioxidative activity. In this way, it can be used as a natural and safe additive in the food industry (Aewsiri et al. 2010). Balti et al. (2011) improved a gelatine extraction process from cuttlefish skin by a protease-aided process using smooth hound crude acid protease extract, which has greater functional properties than bovine skin gelatine, such as foam capacity, foam stability and gel strength. This gelatine extract can be used as an additive in food materials. Alemán et al. (2011b) have hydrolysed gelatines obtained from the skins of jumbo flying squid to produce antioxidant peptides. Hydrophobic amino acids have been obtained in several antioxidant peptide sequences, and, in addition, these authors suggested that the presence of hydrophobic amino acids in the peptide sequences in jumbo squid skin gelatine contributed greatly to its antioxidant properties, mainly inhibition of lipid peroxidation. In this way, hydrolysates of jumbo flying squid skin gelatine had a scavenging effect on radicals, probably because of the presence of proline residues in the peptide sequence.

8.2.4 Calcium and Hydroxyapatite

Cuttlebone from cuttlefish is rich in calcium. Shells are found to be the richest sources of calcium carbonate and have been utilized for various purposes after following appropriate treatments. Due to its similarity with the mineral constituents of bones, hydroxyapatite (HAP; $Ca_{10}(PO_4)_6(OH)_2$) has excellent biocompatibility and has been studied for many years as an implant material (Le Geros and Le Geros 1993). When shells are calcined at a proper temperature, calcium carbonate converts into calcium oxide (CaO), which is a metal oxide. Researchers have found that the CaO prepared from the waste shells can be used as a catalyst in the biodiesel production process. Utilization of waste shells as a source of CaO not only gives an opportunity to use it as catalyst but also adds value to the waste generated (Boroa et al. 2012). Ngamcharussrivichai et al. (2010) investigated the possibility of using various types of natural calcium as the catalysts for the methanolysis of palm kernel oil, which is readily available at a relatively low price from Thailand. Powders of recycled cuttlefish bone and phosphoric acid solutions were used to synthesize HAP. The calcium materials applied to the study included calcite, cuttlebone, dolomite, HAP and dicalcium phosphate. The effects of calcination temperature on the physicochemical properties of the resultant catalysts as well as the effects of reaction conditions on the formation of methyl esters have been investigated (Ngamcharussrivichai et al. 2010). HAP structures for tissue engineering applications can be produced by hydrothermal treatment of the aragonitic form of cuttlefish bone at 200°C for 24 h as processed by Ivankovic et al. (2009).

8.2.5 Functional Peptides

Dietary proteins are a source of biologically active peptides, which are inactive in the parent protein sequence but can be liberated during gastrointestinal digestion, food processing or fermentation. Viscera, digestive organs, brain and endocrine organs are rich in a number of peptides, hormones and neuropeptides, some of which have already been used as therapeutics. This includes generally immunomodulatory, antibacterial, antithrombic and antihypertensive activity (Murray and FitzGerald 2007). Peptides can stimulate blood cells proliferation (Thongthai and Gildberg 2005; Chuang et al. 2000), improve the health and function of the digestive system, positively affect the neurological system (Dorman et al. 1995; Le Poncin 1996a, b), have antioxidative effects (Boukortt et al. 2004) and can inactivate angiotensin I-converting enzyme (ACE; Wako et al. 1996). Moreover, bioactive substances may be derived from the enzymatic hydrolysis of some by-products followed by a qualitative separation of proteins or peptides with specific properties. Depending on the species and by-products, the diversity of potentially bioactive molecules is high. Their properties are numerous: immune stimulation, anti-hypertension, antistress, gastric stimulation and regulation of calcium metabolism (Le Bihan 2006) but they need to be optimized (Guérard 2006). In this way, bioactive peptides can be released by enzymatic proteolysis of food proteins and may act as potential physiological modulators of metabolism during the intestinal digestion of the diet. Bioactive peptides usually contain 3–20 amino acid residues and their activity is based on their amino acid composition and sequence (Le Bihan 2006).

In the last decade, a large number of studies have examined the enzymatic hydrolysis of collagen or gelatine for the production of bioactive peptides. Besides exploring different types of bioactivities, of an antimicrobial, antioxidant or antihypertensive nature, studies have also focused on the effect of oral intake in both animal and human models, revealing the excellent absorption and metabolism of peptides containing hydroxyproline (hyp; Gómez-Guillén et al. 2011). Giménez et al. (2009a) showed that antioxidant capacity of squid gelatines was largely increased by hydrolysis with alcalase for 3 h at 50 °C. Balti et al. (2008) showed that selective enzymatic hydrolysis of cuttlefish product proteins improved their functional and biological activities.

Gelatine produces bioactive peptides by protease hydrolysis which act as inhibitors of ACE (Kim et al. 2001). This enzyme plays an important physiological role in regulating blood pressure. Balti et al. (2010) concluded that the hydrolysate of cuttlefish by-product proteins has an excellent solubility and a high ACE-inhibitory activity. Alemán et al. (2011b) hydrolysed gelatine obtained from giant squid (*D. gigas*) with the aim of producing bioactive hydrolysates. Squid skin gelatines have been reported to give rise to biologically active peptides with high ACE-inhibitory and antioxidant activity, the latter due to its radical-scavenging capacity, metal-chelating effects and reducing power or lipid peroxidation inhibition (Alemán et al. 2011b). Lin and Li (2012) conclude that ACE inhibitors derived from squid by-products could be used to prevent hypertension. In this study, the hydrolysate produced from squid skin gelatine had good ACE-inhibitory activity in vitro (IC50 = 0.33 mg mL^{-1}). Gelatine also produces bioactive peptides by hydrolysis which acts as antioxidants (Lin and Li 2006). Giménez et al. (2009b) concluded in their study that the squid gelatine hydrolysate can be used in food systems as a natural additive with antioxidant properties and foaming and emulsifying functionalities. In another study, Ramasamy et al. (2011) extracted some polysaccharides from cuttlebone and demonstrated that there exists antibacterial activity against different bacterial strains. At the same time, no antifungal activity was detected (Ramasamy et al. 2011).

Le Bihan (2006) has studied the possibility of valorising cuttlefish viscera using a silage method. The author demonstrates that cuttlefish visceral silage can be used in aquaculture for its functional properties. Indeed, the produced silage contains more than 80% of peptides; many of them are functional peptides with numerous bioactivities, such as stimulating the digestion, immune stimulation and growth acceleration (Le Bihan 2006).

8.2.6 Peptones

Peptones are water-soluble protein hydrolysates which are non-coagulating with heat. Their high content in crude protein (resulting from the hydrolysis of proteins) and their natural wealth in fat, salt minerals and vitamins make them an ideal substrate for the microbial industry producing pharmacological substances and starters (lactic acid products, dairy, fungi for the manufacture of cheese, etc.). Thus, it would be very advantageous to convert by-products into a fermentation substrate for microbial growth and production of bio-products. Indeed, growth substrates constitute a major cost in the production of microbial cells and bio-products by the fermentation industry. In most instances, the growth medium accounts for approximately 40% of the production cost of industrial enzymes. Several researchers have explored the potentiality of cephalopod by-products to produce peptones. Peptones are usually obtained from casein, soy protein, gelatine and meat (Reissbrodt et al. 1995). Nevertheless, due to its favourable amino acid balance and high protein content, marine wastes represent a potential source of industrial peptones.

Souissi et al. (2008) studied the cuttlefish powder from *S. officinalis* by-products and tested it as a fermentation substrate for microbial growth and protease production using several species of bacteria. Cuttlefish powder is a complex substrate; notably it contains substances essential to microbial media such as carbon and nitrogen and minerals. This process, which converts underutilized wastes into more marketable and acceptable forms, coupled with protease production, could provide an alternative to the biological treatment of solid and liquid wastes generated by the cuttlefish-processing industry (Souissi et al. 2008). Vázquez et al. (2004) studied various peptones obtained from hydrolysed visceral homogenates of four fishery residues, and showed their suitability for promoting the growth of lactic acid bacteria, microorganisms with particularly complex requirements regarding peptidic nutrients. This work demonstrates that squid hydrolysate can substitute for other peptones in the habitual formulations for culture of lactic acid bacteria, promoting biomass and bacteriocin production that equals or surpasses those obtained on high-cost media traditionally recommended for these purposes. Vázquez and Murado (2008) considered diverse groups of peptones obtained by enzymatic hydrolysis of wastewater from the industrial processing of octopus and showed their effectiveness to promote the growth of lactic acid bacteria and the production of bacteriocins. Le Bihan (2006) explored the potential to use cuttlefish silage as peptones for microorganisms. This study demonstrated the conceivable use of silage as peptone as it can obtain growth kinetics equivalent to that of commercial peptones of various origins (meat, soy and casein).

8.2.7 Enzyme

The fish-producing wastes represent an important source of proteins and enzymes, especially digestive proteases, which are used in many applications in the food industry (Shahidi and Janak Kamil 2001; Gildberg et al. 2000), and other applications, such as the detergent, pharmaceutical, leather and silk industries (Gupta et al. 2002).

The use of proteolytic enzymes is part of many traditional methods of fish treatment as fish sauces, canned, semi-preserved or salted fish. Protease constitutes an important group of industrial enzymes, representing more than 65 % of the total industrial enzyme market (Banik and Prakash 2004). All these processes depend on varying degrees of activity of proteolytic enzymes associated with the viscera themselves (Shahidi and Janak Kamil 2001) with digestive enzymes or other tissues with cathepsins. Nowadays, enzyme-based technology represents an important contribution for many industrial applications. Aquatic invertebrates constitute natural sources of enzymes with colossal interest, such as aspartic pepsin, serine proteases, trypsin, chymotrypsin, collagenase, etc. Due to the prevailing environmental conditions, marine enzymes can effectively operate at low temperatures, below 4°C, and within neutral to alkaline pH values. Enzymatic methods have become an important and essential part of the processes used by the modern food and feed industry, to produce a large and diverse range of products for human and animal consumption. The aquatic environment contains the largest pool of genetic material and hence has enormous potential for the sourcing of different enzymes (Shahidi and Janak Kamil 2001). Cuttlefish wastes constitute an important source of proteolytic enzymes, particularly the digestive gland which contains a high number of proteinases (Hatate et al. 2000). Balti et al. (2009) extracted and purified trypsin from the digestive gland of cuttlefish. A cysteine proteinase was partially purified from the jumbo squid digestive gland (Cárdenas-López and Haard 2009). Hameed and Haard (1985) recovered cathepsin C from the Atlantic short-finned squid digestive gland. Some authors suggested that the salt-tolerant and exopeptidase activity of this enzyme contributes to the pleasant flavour of fermented fish products (Shahidi and Janak Kamil 2001).

8.2.8 Fish Sauce

Some of the potential uses of squid digestive gland proteinases are towards the preparation of fish sauce. The possibility of utilizing squid-processing by-products for low-salt fish sauce production has been investigated since the 1980s (Lee et al. 1982; Raksakulthai et al. 1986). Raksakulthai et al. (1986) showed the high potential of squid digestive gland, containing proteolytic enzymes, which aid the fermentation under optimal conditions.

Xu et al. (2008) concluded that squid by-products could be quickly fermented into low-salt fish sauce with acceptable qualities in terms of aroma and nutrition. Squid viscera have large and wide variety of uses in processed feeds and food supplements because of their nutritional and functional ingredients content. But viscera contain heavy metals, particularly cadmium, so the application of viscera to food is limited (Kawasaki et al. 2008). Some research has been reported, for example, the study of Kawasaki et al. (2008) concerning a purifying treatment for food application, by using microorganisms to remove cadmium from squid viscera. Successful production of fish sauce from male capelin (*Mallotus villosus*) using minced squid digestive gland was reported, with the highest amount of free amino acids and highest taste preference observed for samples treated with squid digestive gland compared to those treated with the fungal protease, pronase, trypsin and chymotrypsin (Shahidi and Janak Kamil 2001).

8.2.9 Organic Fertilizer

Cephalopod by-products, because of their high content of proteins, could be considered, through enzymatic hydrolysis, as a good candidate to be an organic fertilizer, for which there is a growing demand. This could be an interesting development to support the growth in organic agriculture. Some studies have demonstrated that fish protein hydrolysate can have an effect beyond the basic nutrition for plants (Milazzo et al. 1999).

It has been demonstrated that cephalopod waste hydrolysates have potential as organic fertilizers. Squid hydrolysate fertilizer seems to be similar to synthetic fertilizer, in terms of effectiveness and environmental impacts (Fetter et al. 2009). Squid hydrolysate fertilizer can improve the growth of the model plant (Peña-Cortés et al. 2010). Because fertilizer application to lawns can increase nutrient contamination of ground water, NO-N and PO-P have to be quantified. Fetter et al. (2012) studied the nitrate and phosphate leaching with a squid-based organic fertilizer and concluded that it does not appear to be more or less environmentally benign than synthetic fertilizers.

8.2.10 Ink

Cephalopods rely on ejection of dark ink for defence. The ink consists of a suspension of melanin granules and proteoglycans in a viscous and colourless medium. Cuttlefish ink is a natural substance of marine-product processing. Historically, *Sepia* ink has been drawn directly from the ink pouch of the *S. officinalis*. It is a dark and semitransparent colour, which can be used as ink for writing or as watercolour. *Sepia* ink use has been known since the seventeenth century, but it was not until the nineteenth century that it became really common, especially among photographers, artists and painters (López-Montes et al. 2009). Cuttlefish ink is widely used in traditional Chinese medicine, due to its antitumour, immunomodulator and haemostatic effects (Liu et al. 2008). These authors isolated carbohydrates from the cuttlefish (*S. maindroni*) ink, found a novel polysaccharide '*S. maindroni* ink polysaccharide' (SIP) and suggested its potential as an effective natural antimutagenic agent. In another study, a non-sulphated polysaccharide was isolated from the ink sac of squid *Ommastrephes bartrami* after removal of the melanin granules (Chen et al. 2008). Chen et al. (2008) explained that a sulphated glycosaminoglycan-like polysaccharide has been identified in squid ink and is reported to have antibacterial, antitumour and antiretroviral activities.

More recently, some authors have studied the biological activity of ink. Chen et al. (2010) report the preparation, characterization and potential biological activities of a chemically sulphated polysaccharide isolated from the ink of the squid *O. bartrami*. Similarly, a number of naturally sulphated polysaccharides were reported to exhibit diverse biological activities. These authors concluded that an ink polysaccharide is a potential candidate compound for the prevention of tumour metastasis. Also Ding et al. (2011) demonstrated the effect of *Sepia* ink oligopeptides on growth inhibition and could be a potentially useful adjunct in the treatment of cancer. Several active components, including a tyrosinase and an ACE inhibitor, have been identified in cephalopod ink (Chen et al. 2008; Ding et al. 2011). Zong et al. (2013) studied the SIP isolated by Liu et al. (2008) and demonstrated that the sulphated form of SIP, SIP-SII, significantly inhibits the metastase in a melanoma mouse model. The SIP-SII from cuttlefish ink may be used as an anti-metastatic drug.

8.3 Conclusion and Trend

In conclusion, cephalopod by-products represent a diverse array of biomolecules with numerous potential valorisations. Abundant studies have been made concerning the opportunity to develop new products by processing cephalopod by-products. Nevertheless, the major problems to industrialize these developments are:

- Logistical difficulties: Some resources are dispersed and the logistical cost can be excessive.
- Economic weakness with a competition with other raw materials such as vegetables, algae, etc.
- Supply of raw material.
- Freshness of by-products and regulatory requirements.

Cephalopod by-products must be better considered as raw materials than wastes, onboard as well as at the processing plants, with the aim of maintaining freshness of by-products and minimizing the rate of enzymatic degradation and microbial spoilage, which are higher in by-products such as viscera. For use in high-value applications, preservation and storage of by-products are essential.

There are a lot of high-value molecules to extract from cephalopod by-products. Some research is needed to optimize processing methods and limit the economic weakness. In addition, some by-products are used carefully because of their quantities of heavy metals, such as in viscera. Methods to eliminate them are still needed.

In addition, the use of cephalopod by-products needs a regular supply of raw material, and cephalopod populations are sensitive and can fluctuate, which can directly impact the marine fisheries market and cause economic problems for the utilization of cephalopod by-products.

References

Aewsiri T, Benjakul S, Visessanguan W, Wierenga PA, Gruppen H (2010) Antioxidative activity and emulsifying properties of cuttlefish skin gelatin-tannic acid complex as influenced by types of interaction. Innov Food Sci Emerg 11:712–720

Aewsiri T, Benjakul S, Visessanguan W, Wierenga PA, Gruppen H (2011) Surface activity and molecular characteristics of cuttlefish skin gelatin modified by oxidized linoleic acid. Int J Biol Macromol 48:650–660

Al-Sagheer FA, Al-Sughayer MA, Muslim S, Elsabee MZ (2009) Extraction and characterization of chitin and chitosan from marine sources in Arabian Gulf. Carbohydr Polym 77:410–419

Alemán A, Giménez B, Montero P, Gómez-Guillén MC (2011a) Antioxidant activity of several marine skin gelatins. LWT—Food Sci Technol 44:407–413

Alemán A, Pérez-Santín E, Bordenave-Juchereau S, Arnaudin I, Gómez-Guillén MC, Montero P (2011b) Squid gelatin hydrolysates with antihypertensive, anticancer and antioxidant activity. Food Res Intern 44:1044–1051

Balti R, Bougated A, Souissi N, El-Hadj Ali N, Zekri D, Barkia A (2008) Influence of the extent of enzymatic hydrolysis on the functional properties and angiotensin I-converting enzyme inhibitory activity of protein hydrolysates from cuttlefish (*Sepia officinalis*) by-products. In: Nasri M (ed) Recent research developments in food by-products technology and biotechnology. India

Balti R, Barkia A, Bougatef A, Ktari N, Nasri M (2009) A heat-stable trypsin from the hepatopancreas of the cuttlefish (*Sepia officinalis*): Purification and characterisation. Food Chem 113:146–154

Balti R, Bougatef A, El-Hadj Ali N, Zekri D, Barkia A (2010) Influence of degree of hydrolysis on functional properties and angiotensdin I-converting enzyme-inhibitory activity of protein hydrolysates from cuttlefish (*Sepia officinalis*) by-products. J Sci Food Agric 90:2006–2014

Balti R, Jridi M, Sila A, Souissi N, Nedjar-Arroume N, Guillochon D, Nasri M (2011) Extraction and functional porperties of gelatin from the skin of cuttlefish (*Sepia offinalis*) using smooth hound crude acid protease-aided process. Food Hydrocoll 25:943–950

Banik RM, Prakash M (2004) Laundry compatibility of the alkaline protease from *Bacillus cereus*. Microbiol Res 159:135–140

Barwin Vino A, Ramasamy P, Vairamani S, Shanmugan A (2011) Physicochemical characterization of biopolymers chitin and chitosan extracted from squid *Doryteuthis sibogar* adam, 1954, pen. Intern J Pharm Res Dev 2(12):181–190

Barwin Vino A, Ramasamy P, Vairamani S, Shanmugan A (2012) Extraction, characterization and in vitro antioxidatve potential of chitosan and sulfated chitosan from cuttlebone of *Sepia* aculeate Orbigny, 1848. Asian Pac J Tropical Biomed pp 334–341

Boroa J, Dekaa O, Thakurb A (2012) A review on solid oxide derived from waste shells as catalyst for biodiesel production. Renew Sustain Energ Rev 16:904–910

Boukortt FO, Girad A, Prost J, Ait-Yahia D, Bouchenak M, Belleville J (2004) Fish protein improves the total antioxidant status of streptozotocin-induced diabetes in spontaneously hypertensive rat. Med Sci Monit 10:397–404

Cardenas-Lopez JL, Haard NF (2009) Identification of a cysteine proteinase from Jumbo squid (*Dosidicus gigas*) hepatopancreas as cathepsin L. Food Chem 112:442–447

Chen S, Xu J, Xue C, Dong P, Sheng W, Yu G, Chai W (2008) Sequence determination of a non-sulfated glycosaminoglycan-like polysaccharide from melanin-free ink of the squid *Om-*

mastrephes bartrami by negative-ion electrospray tandem mass spectrometry and NMR spectroscopy. Glycoconj J 25:481–492

Chen S, Wang J, Xue C, Li H, Sun B, Xue Y, Chai W (2010) Sulfation of a squid ink polysaccharide and its inhibitory effect on tumor cell metastasis. Carbohydr Polym 81:560–566

Chuang WL, Sun Pan B, Tsai JS (2000) Inhibition of lipoxygenase and blood thinning effects of mackerel protein hydrolysates. J Food Biochem 24:333–343

Chow CK (2000) Fatty acids in foods and their health implications, 2nd edn, New York

Cortizo S, Berghoff C, Alessandrini J (2008) Characterization of chitin from *Illex argentinus* squid pen. Carbohydr Polym 74:10–15

Ding GF, Huang FF, Yang ZS, Yu D, Yang YF (2011) Anticancer activity of an oligopeptide isolated from hydrolysates of *Sepia* ink. Chinese J Nat Med 9(2):0151–0155

Dorman T, Bernard L, Glaze P, Hogan J, Skinnee R, Nelson D, Bowker L, Head D (1995) The effectiveness of *Garum armoricum* (stabilium) in reducing anxiety in college students. J Adv Med 8:193–200

FAO (2010) The state of world fisheries and aquaculture. Fisheries and Aquaculture Department. FAO

FAO (2012) The state of world fisheries and aquaculture. Fisheries and Aquaculture Department. FAO

Fetter J, Brown N, Görres J, Amador JA, Lee C (2009) Effectiveness of squid hydrolysates as a home lawn fertilizer and its environmental impacts. 2009 International Annual Meetings, Pittsburgh

Fetter J, Brown N, Görres J, Lee C, Amador JA (2012) Nitrate and phosphate leaching under turfgrass fertilized with a squid-based organic fertilizer. Water Air Soil Pollut 223:1531–1541

Figueiredo SA, Loureiro JM, Boaventura RA (2005) Natural waste materials containing chitin as adsorbents for textile dyestuffs: Batch and continuous studies. Water Res 39:4142–4152

Ivankovic H, Gallego Ferrer G, Tkalcec E, Orlic S, Ivankovic M (2009) Preparation of highly porous hydroxyapatite from cuttlefish bone. J Mater Sci Mater Med 20:1039–1046

Joseph SM, George MC, Nair JR, Senan VP, Pillai D, Sherief PM (2005) Effect of feeding cuttlefish liver oil on immune function, inflammatory response and platelet aggregation in rats. Curr Sci 88:507–510

Gildberg A, Simpson BK, Haard NF (2000) Use of enzymes from marine organisms. In: Haard NF, Simpson BK (eds) Seafood enzymes. New York

Giménez B, Gómez-Estaca J, Aleman A, Gómez-Guillén MC, Montero P (2009a) Improvement of the antioxidant properties of squid skin gelatin films by the addition of hydrolysates from squid gelatin. Food Hydrocoll 23:1322–1327

Giménez B, Aleman A, Montero P, Gómez-Guillén MC (2009b) Antioxidant and functional properties of gelatin hydrolysates obtained from skin of sole and squid. Food Chem 114:976–983

Gómez-Estaca J, López deLA, Gómez-Guillén MC, López-Caballero ME, Montero P (2009) Antimicrobial activity of composite edible films based on fish gelatin and chitosan incorporated with clove essential oil. J Aquatic Food Product Technol 18:46–52

Gómez-Guillén MC, Turnay J, Fernandez-Diaz MD, Ulmo N, Lizarbe MA, Montero P (2002) Structural and physical properties of gelatin extracted from different marine species: a comparative study. Food Hydro 16:25–34

Gómez-Guillén MC, Giménez B, López-Caballero ME, Montero MP (2011) Functional and bioactive properties of collagen and gelatin from alternative sources: a review. Food Hydro 25:1813–1827

Guérard F (2006) Enzymatic methods for marine by-products recovery. In: Shahidi F (ed) Maximising the value of marine by-products. CRC-Woodhead Publishing, Cambridge

Gupta R, Beg Q, Lorenz P (2002) Bacterial alkaline protease: molecular approaches and industrial applications. Appl Microbiol Biotechnol 59:15–32

Hameed KS, Haard NF (1985) Isolation and characterization of cathepsin C from atlantic short finned squid *Illex illecebrosus*. Comp Biochem Phys B 82(2):241–246

Hatate H, Tanaka R, Suzuki N, Hama Y (2000) Comparison of protease activity in liver among several species of squid and cuttlefish. Fish Sci 66(1):182–183

Hoque S, Benjakul S, Prodpran T (2011a) Effects of partial hydrolysis and plasticizer content on the properties of film from cuttlefish (*Sepia pharaonis*) skin gelatin. Food Hydrocoll 25:82–90

Hoque S, Benjakul S, Prodpran T, Songtipya P (2011b) Properties of blend film based on cuttlefish (*Sepia pharaonis*) skin gelatin and mungbean protein isolate. Int J Biol Macromol 49:663–673

Hwanga L, Liang J (2001) Fractionation of urea-pretreated squid visceral oil ethyl esters. JAOCS 78(5):473–476

Kawasaki KI, Matsuoka T, Satomi M, Ando M, Tukamasa Y, Kawasaki S (2008) Reduction of cadmium in fermented squid gut sauce using cadmium-absorbing bacteria isolated from food sources. J Food Agric Environ 6:45–49

Kim SK, Byun HG, Park PJ, Shahidi F (2001) Angiotensin I converting enzyme inhibitory peptides purified from bovine skin gelatin hydrolysates. J Agric Food Chem 49:2992–2997

Lavall R, Assis O, Campana-Filho S (2007) b-Chitin from the pens of *Loligo* sp.: extraction and characterization. Bioresour Technol 98:2465–2472

Le Bihan E (2006) Valorisation des coproduits issus de la pêche des céphalopodes: application à la seiche *Sepia officinalis*. Université de Caen, p 316

Le Bihan E, Koueta N (2010) Sea by-products as real Material: New ways of application. Transworld Research Network, T.C. 37/661 (2), Fort P.O., Trivandrum-695 023 Kerala, India

Le Geros RZ, Le Geros JP (1993) Dense hydroxyapatite. In: Hench L L, Wilson J (Eds) Introduction to bioceramics. Singapore

Le P (1996a) Experimental study: stress and memory. Eur Neuro Psychopharmacol 6:110-P10-2

Le P (1996b) Nutrient presentation of cognitive and memory performances. Eur Neuro Psychopharmacol 6:187-P19-4

Lee YZ, Simpson BK, Haard NF (1982) Supplementation of squid fermentation with proteolytic enzymes. Food Biochem 6(2):127–134

Liang JH, Hwang LS (2000) Fractionation of squid visceral oil ethyl esters by short-path distillation. JAOCS 77:773–777

Lin L, Li B (2006) Radical scavenging properties of protein hydrolysates from jumbo flying squid (*Dosidicus eschrichitii Steenstrup*) skin gelatin. J Sci Food Agric 86:2290–2295

Lin L, Li B (2012) Angiotensin-I-converting enzyme (ACE)-inhibitory and antihypertensive properties of squid skin gelatin hydrolysates. Food Chem 131:225–230

Liu C, Li X, Li Y, Feng Y, Zhou S, Wang F (2008) Structural characterization and antimutagenic activity of a novel polysaccharide isolated from *Sepiella maindroni* ink. Food Chem 110:807–813

López-Montes A, Blanc R, Espejo T, Navalón A, Vílchez JL (2009) Characterization of *sepia* ink in ancient graphic documents by capillary electrophoresis. Microchem J 93:121–126

Milazzo MC, Zheng Z, Kellett G, Hayneswortg K, Shetty K (1999) Stimulation of benzyladenine-induced *in vitro* shoot organogenesis and endogenous proline in melon (*Cucumis melo* L.) by fish protein hydrolysates in combinaison with proline analogues. J Agric Food Chem 47(4):1771–1775

Murray BA, FitzGerald RJ (2007) Angiotensin converting enzyme inhibitory peptides derived from food proteins: Biochemistry, bioactivity and production. Curr Pharma Des 13:773–791

Muzzarelli RAA, Muzzarelli C (2005) Chitosan chemistry: relevance to the biomedical sciences. Adv Polym Sci 186:151–209

Nagarajan M, Benjakul S, Prodpran T, Songtipya P (2012) Characteristics and functional properties of gelatin from splendid squid (*Loligo formosana*) skin as affected by extraction temperatures. Food Hydrocoll 29:389–397

Navarro J, Villanueva R (2000) Lipid and fatty acid composition of early stages of cephalopods: an approach to their lipid requirements. Aquac 183:161–177

Ngamcharussrivichai C, Nunthasanti P, Tanachai S, Bunyakiat K (2010) Biodiesel production through transesterification over natural calciums. Fuel Process Technol 91:1409–1415

Mizuta S, Tanaka T, Yokoyama Y, Yoshinaka R (2009) Hot-water solubility of mantle collagens in several cephalopod molluscs. Fish Sci 75:1337–1344

Park JH, Lee H, Lee GY, Kim HR, Kim YW, Lee YB (2011) Development of refined cuttlefish (*Toradores pacificus*) oil and its use as functional ingredients. Food Sci Biotechnol 20:389–394

Patil YT, Satam SB (2002) Chitin and chitosan, treasure from crustacean shell waste. Sea Food Export J 33:31–38

Peña-Cortés H, Ramirez I, Simpson R, Urtubia A, Almonacid S (2010) Enzymatic hydrolysate from squid processing waste as an organic fertilizer. J Biotechnol 150:219–220

Raksakulthai M, Lee YZ, Haard N (1986) Effect of enzyme supplents on the production of fish sauce from male capelin (*Mallotus villotus*). Can Inst Food Sci Technol J 19:28–33

Ramasamy P, Barwin VA, Saravanan R, Subhapradha N, Shanmugam V, Shanmugam A (2011) Screening of antimicrobial potential of polysaccharide from cuttlebone and methanolic extract from body tissue of *Sepia prashadi*. Winkworth, 1936. Asian Pac J Tropic Biomed 1(2):244–248

Reissbrodt R, Beer W, Muller R, Claus H (1995) Characterization of casein peptones by HPLC profiles and microbiological growth parameters. Acta Biotechnol 15:223–231

Revathi M, Saravanan R, Shanmugam A (2012) Production and characterization of chitinase from *Vibrio* species, a head waste of shrimp *Metapenaeus dobsonii* (Miers, 1878) and chitin of *Sepiella inermis* Orbigny, 1848. Adv Biosci Biotech 3:392–397

Shahidi F, Janak Kamil YVA (2001) Enzymes from fish and aquatic invertebrates and their application in the food industry. Trends Food Sci Technol 12:435–464

Shahidi F (2006) Maximising the value of marine by-products, 1st edn. CRC-Woodhead Publishing, Cambridge

Shen C, Xie J, Xu X (2007) The components of cuttlefish (*Sepiella maindroni* de Rochebruns) oil. Food Chem 102:210–214

Sikorski A, Kolodziejska M (1986) The composition and properties of squid meat. Food Chem 20:213–224

Sinanoglou VJ, Miniadis-Meimaroglou S (1999) Fatty acid of neutral and polar lipids of (edible) Mediterranean cephalopods. Food Res Int 31(6-7):467–473

Souissi N, Ellouz-Triki Y, Bougatef A, Blibech M, Nasri M (2008) Preparation and use of media for protease producing bacterial strains based on by-products from Cuttlefish (*Sepia officinalis*) and wastewaters from marine-products processing factories. Microbiol Res 163:473–480

Tavakoli O, Yoshida H (2006) Squid oil and fat production from squid wastes using subcritical water hydrolysis: free fatty acids and transesterification. Ind Eng Chem Res 45:5675–5680

Thongthai C, Gildberg A (2005) Asian fish sauce as a source of nutrition. In: Shi J, Ho C T, Shahidi F (eds) Asian functional foods. New York, p 261

Tolaimate A, Desbrières J, Rhazi M, Alagui A, Vincendon M, Vottero P (2000) On the influence of deacetylation process on the physicochemical characteristics of chitosan from squid chitin. Polymer 41(7):2463–2469

Torres-Arreola W, Pacheco-Aguilar R, Sotelo-Mundo RR, Rouzaud-Sández O, Ezquerra-Brauer JM (2008) Partial characterization of collagen from mantle fins, and arms of jumbo squid (*Dosidicus gias*). Cienc Tecnol Alim 6:101–116

Uriarte-Montoya MH, Arias-Moscoso JL, Plascencia-Jatomea M, Santacruz-Ortega H, Rouzaud-Sández O, Cardenas-Lopez JL, Marquez-Rios E, Ezquerra-Brauer JM (2010) Jumbo squid (*Dosidicus gigas*) mantle collagen: extraction, characterization, and potentiel application in the preparation of chitosan-collagen biofilms. Bioresour Technol 101:4212–4219

Vázquez JA, González MP, Murado MA (2004) Peptones from autohydrolysed fish viscera for nisin and pediocin production. J Biotechnol 112:299–311

Vázquez JA, Murado MA (2008) Enzymatic hydrolysates from food wastewater as a source of peptones for lactic acid bacteria productions. Enzyme Microb Technol 43:66–72

Wako Y, Ishikawa S, Muramoto K (1996) Angiotensin I-converting enzyme inhibitors in autolysates of squid liver and mantle muscle. Biosci Biotechnol Biochem 60(8):1353–1355

Wang SL, Wu PC, Liang TW (2009a) Utilization of squid pen for the efficient production of chitosanase and antioxidants through prolonged autoclave treatment. Carbohydr Res 344:979–984

Wang SL, Yang CW, Liang TW, Peng JH, Wang CL (2009b) Degradation of chitin and production of bioactive materials by bioconversion of squid pens. Carbohydr Polym 78:205–212

Xu W, Yu G, Xue C, Xue Y, Ren Y (2008) Biochemical changes associated with fast fermentation of squid processing by-products for low salt fish sauce. Food Chem 107:1597–1604

Yamashita Y, Zhanga N, Nozaki Y (2003) Effect of chitin hydrolysate on the denaturation of lizard fish myofibrillar protein and the state of water during frozen storage. Food Hydrocoll 17:569–576

Zong A, Zhao T, Zhang Y, Song X, Shi Y, Cao H, Liu C, Cheng Y, Qu X, Cao J, Wang F (2013) Anti-metastatic and anti-angiogenic activities of sulphated polysaccharide of *Sepiella maindroni* ink. Carbohydr Polym 91:403–409

Chapter 9
Farming Costs and Benefits, Marketing Details, Investment Risks: The Case of *Octopus vulgaris* in Spain

José García García, Manuel Luaces, Carlos Veiga and Manuel Rey-Méndez

Abstract The common octopus, *Octopus vulgaris,* is an important candidate for marine aquaculture, with optimum conditions for ongrowing such as high growth rate, easy adaptation to captivity and feeding conditions and high market value. The available technology for the culture of the whole life cycle is scarce and has only been achieved on a laboratory scale because there is no commercial diet available. For this reason, the production system is based on the capture of wild subadults, which are kept in different types of cages and fed with species of low commercial value. From the economic point of view, work to date concentrates on cost accounting of octopus ongrowing in floating cages in protected areas, open sea or land-based tanks. Global results show that the production cost in land-based tanks is generally higher than when fattening in sea cages, which may benefit the scale economy. The highest costs in decreasing order correspond to feeding, fixed assets, subadults purchasing and labour costs. Proper development of this activity requires solving two limiting issues: mass production of subadults at an industrial level and a suitable commercial diet.

Keywords Economics · Sea cages · Octopus ongrowing · Costs analysis · Profitability

J. G. García
Consejería Agricultura, Agua y Medio Ambiente de la Región de Murcia, IMIDA,
c/Mayor-Estación Sericícola, La Alberca, 30150, Murcia, Spain
e-mail: jose.garcia21@carm.es

M. Luaces
Arrecifes del Atlántico S.L., c/José Mª del Río 22,
15124, Muxía, A Coruña, Spain

C. Veiga
Samertolameu S. Coop. Galega, Urbanización O Rosal, Portal 9, Entresuelo,
Local 6, 36950, Moaña, Pontevedra, Spain

M. Rey-Méndez
Departamento de Bioquímica y Biología Molecular, Facultad de Biología, CIBUS,
Campus Vida, Universidad de Santiago de Compostela, 15706,
Santiago de Compostela, A Coruña, Spain
e-mail: manuel.rey.mendez@usc.es

J. Iglesias et al. (eds.), *Cephalopod Culture,* DOI 10.1007/978-94-017-8648-5_9,

9.1 Introduction

Industrial octopus ongrowing carried out by fishermen associations in Galicia began in 1996, reaching a production of 12 t in 1997. Interest fundamentally arose from the results achieved through studies about this species' culture by the research teams at the Vigo Coastal Centre of the Spanish Institute of Oceanography (Iglesias et al. 1997) and the Department of Biochemistry and Molecular Biology at the University of Santiago de Compostela (Rama-Villar et al. 1997). In recent years, several small companies have made attempts on an experimental scale, uncovering the big potential that this species ongrowing can have. However, available technology is scarce at present due to the impossibility to complete the artificial biological cycle at an industrial level yet (Iglesias et al. 2004). On the one hand, it is not currently possible to provide a mass production of subadults for ongrowing, and the supply through the fisheries is not always guaranteed; on the other hand, there is no commercial diet available either. These companies reached a maximum production of 49.4 t in 2000, though in 2011 decreased to 2.8 t (press release by the Government of Galicia in 2012), and in 2013 solely one company remains active. Simultaneously, diverse studies of some parameters have been carried out to optimize the production system and identify some factors of interest for developing operations (Rama-Villar et al. 1997; Luaces-Canosa and Rey-Méndez 1999; Tuñón et al. 2000, 2001; Rodríguez et al. 2003; Rey-Méndez et al. 2003).

There is an evident interest in searching new species for aquaculture production, and to increase the diversity of marine culture products it is convenient to select species with a widespread consumption and a high commercial value, as it is the case with the common octopus (Vaz-Pires et al. 2004; Chapela et al. 2006). The fattening studies performed to date have predicted that the common octopus is an important candidate for marine aquaculture, showing optimum ongrowing conditions, high growth rates, easy adaptation for harvesting and feeding up and an appreciable market value (Vaz-Pires et al. 2004; Miliou et al. 2005).

Octopus development at an industrial scale requires mass production of subadults to initiate ongrowing (Navarro and Villanueva 2000; Iglesias et al. 2004, 2007) and specific artificial diets to obtain satisfactory performances (Lee et al. 1991; Vaz-Pires et al. 2004; Cerezo Valverde et al. 2008). The majority of octopus-fattening activities have taken place in Galicia (northwest Spain), though new experiences were also encountered recently in the Mediterranean Sea. However, there are two particular conditioning factors in the Mediterranean area: (1) the fattening activity would be limited by the high temperatures during the summer months (Aguado Giménez and García García 2002) and (2) environmental issues, as well as coastal policy and coastal zone management requirements, which would make open sea (offshore) systems an optimal option.

Work carried out to date concentrates on cost accounting of octopus ongrowing, in floating cages in protected areas (estuaries), open sea or land-based tanks. It is particularly interesting to accomplish economic studies in the field of feasibility and cost analysis, addressing the evaluation of the ongrowing current system and the possible future alternatives. In this sense, the purpose of the econometric models is to estimate production linked to the economic parameters that are not yet defined to the level of a commercial culture at large scale; among them, the final price of the available product, the costs of fry octopus and their feeding and the investment influence on the feasibility or profitability of the commercial activity (García García and García García 2011).

9.2 Exploitation Process

Available technology is scarce at present, due to not having completed the artificial biological cycle at an industrial level. Thus, the production system is based on capturing subadults in the wild, stabling them in different types of cages and tanks and feeding them with distinct species of low commercial value, basically fish discarded from trawling fisheries.

This section can be complemented with information from Chap. 24 on *Octopus vulgaris* ongrowing.

9.2.1 Production Systems

The octopus grown has fundamentally been obtained from two alternative production systems, sea cages and land tanks. The sea-based systems are always developed in cages, although there are some differences among them. They are usually small sized, with little production capacity and built of metallic materials. Octopus shelters are installed inside cylindrical plastic containers compartmentalised individually. The cages are suspended from mussel rafts (*bateas;* Chapela et al. 2006) or floating cages anchored to the sea bottom (García García et al. 2004a; Oltra et al. 2005; Socorro et al. 2005) at different locations, both in the Mediterranean and the Atlantic. The two types of experimental cages tested are shown in Fig. 9.1. Estefanell et al. (2012) tested benthic cages, attaining survival and growth results similar to those in floating cages; García García et al. (2009) have worked with offshore cages. Land-based culture tanks in an open system have also been used (Alemany Sena 2011) by means of a closed system with water recirculation from wells, which permits a better control of the water temperature (García García et al. 2004b).

9.2.2 Subadults Supply

Subadult octopuses are captured through the artisanal fishery that is carried out in the area where the ongrowing facilities are located. The typical catching system uses traps, allowing the octopus to arrive alive and in good condition onboard. The animals designated for ongrowing are individually introduced into a mesh bag having a PVC tube inside (diameter = 15 cm, length = 20 cm), which is used by the

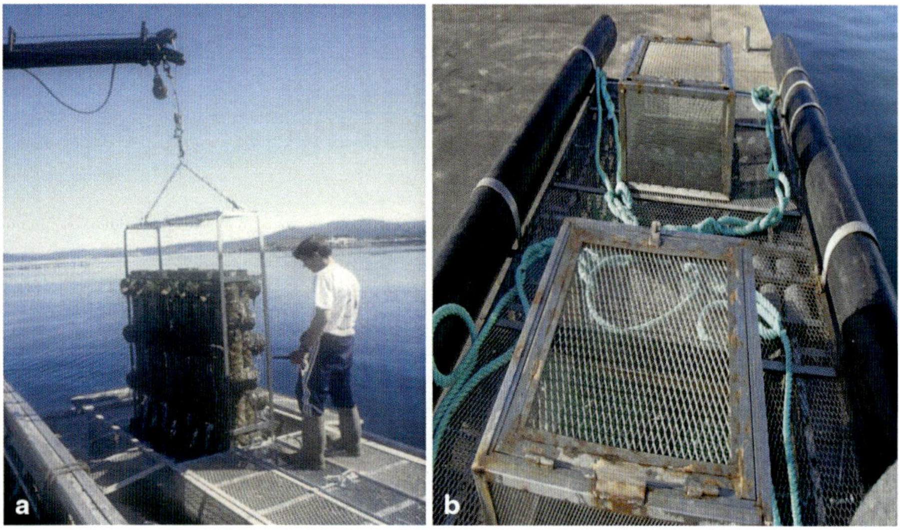

Fig. 9.1 Two types of experimental floating cages that were tested in Spain; **a** in the Atlantic Ocean and **b** in the Mediterranean Sea

octopus as shelter to prevent contact with other congeners and for protection until its entrance into the culture system. Individual bags are also placed inside a bigger mesh bag that is left hanging overboard while the fishing efforts finish. The transfer from the extraction area to the culture system is made with the octopus bag outside the water, for short routes, or submerged in a tank with an open circulation of seawater. The legal minimum size of octopus, as well as the extraction seasons, undergoe a high variability depending on the legal provisions of each removal zone. The octopus-ongrowing process in Galicia is only allowed if it has the legal size for extraction and commercial sale; the minimum size until May of 2007 was 0.75 kg, increasing to 1 kg henceforward, with variable close seasons which in the later years have passed from 1 to 2 months (usually in May and June). This rise of the legal minimum size and the close seasons, added to the fishermen's reluctance to promote the development of an activity that will possibly cause a fall in octopus prices by increased supply, are some of the main causes that make the development of this activity difficult.

9.2.3 Ongrowing Process

As already stated, the ongrowing stock is obtained from the capture of wild subadults. The initial culture density is around 10 kg m^{-3}, with the average of the initial size being about 750–1,000 g per individual. The fattening process takes place over a period of 3–4 months, reaching final weights of around 2,500–3,000 g, and loads of roughly 40 kg m^{-3}, while exhibiting a variable mortality rate. According to the water temperature range and variation, one or two fattening cycles may be completed in the Mediterranean and up to three in open sea in the Atlantic area.

Water temperature is the greatest conditioning factor in octopus growth. Trials in tanks and cages in the Mediterranean have shown that the best performance is obtained between 16 and 21 °C, with mortality notably increasing at temperatures above 22 °C (Aguado Giménez and García García 2002; García García et al. 2009). The open sea locations of the Northwest Atlantic in the Galician coast exhibit ideal conditions for *O. vulgaris,* having a growing temperature within the optimal range indicated. However, this process commonly occurs inside the Galician estuaries, where the annual temperature ranges from 12 to 19 °C, and therefore it is not within the optimum range all year round. In the Mediterranean, the temperature varies widely from 10 to 14 °C in winter and in the range of 25–27 °C in summer; therefore, the fattening activities of this species is restricted to about 7–8 months per year, specifically between October and June (García García et al. 2009).

Survival is one of the production parameters that varies most according to diverse factors, notably the load, size dispersion, type of food supplied, descents in salinity by rainfall or in locations next to river mouths. Under optimum conditions, the survival for a production process could be around 80 % (Rey-Méndez et al. 2003; Alemany 2011; García García et al. 2004b).

9.2.4 Feeding

Most of the ongrowing octopus experiments use similar foodstuff, fundamentally fishery discards, particularly crustaceans, fishes and bivalve molluscs. However, the best performance in terms of growth and food conversion index (CI) has been obtained with mixed diets of crab (*Carcinus mediterraneus*) and bogue (*Boops boops;* García García and Cerezo Valverde 2006). According to the initial and final sizes measured in the ongrowing process at the different stages tested, several survival values and indexes of real conversion were calculated; these results are to be taken into account because of the extra cost that they suppose (Oltra et al. 2005; García García et al. 2004a). In any case, and as already mentioned, to develop the industrial ongrowing of *O. vulgaris* and obtain a satisfactory performance of the whole process, a specific commercial diet is necessary that has not yet been developed (Lee 1994; Vaz-Pires et al. 2004; Cerezo Valverde et al. 2008).

9.3 Costs Analysis

9.3.1 Initial Investment

The initial investment in the octopus culture structures is very high when compared with intensive fish farming, both on land and in floating sea cages. This may be

Table 9.1 Investment and investment/production (K_0/P) according to species, culture system and production capacity

Species	Production system (t year^{-1}) (P)	Investment (€) (K_0)	K_0/P (€ t^{-1} year)	Reference
Octopus vulgaris	Stainless steel cages 45	611,773	13,595	García García et al. (2004a, 2004)
Octopus vulgaris	Land-based 100	1,018,826	10,188	García García et al. (2004b)
Solea senegalensis	Land-based 200	2,593,338	12,967	García García and García García (2006)
Diplodus puntazzo	Floating cages 1,000	1,752,876	1,753	García García and García García (2010)
Octopus vulgaris	Stainless steel cages 131	2,204,160	16,826	García García and García García (2011)

checked by the investment/production indicator, which consists in relativising the investment (K_0 in €) in relation to the exploitation production capacity (P expressed as t year^{-1}; Table 9.1). In a land-based culture with a closed circuit, greater densities may be reached and the K_0/P ratio is high, though lower than for octopus ongrown in sea cages. This fact appears paradoxical, although the explanation lies in the fact that the cages tested up to present have little capacity and the material is very expensive (stainless steel, instead of plastic in order to avoid thefts). Besides, the effect of the inflation in the increase of the investment value as time elapses should be taken into account. It is correct to consider that technological evolution could lead to greater capacity cages and to systems with a much lower relative investment. Moreover, the existence of scale economies, as in fish farming in open sea cages (García García and García García 2010), would indeed confirm a significant decrease in the K_0/P index when reaching greater capacities of exploitation production. As shown in Table 9.1, and because of the aforementioned effect, this index is much lower when growing sharpsnout seabream (*Diplodus puntazzo*), with an annual production of 1,000 t.

9.3.2 Evolution of Octopus Prices

Commercial octopus is classified in categories by weight or size from T7 to T0 in the Spanish market; furthermore, there is another economic variable known as the product price according to size, which undergoes outstanding differences throughout the year. Thus, it may be worth performing one, two or three fattening cycles during the year, following the price according to the size achieved in such cycles (García García and García García 2011). However, both for acquiring subadults and for commercialising the product, it is necessary to analyse the octopus price evolution according to the different commercial sizes, considering that the initial weight categories range from 500 to 1,000 g, whereas the final commercial product size ranges between 2,500 and 3,500 g. In consequence, it is interesting to analyse sizes

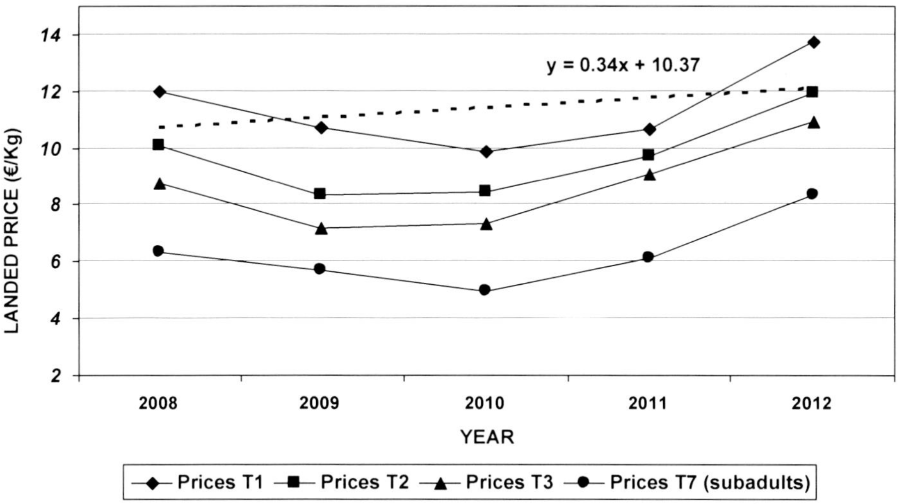

Fig. 9.2 Evolution of the octopus prices in destination during the period 2008–2012

T7 (500–800 g) for subadults and T3 (2,000–3,000 g) as well as T2 (3,000–4,000 g) for final products. Figure 9.2 shows the price evolution of these commercial categories on the target market. The trend line shows the price rising over the period 2008–2012. Data used are the octopus records classified by sizes at the central market of Barcelona (http://www.pesca2.com/informacion/precios_mercabarna.cfm) for this period. The price data of the producer at origin are affected by a reduction coefficient of approximately 15 % in commercialisation costs from origin to destination; as a result, it accounts for half of the price the producer receives and corresponds to 9.9 € kg^{-1} (T1), 8.4 € kg^{-1} (T2) and 7.5 € kg^{-1} (T3), respectively. Likewise, the market purchase price of the subadults amounts to 5.5 € kg^{-1} (T7).

9.3.3 Costs of the Ongrowing Process

An analysis of the octopus ongrowing costs in different experiences and locations has been made. The revised work analyses three distinct systems and presents the cost structures in chronological order, i.e. fattening in cages in protected areas at the Galician estuaries (García García et al. 2004a), in land-based tanks (García García et al. 2004b) and in open sea cages (García García and García García 2011).

The first cost analysis corresponds to the existing floating cages at the Galician estuaries (García García et al. 2004a). An exploitation having 50 floating cages of type C-160 (Fig. 9.1a) with a unitary capacity of 200 octopuses per cage was analysed. The cage section is rectangular, it is built with galvanised steel and it has PVC columns inside consisting of T-shaped attached shelters. The system also has the necessary infrastructure, i.e. a ship, 12 m long with a hydraulic crane of

Table 9.2 Accounting structure variables of the different production systems analysed

Item	Cages (protected areas)	Land-based tanks	Cages (offshore)
Production (kg year^{-1})	45,000	100,000	131,000
Production cost (€ kg^{-1})[a]	7.38	8.97	6.41
Fixed assets cost (%)	11.14	8.03	17.44
Feeding cost (%)	12.42	40.96	40.03
Subadult cost (%)	40.53	18.00	23.02
Labour cost (%)	10.32	18.08	15.60

[a] Production cost updated to 2011

1,100 kg, a truck crane and facilities on land, as well as auxiliary equipment. The ongrowing period consists of two fattening cycles a year, starting with subadults of 800 g, which achieve 2.5–3 kg (size T3) at the end of the process. Feeding is fundamentally based on trawling ship discards comprising of crabs (*Polybius henslowii*), black squat lobster (*Galathea squamifera*), horse mackerel (*Trachurus trachurus*), blue whiting (*Micromesistius poutassou*), etc. The average CI during the ongrowing process is 5.80 and 78 % of final survival is attained.

The total production cost in the year 2004 was of 5.80 € kg^{-1}. If we update it to 2011 considering an average inflation of 3.5 %, a mean updated cost of 7.38 € kg^{-1} would result. This work calculates the break-even point, which is the minimum operational size for costs to equal revenue. The minimum size is 43 cages, providing an annual production of 38.7 t with an average cost of 5.80 € and a maximum acquisition price of subadults and food of 4.58 and 0.155 € kg^{-1}, respectively. The food cost was very low, being sourced from fisheries companies in the area that were related directly to this operation (Table 9.2). The high cost of subadults was due to the fact that large-sized individuals (800 g) with high market price were used as input. The activity was economically feasible, with a low profit rate (profit/cost=4.11 %). Product price evolution rose in all the categories (Fig. 9.2), particularly, the average price of the size T3 in recent years reaches 7.50 € kg^{-1}, slightly higher than the value considered in this chapter (6.01 € kg^{-1}).

The two fattening systems applied in Galicia use metal cages with shelters, either floating or suspended from a raft similar to those used for the mussel culture. A typical outline of work (based on 2007 data from the Arrecifes del Atlántico S.L. company) starts with the subadult octopuses entering the system in two periods, July–August and October–November, leaving for sale in October–December and January–March of the following year, respectively. As a general rule, the first period is the most productive; it shows a greater input of subadults—double that in the second one—which gives a survival rate of 87 % and a final mean weight of 2.96 kg, after remaining in the system between 90 and 100 days. In the second period, the survival is about 69 % and the final mean weight is 2.45 kg, remaining between 100 and 120 days in the system. However, the analysis of data obtained every year (taken by the Arrecifes del Atlántico S.L. company in the period 1997–2010) showed dissimilar results, despite apparently carrying out a similar handling of both the system and the confined individuals.

Among the diverse aspects that should be improved and which are hindering the development of the octopus fattening in Galicia, the dependency on the environment must be especially emphasized, both in obtaining subadults and in their feeding. According to the companies consulted, obtaining subadults is the main problem to be solved due to the supply difficulties faced by the few companies dedicated to the ongrowing process, which also depend on the large annual catch fluctuations in the fishery. A further drawback to secure subadults is the reluctance of many fishermen to develop this activity due to fear of a lowering of prices through increased production. In fact, the sole company that currently remains in business in Galicia is an association of octopus fishermen (Samertolameu S. Coop. Galega), which has guaranteed the provision of subadults through its associates; these types of associations take this activity as an opportunity to obtain further benefits from the extractive fishing. Another aspect is the current fishing dependency on species to be used as food resources for fattening, mainly discards from the industrial trawling fisheries; the number of units dedicated to this type of fishery has decreased in the past years, which has made obtaining suitable food even more difficult and expensive.

To overcome these problems, it is necessary to close the full biological cycle in captivity for obtaining subadults from hatcheries; furthermore, it also becomes crucial to work on a fodder design and manufacture to avoid the dependence on fishing, aiming to improve the survival and growth of the individuals being cultured.

Otherwise, García García et al. (2004b) calculated a production cost of 7.05 € kg^{-1} in an intensive exploitation of land-based tanks, with an annual production of 100 t and commercial size average of 3.5 kg (size T2). When updated to 2011, it would result an average cost of 8.97 € kg^{-1}. The high feeding cost in Table 9.2 is due to fattening the subadults from an initial weight of 500 g to a final average weight of 3,500 g; hence, the feeding cost is noticeably higher than in the previous case. In the same way, it estimates an initial low weight and consistently the subadult cost is relatively lower. The average price of size T2 in recent years is 8.40 € kg^{-1}, much higher (7.60 € kg^{-1}) than the one considered in García García et al. (2004b). The production cost in land-based tanks is noticeably higher than the values obtained by García García and García García (2011) in open sea cages for the same size (T2), which amounted to 6.61 € kg^{-1}. This difference is mainly due to the energy required for pumping and the necessary maintenance of land-based facilities.

The most recent work on octopus fattening in cages at open sea (García García and García García 2011) provides an economic analysis of profitability in the Mediterranean using offshore systems. From an economic point of view, the suitability after completing two ongrowing cycles is analysed. The exploitation was set up to carry out the ongrowing in open sea cages, with a capacity of 30,000 individuals by cycle, which corresponds to 150 cages. They used the cage described in García García et al. (2009) in the Mediterranean. Table 9.2 shows the relevant results of the two ongrowing cycles compared with the two previous works. The items with highest costs, and therefore those with the greatest economic importance, are in order of magnitude the feeding, fixed assets, subadults and labour costs. When comparing the aforementioned costs to those of fish fattening in cages, the cost of subadults has a relatively minor importance in the latter, fundamentally due to the development of a scale economy by the increasing cage capacity in the past years (Merinero et al. 2005; García García and García García 2010).

The option of a unique fattening cycle is more profitable under the current conditions; the profitability threshold shows that this fattening activity would only be feasible when obtaining commercial prices of 6.61 and 6.41 € kg^{-1} for octopus cycles 1 and 2, respectively. Table 9.2 shows that those figures are quite close, although worthy to emphasize that octopus fattened in the one-cycle option achieves the commercial T2 size (3.65 kg) with a noticeably higher market price than in the case of two cycles, which reaches a lower T3 size (2.70 kg).

Global results (Table 9.2) indicate that land-based production cost is generally higher than fattening in sea cages. Moreover, a scale economy exists at sea systems due to greater production capacity, resulting in a lower average cost. Additionally, technological advances may significantly reduce the investment required in sea systems, by improving the cages yield using cheaper materials with greater production capacities, as in fish farming. On the contrary, the capacity to reduce the investment in land-based systems is much lower; further, the land-based location and choice of suitable sites to settle installations is more complex than at sea. Thus, two cycles on-growing with well water would only be possible in areas where the pumping height is low, as reported by García García et al. (2004b).

The wide range of production variables shows the convenience of using econometric tools to analyse possible scenarios that will be economically feasible. Hence, econometric equations can provide a guidance to define the maximum subadults and food prices; in this way, they could ensure that the activity either will be feasible or will have a specific profit rate, providing also information about the balance between the possible operating designs (regarding type and size of cage, anchorages, etc.) and the investment involved. As an example, a 15 % reduction in the variable related to investment, while maintaining the rest of the initial variables, results consequently in an increase of the internal rate of return (IRR) of 22.14 % (IRR = 12.69 %) and 21.03 % (IRR = 14.85 %) for two and one annual cycles, respectively. The results state that from an economic point of view, the profitability (IRR) shows elasticity in relation to the variable of investment.

As shown in the cost analysis, the current situation makes it more profitable to implement one production cycle than two, although the situation may change according to production costs and sale prices. If production costs decrease—particularly those of subadults and food—and the difference in sale prices too, then two fattening cycles a year will become more profitable. Undoubtedly, real development of the octopus-growing activity requires seeking solutions for the two limiting issues still affecting profitability, i.e. the availability of subadults and the existence of a commercial diet providing high efficiency in terms of growth, conversion rate and survival.

In the Mediterranean, this activity would last approximately 6–8 months a year, underutilizing the investment and making fixed asset costs high, as previously noted. A way to decrease production costs would be to integrate the octopus fattening as a complement to fish farming. In gilt-head bream and sea bass farms, operation activities drop off considerably during the winter months as low temperatures cause a major decrease in feeding, becoming perhaps the most important variable in the process. Tuna-fattening operations generally halt the activity precisely during the months when octopus ongrowing could be performed with best results. Thus, establishing integrated octopus- and fish-ongrowing operations in the Mediterranean could contribute to the overall decrease of the production costs, improving the profitability of both types of exploitations.

9.4 Conclusions

Good prospects presented by the octopus *O. vulgaris* have turned it into a target species to develop its cultivation in Spain; diverse business experiences were devoted to ongrowing wild subadults while the complete species cycle was achieved. Cost analysis of this activity in different experiences, locations and systems has permitted to evaluate the importance of the diverse aspects involved in specific real experiences, in order to determine its economic feasibility. Results indicate that production costs in land (8.97 € kg^{-1}) are superior to that of sea cages (6.61 € kg^{-1}); feeding, fixed assets, subadults purchasing and wages are reported as higher costs in descending order. Technological progress related to the confinement systems and the production capacity increase can significantly reduce major costs. The surveyed companies point out that the development of the fattening activity is slowed down by the current dependency on the natural environment (extraction of subadults and food). This dependency is the cause behind the non-full capacity exploitation of some facilities and the activities withdrawal of others. It is thus necessary to solve two crucial limiting factors in order to have a viable development of octopus culture, i.e. securing massive production of subadults in hatchery and provision of a commercial feed rendering high performances in terms of growth, CI and survival.

References

Aguado Giménez F, García García B (2002) Growth and food intake models in *Octopus vulgaris* Cuvier 1797: influence of body weight, temperature, sex and diet. Aquac Int 10:361–377

Alemany Sena F (2011) Aproximación al engorde industrial del pulpo *Octopus vulgaris*. Primeros resultados en una instalación de tanques con circuito abierto en el Delta del Ebro. In: Actas del XIII Congreso Nacional de Acuicultura. Castelldefels, Barcelona, Noviembre de 2011

Cerezo Valverde J, Hernández MD, Aguado-Giménez F, García García B (2008) Growth, feed efficiency and conditions of common octopus (*Octopus vulgaris*) fed on two formulated moist diets. Aquaculture 275:266–273

Chapela A, González AF, Dawe EG, Rocha F, Guerra A (2006) Growth of common octopus (*Octopus vulgaris*) in cages suspended from rafts. Sci Mar 70:121–129

Estefanell J, Roo J, Guirao R, Izquierdo M, Socorro J (2012) Benthic cages versus floating cages in *Octopus vulgaris*: biological performance and biochemical composition feeding on *Boops boops* discarded from fish farms. Aquac Eng 49:46–52

García García B, Cerezo Valverde J (2006) Optimal proportions of crabs and fish in diet for common octopus (*Octopus vulgaris*) ongrowing. Aquaculture 253:502–511

García García J, García García B (2006) An econometric viability model for ongrowing sole (*Solea senegalensis*) in tanks using pumped well sea water. Span J Agric Res 4(4):304–315

García García J, García García B (2010) Econometric model of viability/profitability of ongrowing sharp snout sea bream (*Diplodus puntazzo*) in sea cages. Aquac Int 18(5):955–971

García García J, García García B (2011) Econometric model of viability/profitability of octopus (*Octopus vulgaris*) ongrowing in sea cages. Aquac Int 19:1177–1191

García García B, Cerezo Valverde J, Aguado-Giménez F, García García J, Hernández MD (2009) Growth and mortality of common octopus *Octopus vulgaris* reared at different stocking densities in Mediterranean offshore cages. Aquac Res 40:1202–1212

García García J, Rodríguez González LM, García García B (2004a) Cost analysis of octopus ongrowing installation in Galicia, Spain. J Agric Res 2(4):531–537

García García J, Cerezo J, Aguado F, García García B (2004b) A feasibility/profitability model for common Octopus (*Octopus vulgaris*) ongrowing under intensive conditions on land-based facilities. In: Actas Aquaculture Europe, Barcelona, European Aquaculture Society, 20–23 October 2004, pp 360–361

Iglesias J, Sánchez FJ, Otero JJ (1997) Primeras experiencias sobre el cultivo integral del pulpo (*Octopus vulgaris*) en el Instituto Español de Oceanografía. In: Costa J, Abellán E, García García B, Ortega A, Zamora S (eds) Actas del VI Congreso Nacional de Acuicultura, Cartagena 9–11 de Julio de 1997, MAPA, pp 221–226

Iglesias J, Otero JJ, Moxica C, Fuentes L, Sánchez FJ (2004) The completed life cycle of the octopus (*Octopus vulgaris*, Cuvier) under culture conditions: paralarval rearing using Artemia and zoeae, and first data on juvenile growth up to 8 months of age. Aquac Int 12:481–487

Iglesias J, Sánchez FJ, Bersano JGF, Carrasco JF, Dhont J, Fuentes L, Linares F, Muñoz JL, Okumura S, Roo FJ, van der Meeren T, Vidal EAG, Villanueva R (2007) Rearing of *Octopus vulgaris* paralarvae: presents status, bottlenecks and trends. Aquaculture 266:1–15

Lee PG (1994) Nutrition of cephalopods: Fueling the system. In: Pörtner HO, O'Dor RK, Macmillan DL (eds) Physiology of Cephalopod Molluscs: lifestyle and performance adaptations. Gordon & Brench, Switzerland, pp 35–51

Lee PG, Forsythe JW, DiMarco FP, DeRusha RH, Hanlon RT (1991) Initial palatability and growth trials on pelleted diets for cephalopods. Bull Mar Sci 49:362–372

Luaces Canosa M, Rey MM (1999) El engorde industrial de pulpo (*Octopus vulgaris*) en jaulas: análisis de dos años de cultivo en la Ría de Camariñas (Galicia). In: Actas del VII Congreso Nacional de Acuicultura, Las Palmas de Gran Canaria, mayo de 1999, p 46

Merinero S, Martínez S, Tomás A, Jover M (2005) Análisis económico de alternativas de producción de dorada en jaulas marinas en el litoral mediterráneo español. AquaTIC 23:1–19. http://www.revistaaquatic.com/aquatic/art.asp?t=p&c=188

Miliou J, Fintikaki M, Kountouris T, Verriopoulos G (2005) Combined effects of temperature and body weight on growth and protein utilization of the common octopus, *Octopus vulgaris*. Aquaculture 249:245–256

Navarro JC, Villanueva R (2000) Lipid and fatty acid composition of early stages of cephalopods: an approach to their lipid requirements. Aquaculture 183:161–177

Oltra R, Alemany F, Roig M, Mezquita F (2005) Engorde de pulpo (*Octopus vulgaris* Cuvier, 1797) en jaula flotante en la costa mediterránea de Levante. Bol Inst Esp Oceanogr 21(1–4):187–194

Rama-Villar A, Faya-Angueira V, Moxica C, Rey-Méndez M (1997) Engorde de pulpo (*Octopus vulgaris*) en batea. In: De Costa J, Abellán E, García García B, Ortega A, Zamora S (eds) Actas del VI Congreso Nacional de Acuicultura. Ministerio de Agricultura, Pesca y Alimentación, Madrid, pp 245–250

Rey-Méndez M, Tuñon E, Luaces-Canosa M (2003) Estudio de los efectos del peso inicial y el sexo sobre el comportamiento, la mortalidad y el crecimiento del pulpo (*Octopus vulgaris*, Cuvier 1797) en cultivo industrial. IX Congreso Nacional de Acuicultura, Cádiz del 11 al 16 de mayo de 2003, pp 276–277

Rodríguez C, Carrasco JE, Rodríguez M (2003) Engorde de juveniles de pulpo (*Octopus vulgaris*, Cuvier 1797) en jaulas. IX Congreso Nacional de Acuicultura, Cádiz del 11 al 16 de mayo de 2003, pp 268–269

Socorro J, Roo J, Fernández-López A, Guirao R, Reyes T, Fernández-Palacios H, Izquierdo MS (2005) Engorde de pulpo (*Octopus vulgaris* Cuvier, 1797) en jaulas flotantes alimentado exclusivamente con boga *Boops boops* (L., 1758) de descarte de la acuicultura. Bol Inst Esp Oceanogr 21(1–4):207–212

Tuñón E, Parada A, Caeiro C, Rey Méndez M (2000) Efecto del tamaño inicial en el engorde industrial del pulpo (*Octopus vulgaris*, Cuvier 1797) en jaulas suspendidas en bateas. In: Rey-Méndez M, Fernández Casal J, Izquierdo Rodríguez M (eds) III Foro dos Recursos Mariños e da Acuicultura das Rías Galegas, pp 181–199

Tuñón E, Parada A, Caeiro C, Rey-Méndez M (2001) Estudio comparativo basado en la dieta diferenciada para el engorde de pulpo (*Octopus vulgaris* Cuvier, 1797) en una explotación industrial. In: Rey-Méndez M, Fernández Casal J, Izquierdo Rodríguez M (eds) IV Foro dos Recursos Mariños e da Acuicultura das Rías Galegas, pp 255–269

Vaz-Pires P, Seixas P, Barbosa A (2004) Aquaculture potential of the common octopus (*Octopus vulgaris* Cuvier, 1797): a review. Aquaculture 238:221–238

Part II
Main Cultured Cephalopods

Chapter 10
Nautilus

Gregory J. Barord and Jennifer A. Basil

Abstract Nautiluses are remnants of an ancient lineage that dates back nearly 500 million years. Extant nautiluses still exhibit many traits characteristic of the ancestral species. Nautilus culture systems should therefore take into account both the similarities between nautiloids and modern coleoids and the differences. Nautilus culture systems should be designed to maintain excellent water quality through effective filtration to promote good health. Nautiluses and coleoids differ primarily in their reproductive strategies. Whereas most coleoids are fast growing and semelparous, nautiluses grow slowly, mature later, and are iteroparous. Therefore, nautiluses may necessitate several years of care before becoming sexually mature. Successful reproduction and egg laying by a female yield only a maximum of ten eggs which take up to 1 year to develop and hatch. Currently, nautilus hatchlings have only been reared up to 1 year. The future of nautilus culture systems depends upon a better understanding of both wild and captive reproduction. The success of these culture systems would open up a brand new area of research utilizing different age groups and generations to investigate current and novel questions.

Keywords *Nautilus* culture · *Nautilus* husbandry · *Nautilus* reproduction · *Nautilus* disease · Cephalopod culture

10.1 Background

The nautilid lineage may have existed for more than 500 million years (Ward 1987; Hanlon and Messenger 1996; Strugnell and Lindgren 2007) though the most recent estimation is 416–480 million years (Kröger et al. 2011). Today, nautiluses are grouped under two different genera, *Nautilus* and *Allonautilus* (Ward and Saunders 1997). Throughout the chapter, "nautiluses" will refer to all species of nautiluses

G. J. Barord (✉) · J. A. Basil
Biology Department, Brooklyn College, City University of NY Graduate Center, 2900 Bedford Ave, 11210, Brooklyn, NY, USA
e-mail: gbarord@gc.cuny.edu

J. A. Basil
e-mail: jbasil@brooklyn.cuny.edu

J. Iglesias et al. (eds.), *Cephalopod Culture,* DOI 10.1007/978-94-017-8648-5_10,

Fig. 10.1 External anatomy of *Nautilus pompilius*. (Photo by G. J. Barord)

and specific species will be referred by species name. Six species are commonly referred to in the literature: *Nautilus pompilius, N. stenomphalus, N. belauensis, N. repertus, N. macromphalus,* and *Allonautilus scrobiculatus* (Saunders 2010). However, recent phylogenetic analysis suggests there are only two distinct species: *N. macromphalus* and *A. scrobiculatus* (Bonacum et al. 2011). Genetic analysis of local populations (Sinclair et al. 2007) and DNA bar coding efforts of local and distant populations (Williams et al. 2012) will provide information critical to future management of nautiluses. All nautiluses are found only in the Indo-Pacific region along coral reef slopes, ranging in depths from 0 to 700 m (Ward and Martin 1980; Dunstan et al. 2011a). Their habitat is regulated by both depth and temperature (Saunders and Ward 1987). Shells of nautilus will implode at depths greater than 800 m (Ward et al. 1980; Kanie et al. 1980; Hewitt 2010) which limits their lower depth range. Nautiluses are also sensitive to temperature and do not survive in seawater of 28 °C for longer than 48 h. As such, nautiluses do not inhabit shallow water where temperatures can approach 28 °C.

The general anatomy of nautiluses has been described in detail (Owen 1843; Willey 1902; Budelmann et al. 1997; Sasaki et al. 2010). Nautiluses are the only extant cephalopods with an external shell, as shown in Fig. 10.1. The nautilus shell is striped along the dorsal and lateral sides of the shell which fades away to plain ventrally and towards the aperture of the shell. The shell is composed of several internal chambers which aid the nautilus in buoyancy. A fleshy tube called the siphuncle runs through each of the chambers and regulates the amount of gas and fluid in each of them to maintain neutral buoyancy at any depth. The living animal resides in the largest anterior chamber and the animal grows by adding new chambers. Nautiluses possess up to 90 tentacles that can be extended and retracted into buccal sheaths. Unlike coleoid cephalopods, nautiluses possess a sticky substance on the tentacles for adhesion (von Byern et al. 2012). The eye of nautiluses is less complex than the coleoid eye because it lacks a lens and is referred to as a pinhole camera-type eye. However, it is a large eye, capable of gathering light in dim environments.

10.2 Reproductive Biology

Nautiluses do not fully mature until at least 12–15 years (Saunders 1983; Dunstan et al. 2011b). They have long embryonic periods of 10–12 months (Okubo et al. 1995; Uchiyama and Tanabe 1996) and live for at least 20 years (Dunstan et al. 2011b; Landman and Cochran 2010). Nautiluses reproduce sexually and must locate and identify a mate in deep, dark ocean waters. It is not fully understood how nautiluses locate suitable partners in the deep sea. However, laboratory studies have shown that males are attracted to the scent of other nautiluses, both male and female, and that females are attracted to males, while being repelled by female scent (Basil et al. 2000; Westermann and Beuerlein 2005). Coupling this behavior with their large olfactory organs, it is conceivable that nautiluses detect mates through the ocean currents. It has been demonstrated that they have an acute sense of olfaction and can track patchy, turbulent, and dilute odors from great distances (Basil et al. 2000). When male and female nautiluses are successful in contact, copulation may occur for several hours with the male grasping the female tentacles to tentacles. Actual spermatophore transfer is not well understood, but the male sexual organ, called the spadix, is in contact with the female buccal area during copulation (Mikami and Okutani 1977). After copulation, the female will lay only a few eggs with its labial tentacles (Mikami and Okutani 1977; Arnold et al. 1993). Female nautiluses will only lay 10–20 eggs per year (Okubo et al. 1995; Uchiyama and Tanabe 1996). Based upon captive observations, eggs are laid on a rocky substrate and attached to the substrate with a cement-like substance. There is no information on the exact location of where females lay the eggs, though isotopic analysis suggests the eggs are laid in warmer, shallower water (Arnold et al. 1990; Oba et al. 1992; Landman et al. 1994; Okubo et al. 1995). The eggs take approximately 11 months to develop and hatch (Arnold et al. 1990) at which point the hatchlings have already formed seven to eight internal chambers (Arnold et al. 1987; Oba et al. 1992; Okubo et al. 1995) and migrate to cooler, deeper water (Landman et al. 1994; Landman et al. 2001).

10.3 Embryology

The embryos of nautiluses may provide key insights into cephalopod and molluscan evolution. Nautiluses represent the primitive externally shelled body type of cephalopods and their eggs are ideal for developmental studies because of their large size. The first observations of nautilus embryos occurred in 1985. Arnold and Carlson (1986) confirmed that nautiluses, as in other extant cephalopods, do not have larval stages and develop directly along with many other similarities with extant coleoids. Arnold et al. (1990) described the movement of the embryo within the egg during organogenesis and the formation of the first shell. The movements were attributed to four possible causes: protection, respiration, rearrangement of yolk, and extraembryonic yolk sac respiration. Carlson et al. (1992) reported on the hatching of *N. belauensis* and the distribution of nautiluses. Shigeno et al. (2008) examined *N. pompilius* embryos and the development of the soft parts, and found their developmental trajectory supported the hypothesis that they represent the ancestral cephalopod condition, particularly in the head region. Most notably, Shigeno et al. (2008) support the "arms as foot" hypothesis in nautilus based upon their embryological research.

10.4 Husbandry

The first living nautilus successfully housed in an aquarium occurred in 1958 at the Noumea Aquarium in New Caledonia (Catala 1964). Since then, many more aquariums, zoos, and research institutions have housed nautiluses. Husbandry guidelines were developed over 25 years ago and still serve as the accepted practices (Carlson 1987; Hamada et al. 1987; Spinosa 1987). Though these guidelines still appear to be effective, the increased attention to cephalopod welfare (Moltschaniwskyj et al. 2008) warrants directed attention to follow these standards and do everything possible to improve husbandry.

10.4.1 Source and Capture

Carlson (1987) described the source and capture of live nautiluses though there are several different methods in use. The general collection and acclimatization of nautiluses to captive aquariums are described below. Most of the live nautiluses being imported are collected in the Philippines. Nautiluses are obtained from the wild using baited deep water traps. Though trap design varies, the traps are baited with some type of dead meat, commonly chicken, and set from 200 to 400 m overnight. The traps are retrieved the next morning and nautiluses are recovered. Nautiluses must be transported in chilled oxygenated sea water. Nautiluses received from the wild should be acclimatized to the primary holding system to equalize temperature and pH. This can be done by floating the shipment bags in the holding-tank waters and replacing the shipment water with holding-tank water every 10 min or so. Or, a direct slow drip from the tanks into the holding bags, with the aim of acclimating within about an hour, can be used. After acclimatization has been achieved, the nautiluses should be placed into the holding system with as little contact out of water as possible. Once placed in the primary holding system, each nautilus should be "burped" to remove any possible air bubbles that may have resulted during transit. This can be done by slowly rolling the nautilus in your hand, underwater, with the funnel facing up, for several minutes. Air bubbles trapped within the mantle cavity and eye can cause serious complications down the line. Similar to zebras, nautiluses can be individually identified based upon their striped pattern though permanent markers can be used to label individuals on the shell.

10.4.2 Water

All cephalopods, including nautiluses, require excellent water quality. The epidermis of cephalopods is made up of microvilli that facilitate the uptake of contaminants in their surrounding seawater environment. Nautiluses accumulate trace elements in the wild (Bustamante et al. 2000; Pernice et al. 2009) and are thus susceptible to toxic elements in captive conditions. Water quality can be maintained through the use of strong mechanical, biological, and chemical filtration. The principal parameters to monitor are levels of nitrogenous waste which should be kept between accepted ranges of $<0.10\ \mathrm{mg\ L^{-1}}$ ammonia, $<0.10\ \mathrm{mg\ L^{-1}}$ nitrite, and $<20.00\ \mathrm{mg\ L^{-1}}$ nitrate (Spotte 1979). However, these parameters can and should be maintained near $0.00\ \mathrm{mg\ L^{-1}}$ using adequate filtration methods and consistent water changes. In closed systems, UV scrubbers should be used, as they remove many key bacteria that can cause serious health problems in nautiluses. Nautiluses are sensitive to warm temperatures, so a chilling unit must be installed on the system to maintain temperatures between 15 and 27 °C. Specimens housed in 28 °C for 4 h have shown poor survival (Carlson 1987). In some cases, a rhythmic day/night temperature cycle has been set on aquariums so that the temperature changes throughout the day, from 18 to 21 °C (Carlson 1987), to mimic normal migration patterns in the wild. The use of supplements, such as calcium carbonate and magnesium, has been applied in some systems but there do not appear to be any data to support positive effects in nautiluses. The pH should be maintained at 8.20 through the use of a buffering system or effective water changes. Nautiluses are stenohaline, requiring a small salinity range which should be maintained between 34 and 36 psu (Boyle 1991). When moving nautiluses (for experiments or to another tank, etc.), it is important to keep them submerged in water during transit. Nautiluses are susceptible to mantle infections with repeated exposure to air.

10.4.3 Housing

There is little information to suggest the proper size, shape, and stocking densities of captive nautiluses. However, based upon the vertical migrations of wild nautiluses, nautilus-holding systems should be higher rather than wider to support the vertical movements, as shown in Fig. 10.2. Adding texture to the wall of the tank will make it more attractive to the animal. The average life span of nautiluses in captivity is approximately 3 years though there are occasions of nautiluses living 5–6 years in captivity (Carlson 1987).

10.4.4 Feeding

The gut content from wild nautiluses contains many different prey items, including fish and crustaceans, and crustacean molt shells (Ward and Wickstein 1980).

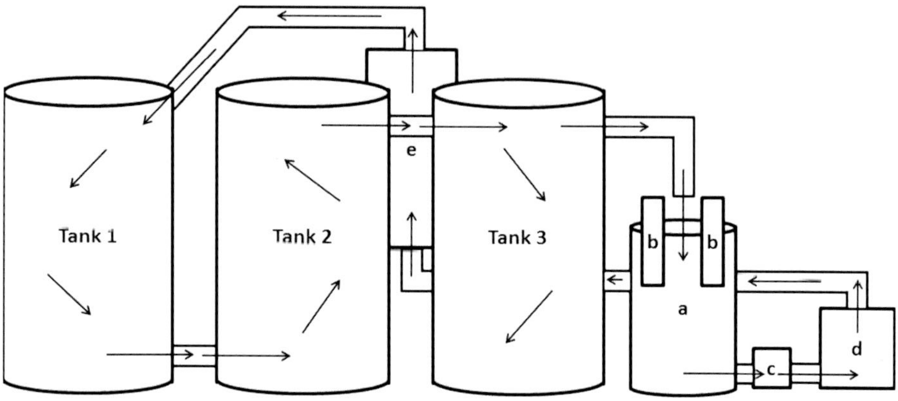

Fig. 10.2 Diagram of nautilus culture system. *Arrows* denote water flow. (Figure by G. J. Barord)

Common prey items for trapping live nautiluses include chicken and tuna though several other different kinds of *meats* have been used. The primary diet of captive nautiluses includes frozen shrimp with the shell on (Carlson 1987), lobster molts, and sometimes fish heads. The addition of live prey items to nautilus systems has shown mixed results. Deep-sea crab species have been consumed by nautiluses in captivity (Carlson 1987) but hermit crabs have not been preyed upon (Barord 2007). The feeding protocols also vary from place to place (Carlson 1987). It is unknown how growth is affected by different feeding protocols. However, a diet with increased calcium resulted in increased growth rates in *N. pompilius* (Barord 2007) though whether this is a positive or negative outcome is unknown. Growth data are available on captive nautiluses and are summarized in Westermann et al. (2004).

10.5 Captive Reproduction

The first nautilus embryos were observed in 1985 at the Waikiki Aquarium (USA); however, the eggs did not hatch (Arnold and Carlson 1986). The first successful hatching occurred in 1988 at the Kagoshima Aquarium (Japan) when three *N. belauensis* hatched after 12 months of incubation at 25 °C (Okubo 1989). In 1990, ten *N. belauensis* eggs hatched at the Waikiki Aquarium (Arnold et al. 1990). The Henry Doorly Zoo (USA) was successful in hatching *N. pompilius* eggs in a closed artificial seawater system (Fields 2006). There have been several other instances of *Nautilus* sp. hatching in captivity but none of the hatchlings have survived past 1 year. From these varied results, it is difficult to ascertain the most successful method of inducing captive reproduction, successful copulation, and egg fertilization, and finally getting the eggs to hatch. One key variable that is difficult to control is the sex ratio of nautiluses in an aquarium. There is no way of predicting the number of males and females that would be received in a shipment. The limiting factor in successful reproduction may in fact be the sex ratio of nautiluses housed in one system. It is unlikely that captive rearing of nautiluses is a viable process, so close monitoring of wild populations and fishing limits are critical to their survival.

10.6 Disease and Anesthesia

Successful maintenance of captive cephalopods depends primarily on excellent water quality but also on constant monitoring (Oestmann et al. 1997). While physical signs of health deterioration should be monitored, behavioral abnormalities may be more telling of health problems. Nautilus disease is not very well understood for a myriad of reasons. Most often, it is difficult to detect the problem before it is terminal and the specimen has expired. However, there are several common reports of health abnormalities. Perhaps the most common are the problems regulating buoyancy. Positive buoyancy occurs in many specimens received from the wild; however, the best treatment regimen appears to be the "sit and wait" approach. Most specimens with positive buoyancy resolve the problem on their own. Animals with positive buoyancy do not seem to suffer any other ill effects, at least in captive populations. Cases of negative buoyancy are less common and even less understood. One case of negative buoyancy in the literature correlated the condition with a severe bacterial infection and a parasitic infestation; however, the treatments were not successful and the specimen expired (Barord and Henderson 2008). Mucodegeneration is perhaps the most common infection in captive nautiluses that appears to be the result of an internal bacterial infection that manifests as significant mucus production around the eyes and tentacles. All reports of this disease had shown it to be fatal until a recent treatment regime of oxytetracycline and iodine on a single *N. pompilius* (Barord et al. 2012). Fungal infestations have been reported, and in one case, the fungus was identified as a *Fusarium* sp. (Scimeca 2006) though no treatment was employed. Anesthetic use in nautiluses is necessary for certain veterinary procedures, such as hemolymph withdrawal, and may also be necessary for specific research methods. Urethane was commonly used as a nautilus anesthetic (Ward et al. 1977) though its inherent carcinogenic effects render this method not viable. Recent trials with ethanol have proven it to be effective in inducing anesthesia for both hemolymph withdrawals and radiographic procedures (Barord and Henderson 2008).

10.7 Conclusions

The large-scale, continuous process of nautilus culture systems does not appear to be feasible at this point. The proxies for captive mating, reproduction, egg laying, egg development, and hatchling survival are not well understood. Therefore, it is difficult to understand what methods should be modified for future success. However, given the declining populations in several areas of the Indo-Pacific (del Norte-Campos 2005; Dunstan et al. 2010; Dunstan et al. 2011c; De Angelis 2012) and that most of the population sizes are still unknown, captive culture systems would nonetheless benefit nautiluses. Perhaps the best line of research is to investigate the initial mechanism of captive mating to begin promoting that aspect of reproduction in captivity. Consistent mating in captivity would enable researchers to identify the mechanisms behind the proceeding steps of reproduction, thereby increasing the likelihood of successful culture systems.

Acknowledgments The authors would like to thank the LIBE, BCR, and AREAC members for their husbandry support throughout the years. The authors would also like to thank John Apolinski, Jean Boal, Bruce Carlson, Paul DiMarco, Frank Grasso, Philip Lee, Heike Neumeister, Janice Simmons, Roxanna Smolowitz, and Peter Ward for all of their advice throughout the years. Thanks also go to Karen Keister and the anonymous reviewers for their insightful comments on the manuscript.

References

Arnold JM, Carlson BA (1986) Living nautilus embryos: preliminary observation. Science 232:73–76

Arnold JM, Landman NH, Mutvei H (1987) Development of the embryonic shell of nautilus. In: Saunders WB, Landman NH (eds) Nautilus: the biologyandpaleobiology of a living fossil, Plenum Press, New York, pp 373–400

Arnold JM, Awai M, Carlson BA (1990) Hatching of *Nautilus* embryos in the Waikiki aquarium. J Ceph Biol 1(2):117–118

Arnold JM, Awai M, Carlson BA (1993) *Nautilus* embryology: egg capsule formation, deposition, and hatching. J Ceph Biol 2:51–56

Barord GJ (2007) An investigation into shell deterioration of the captive chambered Nautilus, *Nautilus pompilius*. Proceedings of the American Association of Zoo Keepers, USA. Vol. 34

Barord GJ, Henderson R (2008) Treatment of negative buoyancy in a captive chambered nautilus, *Nautilus pompilius*. Drum Croaker 39:68–80

Barord GJ, Ju C, Basil JA (2012) First report of a successful treatment of a mucodegenerative disease in the chambered nautilus (*Nautilus pompilius*). J Zoo Wildl Med 43(3):636–639

Basil JA, Hanlon RT, Sheikh SA, Atema J (2000) Three-dimensional odor tracking by *Nautilus pompilius*. J Exp Biol 203:1409–1414

Bonacum J, Landman NH, Mapes RH, White MM, White AJ, Irlam J (2011) Evolutionary radiation of present-day *Nautilus* and *Allonautilus*. Am Malacol Bull 29:77–93

Boyle PR (1991) The UFAW handbook on the care and management of cephalopods in the laboratory. Universities Federation for Animal Welfare London, pp 1–63

Budelmann BU, Schipp R, von Boletzky S (1997) Cephalopoda. In: Harrison FW, Kohn AJ (eds) Microscopic anatomy of invertebrates. Vol 6A. Mollusca II, Wiley-Liss, New York, pp 119–414

Bustamante P, Grigioni S, Boucher-Rodoni R, Caurant F, Miramand P (2000) Bioaccumulation of 12 trace elements in the tissues of the nautilus *Nautilus macromphalus* from New Caledonia. Mar Pollut Bull 40:688–696

Carlson B (1987) Collection and aquarium maintenanceof Nautilus. In: Saunders WB, Landman NH (eds) Nautilus: the biologyandpaleobiology of a living fossil. Plenum Press, New York, pp 563–577

Carlson BA, Awai ML, Arnold JM (1992) Hatching and Early Growth of *Nautilus belauensis* and Implications on the Distribution of the Genus *Nautilus*. Proceedings of Seventh International Coral Reef Symposium, Guam. Vol. 1

Catala RLA (1964) Carnaval sous laMer, Les Editions R. Sicard, Paris, p. 141

De Angelis P (2012) Assessing the impact of international trade on chambered *Nautilus*. Geobios 45(1):5–11

del Norte-Campos AGC (2005) The chambered *Nautilus* fishery of Panay Island, West Central Philippines: fishing practices and yield. Res Bull Phuket Mar Biol Cent 66:299–305

Dunstan A, Alanis O, Marshall J (2010) *Nautilus pompilius* fishing and population decline in the Philippines: A comparison with an unexploited Australian *Nautilus* population. Fish Res 106:239–247

Dunstan AJ, Ward PD, Marshall NJ (2011a) Vertical Distribution and Migration Patterns of *Nautilus pompilius*. Plos One 6(2):e16311. doi:10.1371/journal.pone.0016311

Dunstan AJ, Ward PD, Marshall NJ (2011b) *Nautilus pompilius* Life History and Demographics at the Osprey Reef Seamount, Coral Sea, Australia. Plos One 6:e16312

Dunstan A, Bradshaw CJA, Marshall J (2011c) Nautilus at risk—estimating population size and demography of *Nautilus pompilius*. Plos One 6:e16716

Fields RF (2006) Hatching of *Nautilus pompilius* in a closed system using artificial seawater. Drum Croaker 37:43–47

Hamada T, Mikami S, Okutani T (1987) Chapter: experience with aquarium rearing of *Nautilus* in Japan. In: Saunders WB, Landman NH (eds) *Nautilus*. The biology and paleobiology of a living fossil. Plenum Press, New York, pp 579–583

Hanlon RT, Messenger JB (1996) Cephalopod behaviour. Cambridge University Press, Cambridge, p. 232

Hewitt RAWG (2010) Nautilus shell architecture. In: Saunders WB, Landman NH (eds) Nautilus: biology and paleobiology of a living fossil, 2nd edn. Plenum Press, New York, pp 435–462

Kanie Y, Fukuda Y, Nakayama H, Seki K, Hattori M (1980) Implosion of living Nautilus under increased pressure. Paleobiology 6:44–47

Kröger B, Vinther J, Fuchs D (2011) Cephalopod origin and evolution: a congruent picture emerging from fossils, development and molecules. Bioessays 33:602–613. doi:10.1002/bies.201100001

Landman NH, Cochran JK, Rye DM, Tanabe K, Arnold JM (1994) Early life history of *Nautilus*: evidence from isotopic analyses of aquarium-reared specimens. Paleobiology 20:40–51

Landman NH, Jones DS, Davis RA (2001) Hatching depth of *Nautilus pompilius* in Fiji. Veliger 44:333–339

Landman NH, Cochran JK (2010) Growth and longevity of Nautilus. In: Saunders WB, Landman NH (eds). Nautilus: the biology and Paleobiology of a living fossil, Springer Netherlands, pp 401–420

Mikami S, Okutani T (1977) Preliminary observations on maneuvering, feeding, copulating and spawning behaviors of *Nautilus macromphalus* in captivity. Venus Jpn J Malacol 36:29–41

Moltschaniwskyj NA, Hall K, Lipinski MR, Marian JEAR, Nishiguchi MK, Sakai M, Shulman DJ, Sinclair B, Sinn DL, Staudinger M, Van Gelderen R, Villanueva R, Warnke K (2008) Ethical and welfare considerations when using cephalopods as experimental animals. Rev Fish Biol Fish 17(2–3):455–476

Oba T, Kai M, Tanabe K (1992) Early life-history and habitat of *Nautilus Pompilius* inferred from oxygen isotope examinations. Mar Biol 113:211–217

Oestmann DJ, Scimeca JM, Forsythe JW, Hanlon RT, Lee PG (1997) Special considerations for keeping cephalopods in laboratory facilities. Contemp Topics Lab AnimSci 36:89–93

Okubo S (1989) Hatching of *Nautilus belauensis*in the aquarium. Aquabiology. Tokyo 11:191

Okubo S, Tsujii T, Watabe N, Williams DF (1995) Hatching of *Nautilus-belauensis* Saunders, 1981, in captivity—culture, growth and stable-isotope compositions of shells, and histology and immunohistochemistry of the mantle epithelium of the juveniles. Veliger 38:192–202

Owen R (1843) On the structure and homology of the cephalic tentacles in the pearly Nautilus. Ann Nat Hist XII:305–311

Pernice M, Boucher J, Boucher-Rodoni R, Joannot P, Bustamante P (2009) Comparative bioaccumulation of trace elements between *Nautilus pompilius* and *Nautilus macromphalus* (Cephalopoda: Nautiloidea) from Vanuatu and New Caledonia. Ecotoxicol Environ Saf 72:365–371

Sasaki T, Shigeno S, Tanabe K (2010) Anatomy of living *Nautilus*: reevaluation of primitiveness and comparison with Coleoidea. In: Tanabe K, Shigeta Y, Sasaki T, Hirano H (eds) Cephalopods—present and past. Tokai University Press, Tokyo, pp 33–66

Saunders WB (1983) Natural rates of growth and longevity of *Nautilus belauensis*. Paleobiology 9:280–288

Saunders BW (2010) The species of *Nautilus*. In: Saunders WB, Landman NH (eds) Nautilus: the biology and paleobiology of a living fossil. Plenum Press, New York, p 35–52

Saunders WB, Ward PD (1987) Ecology, distribution, and population characteristics of Nautilus. In: Saunders WB, Landman NH (eds) Nautilus. The biology and paleobiology of a living fossil. Plenum Press, New York, pp 137–162

Scimeca JM (2006) Cephalopods. In: Lewbart GA (ed) Invertebrate medicine. United Kingdom, Blackwell, pp 79–89

Shigeno S, Sasaki T, Moritaki T, Kasugai T, Vecchione M, Agata K (2008) Evolution of the cephalopod head complex by assembly of multiple molluscan body parts: evidence from *Nautilus* embryonic development. J Morphol 269:1–17. doi:10.1002/jmor.10564

Sinclair B, Briskey L, Aspden W, Pegg G (2007) Genetic diversity of isolated populations of *Nautilus pompilius* (Mollusca, Cephalopoda) in the Great Barrier Reef and Coral Sea. Rev Fish Biol Fish 17:223–235

Spinosa C (1987) Chapter: a small, closed aquarium system for *Nautilus*. In: Saunders WB, Landman NH (eds) *Nautilus*. The biology and paleobiology of a living fossil. Plenum Press, New York, pp 585–594

Spotte S (1979) Fish and invertebrate culture. Water management in closed systems, 2nd edn. Wiley, New York, p 179

Strugnell JM, Lindgren AR (2007) A barcode of life database for the Cephalopoda? Considerations and concerns. Rev Fish Biol Fish 17:337–344

Uchiyama K, Tanabe K (1996) Hatching of *Nautilus macromphalus* in Toba Aquarium, Japan. In: Oloriz F, Rodriguez-Tovar FJ (eds) Advancing Research on Living and Fossil Cephalopods. Proceedings of the IV International Symposium on Cephalopods: Present and Past. Kluwer Academic, Grenada, Spain, pp 13–16

von Byern J, Wani R, Schwaha T, Grunwald I, Cyran N (2012) Old and sticky—adhesive mechanisms in the living fossil *Nautilus pompilius* (Mollusca, Cephalopoda). Zoology 115:1–11

Ward PD (1987) The natural history of Nautilus. Allen and Unwin, Boston, p 267

Ward PD, Martin AW (1980) Depth distribution of *Nautilus pompilius* in Fiji and *Nautilus-macromphalus* in New-Caledonia. Veliger 22:259–264

Ward PD, Saunders WB (1997) Allonautilus: a new genus of living Nautiloid cephalopod and its bearing on phylogeny of the Nautilida. J Paleontol 71:1054–1064

Ward P, Wickstein MK (1980) Food sources and feeding-behavior of *Nautilus-macromphalus*. Veliger 23:119–124

Ward P, Stone R, Westermann G, Martin A (1977) Notes on animal weight, cameral fluids, swimming speed, and -color polymorphism of the cephalopod *Nautilus pompilius* in the Fiji Islands. Paleobiology 3:377–388

Ward P, Greenwald L, Rougerie F (1980) Shell implosion depth for living *Nautilus macromphalus* and shell strength of extinct cephalopods. Lethaia 13:182–182

Westermann B, Beuerlein K (2005) Y-maze experiments on the chemotactic behaviour of the tetrabranchiate cephalopod *Nautilus pompilius* (Mollusca). Mar Biol 147:145–151

Westermann B, Beck-Schildwachter I, Beuerlein K, Kaleta EF, Schipp R (2004) Shell growth and chamber formation of aquarium-reared *Nautilus pompilius* (Mollusca, Cephalopoda) by X-ray analysis. J Exp Zool A Comp Exp Biol 301A:930–937

Willey A (1902) Contribution to the natural history of the pearly Nautilus. In: Willey A (ed) Zoological results based on material from New Britain, New Guinea, Loyalty Islands and elsewhere collected during the years 1895, 1896 and 1897, part VI. University Press, Cambridge, pp 691–830

Williams R, Newman SJ, Sinclair W (2012) DNA barcoding in *Nautilus pompilius* (Mollusca: Cephalopoda): evolutionary divergence of an ancient species in modern times. Invertebr Syst 26:548–560

Chapter 11
Sepia officinalis

António V. Sykes, Pedro Domingues and José Pedro Andrade

Abstract This chapter reviews the importance of the European cuttlefish, *Sepia officinalis*, as a potential species for aquaculture and its applications. It provides an overview of cuttlefish culture, its current state of art and future trends. Present cuttlefish culture-related research and recently developed technologies are described. This includes a description of the culture systems for the different life stages, broodstock and egg acclimatization to captivity and management, hatchlings, juvenile and adult-rearing methodologies. Values of fecundity and fertility obtained in different culture conditions (variables include tanks, stocking densities, sex ratios and food); a characterization of different types of cuttlefish egg morphology; growth rates, mortality, feeding rates and food conversions at the hatchling and juvenile stages (including live, frozen and artificial diets); and a comparison between different growout setups are presented. Finally, current bottlenecks are enumerated, prospects for future research are suggested and an overview of whole animal use by the industry is given.

Keywords European cuttlefish · *Sepia officinalis* · Zoo technology · Stocking densities · Diets · Tank and earthen pond culture

11.1 Commercial Value and Capture Methods

The European cuttlefish, *Sepia officinalis* Linnaeus 1758, is mostly found in eastern Atlantic and in the Mediterranean Sea (Boletzky 1983). Cephalopod catches have set a new record in 2008 (FAO 2010) but only contributed to approximately 4% of the world catch from fisheries in 2010 (FAO 2012). Cuttlefish was one of the species attaining high market value in the Mediterranean and Asian markets (FAO 2010);

A. V. Sykes (✉) · J. P. Andrade
CCMAR-CIMAR L.A., Centro de Ciências do Mar do Algarve,
Universidade do Algarve, Campus de Gambelas, 8005-139 Faro, Portugal
e-mail: asykes@ualg.pt

P. Domingues
Centro Oceanográfico de Vigo, Instituto Español de Oceanografía,
Subida a Radio Faro 50, 36390, Vigo, Spain

J. Iglesias et al. (eds.), *Cephalopod Culture*, DOI 10.1007/978-94-017-8648-5_11,

countries such as Spain, Italy and Japan being the main consumers (Boucaud-Camou 1990; FAO 2012).

S. officinalis had an average world production of 16,769 t year^{-1} (2000–2010), showing an increasing trend during this period (FAO-FIGIS 2012). The species is primarily caught by otter and beam trawlers, either as a target species or as catch of demersal finfish fisheries. In southern European countries like Portugal, this species is caught by several fishing methods such as iron traps, trammel nets, bottom trawl, gill nets and purse seine (Sendão et al. 2007), similar to the traditional fishing gears used in other European countries. While trawlers operate in inshore and offshore fishing grounds for both juvenile and adult specimens, artisanal gears, like fish traps, are used to catch spawning animals mainly in inshore areas.

The species potential for aquaculture has been recognized by Barnabé (1996) and Boucaud-Camou (1989). This was due to a set of biological and physiological aspects that were described by Forsythe and Van Heukelem (1987) and which are shared by other cephalopod species. *S. officinalis* has been successfully reared in extensive aquaculture experiments at a medium scale in several EU countries such as Italy, France and Portugal.

The commercial culture of cuttlefish will possibly have a high impact on fisheries production in the near future, as it could allow the sale of undersized individuals (approximately 50 g) that may be produced in only 45–60 days.

11.2 State of the Art

Cultured *S. officinalis* is used as an animal model for biological and biomedical research (e.g. physiology, neuroscience, nutritional biochemistry, ageing, molecular biology or immunology), for aquaculture production and also for public exhibition in aquariums. During the past 20–25 years, a great part of the research on cuttlefish has focused on its use as a new species for aquaculture. When viable, economical and logistic cuttlefish culture is attained, this species will become one of the most relevant marine animal models and make an important impact in other fields of research. The use of cephalopods as animal models has already greatly contributed to the scientific advance of humankind (i.e. the Nobel Prize won by Hodgkin and Huxley in 1963 regarding the squid's giant axon). Thus, research on cuttlefish culture will have further application for the development of culture technology of other cephalopod species and will provide a cultured invertebrate model that shares some biological features with vertebrates.

The most recent revision on the status of cuttlefish culture was performed by Sykes et al. (2006b), in which the major bottlenecks limiting its transition into an industrial scale were identified. Currently, the three main factors still delaying the large-scale culture are (1) the dependence on live prey during the first part of the life cycle, (2) the lack of an adequate artificial diet for all life stages of this species and (3) full control of reproduction in captivity. Due to the low number of laboratories involved in the resolution of these problems, small but steady progress was made since then. Consequently, not only technology has evolved but also cephalopods have been included in the EU animal welfare legislation. Thus, the advances in *S. officinalis* culture technology will be revised also considering a good welfare practice.

11.3 Broodstock Rearing Conditions and Acclimatization to Captivity

Considering that reproduction in captivity is one of the bottlenecks in cuttlefish culture (Sykes et al. 2006b), broodstock maintenance knowledge and conditions have been updated.

11.3.1 Seawater Systems and Conditions

Depending on the location of the husbandry, the three types of seawater systems (open, semi-open or closed) may be applied. Commonly, cuttlefish breeder stocks are obtained from captive cultured populations or through wild captures as described by Sykes et al. (2009). After being checked for illness and skin damage, cuttlefish are conditioned in open seawater systems because of logistics, welfare and economical aspects. Nonetheless, as the technology used in closed and semi-closed systems is becoming cheaper and aquaculture environmental concerns are scaling up, the use of closed seawater systems (RAS), similar to those reported by Hanlon (1990) for cephalopods and recently by Martins et al. (2010) for fish, is expected in the near future. The technology has progressed since 2000, from using 250 L tanks (Correia et al. 2005; Sykes et al. 2006a) to increasing sizes, such as 9,000 L, currently being used after the results of Sykes et al. (2013b, Fig. 11.1).

A typical tank setup may be built indoors or outdoors of the husbandry facilities but it must be located in a low disturbance area (Sykes et al. 2006b). Each tank should have a setup consisting of round-shaped fibreglass tanks comprising enough airlifts, fixed on the tank walls, and air stones in the middle of the tank. These conditions will make water move slowly but like a drain to the outlet pipe, which is located at the centre of the tank. Since cuttlefish is a benthic species (Boletzky 1983), the tank should have large bottom areas (Domingues and Marquez 2010; Sykes et al. 2013b).

If toxic materials (e.g. PVC glue, silicone for marine environments) are used while assembling the tank setup, it should be first filled with tap water for 24 h, followed by a 24 h seawater filling and running, before the animals are placed in the tank. While designing, building and operating this setup, the use of metallic parts, especially copper, should be avoided and kept away from all seawater systems, as it can be toxic in the first stages of development of the cuttlefish (Establier and Pascual 1983) and even in the remaining life stages (Hanlon and Forsythe 1985;

Fig. 11.1 Tanks used at Centro de Ciências do Mar do Algarve (CCMAR) for cuttlefish reproduction. The first experiments (in 2000) were performed in 250 L tanks and the current methods include the use of 9,000 L tanks. (Photos by A Sykes)

Paulij et al. 1990). The use of IP67-electric-certified equipment and material (e.g. illumination) is mandatory if these are installed near the tank.

The use of semi-open or closed systems should comprise a scaled filtration based on the basic setup described by Hanlon (1990) to assure that protein load in the water (resulting from feeding and eventual inking due to reproductive behaviour or stress response) is kept low and spikes or build-up of nitrogenous compounds is avoided. Upgrades on the equipment used in these types of systems are advised to help reduce the area occupied and to attain increased efficiency. The build-up of anoxic areas in the tank, due to low water circulation, should be prevented while designing the system setup.

Airlifts and outflow pipes should be covered with plastic nets of appropriate mesh size to prevent animals and eggs from escaping or being washed out. While designing the seawater system, it should be considered that sharp objects or rough surfaces must not exist inside the tank since they may cause skin damage to the animals. If needed, a wall foam protection system similar to that described by Hanley et al. (1999) should be used. Following a good welfare practice regarding enriched environments and considering the animal's ability to camouflage; this tank should be under low light intensities of 200 Lx or less (measured at the water-air interface of the tank) and under a normal photoperiod that should replicate natural geographical conditions during spawning in the wild. This may be achieved using a combination of natural or natural-resembling artificial light sources and tank colours. If placed outdoors, the 200 lx light intensity may be obtained by using water-repellent shading nets (Fig. 11.1). The use of these nets will prevent excess lighting as well as pH and salinity descent due to rainfall.

In any of the setups described above, animals, tanks, equipment to maintain specific conditions, water flow and aeration should be checked twice a day, in the morning and late afternoon. Water quality parameters should be monitored every day in the morning. Depending on the seawater system used, the determination of different parameters will apply. Open seawater systems will require the measurement of temperature, salinity and dissolved oxygen. It is not possible to fully evaluate the use of semi-open or closed seawater systems, since there are not enough data regarding reproduction and egg quality under these conditions. Nevertheless, if these systems are the only option, determination of previous parameters, plus pH and nitrogenous compounds (ammonia, nitrites and nitrates), is mandatory. In semi-open or closed systems, these parameters should be kept as similar as possible to natural seawater and according to the highest values reported for the culture of this species by Forsythe et al. (1991) and Oestmann et al. (1997). The technology for closed systems has progressed since the end of the 1980s, and there is an increasing pressure to reduce the aquaculture footprint by using these systems (Martins et al. 2010). However, cephalopods are highly sensitive to nitrogenous compounds and the removal of these, to prevent a system overload which will translate into massive death, still requires highly expensive technology, logistics and operational costs. Some trace elements, in particular strontium and calcium, should be kept close to natural seawater values (Hanlon et al. 1989). The use of natural seawater is suggested. However, if, due to logistics, commercial seawater salts are used to produce artificial seawater, its content should comply with these requirements to avoid malformations and the death of hatchlings (Hanlon 1990).

In any case, temperature, salinity, dissolved oxygen, pH and nitrogenous compounds should be maintained according to values reported by Boletzky (1983) for the species. There are reports of culturing cuttlefish beyond the maximum (Domingues et al. 2001a) and minimum (Sykes et al. 2006a) temperature values reported for the species, but a flow-through system was used in both cases, where changes in water quality occurred slowly due to the high volumes of water utilized. Regardless of the seawater system used, water flows should be high enough to maintain water quality and sustain the best reproduction results.

If closed and semi-open systems are used, the use of UV filtration will be necessary to avoid pathologies, as described by Forsythe et al. (1991). Although the effect of ozone on the species is not reported in the literature, its use is not suggested as a way to attain similar results of salubriousness. Nonetheless, if used, measurements of O_3 concentrations and the redox potential of water entering the tank should be enforced. The application of EU Directive 2010/63/EU from January 2013 onwards will impose the use of alarm systems locally and externally, to the person responsible for animal care, through SMS messages or similar systems. Not only should these alarms be set to report a failure of water circulation in tanks or in the marine station but also if a digital solution is applied, a report on these abnormal situations will be created by the system, and later used as experimental data and for reporting animal welfare.

Tank cleaning should be avoided as much as possible to prevent stress, by enforcing an ad libitum feeding scheme. Water and tanks should be kept clean from leftovers, faeces and other debris and removed through water siphoning. Therefore, to avoid problems with cleaning, sand substrates should not be used (Forsythe et al. 1994).

11.3.2 Sex Ratio and Stocking Densities

Reports on stocking densities and sex ratios are scattered in the literature, and experiments dealing primarily with this issue are scarce. Sex ratios and stocking densities are of particular importance since they are thought to have a high impact on cuttlefish fecundity and fertility. This assumption is based on our experience of culturing the species and on previous studies which were not directly related to reproduction (Boal et al. 1999; Forsythe et al. 2002). However, researchers do not agree on which are the best densities and sex ratios that will maximize egg quantity and quality obtained in cuttlefish reproduction in captivity. In addition, most of them do not consider the importance of tank bottom areas with regard to a species that is benthic or the individual contribution of parents that may change depending on available area, stocking densities and sex ratios. For instance, Forsythe et al. (1991) mentions that sex ratios of one male for three females should be used to avoid male aggression and aggressive mating behaviour, and Forsythe et al. (1994) suggests densities of two cuttlefish m^{-2} for breeders.

Table 11.1 summarizes the most relevant fecundity and fertility results obtained in different tanks, stocking densities, sex ratios, temperature and food items (Correia et al. 2005; Domingues et al. 2001b, 2002, 2003b; Sykes et al. 2006a, 2009, 2013b). It is suggested that sex ratios should be maintained at two females for one male and stocking densities kept relatively low when setting up a broodstock. For instance, a 9,000 L tank should have 21 animals, 14 of them females and 7 males, which will correspond to a low stocking density of 4 cuttlefish m^{-2}.

11.3.3 Food Supply

Despite Boletzky's (1983) and Nixon's (1985) descriptions that cuttlefish shifts its food preferences from a predominant crustacean diet (crabs, prawns and shrimps) to a mixture of crustaceans and fish while maturing and reproducing, recent manuscripts report the successful culture of the species using only a frozen grass shrimp (*Palaemonetes varians*)-based diet (Sykes et al. 2006a, 2013b, Table 11.1) during those periods.

The use of different kinds of food throughout the different stages of the life cycle of cuttlefish was never studied, but examples of different feeding strategies are

Table 11.1 Fecundity and fertility results obtained in different tanks, stocking densities, sex ratios, temperature and food

Density (cuttlefish. m−2)	Tank	Animals	Sex ratio	Temperature (°C)	Generation	Food	Fecundity (eggs/♀)	Fertility (%)	Reference
38	250 L round	30	1♂: 1♀	27.0 ± 3.0	F1	Frozen *Carcinusmaenas*	144	50.0	Domingues et al. (2001b)
19	250 L round	15	3♂: 1♀	15.0 ± 4.0	F2	Live P. *varians*	225	41.0	Domingues et al. (2002)
19	250 L round	15	1♂: 1♀	≈ 17.0	F3	Live P. *varians*	150	33.0	Domingues et al. (2003b)
19	250 L round	15	1♂: 1♀			Frozen P. *varians*	411	85.0	
16	250 L round	13	3♂: 1♀	24.5 ± 1.4	NK	Mixture of live *P. varians*, *Carcinusmaenas* and fish	834	35.8 ± 9.4	Correia et al. (2005)
76	250 L round	60	3♂: 1♀				290	62.0 ± 16.9	
9	250 L round	27	1♂: 2♀	17.1 ± 1.7	F2	Live *P. varians*	370	NK	Sykes et al. (2006a)
9	250 L round	30	3♂: 1♀	23.4 ± 1.4	F3		301	16.0	
13	250 L round	35	1♂: 2♀	15.2 ± 3.0	F4		247	47.7	
8	250 L round	50	1♂: 1♀	21.1 ± 2.6	F5		478	67.5	
3	250 L round	60	2♂: 1♀	24.2 ± 1.7	F6		293	0	
15	400 L rectangular	18	1♂: 2♀	19.5 ± 1.1	F2	Frozen *P. varians*	787	n.d.	Sykes et al. (2009)
4	9,000 L round	23	1♂: 2♀	20.5 ± 2.9	F1	Frozen *P. varians*	1,383	72.0	Sykes et al. (2013b)
15	750 L round	23	1♂: 1♀	19.0 ± 2.2	F1	Frozen *P. varians*	223	66.0	
29	250 L round	23	1♂: 1♀	21.2 ± 3.4	F1	Frozen *P. varians*	325	48.0	

Sex ratios are given as observed and not established in the experiments. In a given column, when a given variable is shared by one or more groups, only one value is presented. Fecundity is individual fecundity. Fertility is the percentage of hatching eggs

Values shared by several different groups are presented solely in that column

n.d. not determined, *NK* unknown

found within the literature. For instance, Domingues et al. (2001b) tested the effects of feeding either *Artemia* sp. or *Paramysis nouvelli* during the first 20 days after hatching (DAH), followed by a diet of *P. varians* until 70 DAH and frozen *Carcinus maenas* for the remaining life cycle with similar results of fecundity (Table 11.1). Domingues et al. (2002) used *P. nouvelli* for the first 20 DAH and afterwards live grass shrimp to close the cycle, achieving narrow increased results in individual fecundity but lower fertility (Table 11.1) when compared with the 2001 F1 generation. The effects of using either live or frozen grass shrimp, after being fed during the first 15 DAH, were tested by Domingues et al. (2003b) attaining increased individual fecundity and fertility when using the frozen diet (Table 11.1). A mixture of live *P. varians, C. maenas* and fish used by Correia et al. (2005) attained one of the highest individual fecundity ever in 250 L tanks (834 eggs per female, Table 11.1). We believe that this result is related more to temperature, densities and sex ratio than type of food, but further studies are needed.

Sykes et al. (2006a) studied five consecutive generations of cuttlefish in captivity using a diet of live *P. varians* in 250 L tanks and obtained a maximum individual fecundity of 478 eggs per female and fertility of 67.5% at the F5 generation (Table 11.1). On the other hand, Sykes et al. (2009) confirmed an increased fecundity in a F2 generation (787 eggs per female; Table 11.1), being fed on live grass shrimp for the first 20–30 DAH and frozen grass shrimp onwards (in accordance with the methods described by Sykes et al. 2006b). However, the sex ratio was completely inversed when compared with that obtained by Correia et al. (2005).

Recently, Sykes et al. (2013b) obtained the best values of individual and overall fecundity (1,383 and 16,593 eggs, respectively) and an acceptable individual fertility (72%, Table 11.1) following a similar feed methodology but increasing the bottom area of the tanks.

Based on these results, it seems that the diet per se will not have a direct effect on egg number and quality. However, we do not discard the fact that the use of frozen grass shrimp may be influencing the higher individual and overall fecundity and fertility by promoting lower energy expenditure associated with feeding.

11.4 Spawning Process

11.4.1 Female Conditions

S. officinalis attains sexual maturity at very different sizes/weight (Sykes et al. 2006a) and has an estimated potential fecundity of a maximum of 8,000 eggs in nature (Laptikhovsky et al. 2003). The species displays a very typical behaviour of courtship, mating and agonistic male-to-male interaction, briefly described by Boletzky (1983) and Hanlon and Messenger (1996), where males mature earlier than females. Females receive the spermatophores in the paired seminal receptacles, under the buccal mass, where it may remain and used for as much as 2–5 months

(Hanlon et al. 1999). This is probably due to the biology of the species, with males maturing precociously (Forsythe et al. 1994). Males, before inserting their sperm into this pouch, jet large amounts of seawater inside the seminal receptacles to flush out the sperm from previous males and, in this way, assure that only its genetic contribution will be used (Hanlon et al. 1999).

Females mate repeatedly (Hanlon et al. 1999), display intermittent or chronic spawning (Boletzky 1987) depending on captive conditions and usually will die shortly after laying the eggs. Nevertheless, we have observed an extension of this intermittent spawning when using increased tank bottom areas (based on comparisons between data from Sykes et al. 2006a, 2013b). However, this does not happen repeatedly and in all tanks tested. Therefore, we suspect that the optimal conditions to obtain more and better eggs are related not only to bottom areas but also to other variables, which are being investigated in current projects.

According to Boal (1997), females prefer males that have mated recently instead of choosing males based on a dominance hierarchy, which is promoted by captive conditions (Boal et al. 1999). If this is related to social recognition (Boal 2006), chemoreception (Boal and Marsh 1998; Boal and Golden 1999) or specific pheromones (Zatylny et al. 2000, 2002) remains to be unveiled and are objectives of future studies.

After the copula, it is not advisable to separate females from males since this will not promote a reproduction resembling wild conditions and might have a negative influence on egg quantity and quality. According to Boletzky (1983), bigger females will lay bigger eggs; which does not completely agree with more recent data from Sykes et al. (2013b). By having extended intermittent spawning due to different culture conditions, the amount and quality of eggs should be better. However, what is a quality egg in this species and how can we say that bigger eggs are better eggs, since no results on this subject have been presented?

In the species, no parental care of eggs was ever observed, and senescent females die after spawning. As a senescent species, reproduction will absolutely drain females, which incorporate their little reserves in the eggs. The ability of cuttlefish females to have more than one maturation cycle or continuous maturation during this intermittent spawning is something that will need to be further investigated in the near future, since not all the females or males will die after mating and eggs are laid.

11.4.2 Egg Capture and Handling

Eggs may be obtained from nature or from captive-bred populations, but the latter is recommended. The use of eggs from a known source not only reduces the impact on the environment but also assures that eggs will have a very low probability of introducing contaminants and/or pathologies into the culture facilities. Within the EU, these Animal Resource Centres are regulated by law and a wide network exists (for details, check European Marine Biological Resource Centre—EMBRC—http://www.embrc.eu/).

Fig. 11.2 From copula to incubation in captivity. Copula (**a**), egg laying (**b**), extraction, separation and assessment of eggs (**c, d, e**) and incubation of eggs in 250 L tanks (**f**). (Photos by A Sykes)

In nature, cuttlefish females lay their eggs on hydrodynamic locations, attached to wild flora and fauna or to human-related structures (Boletzky 1983). Egg masses are commonly found on sea grass, seaweeds, polychaetes, buoys, fishnets, traps, boat ropes, etc. Based on this knowledge, Sequi and Palmegiano (1984) and Blanc and Daguzan (1998) described man-made egg collectors that use floating ropes to collect eggs on spawning grounds. However, it is recommended to obtain eggs in the wild directly from the beaches, when they are stranded by storms or other hydrodynamic conditions. The impact of man on wild populations is therefore reduced, considering that these correspond to a percentage of eggs that normally would not hatch because of being exposed to dry conditions.

In captivity and after copulating (Fig. 11.2a), females attach their eggs on ropes, airlines and nets (Fig. 11.2b, c). To make the process of removing the eggs from the tanks and egg collectors easier, it is recommended to use plastic nets. Each breeding tank should have at least one of these collectors that should be checked for eggs every day, preferably in the morning and before feeding. It is very important to avoid disturbing the animals while spawning and proceed accordingly. Afterwards, the egg collector should be removed from the tank and kept humid while eggs are separated and individualized. We recommend the use of a plastic net, with 1 cm × 1 cm net holes (Sykes et al. 2006b), to facilitate egg removal (Fig. 11.2c). The easiest way to perform this operation is using a small knife or scalpel, which will cut the egg lace connecting to the plastic, preventing any harm to the corion. Freshly laid eggs are very soft and gelatinous and should be removed carefully, while those not from that day will be firmer and easy to manipulate. Irrespective of the egg source (nature or captivity) and depending on their numbers, eggs being removed from the

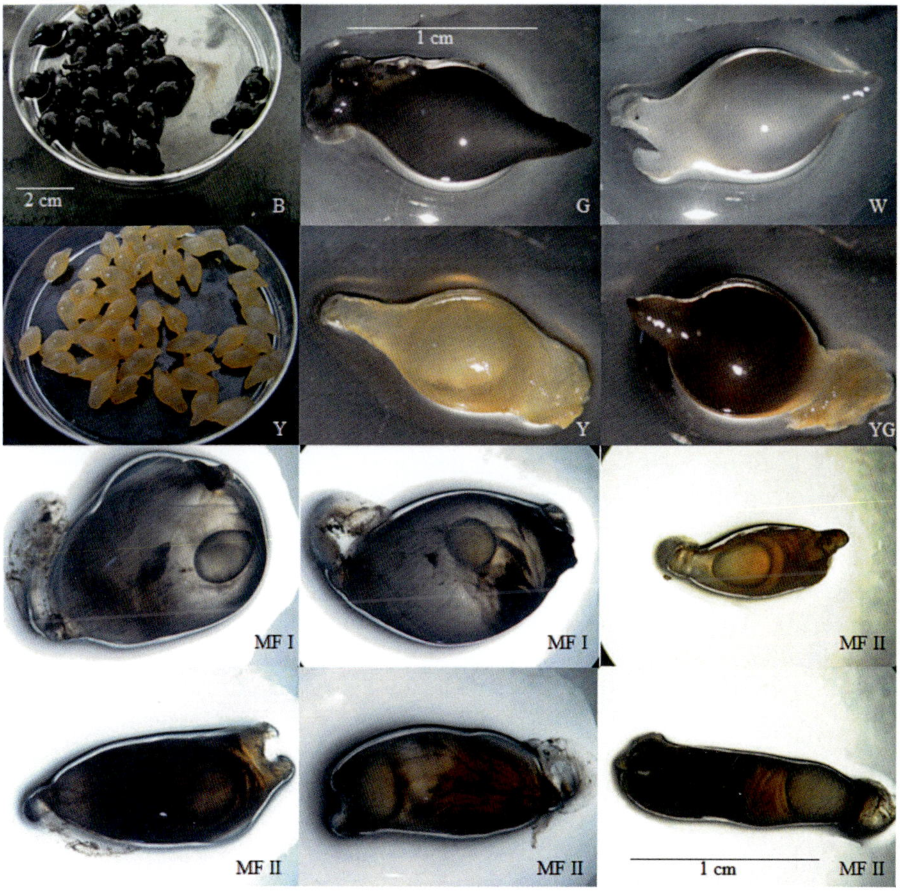

Fig. 11.3 Morphology of different types of cuttlefish eggs obtained in captivity (characterization and identification presented in Table 11.3). (Photos by A Sykes)

natural or man-made collectors should be individually sorted and placed in a bucket with high aeration that will make them move gently (Fig. 11.2d). During egg extraction from both collectors in captivity and egg collection at the beach, eggs should be kept moist and the use of a spray with seawater to accomplish this is recommended. At this time, the eggs should be separated according to their morphology and colour (Figs. 11.2e and 11.3).

Eggs collected from the wild are usually black, present a flask shape and are attached in grape-like clusters to different substrata fixed on the bottom (Boletzky 1983; Boletzky et al. 2006). Eggs will display weights ranging from 0.1 g up to 2.5 g (Sykes et al. 2006a) and a diameter ranging from 1.2 to 1.4 cm. Only black (ink-stained) eggs with a flask-shape morphology should be chosen, while those with a different colour and morphology (Fig. 11.3) should be discarded properly (Sykes et al. unpublished data).

Table 11.2 Characterization of different types of cuttlefish egg morphology

Egg type	ID	Shape	Colour	Transparency
Normal	N	Flask	Black	No
Grey	G	Flask	Grey	No
White	W	Flask	White	No
Orange	O	Flask	Orange	Semi-
Yellow-grey	YG	Flask	Yellow-grey	No
Malformation type I	MF I	Globular	No	Yes
Malformations type II	MF II	Elongated	Dark brown	Semi-

ID—identifies a given egg type in Fig. 11.3

Table 11.2 discriminates and classifies the different egg types ever obtained in captivity.

Despite the recent attempts by Sykes et al. (2013b), there is currently no methodology to assess the quality of eggs at this time and before animals are born. However, in captive conditions, both percentage of rejected eggs and individual egg weight might be indicative of the quality of the egg masses.

After the egg quality assessment and sorting, the eggs should be placed in round-shaped tanks of a flow-through or semi-open system (Fig. 11.2f). If a closed seawater system is used, then the recommendations of Hanlon et al. (1989) regarding strontium should be considered. The tank setup should have airlifts on the walls and air stones in the middle which promote a gentle elliptical movement of the eggs (Fig. 11.2f). This also assures an oxygen-enriched environment, which has proven to prevent necrosis. The amount of eggs should not exceed the tank's carrying capacity, considering that water parameters should remain relatively constant and, therefore, water flows will be small. Water parameters should be set according to the conditions of eggs' geographical location described in the literature (Sykes et al. 2009). So, if needed, seawater should be heated or refrigerated. Nonetheless, according to Palmegiano and Sequi (1984), salinity will increase an egg's viability above 90 % if its values are within 28–50 psu.

Detailed information regarding the embryonic development of the species was reported by Naef (1928), Lemaire (1971) and Boletzky (2003) and was reviewed recently by Boletzky et al. (2006). However, the duration of this stage is dependent on temperature (Koueta et al. 2006), without any linear correlation (Richard 1971) and geographically conditioned (e.g. temperature vs. duration varies considerably between Faro, Portugal and Caen, France (Sykes et al. 2009)). It may range from 40 to 45 days at 20 °C to 80–90 days at 15 °C in the English Channel populations (Boletzky et al. 2006) and from 25 days at 25 °C to 60 days at 15 °C in southern regions, such as Portugal (Sykes et al. 2006b).

It is common that eggs laid during a week by one or more females will display a synchronized hatching on the same day. Whether this synchronized hatching is related to the action of ILME (a waterborne pheromonal peptide released by eggs; Zatylny et al. 2000) or to any other unknown peptide or process remains to be determined.

11.5 Hatchlings Culture

As this stage is considered as one of the bottlenecks in cuttlefish culture (Sykes et al. 2006b), zoo-technology and nutritional knowledge have been updated in this chapter.

11.5.1 Hatchlings Collection and Transfer

According to Paulij et al. (1991), cuttlefish embryos hatch during periods of darkness, so it is important to check the egg-hatching tank in the morning to eventually collect and separate the new offspring.

After hatching, and depending on the purpose, hatchlings may be kept in the same tank or carefully removed to a different one. According to the new animal welfare legislation and cephalopod proposed guidelines, this tank or group of animals will be considered a captive population and assigned an identification, which should include information regarding the generation and source of eggs.

If removed to a new tank, the physical and chemical conditions of the new tank should be similar to those of the hatching tank, although these might be changed, gradually, afterwards. Normally, the procedure of transfer, if the animals remain in the hatchery, is to collect them individually using small movements with an aquarist net. This reduces stress and abrasion against either tank or net. Afterwards, hatchlings are placed in a plastic container of small volume (preferably black) and then submerged in the new tank. The latter will allow cuttlefish to swim freely out of this container into the water of the new tank. If a transfer of the newly hatched animals to other culture facilities is intended, then specific information for the species may be found in the recent animal welfare review by Sykes et al. (2012).

11.5.2 Culture Conditions

11.5.2.1 Seawater Systems

The seawater systems used for the hatchling stage are similar to those used in other life stages of the species and detailed for breeders (see Sect. 11.3.1). Nonetheless, the type of tanks used should be adapted according to the most recently published literature, where factors such as the culture densities of 500 hatchlings m^{-2}, minimum bottom areas of 0.06 m^2 and the avoidance of sand and shelters (Sykes et al. 2003) should be considered. The use of flow-through seawater systems with UV filtration is recommended to avoid problems derived from pathogens present in the water. This stage of the life cycle is the most problematic and where most of the mortality in cuttlefish culture usually occurs.

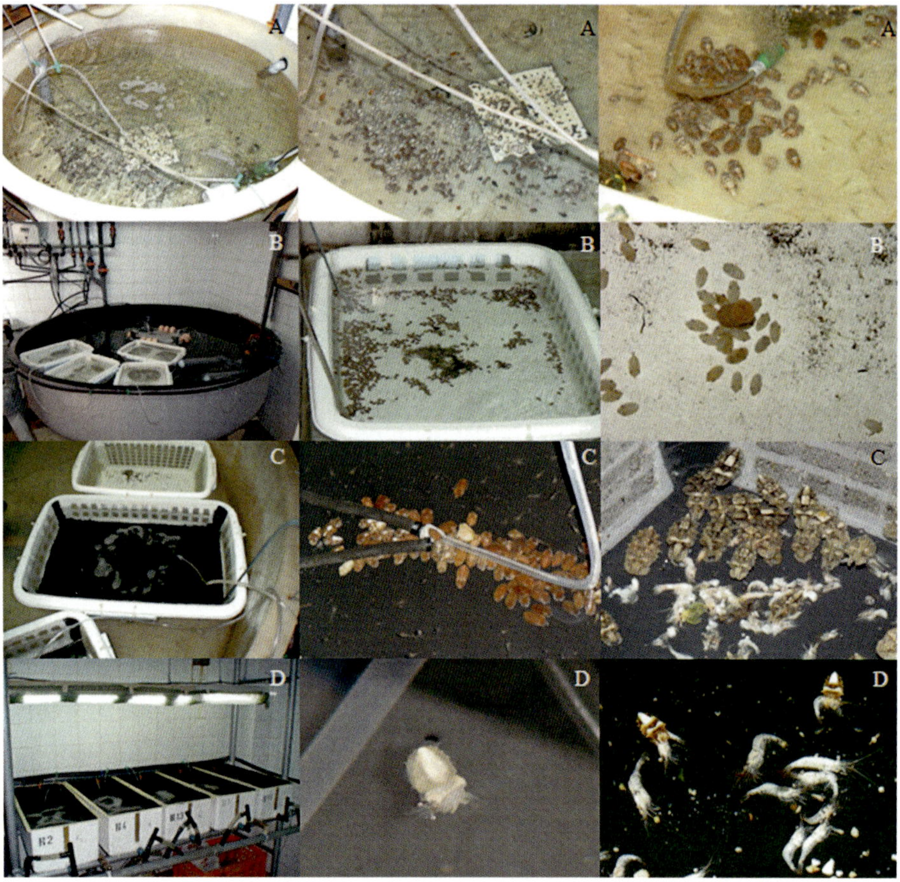

Fig. 11.4 Different hatchling culture systems. **a** In the hatching tank. **b** In a white basket within a 1,500 L tank. **c** In a black basket within a 1,500 L tank. **d** In raceway tanks. (Photos by A Sykes)

Significant progress has been obtained in the optimization of culture conditions at this life stage. Sykes et al. (2006b) previously recommended to rear cuttlefish in the hatching tank (Fig. 11.4a) but currently the use of hatching baskets (Fig. 11.4b), preferably in black (Fig. 11.4c; Sykes et al. 2011), or raceway tanks (Fig. 11.4d) is suggested. The use of low light intensities during this stage is also suggested (Sykes et al. 2013c). These conditions of tank colour, light intensity and similar seawater physical conditions with respect to embryonic development should be maintained to promote appropriate welfare conditions, lower mortality rates and normal growth and development.

11.5.2.2 Physicochemical Conditions

Seawater physicochemical conditions should resemble natural conditions as much as possible, as long as this does not interfere either directly or indirectly with cuttlefish growth and survival. Depending on the seawater system used during this stage, different water parameters should be set and checked (see Sect. 11.3.1 for the list and procedures). Normally, hatchlings' rearing is performed at 20±2 °C, in high dissolved oxygen conditions (>90 %) and high water-flow rates to avoid nitrogenous waste-derived problems in low-volume seawater systems.

11.5.2.3 Light

There is no known minimal amount of light necessary for this stage but recent results recommended 100 lx at the air–water interface of the tank (Sykes et al. 2013c), provided by daylight (450–650 nm) fluorescent bulbs. From published data, only Koueta and Boucaud-Camou (2003) studied the effect of photoperiods in hatchlings and suggested that cuttlefish hunts more during daylight. Photoperiodicity should match habitat and lifestyle at given geographical locations (Sykes et al. 2006b), and daylight should be set with a 12–16 h light cycle. Nonetheless, little is known regarding the daily activity and daily feeding rhythms at this life stage or even for the species (Quintela and Andrade 2002) and this should be researched in the near future.

Up till now, the use of dusk-and-dawn light simulations in the photoperiod or the use of polarization (light or filters) to enhance growth and survival was never reported. Cuttlefish has polarized vision, which it uses to hunt in the natural environment (Shashar et al. 2000, 2002). This characteristic should be further explored to determine if it could enhance conditions in captivity.

11.5.2.4 Diet

Cuttlefish may hatch prematurely, mature or 'over-mature', the differences being characterized by the respective amount of inner yolk reserves remaining. These reserves will determine how long the animal will endure without food and commonly last between 3 and 5 days (Sykes et al. 2004). During this period, the animal must be given the opportunity to feed on an external diet or there will come a point of no return and the cuttlefish will die.

Diets for hatchlings have been one of the most investigated themes in the past 12 years and where research has attained some of the most relevant results. It is known that in nature, cuttlefish hatchlings prey preferably on live crustacean species (Boletzky 1983), and a series of different alternative live diets were tested (DeRusha et al. 1989). For several years, the rearing of cuttlefish during the hatchling stage was obtained using a diet of live mysids followed by live shrimps, when the animals grew bigger (Correia et al. 2005, 2008a; Domingues et al. 2001a, b, 2002, 2003a; Koueta and Boucaud-Camou 1999). Domingues et al. (2004) proposed the use of live *P. varians* as an alternative prey for the rearing of cuttlefish at this life stage, while Sykes et al. (2006a) extended this further by proposing the use of grass shrimp for the culture throughout the life cycle of *S. officinalis*. Irrespective of the prey used, its degree of availability and the use of freshly caught or starved live food affect cuttlefish growth (Correia et al. 2008a, b).

After a series of zoo-technology optimizations for this life stage, Sykes et al. (2013a) managed to rear cuttlefish throughout the hatchling stage performing an early weaning on the first day after hatching with some success (Fig. 11.4d on the right—cuttlefish hatchlings eating frozen grass shrimp). These results are encouraging but further research regarding this issue in terms of nutrition, digestion, etc. is needed.

For the time being, the use of live grass shrimp of appropriate size is still recommended during the hatchling stage and the weaning to frozen grass shrimp should be performed after 20–30 DAH depending on the temperature.

11.5.3 Growth and Survival

Information regarding cuttlefish hatchlings' behaviour, growth and morphological changes during this stage was reported by Nixon and Mangold (1998). Growth rates vary directly with temperature, inversely with size, and play a major role in determining the species' life span (Domingues et al. 2006). Maintaining animals of different age groups together should be avoided at any stage but especially at this stage. It is acceptable, for rearing purposes, to join hatchlings with an age difference of up to 10 DAH. Joining animals with larger age differences will generate increased competition for food, promote the establishment of food hierarchies and, therefore, increase mortality. This stage of cuttlefish life is characterized by the highest growth rates but, at the same time, where mortalities may be higher. Table 11.3 resumes the hatchlings growth rates, mortality, feeding rates and food conversions (during the first 30 days of life, according to type of tank and food) published since 1999.

Normal growth, development and survival depend on the egg content (this content reflects food given to females and the same female may lay eggs with different amounts of yolk depending on the number of spawns and age), food quality, the best quality of seawater and tank setup conditions. Food consumption and feeding rates during this life stage are also extremely high, when compared with the values registered in the subsequent stages.

11.6 Ongrowing of Juveniles and Adults

11.6.1 Tank and Earthen Pond Conditions

Despite the first attempts of rearing cuttlefish in earthen ponds which were performed by obtaining eggs in the wild with egg collectors and placed in these ponds and collecting the juveniles 3 months later (Palmegiano and Sequi 1981), this type of procedure is considered extensive culture. This was, and still is, performed in countries such as Italy, Portugal (Coelho et al. 1989; Gonçalves 1989) and Tunisia (Rodger and Davies 2000). Nevertheless, some authors believe that the potential of cuttlefish culture is able to sustain semi-intensive or intensive aquaculture.

Table 11.3 Growth rates, mortality, feeding rates and food conversions during the first 30 days of life of *Sepia officinalis*

Tank/System	Gen.	T (°C)	S (PSU)	Food/Light Intensity	IGR (% BW/d)	Mortality (%)	FR (% BW/d)	FC	Reference
70 L cylinder tanks/Semi-open	F1	19.0	33.5	Live mysids (*Mesopodopsis slabberi* and *Schistomysis* sp.), *Gammarus* sp., and young shrimps	7.3	3.0	4.6	3.2	Koueta and Boucaud-Camou (1999)
				Live mysids (*Mesopodopsis slabberi* and *Schistomysis* sp.) and frozen mysids (30% BW/d)	4.6	3.0	4.2	2.3	
250 L/Open	F1	27.0 ± 3.0	37.0 ± 3.0	*Artemia* sp. (20 DAH) + *P. varians* afterwards	5.2	33.0	n.d.	n.d.	Domingues et al. (2001b)
				Paramysisnouveli (20 DAH) + *P. varians* afterwards	10.2	10.0	n.d.	n.d.	
				Artemia sp. (10 DAH) + *P. varians* afterwards	8.8	None	n.d.	n.d.	
				P. nouveli (10 DAH) + *P. varians* afterwards	10.0	None	n.d.	n.d.	
Individual baskets /250 L/Open	F1	27.0 ± 3.0	37.0 ± 3.0	*Artemia* sp. (10 DAH) + *P. varians* afterwards	9.9	None	24.3	3.0	Domingues et al. (2001a)
250 L/Open	F2	≈18.0	36.0	*Artemia* sp. (20 DAH) + *P. varians*afterwards	8.7*	None	n.d.	n.d.	Domingues et al. (2002)
Individual baskets /1,000 L/Semi-open	F1	20.0 ± 1.0	35.0	*Crangon crangon* (5 DAH) + live shrimp	8.9	None	n.d.	n.d.	Domingues et al. (2003)
				Crangon crangon (5 DAH) + live fish fry	5.8	None	n.d.	n.d.	
10 L plastic rectangular tanks /Open	F4	20.0 ± 2.0	36.0 ± 1.0	*P. nouvelli*	6.2	9.9	n.d.	n.d.	Domingues et al. (2004)
				P. varians	7.5	26.7	n.d.	n.d.	
				Atherina sp.	2.9	None	n.d.	n.d.	
250 L/Open	F2	17.4 ± 1.8	35.0 ± 2.0	*P. varians*	6.4	14.8	n.d.	n.d.	Sykes et al. (2006a)
	F3	23.7 ± 1.1	37.0 ± 2.0		19.3	20.0	n.d.	n.d.	
	F4	13.3 ± 1.1	35.0 ± 2.0		2.7	14.0	n.d.	n.d.	
	F5	24.0 ± 1.8	35.0 ± 2.0		10.4	None	n.d.	n.d.	
10 L blue plastic rectangular tanks /Open	NK	18.5 ± 0.5	36.0 ± 1.0	*P. nouvelli* (15% BW/d)/535 Lx	7.0	None	13.3	n.d.	Correia et al. (2008a)
				P. nouvelli (30% BW/d)/535 Lx	7.7	None	14.8	n.d.	
45 L raceway/Open/Black	F2	19.3 ± 0.5	37.0 ± 1.0	Mix of *Mesopodopsis slabberi* and *Artemia* sp. (20 DAH) + *P. varians* afterwards/320 Lx	4.6	21.7	n.d.	n.d.	Sykes et al. (2011)
45 L raceway/Open/Sand					4.1	30.0	n.d.	n.d.	
45 L raceway/Open/White					4.3	36.7	n.d.	n.d.	
10 L raceway/Open/Black	F4	24.4 ± 1.3	37.0 ± 2.0	Live *P. varians*	10.4	2.9	n.d.	n.d.	Sykes et al. 2013
				Live *P. varians* (5 DAH) + frozen *P. varians*	5.8	14.3	n.d.	n.d.	
				Frozen *P. varians*	6.2	16.2	n.d.	n.d.	
10 L raceway/Open	F1	23.8 ± 1.1	37.0 ± 1.0	Live *P. varians*/100 Lx	8.0	11.1	104.5	2.0	Sykes et al. 2013c
				Live *P. varians*/350 Lx	7.9	8.9	104.4	2.0	
				Live *P. varians*/1,200 Lx	8.9	38.9	124.4	4.6	

IGR values were estimated when no values were specifically presented. Values shared by several different groups are presented solely in that column

Gen generation, *T* temperature, *S* salinity, *IGR* instantaneous growth rate (% BW/d), *FR* feeding rate (% BW/d), *FC* food conversion, *n.d.* not determined, *NK* unknown. *DAH* days after hatching, *BW* body weight

[a] Represent estimation from data on graphics

Fig. 11.5 Juvenile and adult grow out in tanks. **a** 1,500 L not painted. **b** 1,500 L painted in black and earthen ponds. **c** Pond at Necton, S.A. and cuttlefish juveniles. (Photos by A Sykes)

The conditions presented in this section are based on pilot projects performed in Portugal and Spain with the objective of determining the viability and conditions of a commercial operation. These studies were performed both in aquaculture research stations and in earthen ponds belonging to aquaculture companies, which have expressed interest in cuttlefish culture as an alternative species for diversification.

After going through the hatchling stage in the hatchery, cuttlefish should have attained full development of the digestive system and a mean weight of 5 g. At this point, depending on the facilities and the rearing objectives, cuttlefish may be relocated to tanks of different types (either in fibreglass or in concrete with increased size) or to earthen ponds. If kept in tanks (Fig. 11.5a, b), the seawater systems and conditions used should be similar to those detailed for breeders (see Sect. 11.3.1). Similar to the hatchling stage, the use of black tanks and low light intensities are recommended.

In order to increase the probability of cuttlefish finding the prey and spend the least energy possible while hunting, the water column of the tanks should be low. While cuttlefish grows, the water column should be increased to generate more volume (Forsythe et al. 1994).

If animals are relocated to earthen ponds (Fig. 11.5c), it should be taken care not to place them in a pond with a very different temperature, salinity, pH and dissolved oxygen below 80%. Big differences in these parameters (e.g. 1 °C, 1 psu) will promote immediate mass mortality. Transportation should be performed considering the aspects discussed in Sykes et al. (2012), such as duration and eventual use of anaesthesia. Acclimation after transportation similar to that achieved for fish in an

Table 11.4 Pros and cons of tanks versus earthen ponds

Variables	Tanks	Earthen ponds
Temperature	Only controlled in semi-open and closed seawater systems	Not controlled
Salinity	Only controlled in semi-open and closed seawater systems	Not controlled or hard and expensive to control
Dissolved oxygen	Fully controlled	Controlled through the use of aerators but primary productivity may generate spikes of lower oxygen during the night
Nitrogenous compounds	Controlled in open and semi-open systems	Controlled in ponds with water renewal
Density/biomass	Fully controlled	Hard to control due to water turbidity
Mortality/cannibalism	Fully controlled	Hard to control due to water turbidity
Food/feeding hierarchies	Controlled	Controlled
Predators	None	Require bird nets, filters of appropriate size in inlet water supply, if water renewal is used
System setup and maintenance	Logistics, time consuming and expensive in setup Maintenance is time consuming	Expensive in pond setup. Low expenses during maintenance
Filters inlet/outlet	Inlet filters in water of the facilities are expensive and require logistics and maintenance. Outlet filters used to prevent escape of animals	Inlet filters to prevent entrance of natural predators. Outlet filters to prevent escape of animals. Both are time consuming
Type of culture	Intensive and semi-intensive	Intensive, semi-intensive and extensive
Type of integrated aquaculture	Water used for growth of micro or macroalgae after leaving the tank	Pond used for bivalves and water used for micro or macroalgae after leaving the pond

aquarium should be enforced to obtain the best results. While culturing in ponds, water inlet and outlet filters of appropriate sizes should be used to prevent predator inputs and cuttlefish outputs. In addition, net barriers should be used for preventing cuttlefish capture by birds. Since these ponds are dynamic ecological systems, with proper fauna and flora, the use of oxygenators to efficiently supersaturate the rearing water is recommended. This might be performed either by mechanical movement (paddles) or by oxygen injection and will be particularly important at high water temperatures, in very productive waters and especially during nighttime, when microalgae also consume oxygen.

The pros and cons of using either tanks or ponds are presented in Table 11.4.

11.6.2 Density

Published information provides data regarding density in either open, semi-open and closed systems. As in other life stages, considering not only densities but also the available bottom area is of maximum importance. Forsythe et al. (1994) suggested a density of 20 cuttlefish m^{-2} in closed seawater systems at this ongrowing stage but, despite recognizing that bottom areas are important, no value was provided. A density of 400 cuttlefish m^{-2} is suggested by Forsythe et al. (2002) in 1,800 L circular tanks and closed seawater system. From this study, performed at rearing temperatures of 25 °C, this very high density is on the verge of impacting growth and survival, due to the increase of biomass present in the tanks.

As for densities in flow-through seawater systems, Sykes et al. (2003) suggested the use of 120 cuttlefish m^{-2} and minimum area of about 1,083 cm^2 (in 10 L raceway tanks), when starting a new juvenile tank with individuals of approximately 5 g. According to these authors, these density and bottom area values are valid for animals up to 25 g. Likewise, Domingues and Marquez (2010) studied the effects of both density and bottom areas in open seawater systems (in concrete raceway tanks) and obtained results that support the use of high-density and large bottom areas (33 cuttlefish m^{-2} with an average weight of 9.5 g), registering similar feeding rates (≈ 10 % body weight d^{-1}) but different food conversions. In fact, mortality and growth were similar between high- and low-density tanks using similar large bottom areas, which indicate that the bottom area seems to be more important than the density itself.

Independently from the rearing seawater system, density must be decreased and bottom areas increased while cuttlefish grows. It is suggested that, from 30 DAH to 10 g, the cited values of Sykes et al. (2003) should be used and from this weight and to maturation (which is temperature dependable) the findings of Domingues and Marquez (2010) should be considered.

If cuttlefish is reared in earthen ponds, special attention should be paid to the density and bottom areas, considering the fast growth rates that the species display and the inability to correctly observe what is happening within the ponds. If the pond's carrying capacity and biological limits are reached, cuttlefish mass mortality and loss of total biomass produced will occur. This will be due to the inability to clean the pond which will result in a spike in nitrogenous compounds and a drop in dissolved oxygen. A common observation to detect that this limit is being reached is to find eaten cuttlefish or cuttlebones in the pond as well as cuttlefish floating or swimming in the pond's surface.

11.6.3 Food

According to Warnke (1994), when cuttlefish are fed in a group, individuals hunt three times faster than when isolated, more food is ingested and feeding hierarchies are established.

Currently, growout of cuttlefish juveniles is performed with crustaceans as diet, mainly the grass shrimp—*P. varians* (Sykes et al. 2006a). This is due to the easiness of collection (logistics) and results obtained with this diet (Sykes et al. 2006a). Nonetheless, several different food items have been tested throughout the years, either solely or as mixed diets. Domingues et al. (2001a, b) used the crab *C. maenas*; Domingues et al. (2002, 2003b) and Sykes et al. (2006a) used live or frozen *P.*

varians with success. DeRusha et al. (1989); Domingues et al. (2004) and Almansa et al. (2006) tested the effect of exclusively using fish and reported lower growth than when using grass shrimp.

Despite all this effort, it is not economically viable to produce cuttlefish in large numbers with a growout based on any of those foods. Not only is the amount of food biomass too high but also the availability of these food resources is scarce. Therefore, further research has been performed with the objective of developing alternative diets to cephalopods. Several works regarding trials on artificial diets were performed first by Lee et al. (1991), Castro and colleagues (Castro 1991; Castro et al. 1993; Castro and Lee 1994) and, more recently, by Domingues et al. (2005, 2008) and Ferreira et al. (2010). However, all these attempts to feed cuttlefish on a prepared diet failed to achieve proper growth and survival (for details, see Table 11.5). Fish pellets have also been tried and accepted by juvenile cuttlefish. However, after only 2 days, individuals started to reject this food and cannibalism was observed.

The lack of proper knowledge regarding the physiology and metabolism of the species, at different geographical locations, has hampered the existence of a successful design. Only by having a well-designed, inexpensive and storable artificial diet, cuttlefish aquaculture will reach its maturity and industrial stage (Sykes et al. 2006b). In this way, Domingues et al. (2009) tested the effects of thermal treatment of food given to cuttlefish and obtained interesting growth and survival data that suggest this process affects diet quality, by provoking protein denaturation, washing of proteins and amino acids and lipid oxidation. Current research in Portugal is focused on developing new diets based on the most recent knowledge regarding cuttlefish physiology, metabolism and raw materials. Up to now, similar results to those obtained by Hanlon et al. (1991) and Lee et al. (1991) were registered. Nonetheless, the lack of knowledge regarding cuttlefish physiology and metabolism still remains as the bottleneck to overcome.

11.6.4 Cleaning Techniques

In ponds, periodical cleaning is only performed at the inlet and outlet filters but several cleaning routines should be performed in tanks.

According to Forsythe et al. (1994), no substratum is needed for normal growth and survival, even in cultures performed at high densities. This practice not only facilitates cleaning itself but also prevents pathologies and promotes good welfare under culture conditions (Sykes et al. 2012). Tanks should be siphoned daily with a hose and purged (total time spent will depend on the water quality and volume of the tank). On a weekly basis, airlifts, air stones and water outlet filters should be cleaned with a scrubber with tap water. Afterwards, these elements should be kept for 24 h in Atlantol 914 (Atlantol, Belgium), rinsed with tap water and then placed in VirkonS (DuPont Animal Health Solutions, Europe) for 15 min. Finally, all these materials should once again be abundantly rinsed with tap water. The nets or any material used for animals handling should follow a similar procedure to avoid contaminations, pathologies and spread of diseases.

Table 11.5 Cuttlefish juveniles' growth rates, mortality, feeding rates and food conversions of animals fed live, frozen and artificial diets

Rearing system	Days	T (°C)	S (PSU)	Food	MWWi (g)	GR (% BW/d)	Mortality (%)	FR (% BW/d)	FC	Reference
Closed; 72 L round	57	UK	35.0	Penaeid pellet	13.9 ± 2.6	1.5	35.0	n.d.	n.d.	Lee et al. (1991)
Closed; 500 L round	45	22.0 ± 1.7	35.0	*Palaemon setiferus*	74.5 ± 12.4	2.7	5.0	6.0–8.0	39.0–45.0	Castro et al. (1993)
				P. setiferus pellet		0.3	32.5	< 1.0–3.0	0.0–20.0	
				Catfish surimi		0.5	77.5	8.0–< 4.0	0.0–20.0	
Closed; 500 L round	14	21.0 ± 1.0	35.0	*P. setiferus*	106.0 ± 17.6	2.8	None	7.9	34.6	Castro and Lee (1994)
				Surimi		0.1	None	5.3	0.8	
				Surimi—menhaden oil + egg albumin		−0.2	None	3.2	−8.3	
				Surimi + egg albumin		−0.1	None	3.4	0.1	
				Surimi + casein		0.3	None	4.2	8.6	
				Surimi + whole egg		−0.3	None	2.5	−13.5	
Open; 10 L plastic rectangular tanks	29	24.7	36.5	Live *P. varians*	5.5 ± 0.8	4.4[a]	22.2	11.9[a]	31.0–46.0	Domingues et al. (2003b)
				Frozen *P. varians*	5.6 ± 0.9	4.4[a]	2.2	10.9[a]	34.0–49.0	
Open; 10 L plastic rectangular tanks	28	20.0 ± 2.0	36.0 ± 1.0	Live *P. varians*	0.6 ± 0.1	5.7	None	n.d.	n.d.	Domingues et al. (2004)
				Live *Atherina* sp.		2.4	8.3	n.d.	n.d.	
	28			Live *P. varians*	5.7 ± 0.1	5.8	None	n.d.	n.d.	
				Live *Atherina* sp.		3.5	None	n.d.	n.d.	
	35			Frozen *P. varians*	2.2 ± 0.4	5.1	20.0	12.4	0.4	
				Frozen *Atherina* sp.		3.3	8.1	11.7	0.3	
Closed; 8 L aquaria in 150 L tank	35	19.0	NK	*C. crangon*	2.0–2.8	3.7	None	9.3	2.5[a]	Grigoriou and Richardson (2004)

Table 11.5 (continued)

Rearing system	Days	T (°C)	S (PSU)	Food	MWWi (g)	GR (% BW/d)	Mortality (%)	FR (% BW/d)	FC	Reference
Closed; 1.5 L aquaria in 150 L tank	40	11.0		*C. crangon*	3.0–8.2	0.7	None	2.8	4.0[a]	
		19.0		*C. crangon*		3.5	27.0	8.3	2.4[a]	
Closed; 500 L	30	23 ± 1.5	35.0 ± 2.0	Surimi lysine diet 1	321.8 ± 57.9	−0.02	12.5	2.7	−40.8	Domingues et al. (2005)
				Surimi lysine diet 2		0.2	20.8	2.7	26.7	
				Surimi lysine diet 3		0.1	20.8	2.8	80.9	
				Surimi lysine diet 4		0.3	20.8	2.8	10.7	
Closed; 9 L	21			Surimi lysine diet 1	451.5 ± 103.7	−0.3	NK	1.1[a]	−3.1[a]	
				Surimi lysine diet 2		0.1	NK	2.7[a]	−1.4[a]	
				Surimi lysine diet 3		0.1	NK	1.8[a]	22.1[a]	
				Surimi lysine diet 4		0.4	NK	2.3[a]	6.3[a]	
Open; 250 L	60	21.5 ± 1.5	36.0 ± 1.0	Live *P. varians*	13.3 ± 3.4	3.1	NK	n.d.	n.d.	Almansa et al. (2006)
				Live *Atherina* sp. and *Gobius* sp.	12.9 ± 2.8	1.7	NK	n.d.	n.d.	
				Live *P. varians*	85.6 ± 11.7	2.1	NK	n.d.	n.d.	
				Live *Atherina* sp. and *Gobius* sp.	86.2 ± 8.6	1.0	NK	n.d.	n.d.	
Open; 27 L		21.0 ± 1.0	37.0 ± 1.0	Frozen *Palaemonetes* sp.	44.4 ± 0.5	1.5	None	7.8	18.6	Domingues et al. (2008)
				Frozen *Procambarus clarkii*		1.1	None	8.4	13.7	
	29			Frozen *Sardina pilchardus*		0.3	None	4.4	5.8	
				Artificial diet 1		−0.7	None	2.0	< 1.0	
				Artificial diet 2		−0.5	None	2.4	< 1.0	

Table 11.5 (continued)

Rearing system	Days	T (°C)	S (PSU)	Food	MWWi (g)	GR (% BW/d)	Mortality (%)	FR (% BW/d)	FC	Reference
Open; 40 L	20	20.5 ± 1.0	37.0 ± 1.0	Frozen *P. varians*	12.5 ± 1.0	2.0	None	7.4[a]	23.1	Domingues et al. (2009)
				100 °C boiled *P. varians*	12.4 ± 0.6	0.5[a]	3.3	8.5[a]	3.0	
				60 °C dried *P. varians*	12.6 ± 0.6	−0.2[a]	None	7.7[a]	1.4	
	40	21.5 ± 1.0		Frozen *P. varians*	5.5 ± 0.6	2.1	None	8.6	25.7	
				60 °C dried *P. varians*	5.9 ± 0.3	0.5	8.3	8.7	6.6	
				Freeze-dried *P. varians*	6.0 ± 0.4	1.8	8.3	8.7	20.4	
Open; 40 L	29	21.0 ± 1.0	37.0 ± 1.0	Frozen *Palaemonetes* sp.	13.8 ± 2.3	3.8	3.0	8.8	42.6	Ferreira et al. (2010)
				Frozen *S. pilchardus*		0.8	20.0	6.6	13.6	
				Frozen *P. clarkii*		2.3	13.0	10.5	20.9	
				Artificial diet 1 (fish powder)		−1.8	47.0	4.5	< 1.0	
				Artificial diet 2 (shrimp powder)		−0.3	47.0	2.4	< 1.0	

Values were estimated when no values were specifically presented. Values shared by several different groups are presented solely in that column

Days days of rearing, *T* temperature, *S* salinity, *MWWi* mean wet weight at the beginning of experiment, *GR* growth rate (% BW/d), *FR* feeding rate (% BW/d), *FC* food conversion, *n.d.* not determined, *NK* unknown

[a] Represent estimation from data on manuscript

11.6.5 Growth, Survival and Sampling

Time to marketable size will greatly depend on the requested product itself. For instance, in some Mediterranean countries, such as Portugal and Italy, undersized cuttlefish individuals (5–25 g) are extremely appreciated and their commercial value is higher than that of animals surpassing 100 g. Time to market will also be dependent on culture conditions, especially temperature and food.

S. officinalis is cultured extensively in Portugal, Italy and Tunisia. By definition, this type of culture does not involve human action at the food level, so eggs are caught and left in earthen ponds with naturally occurring prey, where animals are grown and collected a few months later when they reach the marketable size. According to Palmegiano and Sequi (1981), in this type of culture, 0.15–0.30 kg of eggs will produce 800–1,200 kg of cuttlefish with a mean weight of 0.04–0.08 g. The semi-intensive experiments performed in Italy in the 1980s obtained fast growth rates (14.2 g in 60 days, at 21–24 °C in ponds (Palmegiano and Sequi 1981); 25 g in 40 days and 80 g in 100 days, at 21–24 °C in concrete tanks (Palmegiano and Sequi 1984); 80 kg ha^{-1} in 90 days, at 21–24 °C in net cages in ponds (Sequi and Palmegiano 1984)).

Similar high growth rates were obtained in 2004, in earthen ponds of a commercial company at Algarve (Portugal; unpublished data). From an initial biomass of 600 g of cuttlefish juveniles (with mean wet weight of 1.65 ± 0.62 g), 5,000 g of cuttlefish (30.75 ± 11.25 g) were produced in 52 days, with a mean water temperature of 21.3 ± 2.0 °C. Higher growth rates were not obtained due to problems related to initial adaptation of cuttlefish to unfavourable conditions, such as oxygen depletion, lack of food during the fattening and bird predation observed during fishing at the end of the experiment. These factors promoted an overall mortality of 35 %.

Summarizing and comparing data in tanks from Sykes et al. (2006a) with those obtained in earthen ponds, it is possible to see that it takes longer (70–90 days in total) to achieve a similar mean weight in tanks than in ponds, when cuttlefish is reared at a similar temperature. When food is available and other identified conditions are met, cuttlefish will grow faster outdoors than indoors, which is in accordance to the findings of Domingues et al. (2006). So depending on the weight request, it is possible to produce cuttlefish to meet market demands within 2–3 months or even less. To maintain a high-quality product for human consumption and enforce the best welfare practice, cuttlefish should be killed by thermal shock in ice slurry water.

11.7 Trends in Research and Industrial Level

The bottlenecks in the cuttlefish culture identified by Sykes et al. (2006b)—reproduction, feeding and nutrition—are still under research. They need to be solved to improve the current methodology and thereby apply to an industrial scale.

Currently, Centro de Ciências do Mar do Algarve (CCMAR) is the only research centre performing the culture of European cuttlefish in large numbers and they are able to supply cultured animals to research centres and public aquaria. In terms of application, there is potential beyond aquaculture for human consumption, development of refined guidelines and best practice methods for cuttlefish culture. For instance, it is expected that cuttlefish will become the marine laboratory rat, as animals are needed for formation in aquaria or universities and as marine animal models for research. As a matter of fact, cuttlefish is included as one of the model species available at *EMBRC*, a platform created by the EU to facilitate research with animal models, such as specific inbred animals only obtained under culture conditions.

Regarding aquaculture, while commercial retailers are eager to get cultured cuttlefish, with the most appreciated sizes (e.g. cuttlefish under the allowed limit for being captured), the aquaculture industry sees the development of this technology as one with a very low potential, such as something that will never be reached.

At present, gaps in the knowledge of cuttlefish biology, the time expended to develop the culture methods (due to the low number of researchers and research laboratories involved) and the fact that most of the aquaculture investors who are quite unadventurous with regard to business planning are still hampering the excellent opportunity for the diversification of marine aquaculture. The worldwide market prices for the species are high and the short-life cycles will support an easy and short payback time of the investment made. In fact, the payback can be shortened and profit increased if cuttlefish culture is performed in integrated aquaculture. The water of the tanks is extremely rich in nitrogenous compounds which may be used for micro- and macroalgae production. In addition, while performing the experiments in ponds, cuttlefish can be reared at the same time with bivalve species, which will lower the microalgae content in the water and can be sold at the time of harvesting, promoting increased income.

It is also important to acknowledge that, if resources are used efficiently, the whole animal may be sold to be used by different transformation industries. For instance, cuttlefish mantle and ink not only will be used for human consumption but also may be enforced for recycling of by-products for feeds (e.g. cuttlefish viscera is naturally rich in n-3 and n-6 fatty acids and amino acids) and natural products (e.g. cuttlefish bone is made of aragonite and it could be used in the pharmacological industry). For further details, see Chap. 9 'Applications, Uses and By-products from Cephalopods' of this volume.

The species potential is there and it needs to be grasped in short-term future.

11.8 Conclusions

Steady progress in the development of cuttlefish culture technology has been attained since the 1980s, when this species was suggested to have a potential for aquaculture due to the short life cycle and high market prices which translates into short payback time of the investment. This chapter of the book includes a very thorough description of the methods to culture cuttlefish. However, the existing knowledge is still not sufficient to sustain the culture of the species at the industrial level for human consumption. This is due to the existing bottlenecks regarding reproduction, feeding and nutrition that are currently being researched.

The species has an incredible potential regarding its use in integrated aquaculture with bivalves and regarding the use of the whole animal as a resource for different transformation industries, such as recycling of by-products for feeds and other pharmacological products. The species is also seen as a very interesting animal model for several fields of research that include physiology, genetics, etc. For instance, the opportunity to have this animal model culture in the laboratory will allow for inbred samples for upcoming research on the *Sepia* genome. The existing culture protocols allow for small-scale culture for the supply of research centres and public aquaria.

References

Almansa E, Domingues P, Sykes A et al (2006) The effects of feeding with shrimp or fish fry on growth and mantle lipid composition of juvenile and adult cuttlefish (*Sepia officinalis*). Aquaculture 256:403–413

Barnabé G (1996) Bases biologiques et écologiques de l'aquaculture. Editorial Acribia, S.A., Zaragoza

Blanc A, Daguzan J (1998) Artificial surfaces for cuttlefish eggs (*Sepia officinalis* L.) in Morbihan Bay, France. Fish Res 38:225–231

Boal JG (1997) Female choice of males in cuttlefish (Mollusca: Cephalopoda). Behaviour 134:975–988

Boal JG (2006) Social recognition: a top down view of cephalopod behaviour. Vie Milieu 56:69–79

Boal JG, Golden DK (1999) Distance chemoreception in the common cuttlefish, *Sepia officinalis* (Mollusca, Cephalopoda). J Exp Mar Biol Ecol 235:307–317

Boal JG, Marsh SE (1998) Social recognition using chemical cues in cuttlefish (*Sepia officinalis* Linnaeus, 1758). J Exp Mar Biol Ecol 230:183–192

Boal JG, Hylton RA, Gonzalez SA et al (1999) Effects of crowding on the social behavior of cuttlefish (*Sepia officinalis*). Contemp Top Lab Anim Sci 38:49–55

Boletzky SV (1983) *Sepia officinalis*. In: Boyle PR (ed) Cephalopod life cycles. Academic Press, London, pp 31–52

Boletzky SV (1987) Fecundity variation in relation to intermittent or chronic spawning in the cuttlefish, *Sepia officinalis* L. (Mollusca, Cephalopoda). Bull Mar Sci 40:382–387

Boletzky SV (2003) Biology of early life stages in cephalopod molluscs. Adv Mar Biol 44:143–203

Boletzky SV, Erlwein B, Hofmann DK (2006) The Sepia egg: a showcase of cephalopod embryology. Vie Milieu 56:191–201

Boucaud-Camou E (1989) L'Aquaculture des cephalopodes: evaluation et perspectives. Haliotis 19:201–214

Boucaud-Camou E (1990) La Seiche, un animal d'avenir. Peche Maritime 69:321–329

Castro BG (1991) Can *Sepia officinalis* L be reared on artificial food. Mar Behav Physiol 19:35–38

Castro BG, Lee PG (1994) The effects of semi-purified diets on growth and condition of *Sepia officinalis* L (Mollusca, Cephalopoda). Comp Biochem Physiol A Physiol 109:1007–1016

Castro BG, Dimarco FP, Derusha RH et al (1993) The effects of surimi and pelleted diets on the laboratory survival, growth, and feeding rate of the cuttlefish *Sepia officialis* L. J Exp Mar Biol Ecol 170:241–252

Coelho ML, Modesto T, Soares F (1989) *Sepia officinalis:* potential for aquaculture and restocking. In: Boucaud-Camou E (ed) The cuttlefish. Université de Caen, Caen, p 348

Correia M, Domingues PM, Sykes A et al (2005) Effects of culture density on growth and broodstock management of the cuttlefish, *Sepia officinalis* (Linnaeus, 1758). Aquaculture 245:163–173

Correia M, Palma J, Andrade JP (2008a) Effects of live prey availability on growth and survival in the early stages of cuttlefish *Sepia officinalis* (Linnaeus, 1758) life cycle. Aquacult Res 39:33–40

Correia M, Palma J, Kirakowski T et al (2008b) Effects of prey nutritional quality on the growth and survival of juvenile cuttlefish, *Sepia officinalis* (Linnaeus, 1758). Aquacult Res 39:869–876

Derusha RH, Forsythe JW, Dimarco FP et al (1989) Alternative diets for maintaining and rearing cephalopods in captivity. Lab Anim Sci 39:306–312

Domingues P, Marquez L (2010) Effects of culture density and bottom area on growth and survival of the cuttlefish *Sepia officinalis* (Linnaeus, 1758). Aquacult Int 18:361–369

Domingues P, Poirier R, Dickel L et al (2003a) Effects of culture density and live prey on growth and survival of juvenile cuttlefish, *Sepia officinalis*. Aquacult Int 11:225–242

Domingues P, Sykes A, Sommerfield A et al (2003b) Effects of feeding live or frozen prey on growth, survival and the life cycle of the cuttlefish, *Sepia officinalis* (Linnaeus, 1758). Aquacult Int 11:397–410

Domingues P, Sykes A, Sommerfield A et al (2004) Growth and survival of cuttlefish (*Sepia officinalis*) of different ages fed crustaceans and fish. Effects of frozen and live prey. Aquaculture 229:239–254

Domingues P, Ferreira A, Marquez L et al (2008) Growth, absorption and assimilation efficiency by mature cuttlefish (*Sepia officinalis*) fed with alternative and artificial diets. Aquacult Int 16:215–229

Domingues P, Marquez L, López N et al (2009) Effects of food thermal treatment on growth, absorption, and assimilation efficiency of juvenile cuttlefish, *Sepia officinalis*. Aquacult Int 17:283–299

Domingues PM, Kingston T, Sykes A et al (2001a) Growth of young cuttlefish, *Sepia officinalis* (Linnaeus 1758) at the upper end of the biological distribution temperature range. Aquacult Res 32:923–930

Domingues PM, Sykes A, Andrade JP (2001b) The use of Artemia sp or mysids as food source for hatchlings of the cuttlefish (*Sepia officinalis* L.); effects on growth and survival throughout the life cycle. Aquacult Int 9:319–331

Domingues PM, Sykes A, Andrade JP (2002) The effects of temperature in the life cycle of two consecutive generations of the cuttlefish *Sepia officinalis* (Linnaeus, 1758), cultured in the Algarve (South Portugal). Aquacult Int 10:207–220

Domingues PM, Dimarco FP, Andrade JP et al (2005) Effect of artificial diets on growth, survival and condition of adult cuttlefish, *Sepia officinalis* Linnaeus, 1758. Aquacult Int 13:423–440

Domingues PM, Bettencourt V, Guerra A (2006) Growth of *Sepia officinalis* in captivity and in nature. Vie Milieu 56:109–120

Establier R, Pascual E (1983) Effects of cadmium and copper on the development of eggs of *Sepia officinalis* L. Inv Pesquera 47:143–150

FAO (2010) The state of world fisheries and aquaculture 2010. Food and Agriculture Organization of the United Nations, Rome

FAO (2012) The state of world fisheries and aquaculture 2012. Food and Agriculture Organization of the United Nations, Rome

FAO-FIGIS (2012) Fisheries Global Information System (FIGIS). http://www.fao.org/fishery/figis/en. Accessed 15 July 2012

Ferreira A, Marquez L, Almansa E et al (2010) The use of alternative diets to culture juvenile cuttlefish, *Sepia officinalis*: effects on growth and lipid composition. Aquacult Nutr 16:262–275

Forsythe JW, Van Heukelem WF (1987) Growth. In: Boyle PR (ed) Cephalopod life cycles. Academic Press, Inc., London, pp 135–156

Forsythe JW, Hanlon RT, Derusha R (1991) Pilot-scale culture of *Sepia* in biomedical research. In: Boucaud-Camou E (ed) The cuttlefish. Centre de Publications de l'Université de Caen, Caen, pp 313–323

Forsythe JW, Derusha RH, Hanlon RT (1994) Growth, reproduction and life-span of *Sepia officinalis* (Cephalopoda, Mollusca) cultured through 7 consecutive generations. J Zool 233:175–192

Forsythe J, Lee P, Walsh L et al (2002) The effects of crowding on growth of the European cuttlefish, *Sepia officinalis* Linnaeus, 1758 reared at two temperatures. J Exp Mar Biol Ecol 269:173–185

Grigoriou P, Richardson A (2004) Aspects of the growth of cultured cuttlefish *Sepia officinalis* (Linnaeus, 1758). Aqua Res 35: 1141–1148.

Gonçalves JMA (1989) Perspectivas de repovoamento e aquacultura de *Sepia officinalis* (Mollusca: Cephalopoda) em sistemas lagunares costeiros. Açoreana 7:143–152

Hanley JS, Shashar N, Smolowitz R et al (1999) Soft-sided tanks improve long-term health of cultured cuttlefish. Biol Bull 197:237–238

Hanlon RT (1990) Maintenance, rearing and culture of teuthoid and sepioid squids. In: Guilbert DL, Adelman WJ Jr, Arnold JM (ed) Squid as experimental animals. Plenum Press, New York, pp 35–61

Hanlon RT, Forsythe JW (1985) Advances in the laboratory culture of Octopuses for biomedical research. Lab Anim Sci 35:33–40

Hanlon RT, Messenger JB (1996) Cephalopod behaviour. Cambridge University Press, New York

Hanlon RT, Bidwell JP, Tait R (1989) Strontium is required for statolith development and thus normal swimming behavior of hatchling cephalopods. J Exp Biol 141:187–195

Hanlon RT, Turk PE, Lee PG (1991) Squid and cuttlefish mariculture: an updated perspective. J Ceph Biol 2:31–40

Hanlon RT, Ament SA, Gabr H (1999) Behavioral aspects of sperm competition in cuttlefish, *Sepia officinalis* (Sepioidea: Cephalopoda). Mar Biol 134:719–728

Koueta N, Boucaud-Camou E (1999) Food intake and growth in reared early juvenile cuttlefish *Sepia officinalis* L (Mollusca Cephalopoda). J Exp Mar Biol Ecol 240:93–109

Koueta N, Boucaud-Camou E (2003) Combined effects of photoperiod and feeding frequency on survival and growth of juvenile cuttlefish *Sepia officinalis* L. in experimental rearing. J Exp Mar Biol Ecol 296:215–226

Koueta N, Andrade JP, Von Boletzky S (2006) The cuttlefish *Sepia officinalis* a working model in cephalopod research. Vie Milieu 56:67–67

Laptikhovsky V, Salman A, Onsoy B et al (2003) Fecundity of the common cuttlefish, *Sepia officinalis* L. (Cephalopoda, Sepiidae): a new look at an old problem. Sci Mar 67:279–284

Lee PG, Forsythe JW, Dimarco FP et al (1991) Initial palatability and growth trials on pelleted diets for cephalopods. Bull Mar Sci 49:362–372

Lemaire J (1971) Étude du developpement embryonnaire de *Sepia officinalis* L. (Mollusque, Cephalopode). Universite des Sciences et Techniques Lille, France

Martins CIM, Eding EH, Verdegem MCJ et al (2010) New developments in recirculating aquaculture systems in Europe: a perspective on environmental sustainability. Aquacult Eng 43:83–93

Naef A (1928) Die Cephalopoden (Embryologie), Fauna e Flora del Golfo di Napoli. Naples. English edition: Naef A (1992) Cephalopoda Embryology, Fauna and flora of the Bay of Naples (trans: Boletzky Sv). Smithsonian Institution Libraries, Washington, DC

Nixon M (1985) Capture of prey, diet and feeding of *Sepia officinalis* and Octopus vulgaris (Mollusca: Cephalopoda) from hatchling to adult. Vie Millieu 35:255–261

Nixon M, Mangold K (1998) The early life of *Sepia officinalis*, and the contrast with that of Octopus vulgaris (Cephalopoda). J Zool 245:407–421

Oestmann DJ, Scimeca JM, Forsythe JW et al (1997) Special considerations for keeping cephalopods in laboratory facilities. Contemp Top Lab Anim Sci 36:89–93

Palmegiano GB, Sequi R (1981) On the possibility of the introduction of cuttlefish (*Sepia officinalis* L.) in a management strategy for the coastal lagoons. In: Symposium International sur les Lagunes Cotieres. Bordeaux, France 8–14 September 1981, pp 1–9

Palmegiano GB, Sequi R (1984) Allevamento estensivo di *Sepia officinalis* L. Agricoltura e Ricerca 41:49–56

Paulij WP, Zurburg W, Denuce JM et al (1990) The effect of copper on the embryonic-development and hatching of *Sepia officinalis* L. Arch Environ Contam Toxicol 19:797–801

Paulij WP, Herman PMJ, Roozen MEF et al (1991) The influence of photoperiodicity on hatching of *Sepia officinalis*. J Mar Biol Assoc UK 71:665–678

Quintela J, Andrade JP (2002) Diel feeding rhythms, daily ration and gastric evacuation rates of *Sepia officinalis* in the Ria Formosa lagoon (south Portugal). Bull Mar Sci 71:665–680

Richard A (1971) Contribuition à l'étude expérimentale de la croissance et de la maturation sexuelle de *Sepia officinalis* L. (Mollusque, Céphalopode). Université de Lille, Lille, p 264

Rodger GK, Davies IM (2000) Summary of mariculture production in countries neighbouring the European Union. J Appl Ichthyol 16:224–229

Sendão J, Borges TC, Xavier JC (2007) Cephalopoda. In: Borges TC (ed) Biodiversity in the fisheries of algarve (South Portugal). Results of program MARE (22-05-01-FDR-00031). Universidade do Algarve, Faro, p 685

Sequi R, Palmegiano GB (1984) Some alternatives for the use of cuttlefish (*Sepia officinalis* L.) resources in coastal lagoons. In: GFCM Studies and Reviews—Management of Coastal Lagoons Fisheries. FAO, Rome, pp 607–614

Shashar N, Hagan R, Boal JG et al (2000) Cuttlefish use polarization sensitivity in predation on silvery fish. Vision Res 40:71–75

Shashar N, Milbury CA, Hanlon RT (2002) Polarization vision in cephalopods: neuroanatomical and behavioral features that illustrate aspects of form and function. Mar Fresh Behav Physiol 35:57–68

Sykes AV, Domingues PM, Loyd M et al (2003) The influence of culture density and enriched environments on the first stage culture of young cuttlefish, *Sepia officinalis* (Linnaeus, 1758). Aquac Int 11:531–544

Sykes AV, Domingues PM, Andrade JP (2004) Nucleic acid derived indices or instantaneous growth rate as tools to determine different nutritional condition in cuttlefish (*Sepia officinalis*, Linnaeus 1758) hatchlings. J Shellfish Res 23:585–591

Sykes AV, Domingues PM, Andrade JP (2006a) Effects of using live grass shrimp (Palaemonetes varians) as the only source of food for the culture of cuttlefish, *Sepia officinalis* (Linnaeus, 1758). Aquac Int 14:551–568

Sykes AV, Domingues PM, Correia M et al (2006b) Cuttlefish culture—state of the art and future trends. Vie Milieu 56:129–137

Sykes AV, Almansa E, Lorenzo A et al (2009) Lipid characterization of both wild and cultured eggs of cuttlefish (*Sepia officinalis* L.) throughout the embryonic development. Aquac Nut 15:38–53

Sykes AV, Domingues PM, Marquez L et al (2011) The effects of tank colours on the growth and survival of cuttlefish (*Sepia officinalis*, Linnaeus 1758) hatchlings and juveniles. Aquac Res 42:441–449

Sykes AV, Baptista FD, Gonçalves RA et al (2012) Directive 2010/63/EU on animal welfare: a review on the existing scientific knowledge and implications in cephalopod aquaculture research. Rev Aquac 4:142–162

Sykes AV, Gonçalves RA, Andrade JP (2013a) Early weaning of cuttlefish (*Sepia officinalis*) with frozen grass shrimp (Palaemonetes varians) from the first day after hatching. Aquacult Res 44: 1815–1823, doi:10.1111/j.1365-2109.2012.03186.x

Sykes AV, Pereira D, Rodríguez C et al (2013b) Effects of increased tank bottom areas on cuttlefish (*Sepia officinalis*, L.) reproduction performance. Aquacult Res 44: 1017–1028. doi:10.1111/j.1365-2109.2012.03106.x

Sykes AV, Quintana D, Andrade JP (2013c) The effects of light intensity on growth and survival of cuttlefish (*Sepia officinalis*, Linnaeus 1758) hatchlings and juveniles. Aquac Res (in press), doi:10.1111/are.12150

Warnke K (1994) Some aspects of social interaction during feeding in *Sepia officinalis* (Mollusca: Cephalopoda) hatched and reared in the laboratory. Vie Milieu 44:125–131

Zatylny C, Gagnon J, Boucaud-Camou E et al (2000) ILME: a waterborne pheromonal peptide released by the eggs of *Sepia officinalis*. Biochem Biophys Res Commun 275:217–222

Zatylny C, Marvin L, Gagnon J et al (2002) Fertilization in *Sepia officinalis*: the first mollusk sperm-attracting peptide. Biochem Biophys Res Commun 296:1186–1193

Chapter 12
Sepia pharaonis

Jaruwat Nabhitabhata

Abstract The pharaoh cuttlefish, *Sepia pharaonis*, is one of the largest and economically important sepiid cuttlefish in the Indo-Southwest Pacific region. The species became a focus for aquaculture when all previous researches agreed that it is an "easy-to-culture" species due to its reproducibility and high tolerability to culture conditions, hence a high feasibility for commercial scale culture. The cuttlefish benthic habit is one of the advantages for high-density culture which results in high survival. The aquaculture process comprises a collection of live broodstocks, incubation of eggs, nursing of young and growout. The culture methodology is studied in tropical and temperate countries either in closed or in opened seawater systems. Several consecutive generations can be cultured in both systems. Different culture conditions yield different results in growth, final size and longevity of life span. Success in training the pharaoh cuttlefish to feed on dead food yields the success of the culture. The cuttlefish can be cultured as human food, fresh and frozen, ornamental and experimental animals. Various sizes of cuttlefish supplying such various purposes can be produced through difference in culture periods. Innate feeding on specific live prey during the early phase of cuttlefish life is the bottleneck for large-scale culture.

Keywords *Sepia pharaonis* · Large cuttlefish · Benthic habit · High tolerability · Closed and opened seawater systems · Consecutive generations · Various purposes

12.1 Importance of this Species in the Market

The vernacular of the pharaoh cuttlefish, *Sepia pharaonis*, in Thai is "pla muek kradong lai sua" and in Japanese is "torafu-kouika". Both names similarly mean "tiger cuttlefish" after the transverse striped pattern on the dorsum, especially in the male. The pharaoh cuttlefish is a neritic demersal species, occurring in coastal shallows down to a depth of 100 m (Chotiyaputta 1995). The geographical distribution of

J. Nabhitabhata (✉)
Excellence Centre for Biodiversity of Peninsular Thailand, Faculty of Science, Prince of Songkla University, Hatyai, Songkhla 90112, Thailand
e-mail: jaruwat.n5@gmail.com; jaruwat.n@psu.ac.th

J. Iglesias et al. (eds.), *Cephalopod Culture*, DOI 10.1007/978-94-017-8648-5_12,

this species covers the Indian Ocean and the West Pacific, 50–150°E, 40°N–30°S (Reid 1998; Reid et al. 2005). This cuttlefish is one of the largest sepiid species of economic importance (Supongpan 1995). Roper et al. (1984) and Reid et al. (2005) have stated that the maximum size of *S. pharaonis* is between 350 and 420 mm in mantle length and between 4,200 and 5,000 g in weight, respectively. Chotiyaputta (1982, 1995) have reported that the maximum size of this species in Thailand is about 260 mm in mantle length and 1,400 g in body weight, the most common sizes captured being between 40 and 260 mm. In India, the sizes of the mostly captured cuttlefish are between 90 and 230 mm, with a maximum size of 334 mm (Nair et al. 1993; Silas et al. 1985). Females are larger than males, and the male to female sex ratio in the wild is 1:3. The individual fecundity averages 1,400 (800–2,700) eggs in the wild stocks (Chotiyaputta 1979, 1980, 1981, 1982). The spawning is intermittent terminal and occurs all the year round, with two peaks annually, as a consequence of the plasticity in their life cycle, which is less than 1 year either in the short or long mode (Nabhitabhata and Nilaphat 1999).

The total capture of the cuttlefish worldwide is approximately 500,000 t, and has exceeded 250,000 t since 1977, 300,000 t from 1991 to 2008 with a rise to more than 400,000 t during 1995–2002 (Jereb et al. 2005; Food and Agriculture Organization of the United Nations 2010). The major producers of total cuttlefish are China (>50 %) and Thailand (>15 %). The pharaoh cuttlefish, *S. pharaonis*, constitutes about 16 %, or 10,000 t, of the approximately 60,000 t of total cuttlefish that is landed annually by commercial fishing in Thailand. The yield is mainly from offshore otter-board and pair trawlers (>90 %), with some addition from inshore squid trapping, trammel netting, push netting, lift netting and other artisanal fishing activities (Nair et al. 1993; Supongpan 1995). The main target species of squid traps is the bigfin squid, *Sepioteuthis lessoniana* that constitutes 90–95 % of the catch, but *S. pharaonis* is a major by-catch, of approximately 5–10 %, with a maximum of 30 % (Boongerd and Rachaniyom 1990; Chotiyaputta and Yamrungreung 1998; Department of Fisheries 2009; see also Chap. 17 of this volume). For the 2,000–7,000 t annual yield of cephalopods from inshore trap fishing in Thailand since 1981, the cuttlefish yield is about 100–700 t (Department of Fisheries 2012).

The cuttlefish *S. pharaonis* is normally benthic in habit and ritually lies on or burrows into the substratum. The pharaoh cuttlefish is habitually solitary, but is also referred to as being moderately aggregative (Chikuni 1983; Nabhitabhata 1978). Nair et al. (1985) reported that adult *S. pharaonis* in the wild had a strong positive phototaxis, but this is different from observations in captivity, where this species has a moderate positive phototaxis. After 1–3 h of lighting, only about 50 % of the cuttlefish hatchlings move from the dark chamber to the light chamber in an experimental tank (Nabhitabhata, unpublished data). This moderately positive response agrees well with the evidence that the major catch of this species was not from the light-luring fishing but from the trawling, which is in contrast to loliginid squid fishing.

Large-scale aquaculture of the pharaoh cuttlefish has been studied since 1978 (Nabhitabhata 1978) in Thailand, later in India (Anil et al. 2005) and the USA (Minton et al. 2001). The low growth rate and small final weight of about 370 g obtained in captivity are disadvantageous for aquaculture in terms of quantitative production compared to those with larger weight of more than 1,000 g from fisheries

(Nabhitabhata 1995, 2000; Nabhitabhata and Nilaphat 1999). However, the small weight of 370 g turns out to be an advantage, since the small final size is optimal for frozen human food product processing and packaging. In addition, the live cuttlefish of this size are more suitable for ornamental aquarium displays.

12.2 State of the Art

The culture of the pharaoh cuttlefish throughout the life cycle is feasible in either open or closed seawater systems. Broodstocks collected from the wild can reproduce in captivity. Egg capsules are collected and nursed under a controlled environment until hatching after about 14 days at 28 °C. The hatching rate is higher than 90 %. Hatchlings are fed with wild-collected live food for about 30 days (Nabhitabhata and Nilaphat 1999). The volume of water and density are managed to facilitate rapid growth and a high survival in the nursing phase. The ongrowing phase starts after the young are trained to feed on dead food, which is the critical period for aquaculture. The daily growth rate is 1.4 % in length and 3.4 % in weight, throughout the culture period of 210 days. The final size varies depending on the culture conditions, e.g. temperature, nutritional content of food, period of training, etc. Less than 500 g weight of the cuttlefish is required for the frozen food industry (Nabhitabhata 1995). The culture period is recommended to be about 150–210 days with harvest before spawning. Live cuttlefish of similar size are required for the home aquarium trade. Live feed required in the nursing phase is the bottleneck for large-scale aquaculture. However, the large-scale production of live feed and the additional enriched *Artemia* is feasible. The feasibility of producing artificial feed for the ongrowing phase is higher than that for the nursing phase.

12.3 Broodstock Maintenance

Live pharaoh cuttlefish are collected alive from squid traps operating near the shore in Thai waters. The local fisherman maintains live cuttlefish in a partition of the boat where water can flow through (“live chamber”). Cuttlefish can survive for more than 24 h in this chamber. The body weight of the captured individuals that will become spawners range from 400 to 1,500 g, with a mantle length of 200–300 mm. At landing, cuttlefish are transferred into a plastic transportation tank, with a capacity of 500 L and are then transported to the hatchery. They are maintained in 2-m^3 concrete tanks (Fig. 12.1) at a male to female ratio of 1:2, which is similar to the ratio of 1:1.8 that is found for cuttlefish caught in traps (Chotiyaputta and Yamrungreung 1998). Cuttlefish will mate within the first hour after being released into the tank. Pieces of fishing net, with a 25-mm mesh size and sinkers, are put in the tanks as artificial substrates. (For details, see also Chap. 7 “Aquaculture to Restocking”, this volume).

Fig. 12.1 Mating pair of *Sepia pharaonis* broodstock in the maintenance tank (male on *left*). (Photograph of J Nabhitabhata)

Acclimatisation to captivity is fast, since mating and spawning in the tank proceed on the same day or in the next morning (Nabhitabhata 1978; Nabhitabhata and Nilaphat 1999). The females attach their egg capsules in clusters to the artificial substrates. The process of spawning one egg capsule takes about 1 min. In captivity, the wild females are able to spawn 1,600 (500–3,000) egg capsules in total before they die (Nabhitabhata 1978; Nabhitabhata and Nilaphat 1999). The females of cultured consecutive generations produce comparatively lower numbers of eggs. The average individual fecundity is 175 (50–200) eggs per female, with comparatively smaller egg capsules being observed at 28 °C (Nabhitabhata and Nilaphat 1999) and 11–485 eggs at a lower temperature of 25 °C (Minton et al. 2001). Each female, either of wild or of cultured batches, lays eggs in one cluster or more every day or every other day, if they are interrupted while spawning. The species displays intermittent spawning, with mating being observed between spawns. After spawning each cluster, the female will be tired and lies at the bottom, escorted by her mate. However, when she is ready to eat, she immediately seizes the first passing prey or dead feed. The spawning period ranges from 1 to 24 days, depending on the female's size, condition and the number of eggs produced. The male escorts and protects her from other males during spawning (Nabhitabhata 1978; Nabhitabhata and Nilaphat 1999). Only other females are allowed to approach and spawn at the same site and time. The healthy male will turn to form a pair with another selected female after the last spawning of his mate. The mating and spawning are observed early in the morning and late afternoon.

Egg clusters are also available from the wild. The egg clusters are found attached to underwater substrates, i.e. seaweeds and rocks, and can be collected by skin divers (Anil et al. 2005).

Artificial in vitro fertilisation has been studied between mature oocytes from the female oviducts and sperm from the spermatophores in the male Needham's sac. About 40 % of oocytes were successfully fertilised (Nabhitabhata et al. 2001a). The oviducal and nidamental gland jellies are not applied to the eggs in this study. The eggs develop for about 4 h to the blastula stage before ceasing. In the case of naturally fertilised eggs, the embryos in the capsules with a damaged coating tend to be infected by bacteria or fungi after a few days.

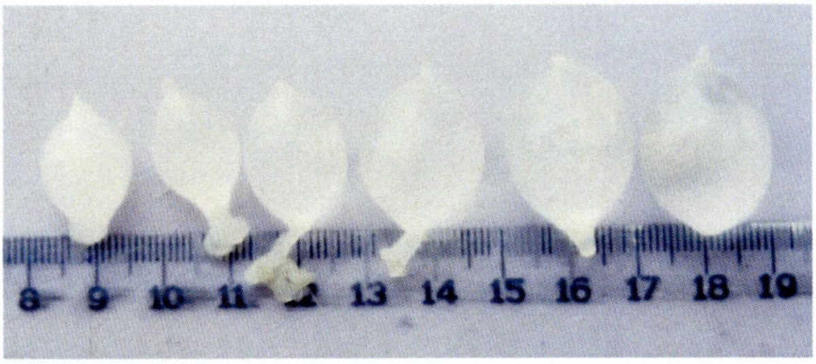

Fig. 12.2 Eggs of *Sepia pharaonis* in sequential stages (*left* to *right*) of embryonic development (0–14 days). (Photograph of J Nabhitabhata)

12.4 Incubation of Egg Masses

12.4.1 Egg Characteristics and Hatching

The single egg is white, opaque and round in shape with a tip and stalk (Fig. 12.2). The egg stalks are twined with or adhered to rod or ribbon-shape or flat substrata. The eggs (approximately 24 mm in length, 14 mm in width) turn larger and become more transparent during the progress of embryonic development, reaching their largest sizes (35 mm in length, 24 mm in width) just before hatching. In extremely rare cases, two embryos embedded in one capsule are observed in one cluster of egg mass. In this case, the embryos are separated from each other by a septum within the capsule. Still, the embryonic development and hatchlings from these eggs masses are normal (Nabhitabhata 2003).

The hatching depends on either biotic or abiotic factors. The average embryonic period is 14 days (9–25) at about 25–28 °C (Anil et al. 2005; Minton et al. 2001; Nabhitabhata 1978; Nabhitabhata and Nilaphat 1999) and can be shortened at higher temperatures (Table 12.1). The hatching period, from the hatching of the first to the last egg of the same cluster, is 3–10 days. More than 50 % of hatching occurs at night on the second and third day after the first hatching. The hatching rate is high, with 99 % being viable eggs from wild spawners and about 50 % by cultured spawners at 28 °C (Nabhitabhata and Nilphat 1999). A low hatching rate of 1.3 % is observed in eggs from cultured broodstocks, reared at 25 °C. Other variables of the hatching rate are the number of unfertilised eggs and abnormally developed embryos in each batch (about 1–2 %).

12.4.2 System Requirements and Management

The eggs are collected daily from the broodstock tanks and transferred to other tanks in the hatchery, which have controlled conditions. The eggs are collected by hand after the females pause their spawning on that day. They are placed and maintained until hatching in plastic baskets with 5 mm mesh size (Nabhitabhata 1978) or in suspended net bags of 10 mm mesh size (Anil et al. 2005). Baskets are inside a concrete tank with a 4,000-L capacity and 1.8 m diameter. In the plastic basket,

Table 12.1 Comparison of culture conditions and production of *Sepia pharaonis* from various studies

	Nair (1985)	Nabhitabhata (1978)	Nabhitabhata and Nilaphat (1999)	Minton et al. (2001)	Anil et al. (2005)
Tank system	Open	Open	Open	Closed	Open
Egg incubation (days)	17–24	16	14.3 (9–25)	13.6 (11–16)	12–17
Temperature (°C)	–	28	28	21–28	27–31
Hatching period (days)	7	2–6	3–10	–	7
Hatchling size (mm, g)	6.5–7	6.8, 0.25	7.7, 0.18	6.4	8, 0.16
Initial density (ind. m^{-2})	–	500	500	–	1
Maturity (first mating, days)	–	110	90	120	120
First spawning (days)	–	110	110	207 (162–309)	–
Size at maturity (mm, g)	–	79, 84	70, 50	171.2–228.9, 612.5–1,317.6 (first spawn)	95, 85
Daily growth (%: ML, W)	–	–	1.37, 3.40	–	1.40, 3.22
Mean final size (mm, g)	–	104, 115	139, 275	214–230, 980–1,320	168, 521
Max size (mm, g)	–	114, 153	16.2, 375	300, 3,045	–
Mean life span (days)	–	–	149.4	267–369	–
Max life span (days)	–	120	271	249 (25–28 °C), 368 (21 °C)	–

ind individual, *ML* mantle length, *W* wet body weight

the capsules are then separated from clusters by scissor cutting. Eggs are visually checked every day and the ones with dead and abnormally developed embryos are discarded in order to prevent the infection—decomposition of the remaining.

In an open system, the change of physical and chemical parameters should be minimised during the embryonic stage. Brief changes in temperature and salinity as well as mechanical stimuli cause premature hatching. To minimise the temperature fluctuation, a low water flow-through of 1 L min^{-1} is used. The fresh seawater is first pumped through a carbon filter and stored in a 200-m^3 concrete tank before being supplied by gravity to the hatchery tanks. This open system is able to maintain an average temperature of 28 °C, with a salinity of 30–33 psu and a pH of 6.0–8.0 (Nabhitabhata et al. 2005). The salinity required for a 50 % hatching rate of eggs is 22–38 psu (Nabhitabhata et al. 1993b, 2001c).

Other important physical factors are turbidity or suspended solids and light. Turbidity should be maintained as low as possible (Nabhitabhata 1995) through filtration and/or prior sedimentation, particularly in open systems. Cephalopod gills have a low efficiency for removing sediment, high levels of suspended solids or high turbidity, which clog the gills and will block oxygen exchange. Any artificial lighting, as well as unnatural daily light–dark periods initiated by any human activity, should be avoided. The growth of algae and fungi on the surface of egg capsules due to excessive light will reduce the hatching rate by blocking the oxygen supply to the embryo. The attachment of algae also initiates fungal growth on the capsule membrane and provokes infection of the embryo. The most convenient way to reduce light exposure is curtaining the hatchery with a camouflage net with a light-reducing efficiency of 80 % (Nabhitabhata et al. 2005).

From the baskets, all hatchlings pass through the basket-mesh into the tank. When the optimum hatchling density in the tank of 500 individuals m^{-2} is achieved, the baskets with the remaining eggs are transferred to another tank of the same water quality. The incubation tank now becomes the nursing tank. The hatchlings are collected and transferred for nursing into other tanks with extreme caution, in a wet condition with gentle movement, in order to avoid skin damage. Such skin damage can subsequently cause mortality, so it could be more appropriate to transfer egg capsules rather than the hatchlings. (For details, see also Chap. 7 "Aquaculture to Restocking", this volume).

12.5 Nursing of Young

12.5.1 Hatchling Characteristics

The overall external morphology and behaviour of the hatchling are similar to that of the adult. Burrowing behaviour in sandy and small gravel substrata is observed at 4 days after hatching. Cuttlefish are solitary in habit, lying or burrowing in the substratum. Gregarious behaviour is occasionally observed when fleeing from a common enemy or in pursuit of prey in the same direction and feeding together in a group after about 10 days after hatching. Such group behaviour is beneficial for the training period during which they feed on dead prey.

12.5.2 System Requirements and Management

Hatchlings of *S. pharaonis* are benthic and the water depth and density settings in this phase have to vary in relation to their growth (cuttlefish size), to suit the management of food. The water depth in tanks is adjusted to encourage aggregation of the live food organisms. Appropriate biomasses of live feed and cuttlefish in a tank reduce excess prey search activity and energy expenditure in food hunting, and hence promote growth. The initial water depth in the egg tanks is lowered from 600 to 300 mm for benthic hatchlings. Instead of lowering the water depth, the egg baskets are floated inside a hanging net cage of a large, fine mesh (1.5 mm) that is placed inside a 3,000-L tank (Minton et al. 2001).

Table 12.2 Expected growth, survival and the managed density of *Sepia pharaonis*, estimation based on Anil et al. (2005), Minton et al. (2001), Nabhitabhata (1978), Nabhitabhata and Nilaphat (1999)

Culture period (days)	ML (mm)	DGRL (%)	W (g)	DGRW (%)	Density (ind. m^{-2})	Water depth (mm)	Survival (%)
0	8	–	0.2	–	500	300	–
30	20	3.0	2	6.0	200	450	90
60	35	2.0	9	4.3	90	600	75
90	60	1.8	35	4.0	40	600>	60
120	85	1.3	75	3.0	20	600>	50
150	110	0.8	130	2.4	15	600>	40
180	130	0.6	275	1.8	10	600>	30
210	155	0.5	400	1.0	5	600>	20

ML mantle length, *DGRL* daily growth rate in mantle length, *W* wet body weight, *DGRW* daily growth rate in weight, *ind.* individual

Every day or every second day, the depth of the nursing tank is increased by 50 mm. The initial density of the animals in the tank is 500 individuals m^{-2} and this is sequentially reduced by 20–30 % after size grading (Table 12.2). The size grading is performed every 10 days due to the rapid growth, and the density will be correspondingly decreased to 375–400, 280–300 and about 200 individuals m^{-2} after 10, 20 and 30 days, respectively. Without size grading and the reduction of the density, after 42 days, there is a decrease in individual size and survival observed in cultured cuttlefish (Barord et al. 2010). These authors reported that, at 28 °C, *S. pharaonis*' average weight increases from 0.01 to 26.4 g in 42 days of rearing at a density of 20 individuals m^{-2} with a survival of 100 %, but the weight increases to only 20.8 g at 100 individuals m^{-2} and 14.9 g at 200 individuals m^{-2} with a survival of 50 and 0.05 %, respectively.

Management of the filtration, flow rate and volume of daily water changes can be varied, depending on the size of the tanks, water depths and cuttlefish density. Anil et al. (2005) increased the initial flow rate in a flow-through system to 2 L min^{-1} in 60-L tanks to 20 L min^{-1} in 500 and 1,000-L tanks, while the volume of water changed daily in this flow-through open system decreased from 80 to 60 % and then to 30 % in 1,000-L tanks, respectively.

The required water-quality criteria in the nursing phase are similar to those in the egg-incubation phase and are managed in a similar manner. In an open seawater system, the change of the physical and chemical parameters is minimised by means of fresh seawater flow-through. Within the range of 20–40 psu, the hatchlings are able to tolerate salinity changes, either briefly or through a gradual change, similar to the egg capsules. The salinity required for 50 % survival of hatchlings is 21–39 psu, with a pH of 6.0–8.4 (Nabhitabhata et al. 1991, 1993b, 2001c; Nabhitabhata and Kbinrum 1984). Cleaning of the tank bottom by siphoning is recommended in order to remove any uneaten prey and avoid food decomposition. Tank water should also be changed daily by at least 50 or up to 80 % if the tanks are smaller. Other routine management of the nursing tanks is similar to those for the egg-nursing tanks.

Nursing of young pharaoh cuttlefish in a floating net cage is feasible. Chiampreecha (1983) reared young cuttlefish from hatching to 63 days in a steel-framed net cage of $600 \times 800 \times 400$ mm. The cage had a water depth of 300 mm. The survival decreased to 63 %, and the initial density of 26 individuals m^{-2} decreased to a final density of 17 individuals m^{-2}. However, the cage in this study is floated in a large concrete tank of unknown volume with a flow-through seawater system.

In closed seawater systems, nitrogenous compound levels can rapidly increase and should be monitored. Although young cuttlefish can tolerate up to 0.6 mg L^{-1} of ammonia (NH_4–N) and 0.02 mg L^{-1} nitrite (NO_2–N) at 28 °C (Chindamaikul et al. 2001), it is recommended to maintain ammonia below 0.1, nitrite below 0.1 and nitrate (NO_3–N) below 20 mg L^{-1} (Minton et al. 2001; Nabhitabhata 2000). Salinity in the system should be monitored daily and be adjusted with synthetic or sterile natural sea salt. In addition, pH should be maintained within the range of 5.9–8.4 for hatchlings (Nabhitabhata et al. 1991).

12.5.3 Feeding

Hatchlings have small external yolk sacs and these are abandoned on the first day. Hatchlings still have some yolk reserves enabling them to survive without food for 3–7 days or even up to 23 days (Nair et al. 1985), and the digestive activities will begin after 3 days, which is similar to what is observed in *S. officinalis* (Boucaud-Camou and Roper 1995).

Various kinds of live food organisms, with corresponding sizes, are fed to the hatchlings. The cuttlefish hatchlings are benthic and begin feeding on the first day, 6–12 h after hatching. The list of accepted food organisms includes live mysid, *Mesopodopsis orientalis* (7 mm in total length and 10 mg body weight) and a mix of fish larvae (10 mm in total length) that are collected from the wild. Other crustacean food organisms can also be used with success, e.g. the mysid *Eurobowmanilla simulans* (4–6 mm), postlarvae of the shrimps *Penaeus indicus*, *Metapenaeus dobsoni* (Anil et al. 2005), the mysid *Mysidopsis* sp., the grass shrimp *Palaemonetes pugio* and guppies *Poecilia reticulata* (Minton et al. 2001). The live food is stocked and provided ad libitum to cuttlefish. The daily feeding ration of the hatchlings is 20–28 % of their body weight, which equals 13 mysids or 10 shrimp postlarvae (Nabhitabhata et al. 1996). Cuttlefish hatchlings prefer crustaceans to fish as food. Newly born hatchlings recognise, attack and eat mysids demonstrating that such behaviour is innate in sepiid cuttlefish as described by Messenger (1977). The attempts to feed young *S. pharaonis* with live food other than crustaceans and fishes have not succeeded (Nabhitabhata 1978; Toll and Strain 1988). Adult *Artemia* spp. are accepted but yield lower growth and survival. They should only be used as a supplementary or reserve food when other crustaceans are in short supply.

Food organisms obtained from commercial hatcheries are more reliable and consistent in supply. The postlarvae of the penaeid shrimp, *Penaeus merguiensis* (8–10 mm) and sea bass, *Lates calcarifer*, fries (7.5 mm) are two of the available species. The cuttlefish consume about 10 mysids or 2 sea bass fries daily (Chankaew

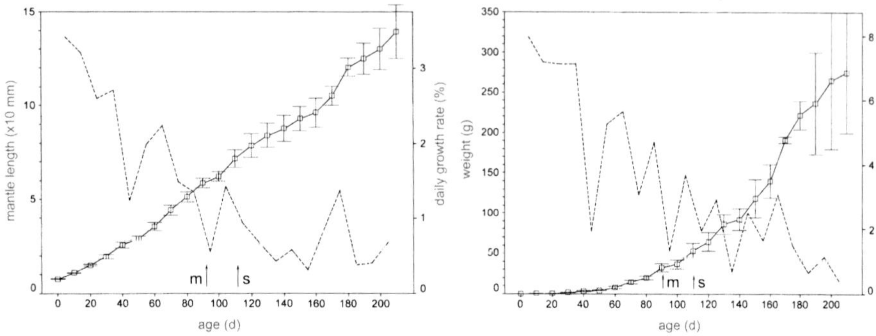

Fig. 12.3 Growth in terms of the relationships between mantle length (mm) (*left*), weight (g) (*right*) and age (days, *full line*), daily growth rate of length and weight (%, *broken lines*). *Arrows* indicate approximate age at mating (m) and spawning (s). (After Nabhitabhata and Nilaphat 1999; Nabhitabhata 2002)

et al. 2003; Nabhitabhata et al. 1996). The survival of cuttlefish hatchlings fed with *P. merguiensis* for 10 days averages 94 % (81–100). The feeding rate is 28 % and the feeding efficiency is 48 %. The unit cost in Thailand is estimated to be US$ 20 for 100 cuttlefish with a feed cost of about 85 % in 1996 (Nabhitabhata et al. 1996), when US$ 1 approximately equal Thai Baht 25 (US$ 1 equal Thai Baht 30 in 2013).

12.5.4 Growth

Young cuttlefish grow rapidly in the first month. In an open system, hatchlings of 7–8 mm in mantle length and 0.2 g wet body weight (Fig. 12.3) grow to 10 mm and 0.40 g, respectively, in the first 10 days at 28 °C. The daily growth rates are 3.4 % in length and 8.0 % in weight, which are the highest rate during their life cycle. Young cuttlefish attain a mantle length of 20 mm and a weight of 2 g after 30 days. The daily growth rates are 2–3 % in length and 6–7 % in weight during the first 40-day period, with a survival rate of about 90 % (Nabhitabhata 1978; Nabhitabhata and Nilaphat 1999).

The growth phase in terms of weight (W, g) and age (T, d; Fig. 12.4) can be separated into two phases. The early phase is from hatching to 60 days of age is exponential (Nabhitabhata and Nilaphat 1999; Nabhitabhata 2002):

$$W = 0.196e^{0.061T}. \quad (12.1)$$

Growth can vary among the cultured batches from within the same population. The range between the maximum and the minimum growth rate in weight can be as large as about four-fold (Nabhitabhata 1978).

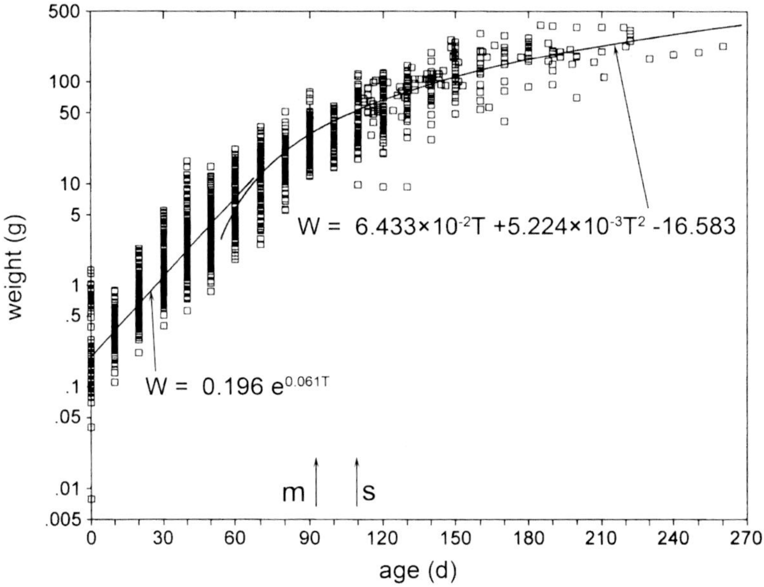

Fig. 12.4 Growth in terms of the relationship between weight (g) and age (d). *Arrows* indicate approximate age at mating (m) and spawning (s). (After Nabhitabhata and Nilaphat 1999; Nabhitabhata 2002)

12.6 Ongrowing

12.6.1 System Requirements and Management

In flow-through open seawater systems, cylindrical tanks of any size can be used. The same tanks used in the nursing phase can be continually used for ongrowing (Nabhitabhata 1978; Nabhitabhata and Nilaphat 1999). In closed seawater systems, two rectangular raceways of 65,000-L volume are used for ongrowing (Minton et al. 2001; Barord et al. 2010). Each raceway is equipped with its own pre-filter tank, airlift and a shared filtration tank. The water is pumped by airlifts from the pre-filter tank into the filtration tank at a flow rate of 228 L min^{-1} before returning to the raceway. The tank volume can be exchanged five times daily. Another set of raceway used has a capacity of 16,000 L with a flow rate of 90 L min^{-1}. The pharaoh cuttlefish can be cultured through three consecutive generations in this system (Minton et al. 2001).

The density during the ongrowing phase is adjusted to correspond to the tank volume and water system used (Table 12.2). In a flow-through seawater system, Anil et al (2005) transferred 30-day-old cuttlefish (approximate 20 mm in mantle length and 1.9 g in weight) from 60-L fibreglass tanks to 500-L tanks where the density is decreased from 1,000 to 100 individuals m^{-3}. After 90 days when the cuttlefish grows 59 mm in mantle length and 34 g in weight, they were transferred

Table 12.3 Effect of biotic and abiotic factors on the final size of *Sepia pharaonis* in culture conditions

Factor	Small final size	Large final size
Dead feed training	Early (30 days)	Late (70 days)
Nutritive value of feed	Low	High
Temperature	High	Low
Density	High	Low
Sex competition	High	Low

again to 1,000-L tanks at a density of 20 individuals m^{-3} and then to 10 individuals m^{-3} after 180 days when the cuttlefish attained a mantle length of 145 mm and a weight of 325 g. In contrast, Nabhitabhata and Nilaphat (1999) use a different rearing methodology to avoid injuring the animals. After 30 days, cuttlefish at about 20 mm in mantle length are not transferred and the ongrowing is continued in the nursing tanks (3,000 L), but their density is decreased from 500 individuals m^{-2} to 200 individuals m^{-2}, with another 20% decrease every 10 days thereafter.

Many abiotic and biotic factors have been reported to affect growth in culture conditions, but not all factors were considered to have obvious effects (Table 12.3). The availability of food and fluctuations of physical and chemical environmental factors are partially controlled or managed in the culture tanks. The most obvious factor is apparently the high density (initial density 500 individuals m^{-2}), which is unavoidable in view of the need to reduce unit costs in large-scale aquaculture. The high culture densities with a high survival result in large numbers of cuttlefish reaching their maturity. This condition can benefit subsequent early maturation (Stearns 1992) by increasing the opportunities for reproductive encounters. The absence of bottom substrates in the culture tanks for the ritualised burrowing behaviour had no effect on normal growth and survival for *S. pharaonis* under high-density conditions. Without the substratum, the cleaning of the tanks is greatly facilitated (Nabhitabhata and Nilaphat 1999).

Since the life cycles of the cuttlefish ends with terminal spawning, the harvest has to be performed as soon as the pair formation or mating is observed, prior to spawning, to gain maximum production.

12.6.2 Feeding

Training the pharaoh cuttlefish to feed on dead food after about 30 days of age is the critical stage for the success of aquaculture. Thirty days after hatching, the cuttlefish will be trained to feed on dead fish meat (*Caranx leptolepis*). The fish are available in a large quantity with low cost, storable for at least a week in a refrigerator and are available all the year through commercial fisheries. Thus, they can be supplied constantly, consistently and reliably. Cuttlefish starve to death if they do not favour and accept the dead food as a result of the lack of the live prey. Survival, hence production, may highly decrease during this period (Nabhitahata 1978) if training is not successful. The period from 30 to 50 days, or 20 to 30 mm mantle length, is appropriate for training. The older and larger they are, the more easily they can be trained to accept dead food and a varied diet. Anil et al. (2005) began training cuttlefish to accept whole mysids, *Acetes* sp., and fish after 50 days, while Minton et al. (2001) began with thawed shrimps, *Penaeus* sp., after 70 days. Overall, the success of training depends on the right timing as well as the experience of human trainers. Trained cuttlefish are fed ad libitum, twice daily.

Late training for accepting dead food at an older age, from 30 to 70 days, is considered to affect the subsequent higher growth and final size (Minton et al. 2001). The advantage is the larger size of the cuttlefish at training, but the extension of the nursing phase to 40 days with additional costs regarding live food should also be a concern in large-scale cultures.

Preliminary attempts to train *S. pharaonis* to feed on artificial diets instead of fish meat did not succeed (Sangpradab et al. 1984), and no further studies have been reported.

Consequent changes in feeding behaviour can be seen. Feeding behaviour changes from using tentacles to using only arms for the seizing of food, without the positioning step (Messenger 1977) when they are fed on dead feed. The feeding behaviour of one cuttlefish also visually stimulates others in the same tank to behave similarly, as also observed in *S. officinalis* (DeRusha et al. 1989). Such stimulation accelerates the acceptance of dead food and reduces the time required for training.

12.6.3 Growth

Growth can be observed from changes in colour patterns and morphology. After about 60 days of age, the normal colour pattern of the cuttlefish turns to dark brown with scattered small black rings on the dorsum and a dark brown colour on both fins. The fin length is about 95 % of the mantle. After about 90 days of age, sexual dimorphism may be recognised by the colour pattern (Nabhitabhata 1978). The male displays brown-white transverse stripes or a tiger pattern on the dorsum. The female colour pattern does not obviously change with growth. The tiger pattern of the male gradually turns clearer with growth to 120 days of age.

The daily growth rates in weight increase with age and then decrease corresponding to the period of sexual maturity and mating. In an open system, the daily growth rates in terms of weight were 7.2 % at 60 days after hatching and these increase to 7.9 % at 70 days after hatching and decrease to 6.2 and 4.5 % after mating at about 90 days after hatching and spawning at about 110 days after hatching, respectively (Nabhitabhata and Nilaphat 1999). The first spawning is observed at a mantle length of about 70 mm (50 g) and generally to about 100 mm (100 g). On and after a 10-day period, the daily growth tends to decrease with the increase in age. For the overall life cycle of 210 days, the average daily growth rate is 1.4 % in mantle length and 3.4 % in body weight at 28 °C.

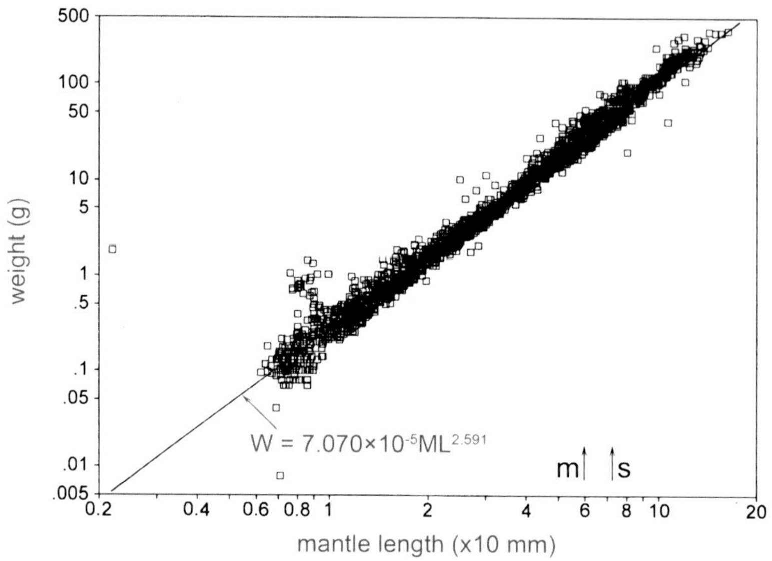

Fig. 12.5 Growth in terms of the relationship between mantle length (mm) and weight (g). *Arrows* indicate approximate age at mating (m) and spawning (s). (After Nabhitabhata and Nilaphat 1999; Nabhitabhata 2002)

The culture conditions cause variations of the final size and the life span of *S. pharaonis* in captivity (Table 12.1). Lower temperatures can extend sexual maturity, the longevity of life span and the subsequent larger final size. Nabhitabhata and Nilaphat (1999) reported that sexual maturity was reached at about 90 days of age and a mean final size of 140 mm and 275 g at 210 days of age with an average life span of 150 days at 28 °C from the open system. At similar temperatures, Anil et al. (2005) reported later mating at 120 days after hatching but cuttlefish displayed a larger size (168 mm) and increased weight (521 g) at 210 days after hatching. From a closed system at 25–28 °C, further growth seems to occur. Minton et al. (2001) reported that the sexual maturity postponed to 150–240 days after hatching and to a mean final size of 214–230 mm, 980–1,320 g with an average life span of less than 270 days. The differences in the two studies reflect the plasticity of the strategies of the life cycle in this species, such as the short and the long generation mode, as adaptations to the environment.

Models of growth from hatching have been reported by Nabhitabhata and Nilaphat (1999) and Nabhitabhata (2002). The relationship between mantle length (ML, mm) and weight (W, g) of the pharaoh cuttlefish from hatching to 260 days can be expressed in a single model as follows (Fig. 12.5):

$$W = 7.070 \times 10^{-4}\, ML^{2.591}. \tag{12.2}$$

The growth in terms of the mantle length–age (T, d) relationship (Fig. 12.6) can also be expressed in a single model:

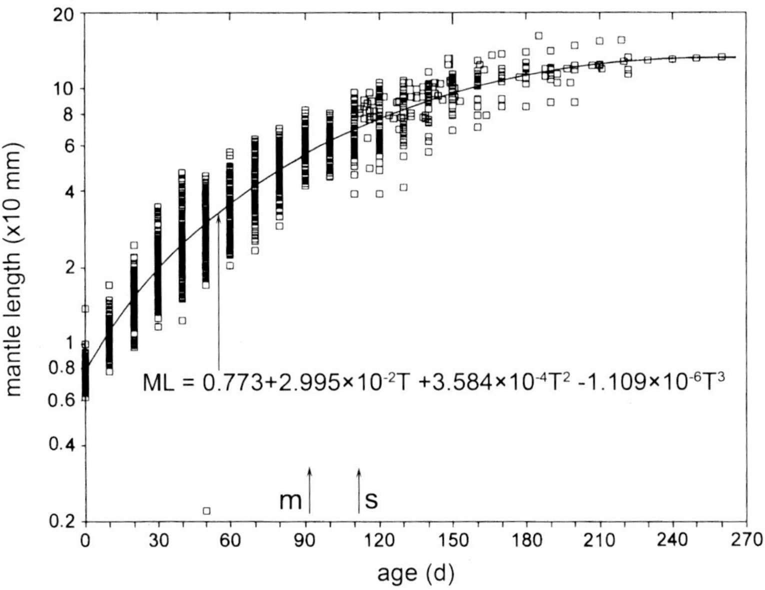

Fig. 12.6 Growth in terms of the relationship between mantle length (mm) and age (d). *Arrows* indicate approximate age at mating (m) and spawning (s). (After Nabhitabhata and Nilaphat 1999; Nabhitabhata 2002)

$$ML = 0.077 + 2.995 \times 10^{-3}T + 3.584 \times 10^{-5}T^2 - 1.109 \times 10^{-7}T^3. \qquad (12.3)$$

The second growth phase in terms of the weight–age relationship is from 50 to 260 days of age, with the inflection point at about 60 days, and this can be fitted by the following equation (Fig. 12.4):

$$W = 6.433 \times 10^{-2}T + 5.224 \times 10^{-3}T^2 - 16.583. \qquad (12.4)$$

12.7 Trends in Research and Industrial Level

The low growth rate and small final weight of about 400 g at 210 days (7 months) are disadvantageous for aquaculture in terms of quantitative production compared with the larger weight of more than 1,000 g from the fisheries ground (Nabhitabhata 1995, 2000). However, this happens because the technology might need optimization, and more knowledge regarding the species physiology is required. Culture for a larger final size is feasible at low density, low temperature, longer periods and high nutritive feeding. Such production comes at a cost that does not attract any investment for commercial production. Another point of concern is the feasibility of decrease in the final size after consecutive-generation cultures due to inbreeding, and the fertility in cultured batches is much lower than in wild stocks (Nabhitabhata and Nilaphat 1999; Minton et al. 2001). From the aquaculture point of view, it means that seed production cannot rely solely on cultured broodstocks and has to depend, at least, partially on captured spawners. However, the small size at yielding is an advantage, whereas a weight of 100–400 g is optimum and required by the seafood industry for frozen food product processing and packaging. It takes only 150 days to produce the cuttlefish of the mentioned size. The short rearing period facilitates a higher production and a more reliable supply (through higher survival) and a lower cost of production.

Live cuttlefish products can be distributed to the ornamental aquaculture trade for which there is a rapidly growing demand. Live cuttlefish of this size are suitable for the small volumes of home aquaria. Their benthic habits and high tolerance to different culture conditions are undeniable advantages.

An increased rearing period of more than the suggested 150 days, obtained at high-temperature conditions, is feasible and will postpone the maturity of the cuttlefish. The high temperature that accelerates the growth of *S. pharaonis* and shortens the life span through the early maturity is unavoidable in flow-through open systems (operated at a lower cost compared to closed systems) used in tropical countries. A possible solution to this problem may involve decreasing sexual stimuli and activities by separation of males and females, hence the mono-sex culture.

The main focus of future research should be devoted to developing feeds for large-scale culture, both live and artificial. Live feed is required in the nursing phase, since the young innately feed on a certain prey. The development of mass culture of mysids as well as any other crustaceans is necessary. The use of commercially produced penaeid shrimp larvae is reliable, but their own commercial value discourages production. Another trend of research on the live food area is the enrichment of *Artemia* sp. as a supplementary and substitute feed for young cuttlefish in the nursing phase, following studies in *S. officinalis* (Koueta et al. 2002) and *Sepiella inermis* (Muthuwan et al. 1993; see also Chap. 13 of this volume). The innate feeding on live feed is really the bottleneck of the nursing phase. The artificial feed is feasible through behaviour-approached research on the acceptable form of the food pellets. The making of artificial feed for the ongrowing phase is feasible in the short term, because of the higher probability of acceptance of an artificial diet by juveniles than hatchlings since their ability to learn and their lower degree of food preference are known. The acceptable form of feed is the first step to be studied through behavioural approaches (Nabhitabhata et al. 2001b). The form has to mimic the appearance (shape, colour, brightness, locomotion) and characters (texture, "odour") of natural feed. It will be of great benefit to consider some details, i.e. the acceptable form, from the studies previously held on *S. officinalis* (Castro 1991; Castro et al. 1993).

The difference in growth rate among different morpho-types and molecular types, thus inter-populations and intra-populations, could be very interesting in view of maximising the aquaculture production. The *S. pharaonis* is probably a complex species (Reid et al. 2005) of three morphological types (Norman 2000) and five molecular clades (Anderson et al. 2011). The cultured offspring from spawners collected from the Andaman Sea (Indian Ocean) tended to grow much faster, at similar temperature (28 °C) and ad libitum feeding, than those from the Gulf of Thailand (Pacific Ocean) by about two-fold in length and four- to five-fold in weight (Nabhitabhata and Nilaphat 1999).

12.8 Conclusions

The culture process of *S. pharaonis* comprises broodstocks collected from the wild, incubation of eggs, nursing of the young in the hatchery and growout phase. Cuttlefish broodstock collected from the wild and from cultured batches can reproduce in captivity. Collected egg masses and young cuttlefish are nursed in concrete tanks. Water quality is fully controlled in closed seawater systems and partially controlled in open seawater systems. The young are fed with live prey organisms either collected from the wild (mysids) or produced from a hatchery (penaeid shrimp postlarvae) from the time of hatching to about 30 days. Density is area oriented because the ritual habit of the cuttlefish is lying at the tank bottom. Initial density is 500 individuals m^{-2} with a 25 % decrease through regular size grading over a 10-day period. Survival is more than 90 % after 30 days. Training the pharaoh cuttlefish to feed on dead food after about 30 days of age is the critical stage. The success in such training will start the growout phase and also indicates the consequent success of culture. The growout phase starts after the young are able to accept dead feed. Overall, daily growth rate is approximately 1.4 % by length and 3.4 % by weight. High temperature (>25 °C) plays a key role in affecting higher growth, smaller final size and shorter life span in culture conditions. The mean final weight and life span is approximately 300 g, 150 days in open seawater systems at 28 °C and 1,000 g, 300 days in closed systems at 21–25 °C. Cultured cuttlefish have lower fecundity compared to wild ones. Innate feeding on specific live prey is the bottleneck for large-scale aquaculture as in other cephalopods. Development of artificial feed could be a solution and should be able to reduce the cost of production.

References

Anderson FE, Engelke R, Jarrett K, Valinassab T, Mohamed KS, Asokan PK, Zacharia PU, Nootmorn P, Chotiyaputta C, Dunning M (2011) Phylogeny of the *Sepia pharaonis* species complex (Cephalopoda: Sepiida) based on analyses of mitochondrial and nuclear DNA sequence data. J Mollusc Stud 77:65–75

Anil MK, Andrews J, Unnikrishnan C (2005) Growth, behavior and mating of pharaoh cuttlefish (*Sepia pharaonis* Ehrenberg) in captivity. Israel J Aquacult 57:5–31

Barord GJ, Keister KN, Lee PG (2010) Determining the effects of stocking density and temperature on growth and food consumption in the pharaoh cuttlefish, *Sepia pharaonis*, Ehrenberg 1890. Aquacult Int 18:271–283

Boongerd S, Rachaniyom S (1990) Squid trap fishing. Tech Pap 1/1990, Fish Tech Subdiv, Mar Fish Div, Dept Fish

Boucaud-Camou E, Roper CFE (1995) Digestive enzymes in paralarval cephalopods. Bull Mar Sci 57:313–327

Castro BG (1991) Can *Sepia officinalis* L. be reared on artificial food? Mar Behav Physiol 19:35–38

Castro BG, DiMarco FP, DeRusha RH, Lee PG (1993) The effects of surimi and pelleted diets on the laboratory survival, growth, and feeding rate of the cuttlefish *Sepia officinalis* L. J Exp Mar Biol Ecol 170:241–252

Chankaew K, Nabhitabhata J, Sanein N, Pattanakul S (2003) Management of sea bass fries as live feed of hatchlings of bigfin squid and pharaoh cuttlefish. In: Nabhitabhata J (ed) Report on cephalopod culture research and development programme, 2002–2004. Rayong Coast Fish Res Contrib 27

Chiampreecha B (1983) Survival, growth and length-weight relationship of cuttlefish, *Sepia pharaonis* from rearing experiment in floating net-cage. Tech Pap 45, Pelagic Fish Sect, Mar Fish Div, Dept Fish

Chikuni S (1983) Cephalopod resources in the Indo-Pacific region. In: Caddy JF (ed) Advances in assessment of world cephalopod resources. FAO Fisheries technical paper 231. FAO, Rome

Chindamaikul T, Kaoian K, Chainak P, Nabhitabhata J, Samrarnrat N (2001) Effect of nitrite-nitrogen and ammonia-nitrogen on survival of hatchlings of pharaoh cuttlefish, *Sepia pharaonis* Ehrenberg, 1831. In: Nabhitabhata J (ed) Report on cephalopod culture research and development programme, 2002–2004. Rayong Coast Fish Res Contrib 27

Chotiyaputta C (1979) Length-weight relationships and sex ratios of cuttlefishes *Sepia pharaonis* Ehrenberg and *Sepia brevimana* Steenstrup. Tech Pap 15/1979, Invert Sect, Mar Fish Div, Dept Fish

Chotiyaputta C (1980) Biological study on cuttlefish, *Sepia pharaonis* Ehrenberg in the Gulf of Thailand. Tech Pap 16/1980, Invert Sect, Mar Fish Div, Dept Fish

Chotiyaputta C (1981) Biology of the cuttlefish, *Sepia pharaonis* Ehrenberg. Ann Rep 1981, Invert Sect, Mar Fish Div, Dept Fish

Chotiyaputta C (1982) Biological study on cuttlefish, *Sepia pharaonis* Ehrenberg. Tech Pap 4/1982, Invert Sect, Mar Fish Div, Dept Fish

Chotiyaputta C (1995) Biology of cephalopods. In: Nabhitabhata J (ed) Biology and culture of cephalopods. Rayong Coast Aquacult St, Rayong

Chotiyaputta C, Yamrungreung A (1998) Trap fisheries for squid and their impact on spawning. S Afr J Mar Sci 20:285–291

Department of Fisheries (2012) Fisheries Statistics 2009. www.fisheries.go.th/it-stat/. Accessed 15 February 2012

DeRusha RH, Forsythe JW, DiMarco FP, Hanlon RT (1989) Alternative diets for maintaining and rearing cephalopods in captivity. Lab Anim Sci 39:306–312

Food and Agriculture Organization of the United Nations (2010) FAO Yearbook 2008. Fishery and aquaculture statistics. FAO, Rome

Jereb P, Roper CFE, Vecchione M (2005) Introduction. In: Jereb P, Roper CFE (eds) FAO Species catalogue for fishery purposes, no 4, vol 1. Cephalopods of the world. An annotated and illustrated catalogue of cephalopod species known to date. Vol 1: Chambered nautiluses and sepioids (Nautilidae, Sepiidae, Sepiolidae, Sepiadariidae, Idiosepiidae and Spirulidae). FAO, Rome

Koueta N, Boucaud-Camou E, Noel B (2002) Effect of enriched natural diet on survival and growth of juvenile cuttlefish *Sepia officinalis* L. Aquacult 203:293–310

Messenger JB (1977) Prey-capture and learning in the cuttlefish, *Sepia*. In: Nixon M, Messenger JB (eds) The biology of cephalopods. Academic Press, London

Minton JW, Walsh LS, Lee PG, Forsythe JW (2001) First multi-generation culture of the tropical cuttlefish *Sepia pharaonis* Ehrenberg, 1831. Aquacult Int 9:375–392

Muthuwan V, Sawatpeera S, Mamat P, Kuandee P (1993) Rearing of spineless cuttlefish, *Sepiella inermis* Fer.& d'Orb. with adult brine shrimp, *Artemia salina* fed with various kinds of feed. Res Pap 53/2536, Inst Mar Sci, Burapha Univ

Nabhitabhata J (1978) Rearing experiment on economic cephalopod-II: cuttlefish *Sepia pharaonis* Ehrenberg. Tech Pap 1978, Brackish Fish Sect Rayong Brackish Fish Stn, Brackish Fish Div, Dept Fish

Nabhitabhata J (1995) The culture of cephalopods: commercial scale attempt in Thailand. In: Nambiar KPP, Singh T (eds) Proc INFOFISH-AQUATECH' 94, Int Conf Aquacult. INFOFISH, Columbo

Nabhitabhata J (2000) Cephalopod culture: present and future status. In: Aryuthaka C (ed) Proc Wantana Yoosuk sem moll stud year 2000. Fac Fish Kasetsart Uni, Bangkok

Nabhitabhata J (2002) Life history and behaviour of reared neritic cephalopods from Thai Waters. PhD Thesis, Tokyo University of Fisheries

Nabhitabhata J (2003) Double Eggs of Pharaoh Cuttlefish, *Sepia pharaonis* Ehrenberg, 1831. Veliger 46:97–98

Nabhitabhata J, Kbinrum S (1984) Biological studies on economic cephalopods-III: note on survival of young cephalopods during continuous change of salinity. Tech Pap 6/1984, Rayong Brackish Fish Stn, Brackish Fish Div, Dept Fish

Nabhitabhata J, Nilaphat P (1999) Life cycle of cultured pharaoh cuttlefish, *Sepia pharaonis* Ehrenberg, 1831. Phuket Mar Biol Cent Spec Pub 19:25–40

Nabhitabhata J, Kbinrum S, Soodmee Y (1991) Effect of pH on survival rate of cephalopods hatchlings. Tech Pap 17/1991, Rayong Coast Aquacult Stn, Coast Aquacult Div, Dept Fish

Nabhitabhata J, Nilaphat P, Kongrot J, Ratanasoot S (1993a) Effect of salinity on survival rate of pharaoh cuttlefish hatchlings, *Sepia pharaonis* Ehrenberg. Tech Pap 11/1993, Rayong Coast Aquacult Stn, Coast Aquacult Div, Dept Fish

Nabhitabhata J, Phudtaraksa W, Nilaphat P, Phetrasatian G (1993b) Effect of salinity on hatchling rate of pharaoh cuttlefish, *Sepia pharaonis* Ehrenberg. Tech Pap 10/1993, Rayong Coast Aquacult Stn, Coast Aquacult Div, Dept Fish

Nabhitabhata J, Kakhai N, Suriyawarakul J, Nilaphat P (1996) Feeding and cost of seed production of cephalopods, *Sepioteuthis lessoniana* Lesson; *Sepia pharaonis* Ehrenberg; *Sepiella inermis* Orbigny. Tech Pap 1/1996, Rayong Coast Aquacult Stn, Coast Aquacult Div, Dept Fish

Nabhitabhata J, Nilaphat P, Promboon P (2001a) Artificial fertilisation and embryonic development of pharaoh cuttlefish, *Sepia pharaonis* Ehrenberg, 1831. In: Nabhitabhata J (ed) Report of cephalopod culture research and development programme, 2002–2004. Rayong Coast Fish Res Contrib 27

Nabhitabhata J, Nilaphat P, Promboon P (2001b) Behavioural approaches on the acceptable form of artificial feed of pharaoh cuttlefish, *Sepia pharaonis* Ehrenberg, 1831. In: Nabhitabhata J (ed) Report of cephalopod culture research and development programme, 2002–2004. Rayong Coast Fish Res Contrib 27

Nabhitabhata J, Asawangkune P, Amornjaruchit S, Promboon P (2001c) Tolerance of eggs and hatchlings of neritic cephalopods to salinity changes. Phuket Mar Biol Cent Spec Pub 25:91–99

Nabhitabhata J, Nilaphat P, Promboon P, Jaroongpattananon C, Nilaphat G, Reunreng A (2005) Performance of simple large-scale cephalopod culture system in Thailand. Phuket Mar Biol Cent Res Bull 66:337–350

Nair KP, Thomas PA, Gopakumar G, Vincent SG, Omana TA (1985) Some observations on the hatching and post-hatching behaviour of the cuttlefish *Sepia pharaonis* Ehrenberg. In: Silas EG (ed) Cephalopod bionomics, fisheries and resources of the exclusive economic zone of India. Cent Mar Fish Res Inst Bull 37

Nair KP, Srinath M, Meiyappan MM, Rao KS, Sarvesan R, Vidyasagar K, Sundaram KS, Rao GS, Lipton AP, Natarajan P, Radhakrishnan G, Sunilkumar K, Mohamed, Narasimham KA, Balan K, Kripa V, Sathianandan TV (1993) Stock assessment of the pharaoh cuttlefish *Sepia pharaonis* Ehrenberg. Indian J Fish 40:85–94

Norman M (2000) Cephalopods, a world guide. Conchbooks, Hackenheim

Reid AL (1998) Sepiidae Cuttlefishes. In: Carpenter KE, Niem VH (eds) FAO Species Identification Guide for Fishery Purposes. The living marine resources of the western central Pacific. Vol 2: Cephalopods, crustaceans, holothurians and sharks. FAO, Rome

Reid AL, Jereb P, Roper CFE (2005) Cuttlefishes. In: Jereb P, Roper CFE (eds) FAO Species catalogue for fishery purposes, no 4, vol 1. Cephalopods of the world an annotated and illustrated catalogue of cephalopod species known to date. Vol 1: Chambered nautiluses and sepioids (Nautilidae, Sepiidae, Sepiolidae, Sepiadariidae, Idiosepiidae and Spirulidae). FAO, Rome

Roper CFE, Sweeney MJ, Naun CE (1984) FAO Fisheries synopsis, no 125, vol 3. FAO species catalogue. Vol 3: Cephalopods of the world. An annotated and illustrated catalogue of species of interest to fisheries. FAO, Rome

Sangpradab S, Veschasatra V, Kbinrum S (1984) Experiments on feeding of cuttlefish (*Sepia pharaonis* Ehrenberg) with four different kind of feed. Tech Pap 2/1984, Rayong Brackish Fish Stn, Brackish Fish Div, Dept Fish

Silas EG, Sarvesan R, Nair KP, SastriYA, Sreenivasan PV, Meiyappan MM, Vidyasagar K, Rao KS, Rao BN (1985) Some aspects of the biology of cuttlefishes. In: Silas EG (ed) Cephalopod bionomics, fisheries and resources of the exclusive economic zone of India. Cent Mar Fish Res Inst Bull 37

Stearns SC (1992) The evolution of life histories. Oxford University Press, London

Supongpan M (1995) Cephalopod resources in the Gulf of Thailand. In: Nabhitabhata J (ed) Biology and culture of cephalopods. Rayong Coast Aquacult St, Rayong

Toll RB, Strain CH (1988) Freshwater and terrestrial food organisms as an alternative diet for laboratory culture of cephalopods. Malacologia 29:195–200

Chapter 13
Sepiella inermis

Jaruwat Nabhitabhata

Abstract The spineless cuttlefish, *Sepiella inermis,* is an economic species of the Indian Ocean. *S. inermis* habit is bentho-nektonic but active in a higher degree compared to *Sepia* cuttlefish. This species can tolerate environment fluctuations and culture conditions very well, which favour aquaculture. *S. inermis* can be cultured through several consecutive generations in open seawater systems. The culture methodology is comparable to other sepiid cuttlefish, particularly *Sepia pharaonis,* comprising collection of live broodstocks, incubation of egg masses, nursing of young and growout phase. The planktonic phase of hatchling is different from *Sepia* cuttlefish. The moderate final size of 50–100 g is appropriate to frozen food product packaging and maintenance in home aquarium.

Keywords *Sepiella inermis* · Moderate size · Planktonic hatchling · Bentho-nektonic · Open seawater systems · Consecutive generations

13.1 Importance of this Species in the Market

The common name "spineless cuttlefish" comes from the unique character of the cuttlebone which lacks a spine at the posterior tip (Reid 1998; Reid et al. 2005; Roper et al. 1984). The Thai vernacular "Pla Muk Kra-dong Hang Mai" arises from the black colour of the anal gland and pore at the posterior tip of the mantle, which is another unique characteristic. The function of this gland is still unknown. Nabhitabhata and Polkhan (1983b) observed the release of brown fluid from the anal pore after an abrupt change to low salinity, suggesting the functions are related to the osmoregulation process. This species is the only sepiid cuttlefish known to inhabit the estuarine areas (Roper et al. 1984). The geographical distribution covers the Indo-West Pacific region, 30–120°E, 30°N–20°S (Reid 1998; Reid et al. 2005; Roper et al. 1984). *Sepiella inermis* is a demersal, benthic species, with distribution from shallow water to depths of about 40 m. The maturity in the wild stocks is at a mantle length of about 50 mm, with fecundity of between 200 and 1,000

J. Nabhitabhata (✉)
Excellence Centre for Biodiversity of Peninsular Thailand, Faculty of Science,
Prince of Songkla University, HatYai, Songkhla 90112, Thailand
e-mail: jaruwat.n5@gmail.com; jaruwat.n@psu.ac.th

J. Iglesias et al. (eds.), *Cephalopod Culture,* DOI 10.1007/978-94-017-8648-5_13,

eggs (Chotiyaputta 1995; Sundaram and Khan 2011; Unnithan 1982). Spawning occurs throughout the year with seasonal peaks varying on geographical localities, environment and habitats. The maximum length is about 125 mm with an estimated lifespan of 720 days (Reid et al. 2005; Roper et al. 1984). The final size and lifespan tend to be shorter in the estuarine populations.

This is an economically important species of the region, especially in India and Sri Lanka, with captures representing between 2 and 4% of the total cephalopod catch in India (Sundaram and Khan 2009, 2011), and 3% (approximately 2,000 t) of total cuttlefish catch in Thailand (Supongpan 1995). They are mostly captured by trawlers offshore and by push nets inshore and in the estuary (Supongpan 1995). Average capture sizes range between 40 and 60 mm in India and between 20 and 80 mm in Thailand. The estimated age to reach this size is 180 days with a growth rate of 2.5 mm per month (Sundaram and Khan 2009; Unnithan 1982).

The spineless cuttlefish has a more active behaviour than other sepiid cuttlefish. They ritually hover in the water column like a pelagic squid and dwell on the bottom like other cuttlefish. The behaviour is aggressive to a higher degree than the pharaoh cuttlefish. The spineless cuttlefish sometimes bite others in the same school, as well as humans when handled with bare hands (Nabhitabhata et al. 1984, 1985). Because of their benthic–pelagic habit, spineless cuttlefish require less interindividual space or territory compared with pelagic squid of the same size, and can be maintained at a higher density. Considering the unique life history and behaviour, this species would have to be named the "active cuttlefish".

Aspects related to aquaculture potential of this species were reported at both laboratory scale (Boonprakob et al. 1977a, b; Anil 2003) and large scale (Nabhitabhata 1997, 2000; Nabhitabhata et al. 1984, 1985; Sivalingam 1999). The small final size turns out to be an advantage for frozen human food product processing and packaging and ornamental aquarium displays, as in *Sepia pharaonis* (see Chap. 12, this volume). The high tolerance to variation of environmental conditions makes the spineless cuttlefish biologically interesting from a large-scale culture point of view. Furthermore, this species is a less selective feeder, and can be trained to accept dead feed easier than pelagic squid in culture tanks.

13.2 State of the Art

The spineless cuttlefish is highly tolerable to fluctuations of the environment and has a high adaptability to culture conditions. This species can be cultured through several consecutive generations in tanks in flow-through systems. Broodstocks collected from the wild can spawn in captivity. Egg masses are collected and nursed under controlled environment until hatching after about 13 days at 28°C. The hatching rate and survival of the young at 30 days are higher than 95%. Hatchlings are fed with wild-collected live feed for about 30 days, but training them to accept dead feed can be as early as 20 days. Water volume and culture density are managed to correspond with their rapid growth and high survival during the nursing phase.

Fig. 13.1 Mating of *Sepiella inermis* broodstock in the maintenance tank (male is on the *left*). (Photograph of J Nabhitabhata)

After enrichment with fatty acids, mass-produced *Artemia* can be used for nursing whenever other live feed is unavailable. The ongrowing phase starts after the young accept dead food. The daily growth rate is 2.4% in length and 5.5% in weight through the culture period of 130 days. The high tolerability to changing environments enables their ongrowing in saltwater earthen ponds.

13.3 Broodstock Maintenance

Broodstocks of the spineless cuttlefish are collected alive with push netting near the shore and in the estuarine area. Onboard, they are maintained in circular polyvinyl chloride (PVC) tanks of 50-L capacity equipped with portable aeration devices. The cuttlefish size ranges between 40 and 60 mm in mantle length and 30–50 g in weight. The tanks and the live cuttlefish are transported with an air supply to the hatchery and then maintained in 2 m^3 concrete tanks at a male to female ratio of 1:2. The broodstock tanks must be previously prepared before introducing the cuttlefish, and artificial substrates for attaching eggs, made from pieces of fishing net of 25 mm mesh size with sinkers, have to be provided (Nabhitabhata et al. 1984; Nabhitabhata 1997).

The acclimatisation period to captivity is short; mating and spawning in the tank occur within 12 h. The male selects his mate by displaying a dark brown colour with white and orange spots on its fins, raising a pair of dorsal arms (arms I) and spreading out the other pairs of arms, while the consenting female displays small white spots along the base of the fins. During and after this stage, the male escorts his mate by swimming alongside her all of the time. The mating position is head-to-head (Fig. 13.1). The females spawn in the tank and attach their egg capsules in clusters to the artificial substrates (Fig. 13.2). In the wild, the females prefer branch-like substrates (e.g., tree branch, sea fan) for attachment of its egg capsules. Egg laying occurs the morning after, in the same period of the day as mating (Nabhitabhata et al. 1984; Nabhitabhata 1997).

The number of eggs laid by one female is approximately 400. The female lays 1–5 batches of eggs and does not eat during the interval of each batch release. The

Fig. 13.2 Female *Sepiella inermis* attaching her egg to an artificial substrate. (Photograph of J Nabhitabhata)

number of batches laid by one female depends upon the number of eggs in one batch which ranges from 20 to 500 eggs (Nabhitabhata 1997). Any interruption during egg laying will delay the process for at least 24 h. The female does not take care of her eggs. The male will turn to form a pair with another female after the last spawning of his mate. The healthy male switches to mate with another selected female after the final egg laying of his mate.

13.4 Incubation of Egg Masses

13.4.1 Egg Characteristics and Hatching

The characteristics of the eggs are similar to *S. pharaonis,* except for the capsule colour. The egg capsule of spineless cuttlefish contains a single egg that is attached in clusters to the underwater substrate. The capsules are stained black in colour with ink from the female. They are round in shape with a tip and a short stalk (Fig. 13.3). The healthy female spawns dark black egg capsules, while the weak female spawns egg capsules that are paler in colour. Paler and white-coloured egg capsules are the result of an abnormal coating without ink, revealing that the female is weak or not in good condition (Nabhitabhata et al. 1984; Nabhitabhata 1997). The offsprings that hatch from the pale and white capsules usually have low survival. The egg capsules with a pale black colour are also pilot eggs which are first laid by the female in order to test the substrates. The eggs become comparatively larger and more transparent during the progress of their embryonic development compared to the freshly laid ones and reach their largest size just before hatching.

The length of the embryonic period and hatching pattern are also similar to that of *S. pharaonis*. The embryonic period is between 8 and 19 days, on a mean of 13 days at about 28 °C (Table 13.1). Hatching occurs mostly at night and the hatching

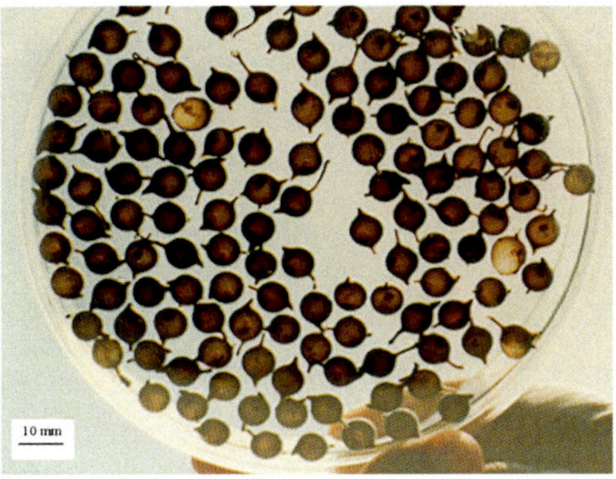

Fig. 13.3 Eggs of *Sepiella inermis* at near hatching. (Photograph of J Nabhitabhata)

rate is normally more than 90 % in controlled conditions. Brief changes of temperature and salinity and mechanical stimuli cause premature hatching and yield hatchlings of comparatively smaller size. The smaller and weaker the hatchlings, the lower is their survival rate.

13.4.2 System Requirements and Management

Eggs are collected in the morning and the nursing is performed in other concrete tanks of similar size and water quality. Small clusters of eggs may be nursed in a small 50-L tank. In the egg cluster, the egg stalks are twined to the substratum and also to stalks of other eggs and each other at the base of the cluster. The eggs are separated from the cluster by hand or by scissor cutting. They are aerated and maintained until hatching in plastic baskets for better ventilation. The egg baskets of 2 mm mesh size are floated and aerated in the tanks. Flow-through systems are employed in *S. inermis* culture, to maintain optimum water quality. Temperature variations are minimised at a rate of 1 L min^{-1}. This maintains an average temperature of 28 °C, a salinity of 30–33 psu and a pH of 6.0–8.0 (Nabhitabhata et al. 2005). Light is reduced with a camouflage net to prevent algal growth. All other details of the method are similar to those described for *S. pharaonis* (Chap. 12, this volume). The optimum temperature for 99 % hatching rate is 24–32 °C (Junpramuk 1981) and the salinity for 50 % hatching ranges from 20 to 36 psu (Boonprakob et al. 1982).

13.5 Nursing of Young

13.5.1 Hatchling Characteristics

Hatchlings are miniature adults in their morphology, but with a planktonic habit. This habit is different from benthic habit of *Sepia* hatchling, since the life mode of

Table 13.1 Comparison of culture conditions and production of *Sepiella inermis* from various studies

Aspect	Boonprakob et al. (1977a, b)	Nabhitabhata et al. (1984); Nabhitabhata (1997)	Sivalingam (1999)	Anil (2003)
Tank system	Open	Open	Open	Open
Egg incubation (days)	12.5 (11–20)	12.6 (8–19)	13	13–20
Temperature (°C)	28	28	–	27–32
Hatching period (d)	7	3–10	10	7
Hatchling size (mm, g)	3.3, 0.01	4.3, 0.04	2	4, 0.02
Initial density	–	100 ind m^{-2}	–	500 ind m^{-3}
First spawning (d)	98.7	87 (79–96)	75	86
Size at maturity (mm, g)	69.5, 61.8	61, 36.6	57.2, 37.9	64, 46.5
Daily growth (%: ML, W)	2.4, 6.8	2.4, 5.5	–	2.5, 6.0
Mean final size (mm, g)	69.5, 61.8	68.2, 48.2	–	69, 54.7
Max size (mm, g)	–	77, 52.6	–	–
Mean lifespan (days)	109.7–114.6	117–115.1, 116.2 (84–149)	–	–
Max lifespan (days)	138	149	–	110

most young cephalopods, particularly sepiid cuttlefish, largely corresponds to the benthic mode of the adult, as in *S. pharaonis* (Fig. 13.4). They usually hatch out with an external yolk sac, which is smaller than their own head. The yolk sac is abandoned on the first day after hatching. Hatchlings swim hovering at an angle of about 60–80° to the tank bottom. Chromatophores function and their normal colour pattern is dark brown, white or transparent, with a dark red spot of the anal gland at the posterior tip of the dorsum. The size of the anal gland is comparatively large with a diameter of more than about 20 % of their own mantle length (Nabhitabhata 1997). Fins are separated and projected to the posterior and their length is about 80 % of the mantle length.

At 3 days old, hatchlings dive to the floor, entering their benthic juvenile stage. However, the period of the planktonic stages varies among hatchlings from different batches. Such periods range from less than 12 h to 5 days, probably due to slight differences of their embryonic stages at hatching. At this early benthic stage, the young cuttlefish hold on to the floor with the tips of their arms. They touch the floor while their head and mantle are still hovering at an angle of about 20–30° to the floor (Nabhitabhata et al. 1984). Their colour is dark brown with the dark red spot at the anal gland. After 5 days, the cuttlefish completely adopt the benthic phase, lying on the floor. At this age, the cuttlefish do not burrow into the substratum but prefer lying under shelters. The tentacles begin to function for attacking prey. Their arms are still short but strong enough to catch prey up to twice their mantle length in size.

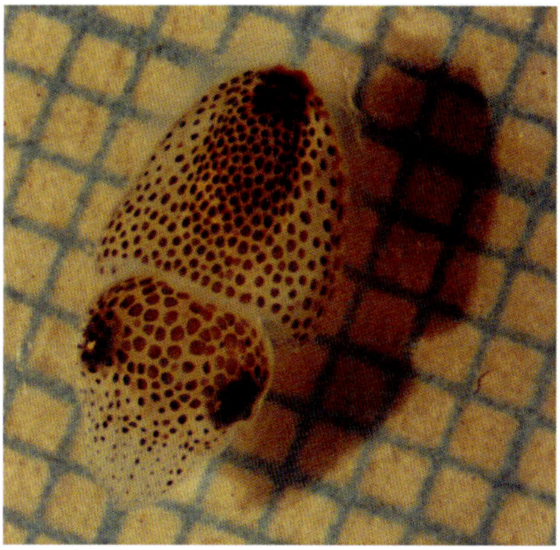

Fig. 13.4 Planktonic hatchling of *Sepiella inermis*. (Photograph of J Nabhitabhata)

13.5.2 System Requirements and Management

The planktonic habit of hatchlings requires a directed water flow in the tanks which enables them to suspend in the water column. Aeration should be managed to serve another function, as the current generator in the tanks. Adjusting the current velocity to an optimum velocity can reduce stress and energy consumption in countercurrent swimming. The optimum current velocity which enhances survival, growth and hence production can be observed from the angle of hovering that has to be more than 45° to the tank bottom at all times (Nabhitabhata et al. 1984; Nabhitabhata 1997). A simple air-lift system is used to generate the horizontal flow in the cylindrical tank. A PVC pipe of 50 mm diameter and 800 mm length is longitudinally cut into two halves and each half is drilled at one end. The air pipe and air stone are attached to the interior face of the half-pipe at one end through the drilled hole. The set of prepared half-pipe is placed upright at 45° with the air stone at the bottom of the tank. At least two sets are placed in one circular tank. The air stones and half-pipe sets face the same direction to generate a continuous flow. The current velocity is adjusted by adjusting the aeration rate. The numbers of the flow-generating pipe sets can be increased, depending on the size and shape of the tanks. A rectangular tank requires one set for each corner.

The criteria of the required water quality and tank management in the nursing phase are similar to that in the egg-nursing phase. The optimum temperature is below 34°C (Junpramuk 1981), while the tolerable salinity range of hatchlings of *S. inermis* is very wide, between 16 and 44 psu. The salinity required for 50% survival, after a brief change, is 18–44 psu (Anil 2003; Boonprakob et al. 1977b; Nabhitabhata and Darunchoo 2001). The young cuttlefish can tolerate high levels of ammonia (NH_4-N) of up to 0.6 mg L^{-1} for at least 24 h and 0.12 mg L^{-1} for at least 48 h (Wongwiwathanawut et al. 2001).

Table 13.2 Expected growth, survival and managed density of *Sepiella inermis*. (Estimation from Anil 2003; Boonprakob et al. 1977a, b; Nabhitabhata et al. 1984 and Nabhitabhata 1997)

Culture period (days)	ML (mm)	DGRL (%)	W (g)	DGRW (%)	Density (ind m^{-2})	Water depth (mm)	Survival (%)
0	4	–	0.03	–	500	300	–
30	22	4.6	2.20	6.5	200	450	95
60	42	2.2	11.0	4.4	90	550	90
90	62	1.3	43.0	3.8	40	600	85
100	67	1.1	50.0	3.6	20	600	80
130	70	0.2	60.0	0.8	10	600	75

ML mantle length, *DGRL* daily growth rate in mantle length, *DGRW* daily growth rate in weight, *W* wet body weight

Since the hatchlings are planktonic, the initial density depends on the water volume, not on the area as in benthic *S. pharaonis*. The initial water depth in tanks is 300 mm with an increase of 50 mm every 10 days. The initial density is about 500 individuals m^{-3}, which corresponds to the water depth (Tables 13.1, 13.2). At 3 days old, hatchlings enter the benthic juvenile stage. After 10 days, the density turns from a volume basis to an area basis, decreasing to 100 individuals m^{-2}. The size grading for a suitable management of the food has to be performed every 10 days, and this also starts in this period.

13.5.3 Feeding

Hatchlings are instantly fed an excess quantity of live food, and the stocking density and water level are managed to promote feeding. Hatchlings of the spineless cuttlefish are 43 mm in mantle length and 0.04 g in body weight and the live food is of a corresponding size. The food organisms are mysids (*Mesopodopsis orientalis*) of about 7 mm in total length, and the postlarvae of penaeid shrimp (*Penaeus merguiensis*) of 5 mm in total length (Nabhitabhata 1997, 2002; Nabhitabhata et al. 1984). Sivalingam (1999) and Anil (2003) also fed the spineless cuttlefish with mysids and shrimp postlarvae.

Young spineless cuttlefish can feed on live adult *Artemia*. Before being fed to the cuttlefish, *Artemia* have to be enriched with the Omega-3 group of highly unsaturated fatty acid (n3HUFA) through feeding. Muthuwan et al. (1993) reported that young fed on adult *Artemia,* enriched with a rice bran suspension, fish liver oil or 12%-fat microencapsulated feed, could maintain survival of more than 44% up to 31 days after hatching compared to 84% of those fed on mysids, *Mesopodopsis* spp. Adult *Artemia* enriched with plankton, live *Tetraselmis* sp. and a dry powder of *Spirulina* sp. are also consumed by cuttlefish, but yield lower growth and survival of 25–27%. Growth among the cuttlefish that are fed on *Artemia* is not significantly different but is 4–16 times lower than those fed on mysids. Enrichment with the rice

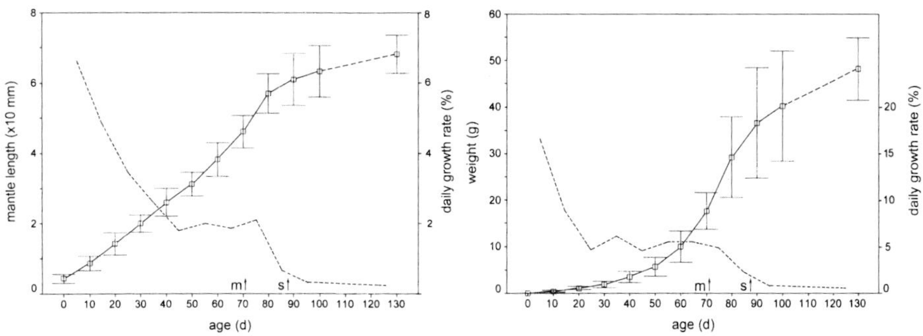

Fig. 13.5 Growth in terms of the relationships between mantle length (mm) (*left*), weight (g) (*right*) and age (d, *full line*), daily growth rate (%, *broken lines*) of length and weight (after Nabhitabhata 2002). *Arrows* indicate approximate age at mating (m) and spawning (s)

bran suspension is the most appropriate method according to economic viability and the most convenient process compared to the encapsulated diet and fish liver oil. From a mass culture point of view, the enriched *Artemia* can be used as a substitutional feed for a short period when the main feed (e.g. live mysids) is in short supply. It is notable that the cuttlefish eats only a small quantity of each *Artemia* body and the rest are left uneaten. According to such habits, it is not appropriate to use live *Artemia* as additional feed (e.g. night feed) since the cuttlefish will learn to wait for their favourable feed in the subsequent feeding (Nabhitabhata et al. 1984).

13.5.4 Growth

The highest growth rate of spineless cuttlefish is in the first 30 days after hatching, during the nursing phase. The hatchlings of 4.3 mm mantle length and 0.04 g body weight grow to 8.6 mm and 0.44 g after 10 days (Fig. 13.5). During this period, the daily growth rate in the mantle length is 6.7% and in the weight is 16.7%. Daily consumption is of 3 mysids per individual and the food conversion efficiency is about 60% at a 20% ration which is 12% higher than that for *S. pharaonis* (Nabhitabhata et al. 1996). Survival is more than 90% for the nursing phase (30 days). The cost of production in Thailand is estimated to be about US$ 3.5 for 100 individuals in 1996 (Nabhitabhata et al. 1996), when US$ 1 equal about Thai Baht 25 (US$ 1 equal Thai Baht 30 in 2013).

Growth of the spineless cuttlefish is allometric and can be separated into 2 phases (Nabhitabhata 2002). The early phase is from hatching to 30 days of age, during the nursing phase. The relationship between mantle length (ML, mm) and weight (W, g) (Fig. 13.6) in the early phase is:

$$W = 4.288 \times 10^{-3} ML^{2.032} \tag{13.1}$$

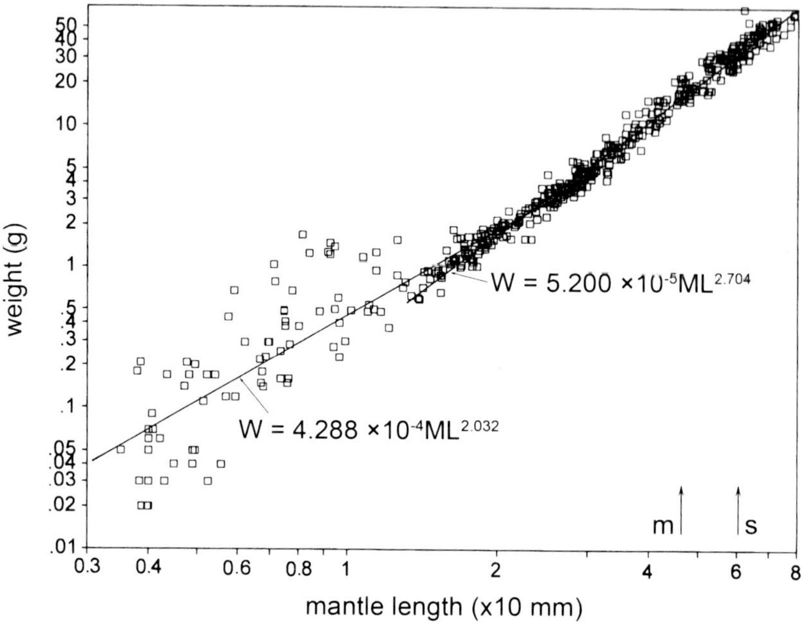

Fig. 13.6 Growth in terms of the relationship between mantle length (mm) and weight (g) (after Nabhitabhata 2002). *Arrows* indicate approximate age at mating (m) and spawning (s)

The early growth phase of the cuttlefish from hatching to 30 days of age in terms of mantle length–age (T:d) relationship (Fig. 13.7) is expressed with the exponential equation:

$$ML = 0.0461e^{0.0482T} \tag{13.2}$$

The growth phase in terms of weight–age relationship (Fig. 13.8) in the early phase is exponential:

$$W = 8.867 \times 10^{-2} e^{0.102T} \tag{13.3}$$

13.6 Ongrowing

13.6.1 System Requirements and Management

Twenty days after hatching, the cuttlefish are trained to feed on dead prey. They are approximately 14 mm in mantle length and their normal colour turns darker. Young cuttlefish seize prey in the water column and eat them while hovering above

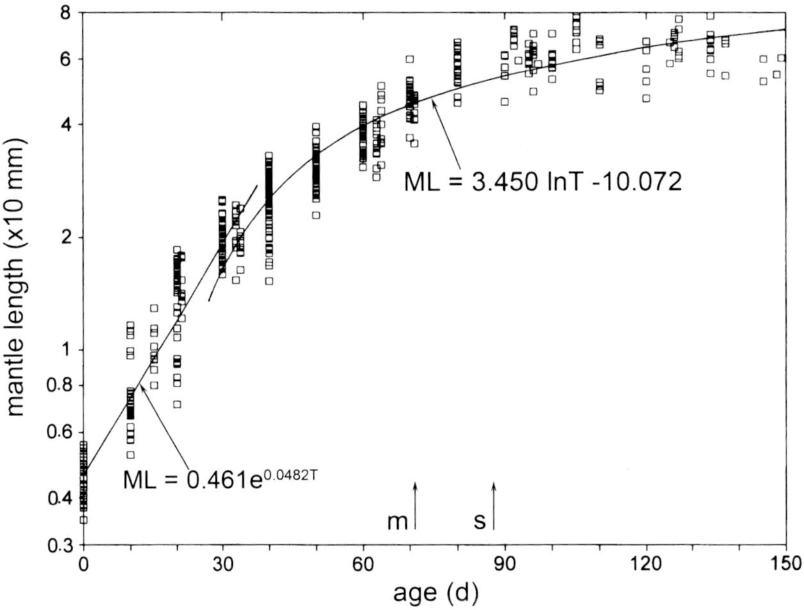

Fig. 13.7 Growth in terms of the relationship between mantle length (mm) and age (d) (after Nabhitabhata 2002). *Arrows* indicate approximate age at mating (m) and spawning (s)

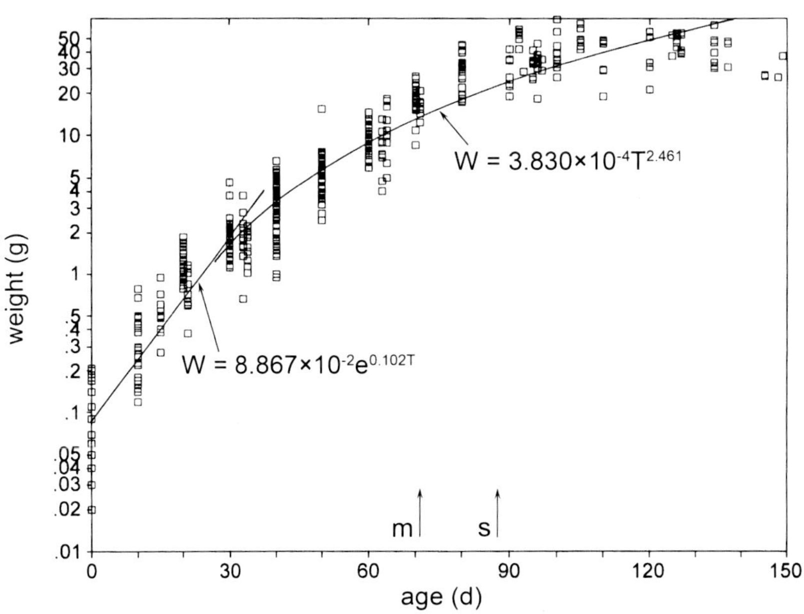

Fig. 13.8 Growth in terms of the relationship between weight (g) and age (d) (after Nabhitabhata 2002). *Arrows* indicate approximate age at mating (m) and spawning (s)

or lying on the substrate. The prey is seized with the tentacles and kept within their arms. When the cuttlefish are fed on dead food, they change their feeding behaviour to use only arms to seize prey without the positioning step in the feeding behaviour. Tang and Khoo (1974) also reported the same change and suspected it was a sign of weakness. This change indicates the success of training, and that the spineless cuttlefish are accustomed to the feed and culture conditions (human appearance with feed and the environment of the culture tank). Another reason for the change in behaviour is the lack of motion of dead feed compared to the active live food. It is unnecessary for the spineless cuttlefish to use a rapid and energy consuming motion like stretching out the tentacles to capture the dead feed. This kind of behavioural change is similar to that seen in cultured pharaoh cuttlefish, *S. pharaonis*. Aggressive behaviour is prominent at this age and cannibalism is observed whenever the food is in short supply. Males sometimes bite other males in connection with their sexual activity (Nabhitabhata et al. 1984).

Tanks of any size can be used for the ongrowing phase of the spineless cuttlefish. The initial stocking density of this phase, after 30 days, is 30 individuals m^{-2}. Size grading is performed every 10–20 days until ceasing at the beginning of reproduction, at about 70 days. Feeding is twice per day at the ration of 10 % of body weight. Food with high nutritional value, i.e. whole shrimps, large mysids, with additional cost, may be added as a supplement. Survival is cumulative, and of 80–90 % during ongrowing (Table 13.2).

Ongrowing of *S. inermis* in an earthen pond has been studied in Thailand (Nabhitabhata et al. 1985). Cuttlefish accustomed to the dead feed, of about 38 mm mantle length and 11 g weight (50–60 days) were released into a pen covering an area of 45 m^2 in a large (25,600 m^2) earthen pond at 3–6 individuals m^{-2}. Survival was of 40 % after 50 days. The production was about 4,000 g and the feed conversion efficiency was of 13–18 %. One explanation for this low survival was the low quality of the benthic substrate in the pond, which lowered the water quality and initiated bacterial infections. Another problem was a competing fish, *Therapon* spp. that had infiltrated the pen through the net fence. Those fish were more active than the cuttlefish, interfering in the ritual feeding and resting of the cuttlefish.

13.6.2 Growth

The relationship between mantle length and weight in the ongrowing phase after 30 days is also expressed with the power equation (Nabhitabhata 2002) (Fig. 13.6):

$$W = 5.200 \times 10^{-4} ML^{2.704} \tag{13.4}$$

The slope of the length–weight relationship equation obtained from cultured batches is 2.0–2.7 which is within the range of 1.9–2.7 obtained from natural stocks (Chotiyaputta 1981; Unnithan 1982; Sundaram and Khan 2011). The second growth phase in terms of mantle length–age relationship of the cuttlefish from 30 days old, the inflection point (Fig. 13.7), is expressed with the logarithmic regression (Nabhitabhata 2002):

$$ML = 0.345 \ln T - 1.007 \tag{13.5}$$

The growth phase in terms of weight–age relationship (Fig. 13.8), in the second phase, also starts at the inflection point of 30 days, and can be expressed by the power model:

$$W = 3.830 \times 10^{-4} T^{2.461} \tag{13.6}$$

After about 60 days, or 40 mm and 10 g, the colour pattern allows sex recognition. The male colour pattern is dark brown with white and orange spots on the dorsum along the base line of the fins. The female colour pattern is dark brown, but a lighter shade than that of the males, with smaller white spots and without orange spots. *S. inermis* begin to form mating pairs on and after the age of 60 days, and the first mating is observed at 70 days old, or 50 mm and 20 g (Table 13.1).

The growth rate of the spineless cuttlefish decreases after reproduction. During the reproductive period from 80 to 130 days of age, growth rates of 0.2–0.7% in length and 0.6–2.0% in weight are reported, compared to those in the previous phase of 30–80 days of age, 2.0–2.6% in length and 5–6% in weight (Nabhitabhata 1997). Most of the mating and spawning of the spineless cuttlefish is observed at the size of 60 mm mantle length and 40 g weight (90 days). The daily growth rate in length decreases to lower than 1.0% on and after 90 days and that in weight also decreases after 100 days of age. On average, the overall daily growth rate from hatching to 130 days of age is 2.4% in length and 5.5% in weight.

The final size before senescence of this species is mostly attained during 100–130 days and the females are larger than the males. After 130 days, the average mantle length is approximately 70 mm, while average weight is 50 g. The maximum size recorded in culture conditions is about 70 mm mantle length and 80 g weight in a male (126 days) and 80 mm and 65 g (105 days) for a female (Tables 13.1, 13.2) (Boonprakob et al. 1977a; Nabhitabhata et al. 1984). The lifespan is approximately 120 days, with a maximum of 150 days.

Growth tends to be much faster in culture condition compared to in the wild. Silas et al. (1985) reported the maximum mantle length of the spineless cuttlefish to be 125 mm based on the research on the open sea population. Animals attain 29–35 mm mantle length after 6 months, 53–61 mm after 12 months and 74–82 mm after 18 months. The lifespan is six times longer and growth is approximately four-fold slower than that obtained from culture batches. The culture conditions, i.e. sufficient food availability and stable environmental factors, tend to induce fast growth and early maturation, as observed in *S. pharaonis* (see Chap. 12, this volume).

13.7 Trends in Research and Industrial Level

Inbreeding has no negative effect on growth rate and the final size after culture through three consecutive generations (Nabhitabhata et al. 1984; Nabhitabhata 1997). The indifferent growth could, in the short term, provide a reliable and continuous supply of domesticated cultured broodstocks for hatchling production at a certain degree. However, there would be an unknown and unavoidable risk that culture through several (more than three) generations could result in some inbreeding effects. Therefore, it would be better to obtain additional broodstocks from different populations (cultured or wild) in order to avoid a potential decrease of growth due to inbreeding. From an aquaculture-for-restocking point of view, this matter should be considered in order to maintain a gene pool and genetic variation of natural populations.

The small final size turns out to be the advantage, being required by the seafood industry for frozen food product processing and packaging as well as live cuttlefish products that can also be distributed to the ornamental aquaculture trade, in a similar manner to *S. pharaonis* (see Chap. 12, this volume).

The main focus of future research is the developing of feed, both live and artificial, for large-scale culture. However, the feasibility is at a higher level due to the active habit compared to *S. pharaonis*. The acceptability of the enrichment of *Artemia* sp. (Muthuwan et al. 1993) as supplementary and substitute feed for young cuttlefish in the nursing phase is a promising trend for the solution of the bottleneck during the nursing phase.

The differences in growth rate probably also depend on the environmental history of the broodstocks between estuarine and open sea populations. Early maturation, a smaller final size and a shorter lifespan tend to occur in the estuarine population due to the effects caused by the diversity of environmental fluctuations in estuaries. The source for the collection of wild broodstocks has to be known in order to have a proper management. Nabhitabhata and Polkhan (1983b) reported that the young *S. inermis,* offspring of estuarine broodstocks in Thailand, grew faster with a higher survival in salinity of 28 psu than in 32 psu which is the normal salinity of open waters. The young also demonstrated their preference for the mud substrate to a similar degree to the sand substrate, in contrast to the obvious preference for sand by *S. pharaonis* (Nabhitabhata and Polkhan 1983a; Nabhitabhata and Nilaphat 2000). In view of maximising the aquaculture production, the differences in the growth rate and reproductive products among different populations should be identified and studied in detail to determine whether it is the result of the plasticity of life cycles in different habitats or the genetic difference. Reid et al. (2005) suggested that *S. inermis* is possibly a species complex. Overall, this species has good adaptability to culture conditions that makes them easy to culture.

13.8 Conclusions

Culture process of *S. inermis* is comparable to that of other sepiid cuttlefish, which comprises a collection of broodstocks from the wild, incubation of egg masses, nursing of the young in the hatchery and growout phase. Egg masses and hatchlings are nursed in concrete tanks. Water quality is partially controlled in open seawater systems. The short period of planktonic habit of hatchling requires a directed water flow in nursing tanks. Density is area oriented due to benthic habit. Initial density is 500 individuals m^{-2} which a 25 % is decreased through regular size grading, over a 10-day period. The young are fed from hatching to about 30 days with live prey organisms either collected from the wild (mysids) or produced in hatchery (penaeid shrimp postlarvae). The active habit of the young causes this species to be able to accept dead feed in a shorter period, compared to *S. pharaonis*. Survival is more than 90 % 2 months after hatching. Growout in an earthen pond is feasible. Overall daily growth rate is approximately 2.4 % in length and 6.0 % in weight. The mean final size and lifespan is approximately 50 g, for 115 days in open seawater systems at 28°C. Innate feeding on specific live preys is the bottleneck and development of artificial feed is required in the long term.

References

Anil MK (2003) Laboratory rearing and breeding of spineless cuttlefish *Sepiella inermis* Orbigny (Mollusca: Cephalopoda). J Aquacult Trop 18:35–44

Boonprakob P, Satayalai O, Sithigorngul P, Siripoonya P, Yodyingyuad U (1977a) Biological study of cuttlefish *Sepiella inermis*. Tech Paper Silver Jubilee Res Fund, Chulalongkorn University, Bangkok

Boonprakob P, Siripoonya P, Yodyingyuad U, Satayalai O, Sithigorngul P (1977b) Studies on embryonic development, karyotype, effects of physical factors on development and behavioural response to physical factors of cuttlefish *Sepiella inermis*. Rep National Research Council of Thailand

Boonprakob P, Siripoonya P, Satayalai O (1982) Effect of salinity on hatching period and development of cuttlefish *Sepiella inermis*. Rep Department of Biology, Chulalongkorn University, Bangkok, Thailand

Chotiyaputta C (1981) Length-weight relationships and sex ratios of cuttlefish *Sepia lycidas* and *Sepiella inermis* Ferrussac & d' Orbigny. Tech Paper 11/1981, Invertebr Sub-Div, Mar Fish Div, Dept Fish

Chotiyaputta C (1995) Biology of cephalopods. In: Nabhitabhata J (ed) Biology and culture of cephalopods. Rayong Coast Aquacult Station, Rayong

Junpramuk J (1981) Effect of temperature on hatching and lethal temperature of young cuttlefish (*Sepiella inermis* Ferussac & d'Orbigny). MS thesis, Chulalongkorn University, Bangkok

Muthuwan V, Sawatpeera S, Mama P, Kuandee P (1993) Rearing of spineless cuttlefish, *Sepiella inermis* Ferussac & d'Orbigny with adult brine shrimp, *Artemia salina* fed with various kinds of feed. Research Paper 53/2536, Institute of Marine Science, Burapha University, Chon Buri, Thailand

Nabhitabhata J (1997) Life cycle of three cultured generations of spineless cuttlefish, *Sepiella inermis* (Ferussac & d'Orbigny, 1848). Phuket Mar Biol Cent Spec Pub 17:289–298

Nabhitabhata J (2000) Cephalopod culture: present and future status. In: Aryuthaka C (ed) Proceedings of the Wantana Yoosuk seminar on mollusc studies in the year 2000. Faculty of Fisheries, Kasetsart University, Bangkok

Nabhitabhata J (2002) Life history and behaviour of reared neritic cephalopods from Thai Waters. Ph D Thesis, Tokyo University of Fisheries, Tokyo

Nabhitabhata J, Darunchoo L (2001) Effect of different levels of salinity on survival of hatchlings of spineless cuttlefish, *Sepiella inermis* Ferussac & d'Orbigny, 1948. Rayong Coast Fish Res Contrib 27:17

Nabhitabhata J, Nilaphat P (2000) Behaviour of juvenile cephalopods: preference for texture and brightness of substrata. Phuket Mar Biol Cent Spec Pub 21:103–112

Nabhitabhata J, Polkhan P (1983a) Biological studies on economic cephalopods-I: preference on types of substratum of sepiid cuttlefish (Cephalopoda: Sepioidea). Tech Paper 2/1983, Rayong Brackish Water Fisheries Station

Nabhitabhata J, Polkhan P (1983b) Biological studies on economic cephalopods-II: survival and growth of cuttlefish, *Sepiella inermis* Ferrussac & d' Orbigny in various salinity conditions. Tech Paper 4/1983, Rayong Brackish Water Fisheries Station, Brackish Water Fisheries Division, Department of Fisheries

Nabhitabhata J, Polkhan P, Kbinrum S (1984) Culture, growth and behaviour of spineless cuttlefish, *Sepiella inermis* Ferussac & d'Orbigny. Tech Paper 5/1984, Rayong Brackish Water Fisheries Station, Brackish Water Fisheries Division, Department of Fisheries

Nabhitabhata J, Kbinrum S, Kumpang P (1985) On earthen pond culture of spineless cuttlefish, *Sepiella inermis* Ferussac & d'Orbigny. Tech Paper 5/1985, Rayong Brackish Water Fisheries Station, Brackish Water Fisheries Division, Department of Fisheries

Nabhitabhata J, Kakhai N, Suriyawarakul J, Nilaphat P (1996) Feeding and cost of seed production of cephalopods, *Sepioteuthis lessoniana* Lesson; *Sepia pharaonis* Ehrenberg; *Sepiella inermis* Ferussac & d'Orbigny. Tech Paper 1/1996, Rayong Coastal Aquaculture Station, Coastal Aquaculture Division, Department of Fisheries

Nabhitabhata J, Nilaphat P, Promboon P, Jaroongpattananon C, Nilaphat G, Reunreng A (2005) Performance of simple large-scale cephalopod culture system in Thailand. Phuket Mar Biol Cent Res Bull 66:337–350

Reid AL (1998) Sepiidae Cuttlefishes. In: Carpenter KE, Niem VH (eds) The living marine resources of the western central Pacific vol 2 cephalopods, crustaceans, holothurians and sharks. Food and Agriculture Organization of the United Nations, Rome

Reid AL, Jereb P, Roper CFE (2005) Cuttlefishes. In: Jereb P, Roper CFE (eds) Cephalopods of the world an annotated and illustrated catalogue of cephalopod species known to date vol. 1 chambered nautiluses and sepioids (Nautilidae, Sepiidae, Sepiolidae, Sepiadariidae, Idiosepiidae and Spirulidae) FAO Species Catalogue for Fishery Purposes no.4, vol 1

Roper CFE, Sweeney MJ, Naun CE (1984) FAO species catalogue vol. 3 cephalopods of the world an annotated and illustrated catalogue of species of interest to fisheries. FAO fisheries synopsis no. 125 vol. 3. FAO, Rome

Sundaram S, Khan MZ (2009) Stock assessment of *Sepiella inermis* (Orbigny, 1848) from Mumbai waters. J Mar Biol Ass India 51:178–183

Sundaram S, Khan MZ (2011) Biology of the spineless cuttlefish *Sepiella inermis* (Orbigny, 1848) from Mumbai waters. Indian J Fish 58:7–13

Silas EG, Sarvesan R, Nair KP, Sastri YA, Sreenivasan PV, Meiyappan MM, Vidyasagar K, Rao KS, Rao BN (1985) Some aspects of the biology of cuttlefishes. In: Silas EG (ed) Cephalopod bionomics, fisheries and resources of the exclusive economic zone of India. Cent Mar Fish Res Inst Bull 37

Sivalingam D (1999) Seed production of spineless cuttlefish (*Sepiella inermis*) a breakthrough in cephalopod culture. Seafood Expt J 30:3–5

Supongpan M (1995) Cephalopod resources in the Gulf of Thailand. In: Nabhitabhata J (ed) Biology and culture of cephalopods. Rayong Coast Aquacult Station, Rayong

Tang I, Khoo H (1974) The food and feeding habits of the cuttlefish *Sepiella inermis* (Ferussac & d'Orbigny). Veliger 16:405–410

Unnithan KA (1982) Observations on the biology of cuttlefish *Sepiella inermis* at Mandapam. Indian J Fish 29:101–111

Wongwiwathanawut J, Darunchoo L, Nabhitabhata J (2001) Acute toxicity of ammonia on young spineless cuttlefish, Ferussac & d'Orbigny, 1948. Rayong Coast Fish Res Contrib 27:26

Chapter 14
Sepiella japonica

Xiao-Dong Zheng, Xiang-Zhi Lin, Zhao-Sheng Liu, Zhao-Kai Wang and Wei-Bing Zheng

Abstract The spineless cuttlefish *Sepiella japonica* has been cultured in China in recent years. After acclimatization to captivity, spawning induction and mating of the broodstock, the females lay eggs attaching to the artificial substratum. Seawater temperature and salinity of hatching are 20–26 °C and 25.0–32.0 psu, respectively. The hatching period lasts for 17–26 days. Hatchlings are planktonic, with approximately 4 mm mean mantle length (ML). The initial feedings are live cladoceran, copepods and enriched *Artemia* nauplii. The spineless cuttlefishes are sexually mature after 90–120 days, and mating is observed after that. Twenty-thousand mature individuals are obtained, and survival rate from the juvenile to adult phase is about 60 %. The mean weight of adult is 100 g. ML is 50 mm in the smallest sexually mature individual. Spawning occurs as early as day 120. The amount of spawned eggs is 300–500 per individual.

Keywords *Sepiella japonica* · Embryonic development · Ongrowing · Culture · Life cycle

14.1 Importance of *Sepiella japonica* in the Market

Sepiella japonica Sasaki, 1929 belongs to Family Sepiidae. Russian Federation, Possiet Bay in southern Primorye and in Japan, from the Kanto region of Honshu to South Korea, in the East China Sea as well as in the South China Sea (Jereb and Roper 2005, p. 133). *S. japonica* plays an important role in the fisheries of Japan, South Korea and China. It is caught in large numbers in bottom trawls in China, and reported under the erroneous name of *Sepiella maindroni* de Rochebrune, 1884 (Dong 1988; Lu et al. 2012). A great deal of fundamental and implicational research about the species has been carried out since the 1930s, with emphasis on the following subjects: faunal surveys, systematics, embryology and development, ecology, genetics as well as fishery resources (Ohshima and Choe 1961; Choe 1966a, b; Adam and Rees 1966;

X.-D. Zheng (✉) · Z.-S. Liu
Fisheries College, Ocean University of China, Qingdao 266003, China
e-mail: xdzheng@ouc.edu.cn

X.-Z. Lin · Z.-K. Wang · W.-B. Zheng
Third Institute of Oceanography, SOA, Xiamen 361005, China

J. Iglesias et al. (eds.), *Cephalopod Culture*, DOI 10.1007/978-94-017-8648-5_14,

Yamamoto 1982; Wu and Tang 1990; Dong 1993; Zhang et al. 1997, 2001, 2005, 2010; Wu et al. 2010).

The major stocks of *S. japonica* live in warm water. Fishing season comes earlier in the south than in the north: February–March in Guangdong and Fujian provinces in the South China Sea, April–June in Zhejiang Province in the East China Sea and June–July in Shandong Province in the Yellow Sea. Breeding migration starts in spring from the deep waters to shallow waters. The breeding sites are around the islands close to the open seas with clear water, gentle current, high salinity and convergence of Kuroshio Current and coastal waters (Dong 1988).

The time of breeding migration is related to the movement of the isotherm. The appropriate water temperature for spawning population in the northern East China Sea is 18–22 °C. They mate and lay eggs along the island and reefs. The eggs are oval, with flat and forked stalk with which the eggs are attached onto substrates such as bamboo, tree branches, plastic rope, wire, straw, coral and seaweeds. *S. japonica* attaches one egg to the substratum and then winds the egg onto the fixed stalk and continues winding in the same manner to make a string of up to 200 eggs like grapes, hence the nickname 'sea grape' referring to the egg clusters by the fishermen. The eggs require 28–35 days of development before hatching at the temperature of 20–26 °C. The juvenile can live in 10–30 °C, and can adapt to a wide range of salinity within 20–35 psu, while the adults live in salinity of 30–35 psu generally.

With 92 % of its body being edible, *S. japonica* is delicious and nutritious, and rich in protein, vitamins and trace elements (Wang and Wang 2008). Meanwhile, it is a popular traditional dish of Chinese people, and can be canned or dehydrated for market. Most of the catch is marketed as 'shiriyakeika' in Japan.

S. japonica is the dominant species of cuttlefish in China, and its output has been up to 60 % of the total amount of cephalopods caught in the 1980s. The resource and output have been depleted dramatically in Bohai Sea, Yellow Sea and East China Sea since the early 1990s due to overfishing, and the fishing has even disappeared during recent years. Studies on culture of *S. japonica* have become an imperative task to resource restocking and protection.

14.2 Broodstock Acclimatization to Captivity

S. japonica broodstock are captured by traps along the coastal waters of Fujian Province during March and April every year, and then transported alive to the nursery factory directly for maintenance. Meanwhile, egg clusters can be collected from the same area, and transported directly indoors for hatching and cultivation.

Big and vigorous *S. japonica* individuals with well-matured gonads should be selected as broodstock, and be transported to the indoor rearing pools for life-cycle culture, including broodstock maintenance, spawning and hatching, nursing of juveniles, ongrowing and post-spawning natural dying. For hatchling rearing and ongrowing, 1–10 m^2 round or rectangular concrete pools can be used. The entire

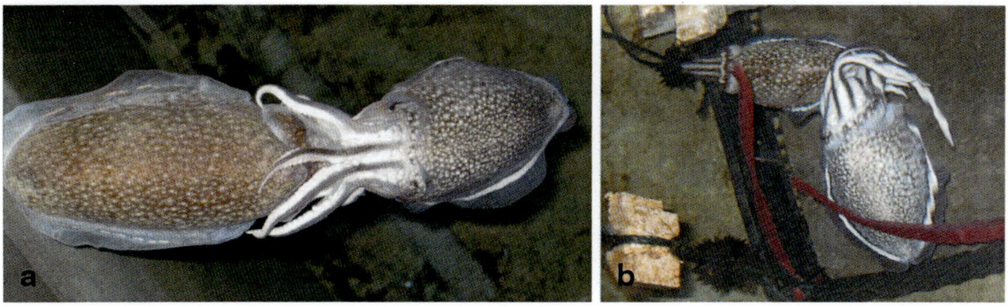

Fig. 14.1 Mating and spawning of *Sepiella japonica*. **a** Mating. **b** Spawning

process can be done under the ambient water temperature with illumination intensity below 1,000 Lx and salinity over 20 psu.

S. japonica is very sensitive to environmental changes, which will result in stress reaction and ink ejecting. Therefore, before introducing them to the new environment, the water temperature and salinity should be adjusted to almost the same as under the transit conditions. Besides, the density should be kept at 2–3 individuals m^{-3}. The top of the pools should be shaded and the illumination intensity on the water surface should be 200–800 Lx. The water should be sand-filtered with salinity of 18–32 psu, dissolved oxygen (DO) >4 mg L^{-1} and pH 7.8–8.4. The daily water renewal should be one to two times of the total quantity. Sex ratio of broodstock should be 1:1.

The feed can be fresh shellfish, fish or shrimps, such as *Sardinella lemuru* and *Sinonovacula constricta*. The feeding rate should be about 10% of the broodstock total weight. Observation must be done every day and record should be maintained to get a clear knowledge of their activities, ingestion and growth.

14.3 Spawning Process

14.3.1 Mating

The broodstock can be forced to mature by increasing the water temperature little by little, i.e. 1 °C in 2–4 days. After some days, they become agitated swimming or even fighting violently, indicating that they are sexually matured. A male only approaches one female, and where there are more males than females, there are usually fierce fights between the males. Only the winners can get access to the female. When mating, the male holds the female tightly by its arms crossed and in a head-to-head fashion (shown in Fig. 14.1a). The entire mating process lasts for 5–15 min. The male uses the hectocotylized arm to deposit the spermatophores to the sperm receptacle in the buccal region of the female.

14.3.2 Spawning

The sperm live for a certain period in the sperm receptacle near the buccal membrane of the female. When a suitable breeding site is found, the eggs mature in batches and are transferred to the sperm receptacle, fertilized and then laid one by one (Zheng et al. 2010). The fertilized eggs are wound onto seaweed, branches or ropes. Most of the parents die after spawning. The main spawning period is from mid-April to late June under the conditions of 5–50 m depth with abundant seaweed and hard bottom; however, there are two spawning periods along the coastal waters of Fujian Province: April–May in spring and September–October in autumn. As for the rearing indoors, the settled substrata such as ropes and pieces of mesh should be treated to have rough surfaces, and put into the cultivating pools. After 3–5 days, beneficial bacterium will develop on the substrata surface, which will help the eggs to attach (Zheng et al. 2010). During the spawning period, the male always swims around to protect the female from other males and mates with the female (Fig. 14.1b). In cases where there are other males, the male usually fights fiercely with them. The mated female can lay eggs without a male. The spawning period usually lasts from several days to 10 days. *S. japonica* prefers substrata with egg clusters, and usually there is more than one female laying eggs together. Putting the substrata with eggs together can induce the females to lay eggs continuously and attract more females to come and lay eggs. Parents do not take care of the eggs after spawning; instead, they die very soon.

14.3.3 Nursing of Eggs

In general, one female can lay 1,000–2,000 eggs. The fertilized eggs should be hatched under stable room temperature, and should be checked regularly. The seawater in the cement pools keeps flowing all day, and the salinity should be above 20 psu.

The number of days required for hatching depends on the water temperature—the higher the temperature, the shorter the hatching period (Table 14.1). The newly laid eggs are soft, black in colour and look like grapes (Fig. 14.2a), with an average diameter of 11.0 mm × 7.6 mm and wet weight of 0.31 g. After 10 days, the egg becomes larger, and the embryo can be seen inside (Fig. 14.2b). Figure 14.2c shows a 20-day embryo under the temperature of 22 °C. There are yellowish pigment granules on the arms and at the ventral edge of the mantle, and an initially developed ink sac. On the 24th day, a well-developed embryo hatches, still with a little external yolk and 4 mm long in mantle. The hatching rate can reach 80 % or over. The hatchlings can swim, and their morphology is similar to the adults. The yolk is consumed in 1–2 days, and the hatchling starts to look for food (Fig. 14.2d). It is capable of moving actively by water ejecting.

Table 14.1 The relationship between temperature and time of hatching in *Sepiella japonica*

Hatching temperature (°C)	16	18	20	22–24	26	28
Hatching (days)	33–38	28–31	24–27	20–22	15–18	12–14

Fig. 14.2 Embryonic development of *Sepiella japonica*. **a** Newly laid eggs. **b** 10-days eggs. **c** 20-days embryo. **d** Hatchling

14.4 Hatchling and Juvenile Nursing

Once the fertilized eggs are hatched, they are then transferred to the nursing pools. The hatchlings with about 4 mm ML usually swim near the water surface. They can be nursed in round pools with rotating current to avoid mortality by collision with the walls (Fig. 14.3). The current and water level can be adjusted by valves outside the pool so as to make the hatchling distribution more uniform to effectively utilize the water volume and increase density. During the planktonic phase, the hatchlings should be closely observed for their growth, ingestion and condition of the feeds, so as to discover and isolate the sick ones at an early time.

14.4.1 Feed

Feed is a key factor for hatchling growth and development. The size, buoyancy, density and nutrition value of feed should be taken into consideration. The hatchling has yolk in its mouth which may disappear in 1 or 2 days. Most of them take feed soon after hatching. *S. japonica* is a typical carnivorous cuttlefish. As soon as the hatchlings are placed into the pools, free-swimming copepods, *Artemia* nauplii or *Miona mongolica*, should be added at a density of 0.1–0.2 individuals L^{-1}, 1–2 times per day. The hatchling survival rate can be higher (83.3 %) when initial feedings are cladoceran, copepods, shrimp or *Artemia* (Li et al. 2007).

Fig. 14.3 Hatchling nursing of *Sepiella japonica*

The juveniles start their demersal life when the ML reaches 8–10 mm. As for above 10 mm ML juveniles, shrimp (e.g. *Neomysis awatschensis*) or crab (e.g. *Portunus trituberculatus*) larvae can be used as feed. The feed density is up to 0.3–0.5 individuals L^{-1}. When the ML of the juveniles reaches 12–15 mm, dried shrimp (*Acetes* sp.) or crushed fish (*Pseudorasbora parva*) at a body length of 10 mm can be used as supplement, and gradually, be used as the only diet, at the rate of 10–20 % of the body weight.

14.4.2 Water Quality, Illumination, Aeration and Nursing Density

The suitable temperature for juveniles is 19–30 °C, with 23–26 °C being the best range (Zheng et al. 2010). Within this range, the warmer the water is, the faster the hatchlings grow. The standards regarding the water quality are: pH 7.6–8.6, DO $\geq$5 mg L^{-1} and ammonia nitrogen $\leq$0.2 mg L^{-1}. These conditions should remain stable. The juveniles have good adaptability to low salinity, so long as it is above 20 psu, with 25–32 psu being the best.

The juvenile cuttlefishes prefer dim light and excessive illumination can reduce their ingestion, which has a negative impact on the growth and survival rate. Bright light may cause them to eject ink. The nursing pools should be covered with black cloth sheeting, and dim light ($\leq$500 Lx) should be maintained during the entire nursing period. Oxygen demand by the juveniles increases along with their growth.

Low aeration rate can be carried out in the earlier period, which should be intensified in the later period.

At the initial period, the culture density is maintained at 1,000–1,500 individuals m^{-3}, which should be reduced to 300–500 individuals m^{-3} once the ML reaches 15–20 mm. The crowding and insufficient feed may cause the juveniles to fight for feed or even eat each other.

Thirty to forty days after hatching, the ML of the juvenile *S. japonica* increases to 20 mm and total wet weight reaches 2–3 g.

14.5 Ongrowing

When the ML of the juveniles reaches 20 mm or when the hatchlings start demersal life, they can be transferred to indoor pools, outdoor ponds as well as cages for further growing.

14.5.1 Indoor Ongrowing

Indoor round pools of 10 m^2 are used. Indoor ongrowing can guarantee the physical and chemical conditions of the seawater and allows fine management. The pools should be covered with black cloth sheeting, and water inside it should be flowing constantly. There should be fresh air in the room and optimal quality water with appropriate DO in the pools. The room temperature should be controlled at 15–20 °C. There should be good drainage, the noise from the pump should be low and the feeding operation should be convenient to conduct. The main parameters for the ongrowing period are as follows:

Feed There must be change of feed at the beginning of this period. Small shrimps (e.g. *N. awatschensis, Acetes* sp.), *Artemia urmiana* and fish (*P. parva*) captured from the wild can be used. Feeding should be done two to three times a day, with the quantity equal to 8–10 % of the main total body weight of juveniles. Gradually, inexpensive fresh fish (*Ammodytes personatus*) and shellfish (*Ruditapes philippinarum*) can be introduced as interim or transition feed, and then decreased gradually through the middle and later stage. Finally, frozen fish (*S. lemuru*) and shellfish (*S. constricta*) muscles will become main diet for adult cuttlefishes.

Water temperature and salinity The temperature is controlled at 10–28 °C, salinity at 20–32 psu, and the water depth should be at least 0.5 m.

Culture density It depends on the size of the individuals. For the juvenile with 12–20 mm ML, the density ranges from 600 to 800 individuals m^{-3}, whilst with ML over 60 mm, less than 100 individuals m^{-3} (Fig. 14.4a).

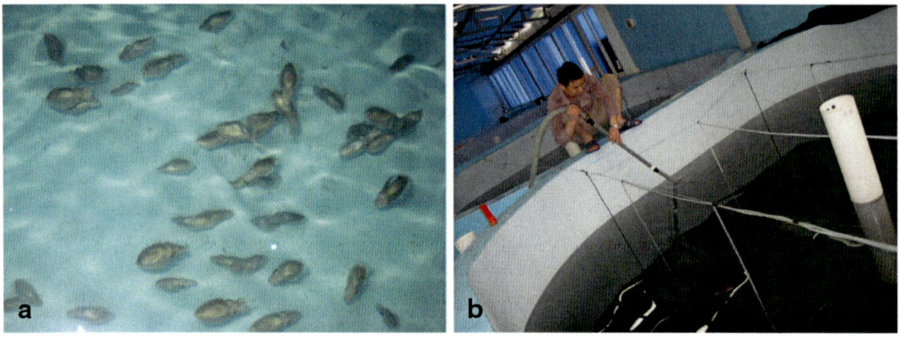

Fig. 14.4 Ongrowing of *Sepiella japonica*. **a** Growing density. **b** Management

Management The daily water renewal should be equal to the total quantity generally, or half of the total quantity in the hot season, or keeping the water constantly flowing. In the cold season, 10–20% of the water should be renewed per day. The water in the cement pools should always be clean and fresh, with sufficient DO and standard pH value (Fig. 14.4b).

The spineless cuttlefishes are sexually mature after 90–120 days of growth, and mating has been observed since then. Twenty-thousand mature individuals are obtained, and survival rate from the juvenile to adult phase is about 60%. The adult mean weight is 100 g. The ML is 50 mm in the smallest sexually mature individual. Spawning occurs as early as day 120. The amount of spawned eggs is 300–500 per individual (Zheng et al. 2010).

14.5.2 Cage Culture

The plate and frame type cages, 4 m × 4 m × 4 m in size, covered with polyethylene knotless nets are used in the southeast China Sea for raising *S. japonica*. The mesh diameter of the net ranges from 10 to 17 mm. The cages have many advantages such as easy management, convenient operation, low price and easy to use widely.

14.5.2.1 Selection of the Farm Location, Cage Size and Layout

Usually, good locations for *S. japonica* rearing are the bays or waters with active water exchange, less pollution, gentle waves, suitable salinity, small change in salinity after heavy rains, over 5 m depth in low tide, perennial water temperature between 7 and 30 °C, DO >4 mg L^{-1}, pH 6.8–8.5 and quiet surrounding environment.

Floating and deepwater cages of various sizes and shapes are suitable. The cages should be arranged in a pyramid pattern along the tide. The mesh specification depends on the size of the juveniles. For individuals with 15–20 mm ML, 4 mm diameter mesh is suitable, and when the individuals are growing larger, the mesh size should be adapted accordingly. The cages should be shaded by black mesh on the top to avoid stress by bright light, which may cause the individuals unrest, or even eject ink and die. The height of the net should be convenient for feeding operation. In order to avoid attacks by the birds, net cover should be provided.

14.5.2.2 Juvenile Size and Density

Active, healthy juveniles of 15–20 mm ML can be introduced to the ongrowing process in 13–32 °C, optimally 27 ± 2°C, and salinity of 20–35 psu. Stocking density depends on the size of the individuals. A 3 m × 3 m × 3 m cage can accommodate 2,000–2,500 juveniles of 0.7–1 cm total length (TL), or 1,200–1,500 juveniles of 3–5 cm TL, or 800–1,000 juveniles of 6–8 cm TL or 300–500 (>9 cm TL).

14.5.2.3 Feeding

Wild fresh small shrimp (*N. awatschensis*) and fish (*P. parva*) are the major feed for the juveniles during the ongrowing period. Daily feed quantity should be 10–15 % of total body weight of juveniles. Besides, lamp light can be used to attract wild preys, or if conditions permit, use lamps on top of the cages to attract zooplankton (such as copepods, cladoceran) as supplement

14.5.2.4 Daily Management

During the ongrowing period, the fouling on the cage should be cleared and mesh of suitable size should be changed periodically. Water flow in the cage should be maintained. Water exchange must be guaranteed to ensure that there is always fresh water in the cage, but the mesh size should be appropriate to prevent individuals from escaping.

During the period 2006–2007, a total of 50,000–60,000 *S. japonica* adult individuals were reared in cages in the seas off the coast of Ningde and Putian counties, Fujian Province, at a total survival rate of 60 %. Meanwhile, the outdoor ponds were also used to *S. japonica* ongrowing in East China Sea (Wang et al. 2006); the survival rate of the species was 26.7 % in 2005, and mean wet weight was 250 g; the survival rate was up to 47.4 % in 2006, and the maximum weight reached was 400 g.

14.6 Harvest

After 3–4 months of ongrowing, *S. japonica* reaches commercial size and can be harvested for marketing. The spineless cuttlefishes are active when water temperature is over 15 °C. Cuttlefish fishing traps which are cylinders of 20–30 cm height by 30 cm diameter, covered with the polyethylene net of 10–20 mm mesh diameter with reversed whisker, can be used to capture them. Firstly, baits (fresh fish or shrimp) are deposited into them, and then the traps are hung into the water. The traps will be lifted when there are a sufficient number of adult cuttlefishes. It is not advisable to take them out of the traps by hands; instead, they should be taken out with big spoons with water inside. However, when water temperature is below 15 °C, *S. japonica* is not active and the trap capture is not feasible; instead, purse seine or draining-out-the-pond method can be adopted.

14.7 Trends in Research and Industrial Level

Optimization of the feed supply is a key to commercial culture which still requires further research. Furthermore, the production of artificial feed is still a scientific blank at the moment.

The culture of *S. japonica* has not caused loss of meat quality so far, and cultured production can fully substitute wild products to be an ideal healthy food (Chang et al. 2008). But the mean size of the stocks reared indoors is 20 % smaller than the wild ones. Whether this is caused by genetic degeneration, the artificial environment or culture technique is still unknown. *S. japonica* spends at least 6–8 months to reach sexual maturity in the wild; however, under the artificial environment, it requires only 3–4 months. The authors suggest that early sexual maturity and shorter growth period should be the main reasons why the cultured individuals are significantly smaller than their wild counterparts. Although the maturation phase of the species has been lengthened from 3–4 months to 6 months by adjusting the light intensity indoors, related problems still have to be solved. It is worth noting that the wild resource restocking of the species has been achieved by releasing of artificial hatchlings in recent years (Wu et al. 2010).

14.8 Conclusions

The process of spineless cuttlefish culture includes broodstock capture, acclimatization to captivity, spawning, hatching, nursing, ongrowing and harvest. The main conclusions are as follows:

1. In general, each wild female of *S. japonica* lays 1,000–2,000 eggs. Water temperature and salinity of hatching are 20–26 °C and 25.0–32.0 psu, respectively. The hatching period is 17–26 days. Hatchlings are planktonic, and mean ML is approximately 4 mm. The highest hatchling survival rate (83.3 %) is obtained when initial feeding is based on cladoceran, copepods, shrimp or *Artemia*.
2. When the ML of juveniles reaches 20 mm or when the hatchlings start benthic life, they can be transferred to indoor pools, outdoor ponds as well as cages for growing. The survival rate from the juvenile to adult phase is about 60 %. The

mean weight of adults is 100 g. After 3–4 months of ongrowing, *S. japonica* reaches commercial size and can be harvested for marketing.

3. The spineless cuttlefishes show early sexual maturity after 90–120 days growth indoors. Spawning occurs as early as day 120. And the number of spawned eggs from cultured females ranges from 300 to 500 per individual. The maturation phase of the species is delayed by adjusting the light intensity indoors.
4. *S. japonica* is the dominant and commercial cuttlefish in China, especially in the southeast China Sea. The broodstock do not take care of their eggs. Instead, they leave them to develop alone on the yolk's nutrition. Artificial breeding can greatly enhance hatching rate. So its culture has been developing prosperously in the current decade, and will be available to become an emerging mariculture industry in the coastal waters of China.

Acknowledgements The authors thank Dr. Chung-Cheng Lu and Mrs. Mairi Lu from National Chung Hsing University for language checking and Hai-Bin Zhang from Monterey Bay Aquarium Research Institute for comments. The study was supported by the grants from the National High Technology Research and Development Program of China (2010AA10A404) and National Marine Public Welfare Research Project (200805069, 201305043).

References

Adam W, Rees WJ (1966) A review of the cephalopod family Sepiidae. Sci Rep John Murray Exped 11:1–165

Chang KM, Wu CW, Lu ZM, Zhu AY, Zhang JS (2008) Comparison in biochemistry of tissues of wild and cultured *Sepiella maindroni*. Oceanol Limnol Sin 39(2):145–151. (In Chinese with english abstract)

Choe S (1966a) On the egg, rearing, habits of the fry, and growth of some cephalopoda. Bull Mar Sci 16:330–348

Choe S (1966b) On the growth, feeding rates and the efficiency of food conversion for cuttlefishes and squids. Korea J Zool 9:12–20

Dong ZZ (1988) Fauna sinica. Phylum Mollusca. Class Cephalopoda. Science Press, Beijing, pp. 181–182. (In Chinese)

Dong ZZ (1993) Study of cephalopods. In: Zeng CK, Zhou HO, Li BC (eds) Marine science study and its prospect in China. Qingdao Publishing House, Qingdao, pp 23–26

Jereb P, Roper CFE (2005) Cephalopods of the world. An Annotated and Illustrated catalogue of Cephalopod species known to date. Vol. 1. Chambered nautiluses and sepioids (Nautilidae, Sepiidae, Sepiolidae, Sepiadariidae, Idiosepiidae and Spirulidac). FAO Spcc Cat Fish Purp 4(1):132. FAO, Rome

Li Z, Jiang XM, Wang CL (2007) Effects of diets on growth, survival and nutrient composition of cuttlefish *Sepiella maindroni*. J Dalian Ocean Univ 22(6):436–441. (In Chinese with english abstract)

Lu CC, Zheng XD, Lin XZ (2012) Diversity of Cephalopoda from the waters of the Chinese mainland and Taiwan. In: Lin M, Wang CG (eds) Proceedings of the 1st Mainland and Taiwan Symposium of Marine Biodiversity Studies. China Ocean Press, Beijing, pp 76–87

Ohshima Y, Choe S (1961) On the rearing of young cuttlefish and squid. Bull Jap Soc Sci Fish 27(11):979–986. (In Japanese with english abstract)

Wang RC, Wang ZP (2008) Science of marine shellfish culture. Ocean University Publishing House, Qingdao, pp 484–498

Wang CL, Jiang XM, Qiu YG (2006) The pond culture of *Sepiella japonica*. Chin Aqua 8:50–51. (In Chinese)

Wu YQ, Tang ZC (1990) Population composition and migratory distribution of cuttlefish (*Sepiella maindroni* de Rochebrune) in the Huanghe estuary and the Laizhou Gulf. J Fish Chin 14(2):149–152. (In Chinese with english abstract)

Wu CW, Dong ZY, Chi CF, Ding F (2010) Reproductive and spawning habits of *Sepiella maindroni* off Zhejiang, China. Oceanol Limnol Sin 41(1):39–46

Yamamoto M (1982) Normal stages in the development of the cuttlefish, *Sepiella japonica* Sasaki. Zool Mag 91:146–157

Zhang BL, Sun DW, Bi HS, Wu YQ, Huang B (1997) Growth and seasonal distribution of *Sepiella maindroni* in the Jiaozhou Bay and adjacent waters. Mar Sci 5:61–64. (In Chinese with english abstract)

Zheng XD, Wang RC, Wang XF, Xiao S, Chen B (2001) Genetic variation in populations of the common Chinese cuttlefish *Sepiella maindroni* (Mollusca: Cephalopoda) using allozymes and mitochondrial DNA sequence analysis. J Shellfish Res 20(3):1159–1165

Zheng XD, Li Q, Wang ZP, Yu R, Tian C (2005) Genetic diversity in populations of *Sepiella japonica* based on mtDNA sequence analysis. Phuket Mar Bio Cent Res Bull 66:195–201

Zheng XD, Lin XZ, Wang ZK, Yu RH, Tian CY, Li Q (2010) Studies on life cycle of cultured spineless cuttlefish *Sepiella japonica*. Trans Oceanol Limnol 3:24–28. (In Chinese with english abstract)

Chapter 15
Euprymna hyllebergi and *Euprymna tasmanica*

Jaruwat Nabhitabhata and Michelle K. Nishiguchi

Abstract Bobtail squids of the genus *Euprymna* are small in size with a benthic habit. Such small size results in their insignificance in fisheries and aquaculture focused for human consumption. The unique ability of the voluntary adhesion system and symbiotic bacteria used for bioluminescence is now a primary research focus with potential industrial and biomedical applications. Their small size is well suited for the home aquarium with small volume. Culture of this cephalopod group can therefore serve both research and recreational purposes. Aquaculture in the laboratory provides valuable information for culture methodology that is utilized throughout the entire life cycle of several consecutive generations. This small size and benthic habit of *Euprymna* are advantageous for small-scale closed or open seawater culture systems. Major trends for culturing *Euprymna* are similar to other cephalopod groups, particularly benthic octopus that also produce planktonic hatchlings. Reduction of the cost of production is necessary for future large-scale production, with novel protocols for live feed requirements of planktonic young in the nursing phase.

Keywords *Euprymna* · Small size · Benthic habit · Small-scale culture · Closed and open seawater systems · Research and recreational purposes

15.1 Importance of the Species

The Thai bobtail squid, *Euprymna hyllebergi* Nateewathana 1997, is a common species occurring in the Andaman Sea of Thailand (Indian Ocean) and the Gulf of Thailand (Pacific Ocean; Nateewathana 1997; Nateewathana et al. 2001; Aungtonya et al. 2011). This species is small (20–40-mm mantle length, ML),

J. Nabhitabhata (✉)
Excellence Centre for Biodiversity of Peninsular Thailand, Faculty of Science,
Prince of Songkla University, Hatyai, Songkhla 90112, Thailand
e-mail: jaruwat.n5@gmail.com; e-mail: jaruwat.n@psu.ac.th

M. K. Nishiguchi
Department of Biology – MSC 3AF, New Mexico State University,
PO Box 30001, Las Cruces, NM 88003-8001, USA
e-mail: nish@nmsu.edu

J. Iglesias et al. (eds.), *Cephalopod Culture*, DOI 10.1007/978-94-017-8648-5_15,

Fig. 15.1 Partial sand-coated *Euprymna hyllebergi* shortly after emerging from its burrowing site. (Photograph of J Nabhitabhata)

neritic and strictly nektobenthic, inhabiting coastal waters in a similar manner to its congeners that occur in the Indo-west Pacific region (Summers 1985; Norman and Lu 1997; Reid and Jereb 2005). *E. tasmanica* (Pfeffer 1884), otherwise known as the "southern dumpling squid", is a species that resides around the continent of Australia and is found in similar habitats as *E. hyllebergi* (Reid and Jereb 2005). *E. tasmanica* is somewhat larger in size than *E. hyllebergi,* with an approximate ML of 30–40 mm (Norman and Lu 1997). In Thailand, *E. hyllebergi* and congeners are captured as by-catch of commercial fishing, particularly push netting and trawling (Nateewathana 1997). The yields are discarded as trash fishes due to their small size and low economic value. Because of this, fishing statistics of both species are not available (Nateewathana et al. 2001; Reid and Norman 1998; Reid and Jereb 2005).

Although the economic value of *Euprymna* as human food is low, there is a growing importance of the animals as scientific experimental models. The unique behaviour of *Euprymna* is its capability to retain a "coat" of sand or other debris on its dorsum (Fig. 15.1) when it emerges from its daily buried state to hunt prey at night (Anderson et al. 2002). The function of the sand coat is presumably for camouflage, making the squid difficult to be detected visually by its predators and prey (Anderson and Mather 1996; Shears 1988). The stickiness of the sand coat depends upon secretions of the ectodermal epithelium (Moynihan 1982). Choice between being sticky and nonsticky is voluntary and variable (Moynihan 1982). This indicates that the ability to use a sand coat might have evolved from the initial use of the behaviour for sand consolidation when the squid buries (Shears 1988). The ability to begin sand coating starts 5–7 days after hatching, simultaneous to their burrowing ability (Nabhitabhata et al. 2005). Additionally, bobtail squids are presently used as a model organism to identify a new generation of biomimetic adhesives and marine antifouling compounds with potential industrial value (Byern and Grunwald 2010).

Symbiotic associations between *Euprymna* and the bioluminescent bacterium *Vibrio fischeri* has been a recent focus as a model for investigating the process of bacterial colonization of host tissues and its effect on host development (Ruby 1999, Ruby and Lee 1998). *V. fischeri* and other luminous bacteria form a variety of pathogenic and cooperative associations with marine animals; more recently, they are being increasingly recognized as causes of invertebrate diseases (Guerrero-Ferreira and Nishiguchi 2011; Guerrero-Ferreira et al. 2012). Since the process of bacterial colonization of the squid light organ begins immediately after hatching (Ruby and McFall-Ngai 1992), independent aquaculture of the squids and their luminous bacterial partners could yield valuable results for biotechnological and biomedical sciences (McFall-Ngai et al. 2012; Nyholm and Nishiguchi 2008). Because of these newly developed models for basic research, the advancement of culturing techniques for species of *Euprymna* has been especially important for monitoring fitness between generations, effects of inbreeding and, more importantly, diet and stress under laboratory conditions (Nabhitabhata et al. 2005; Sinn 2005; Sinn et al. 2008; Moltschaniwskyj and Carter 2010). Additionally, more information on their development, growth and time to reproduction can indicate whether all species have similar life-history strategies, and if this is dependent upon habitat or other abiotic factors.

15.2 State of the Art

Both *E. hyllebergi* and *E. tasmanica* are small in size and found living within benthic habitats. This is an advantage to provide culture conditions on a smaller scale with lower cost and less requirement of facilities compared to those used for pelagic and large-sized species. *Euprymna* can be cultured through several consecutive generations ensuring a supply of broodstocks. Broodstocks collected from the wild can spawn in captivity and are maintained throughout the life history of the animal. *E. hyllebergi* hatchlings are fed with wild-collected live feed for approximately 30 days, but later can be trained to accept dead feed. *E. tasmanica* hatchlings are fed small mysid shrimp two times a day for approximately 6 weeks, and then moved to a diet of ghost shrimp for the duration of their lives while in captivity. Interestingly, *E. tasmanica* adults were not trained to feed on dead material, and prefer not to eat food items that do not move. These same facilities for raising juvenile squids can be used for culture throughout the squid's entire life cycle. The daily growth rate for *E. hyllebergi* is 2.4 % in length and 7.5 % in weight through the culture period of 100 days. Growth rates for *E. tasmanica* were approximately 3.5 % in length and 10 % in weight for approximately 60 days. Growing demands for these squid for use as both biotechnological and biomimetic experimental models as well as ornamental animals for home aquaria and teaching laboratories are beneficial for aquaculture and biomedicine, e.g. Nabhitabhata et al. (2005), Moltschaniwskyj et al. (2007), Sinn and Moltschaniwskyj (2005) and Sinn et al. (2008).

15.3 Broodstocks Maintenance

Broodstocks of the Thai bobtail squid, *E. hyllebergi,* are collected live from otter board trawlers and beam trawlers, operating along the eastern part of the Gulf of Thailand, South China Sea and Pacific Ocean. Onboard, the squids are maintained in cylindrical fibreglass tanks of 50-L capacity containing 30 L of fresh seawater with aeration and then, upon landing, are transported to the cephalopod hatchery. The broodstocks are maintained in an open system of cylindrical concrete tanks of 2 m^3 with flow-through filtered seawater (for the seawater supply system in this chapter, see also Chap. 7 "Aquaculture to Restocking"). Artificial substrates, made from pieces of longitudinal-cut polyvinyl chloride (PVC) pipe (50-mm diameter, 150-mm length), are previously placed on the tank bottom as shelters or "dens".

Broodstocks of southern dumpling squids, *E. tasmanica,* are collected by seine net in shallow waters of Botany Bay, New South Wales, Australia. Adult animals are transported to running open seawater facilities located at the Sydney Institute of Marine Sciences (SIMS) at Chowder Bay, NSW. Adults are acclimated to the conditions in the tanks (34 psu, 18 °C) and transported to New Mexico State University within 2–3 days. Transport of the squids takes approximately 36 h tank to tank in aquaria bags with less than 1 L of seawater. Animals are then acclimated to the contained recirculating artificial seawater tanks (100 L) at New Mexico State University under the same culturing conditions. Each tank is divided into eight cubic sectionals (each 0.3 m^2), which holds three to four adult individuals. Sexes are continuously kept separate, since the presence of males can stress female behaviour. The only time males are placed with females is when a planned mating is scheduled. In this manner, we can document which particular male has mated with which female (and therefore, track fecundity of each female). Males are placed in the female cubical (usually at a 1:1 or 1:2 male to female ratio) and are removed anytime between 4 and 10 h. PVC pipe cut longitudinally is placed in the female cubical and used as artificial substrates for the females to lay their eggs after they have been mated.

Squids will mate and spawn in the tanks. Mating occurs without prior pair formation for *E. hyllebergi,* and controlled conditions (noting which pairs are mated, and how many times) are maintained for *E. tasmanica*. Spawning behavioural pattern is similar for both species. The male responds to the presence of a swimming female by initially approaching and then grasping her from below in a male to female neck position. The female is pulled down to the bottom where copulation takes place. Copulation takes 7–10 min and then the pair separate. Spawning is observed at dawn, 2–3 days after mating. Prior to spawning, the female investigates substrates for attaching her eggs by swimming around, and touching the substrata with the tip of her arm cone. In the tanks, the female attaches her eggs in clusters to the inner surface of the artificial substrates (Fig. 15.2). The time period for attaching is 40–60 s for one egg. Intervals between each egg attachment lengthens as the number of eggs increases, up to 2–3 min. Spawning is intermittent and irregular and may be extended over several weeks. The total number of eggs per female is about 100–470 with an average of 200 eggs. Females can spawn up to three to four clutches in her lifetime, with the number of eggs decreasing as the female becomes older (Steer et al. 2004; Nabhitabhata et al. 2005). For *E. tasmanica,* adults reared in captivity (F1 generation) live longer (2–3 months) and produce larger and more clutches per female. Wild-caught adult *E. tasmanica* at maturity produce approximately 3 clutches while in captivity, ranging from 25 to 100 eggs per clutch (with one exceptional female, which laid approximately 500 eggs in one clutch). F1

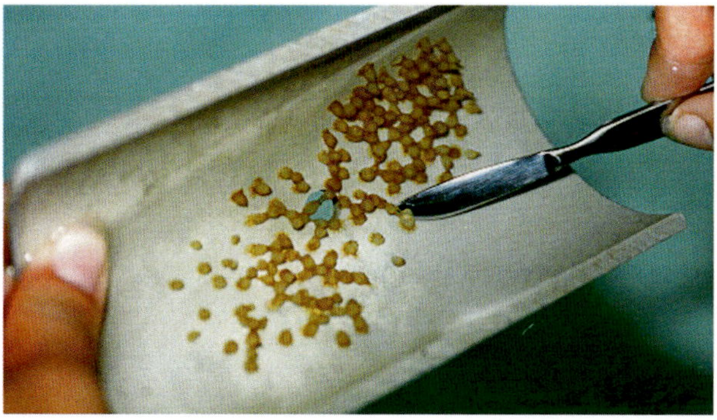

Fig. 15.2 Egg capsules of *Euprymna hyllebergi* attached to the inner side of the artificial substratum, a piece of cut polyvinyl chloride (*PVC*) pipe. (Photograph of J Nabhitabhata)

generation *E. tasmanica* females lay up to 5 clutches/lifetime, with sizes ranging from 100 to 250 eggs per clutch. Hatching rate from the F2 clutches is approximately 99 % for the first clutch, with a decrease leading up to approximately 20 % (~80 % hatching rate) for later clutches. F1 adults are larger and thus far have lived for 1 year in captivity (Nishiguchi, unpublished). F2 adults have similar longevities, but hatching rates for the F3 generation was somewhat lower (70–80 %)

15.4 Nursing of Eggs

15.4.1 Egg Characteristics

Eggs are single, stalkless and opaque white, having a droplet shape and calcareous leather-like coating capsule (Fig. 15.3). The size of each egg is about 4 mm along its major axis, 3 mm in its minor axis and weighs about 0.02 g. About 2 h after being laid, the outer coat (or capsule) turns brown, leathery and rigid in *E. hyllebergi* (Nabhitabhata et al. 2005), whereas in *E. tasmanica* the egg is orange from wild-caught specimens. In F1 and subsequent generations of *E. tasmanica,* egg capsules are white to opaque and remain so during development. This solid protection allows the developing embryo to become a "sessile organism" during the extended period of development (Boletzky 1998). Eggs are telolecithal. Asymmetric eight-cell cleavage occurs 10 h after fertilization. Clockwise rotation of the embryo occurs from days 3 to 8, at 28 °C. Organogenesis occurs from day 4. The unique bilobed character of the external yolk sac appears after day 5 when the capsule becomes more transparent and the embryo is now visible. Chromatophores develop from day 6, and four diverticula of the internal yolk sac from day 8. The first hatching occurs at day 12 (Fig. 15.4) for *E. hyllebergi,* and day 32 for *E. tasmanica.* The embryonic phase is about 12–18 days, after approximately 14 days at 28 °C for *E. hyllebergi,* and 21–28 days at 18 °C for *E. tasmanica.* The hatching period of eggs in the same clutch takes 5 days from first to the last eggs and primarily occurs on the third day. Average hatching rate is about 94 % (82–100 %) for both species.

Fig. 15.3 Egg capsule of *Euprymna hyllebergi*; surface is colored brown by attached diatoms (40x). (Photograph of J Nabhitabhata)

Fig. 15.4 Hatchling of *Euprymna hyllebergi* (dorsum, 17x). (Photograph of J Nabhitabhata)

15.4.2 System Requirements and Management

Artificial substrates with egg clusters are transferred to hatch in fibreglass tanks of 50-L capacity containing 40-L filtered seawater. Two pieces of longitudinal-cut PVC pipe (25-mm diameter, 400-mm length) equipped with aeration devices are placed in each tank, facing in the same direction, to generate and direct an artificial current (Fig. 15.5). Tanks are cleaned by siphoning out the old water and replaced by a volume of 50 %. Temperature change is minimised by means of outside running water around the tank base (Nabhitabhata et al. 2005). The average temperature

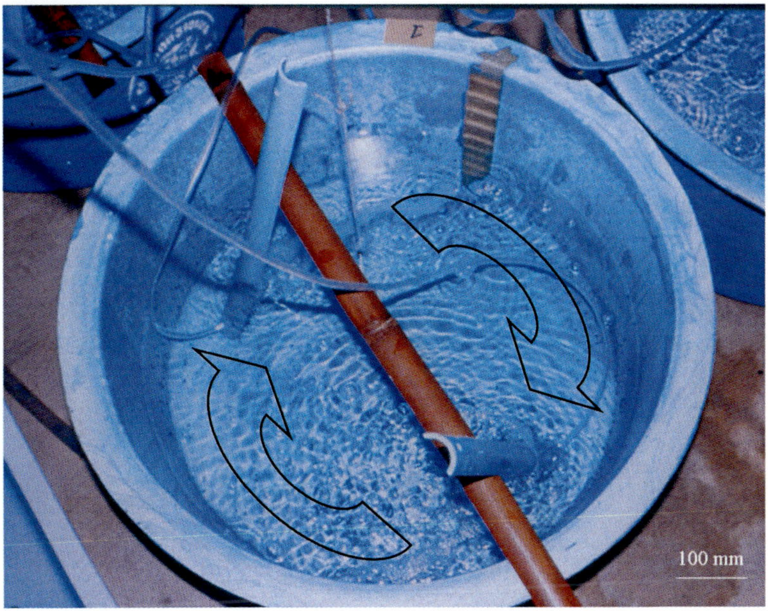

Fig. 15.5 A culture tank of *Euprymna hyllebergi* equipped with current generator devices (*arrows* indicate current direction). (Photograph of J Nabhitabhata)

can be maintained at approximately 28.2 °C, pH 8.0 and salinity at 32.5 psu. For *E. tasmanica,* clutches are placed in a 100-L polycarbonate tank with ultraviolet (UV)-filtered artificial seawater. Each individual clutch is placed in a cage made out of 2 pieces of PVC tubing, cut longitudinally and then glued back together with mesh netting in between. The clutches are placed in these cages so as not to have hatchlings mix with other clutches, as well as receiving enough aeration from below during development. Each cage is aerated with water from an individual spout that provides oxygenated seawater (Fig. 15.6). Water is kept at constant temperature (20 °C) with pH 8.0 and salinity 34.0 psu. Water changes are completed every other day to maintain salinity due to evaporation.

15.5 Nursing of Young

15.5.1 Hatchling Characteristics

The living mode of the hatchling includes a planktonic phase lasting from 6 to 8 h before the hatchling gradually adopts a benthic habit. The settling stage is approximately 5 days. Juvenile squids still enter the water column on a regular basis, alternatively planktonic and benthic, until 25–30 days after hatching. The internal yolk sac is still visible through the transparent mantle from hatching until the third day (Fig. 15.4).

Fig. 15.6 Culture facilities for clutches/hatchlings for *Euprymna tasmanica*. (Photograph of MK Nishiguchi)

15.5.2 System Requirements and Management

Nursing of young is performed using the same system as for nursing of eggs for both species.

15.5.3 Feeding

The general task is to feed planktonic food to juvenile squids before the settling stage, at which time the squids are column feeders. Subsequently, benthic food is provided after the settling stage, when the juvenile squids settle to the bottom (Hanlon 1990; Hanlon et al. 1997; Nabhitabhata et al. 2005). *E. hyllebergi* hatchlings are fed with live, hatchery-produced penaeid shrimp larvae (*Penaeus merguiensis, P. monodon*) of the protozoea and mysis stages for 5 days after hatching (Fig. 15.7). Postlarvae of penaeid shrimps of the same species as well as wild mysids (*Mesopodopsis orientalis*) are also fed to the squids from hatching to 40 days. The planktonic young seize and eat its prey in the water column while hovering. After 25 days, the juvenile gradually changes to a benthic feeder, seizing its prey in the water column and then consuming it on the bottom substrate.

After 30 days, supplementary prey organisms for *E. hyllebergi* are palaemonid shrimps (*Palaemon styliferus*) and wild mysids (*Acetes* spp.). Training the squids to feed on dead fish meat (*Caranx leptolepis*) begins during this

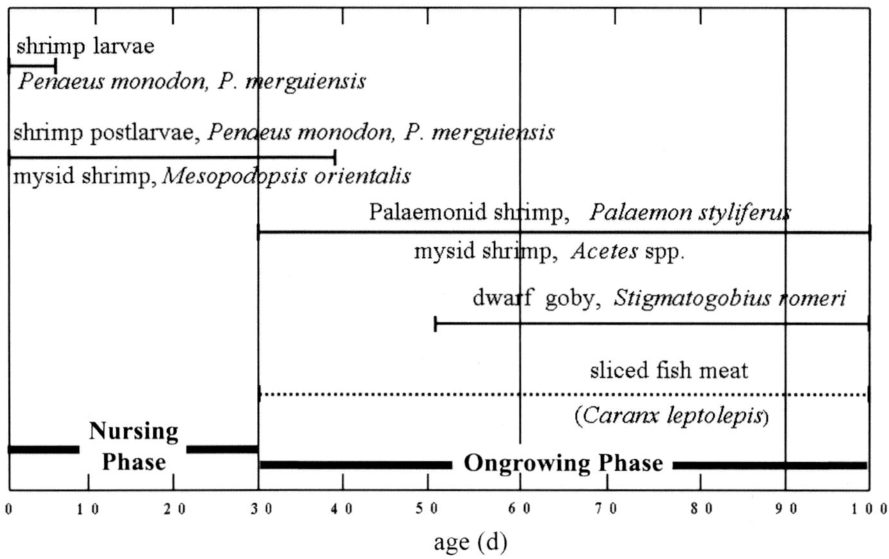

Fig. 15.7 Diagram of feeding of cultured *Euprymna hyllebergi* in the nursing phase (0–30 d) and ongrowing phase (after 30 d settlement) with live feed (*full line*) and dead feed (*dotted line*). (After Nabhitabhata et al. 2005)

period. Size grading is also initiated and continued every 10 days. The density is reduced from the initial 2–6 individuals L^{-1} by at least 25% after each grading (Nabhitabhata et al. 2005). *E. tasmanica* juveniles are fed on live, mysid shrimp for the first month (~30 days), i.e. *Tasmanomysis oculata, Paramesopodopsis refa,* and then are moved to smaller, post-larval panaeid shrimp (*Penaeus* sp.). Since *E. tasmanica* are larger when hatched, they are capable of obtaining bigger prey items earlier in their development than *E. hyllebergi*. Enriched brine shrimps (*Artemia parthenogenetica* and *A. franciscana*) can be used as substituted food when mysids are unavailable (Sinn 2005; Sinn and Moltschaniwskyj 2005; Sinn et al. 2008) although they are generally less preferred. *E. tasmanica* does not take dead prey, although there is no attempt to train juveniles to feed on this type of material.

15.5.4 Euprymna hyllebergi Growth

Hatchlings of Thai bobtail squid grow from about 2-mm ML and 0.004-g weight to 7-mm ML and 0.26 g in the first month (Fig. 15.8a, b; Nabhitabhata et al. 2005). The daily or instantaneous relative growth rate (IGR) is the highest between 10 and 20 days after hatching, about 5% in length and 17% in weight (Fig. 15.8b). The survival in the nursing phase from hatching to settling stage (0–30 days) is approximately 30%.

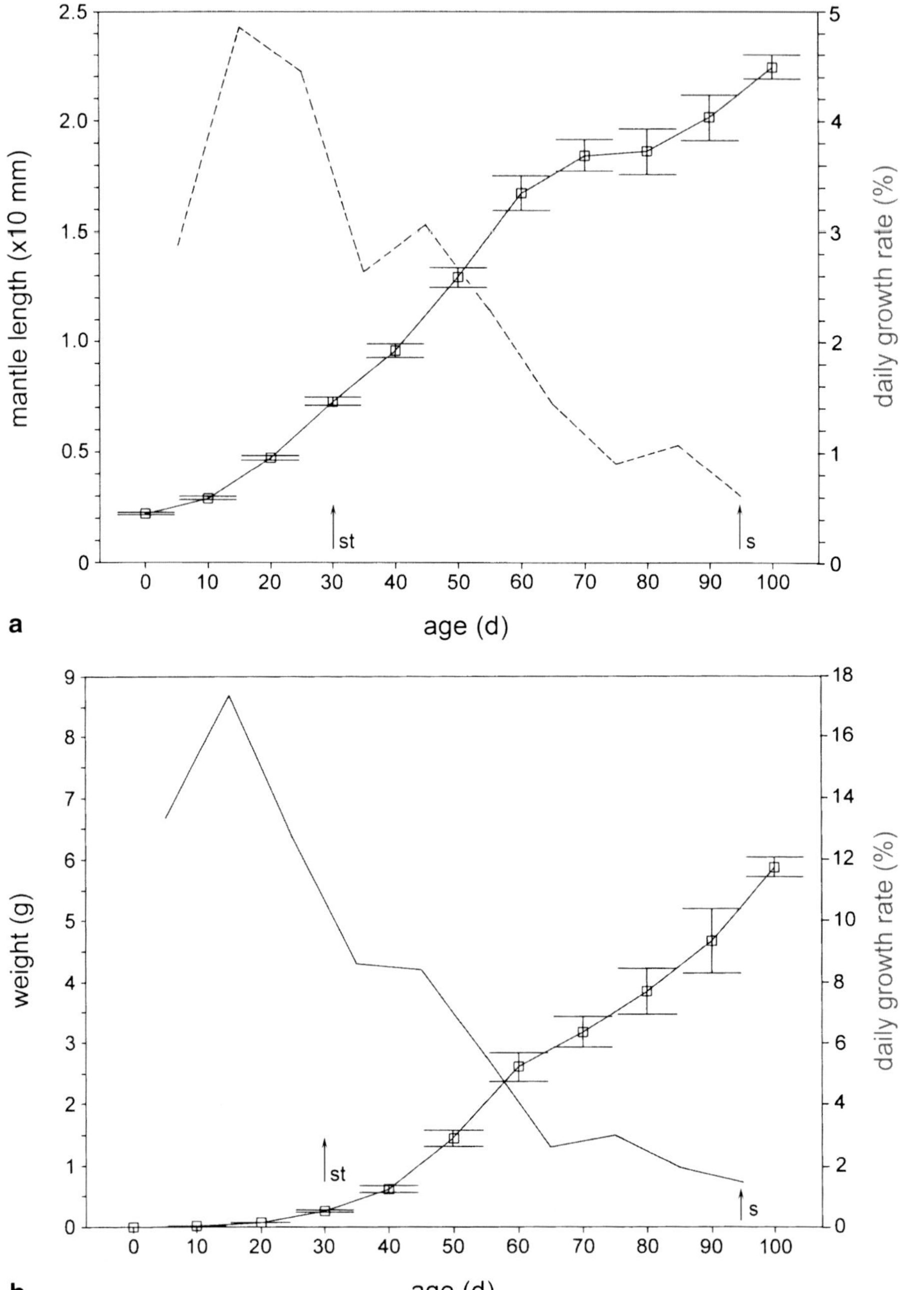

Fig. 15.8 Growth of *Euprymna hyllebergi* in terms of (*above*) mantle length (x10 mm), instantaneous relative growth rate (*IGR*: %) and age (*d*) after hatching and (*below*) weight (*g*), *IGR* (%) and age (*d*) after hatching. *Arrows* indicate spawning (*s*) and settling stage (*st*). (After Nabhitabhata et al. 2005)

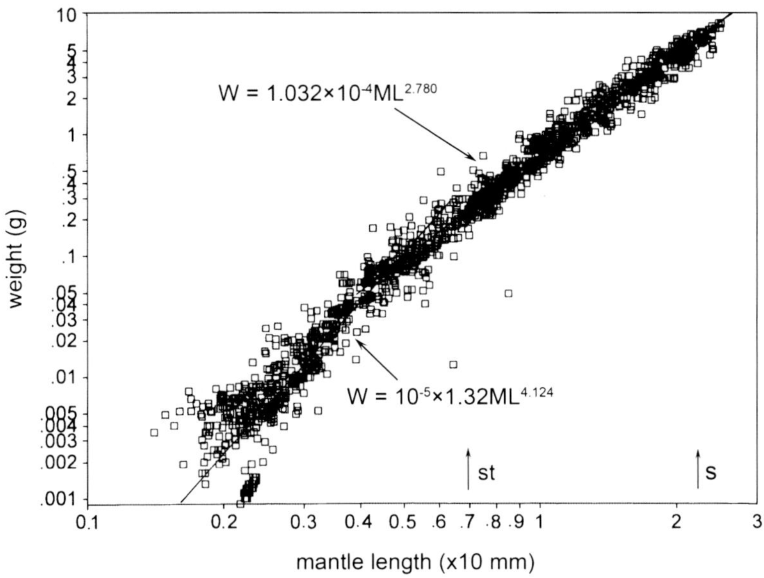

Fig. 15.9 Relationships between mantle length (x10 mm) and weight (*g*) of *Euprymna hyllebergi*; intercept of the two regressions at 5.5 mm mantle length. *Arrows* indicate spawning (*s*) and settling stage (*st*). (After Nabhitabhata et al. 2005)

Growth models demonstrate two phases of growth. The early phase was from hatching to 30 days where the relationships between the ML (mm) and weight (g) can be expressed as a power regression model (Fig. 15.9; Nabhitabhata et al. 2005):

$$W = 1.230 \times 10^{-4}\ ML^{4.124}. \tag{15.1}$$

The relationships between ML and age (d: days after hatching, Fig. 15.10; Nabhitabhata et al. 2005) and between weight (g) and age (d, Fig. 15.11; Nabhitabhata et al. 2005) can be expressed as the exponential models:

$$ML = 1.988e^{4.205\times10*(-2)A} \tag{15.2}$$

$$W = 2.750 \times 10^{-3} e^{0.153A}. \tag{15.3}$$

15.6 Ongrowing

15.6.1 System Requirements and Management

For *E. hyllebergi,* ongrowing phase starts after the benthic young are able to accept dead fish meat. Tanks for ongrowing are the same tank used for the nursing phase and with similar management for both species. The density of *E. hyllebergi*

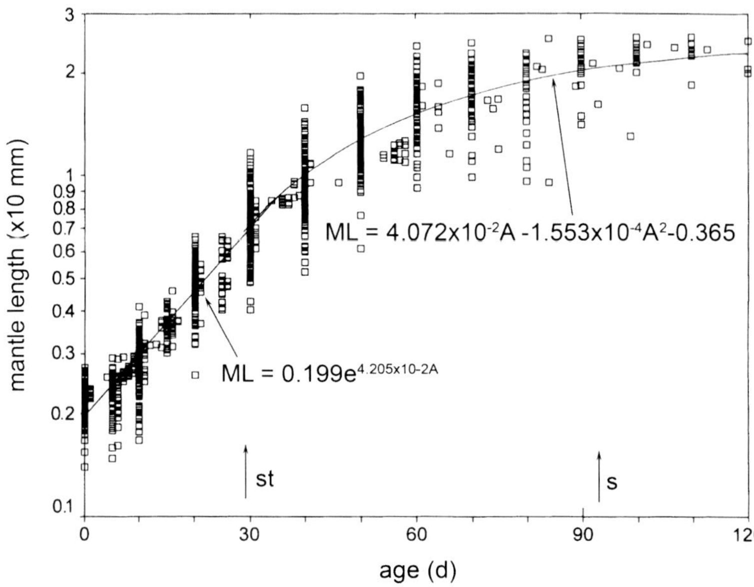

Fig. 15.10 Relationships between mantle length (x10 mm) and age (*d*) of *Euprymna hyllebergi*. *Arrows* indicate spawning (*s*) and settling stage (*st*). (After Nabhitabhata et al. 2005)

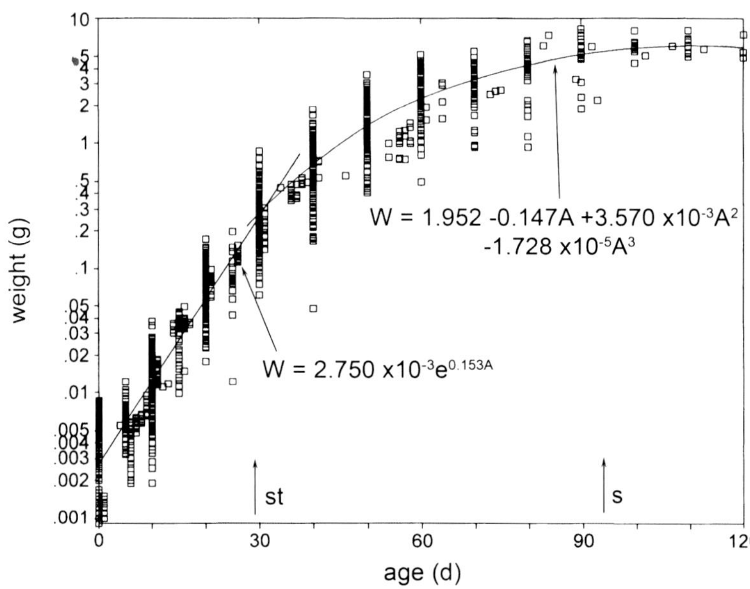

Fig. 15.11 Relationships between weight (*g*) and age (*d*) of *Euprymna hyllebergi*. *Arrows* indicate spawning (*s*) and settling stage (*st*). (After Nabhitabhata et al. 2005)

is changed from a water volume oriented to an (bottom) area oriented as 4–5 individuals m^{-2}. *E. tasmanica* juveniles are initially raised in round glass bowls (approximately 2 L) with sand at the bottom so the juveniles can settle. Unlike *E. hyllebergi* juveniles, *E. tasmanica* immediately settle on the bottom once they are hatched. Water is changed daily since the volume is small and there is greater evaporative loss from this volume. At approximately 2–3 weeks, juvenile squids are transferred to 40-L aquaria with sand on the bottom and raised until sexually mature (2 months). Generally, 20 squids are kept in an aquarium this size due to space limitations, but this number does not seem to affect their behaviour with any visible signs of stress. Since *E. tasmanica* F1 and F2 generations have higher growth rates than those caught in the wild, these individuals are moved earlier to the adult cubicals.

15.6.2 **Euprymna hyllebergi** *Growth*

The growth rate from hatching to 100 days of age for *E. hyllebergi* is approximately 2.4% in length and 7.5% in weight. At 60 days after hatching, the squid had grown to 17-mm length and 2.6-g weight and 22 mm and 6 g at 100 days. Food consumption of about 0.2 g d^{-1} or 37% body weight d^{-1} enables calculation of the food conversion efficiency of about 37% (range 14–99) from hatching to 100 days. This rate increases from 30 to 40% after hatching to 60–70% during 40–60 days with a peak of about 64% between 50 and 60 days (Fig. 15.12). These values potentially relate to the storage of energy for the consequent reproductive period (Nabhitabhata et al. 2005). At 90 days after hatching, the survival from hatching is approximately 10% and from settlement is 70%.

Transition in growth phases is reflected in the nature of the growth models. The stage where the models shifted to a higher elevation is at about 30 days after hatching, and this corresponds to the observed settlement stage (Figs. 15.8–15.11). The second growth phase is from 30 to 122 days. The relationships between ML (mm) and weight (W, g) can also be expressed as a power regression model (Nabhitabhata et al. 2005) as happened in the early phase (Fig. 15.9):

$$W = 1.032 \times 10^{-3}\ ML^{2.780}. \tag{15.4}$$

The relationships between ML and age (d, days after hatching) and between weight (g) and age can be expressed as the quadratic equation (Fig. 15.10; Nabhitabhata et al. 2005) and a cubic regression model (Fig. 15.11; Nabhitabhata et al. 2005):

$$ML = 0.407A - 1.553 \times 10^{-3}\ A^2 - 3.648 \tag{15.5}$$

$$W = 1.952 - 0.147A + 3.570 \times 10^{-3}A^2 - 1.728 \times 10^{-5}A^3 \tag{15.6}$$

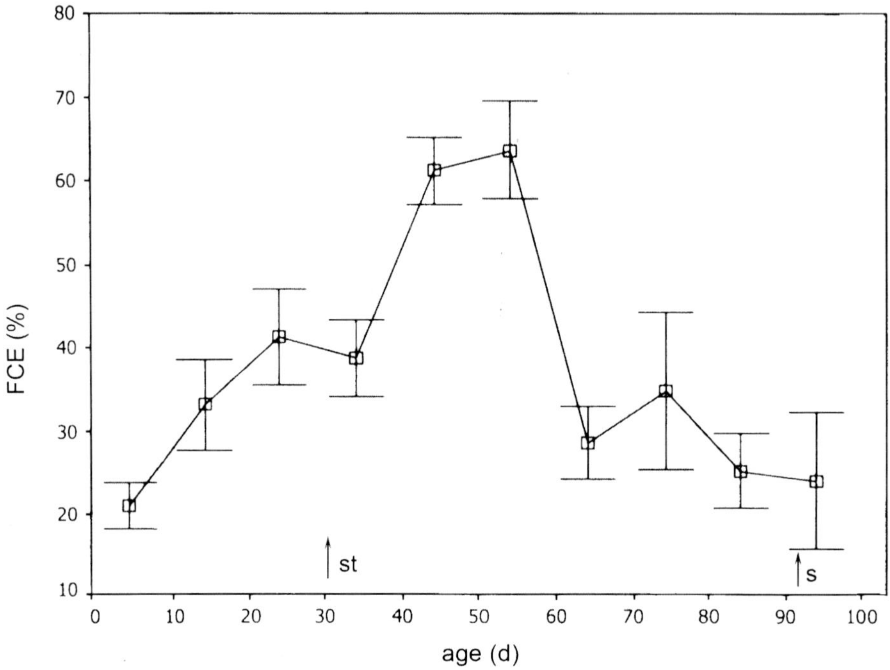

Fig. 15.12 Food conversion efficiency (*FCE*: %) of cultured *Euprymna hyllebergi* during growth (age: d). *Arrows* indicate spawning (*s*) and settling stage (*st*). (After Nabhitabhata et al. 2005)

15.6.3 Euprymna tasmanica *Growth*

Detailed studies on growth of *E. tasmanica* are scarce. Growth rate of *E. tasmanica* is rapid, with hatchlings reaching adult size in 2 months at 18 °C. *E. tasmanica* can grow from a ML of approximately 1.7 mm and a weight of 0.012 g at hatching (Steer et al. 2004) to 0.06 g at 21 days and to 6.8 g at 112 days (Sinn et al. 2008). The daily growth rate from day 21 to 63 is 7–9 %, from 63 to 84 days decreases to 2–4 % and from 84–112 days 1–2 % at 18 °C (Sinn 2005).

The relationships between weight and age are exponential from 7 to 44 days after hatching and linear from 58 to 140 days (Moltschaniwskyj and Carter 2010) which can be expressed as:

$$\ln W = 0.069A - 4.06 \qquad (15.7)$$

$$W = 0.07A - 3.28. \qquad (15.8)$$

15.7 Trends in Research and Industry

The main purpose of culture *Euprymna* is obviously not for human consumption. Present research in the fields of marine pharmacology, biotechnology and mimetic engineering requires small and "easy to culture" squids. Rapidly growing demand in the ornamental aquaculture trade also requires organisms of similar characters, specifically size and ease of care. The small adult body size, benthic habit and good adaptability to culture conditions of *Euprymna* are prominent character suites that are well adapted to the aforementioned purposes. Based on such qualities, bobtail squids should be cultured on a small scale in order to reduce the cost of production. Additionally, a small-scale culture has advantages of the reduced size and benthic habits of the squids. The variety of flow-through open or closed seawater systems can yield different results and should be further studied to better quantify which systems are best for maximising production and those appropriate for each species.

Culture of *Euprymna* similarly encounters a bottleneck during the nursing phase similar to other cephalopods, since young innately feed on live feed. Future research should focus on developing feeds, both live and artificial. However, small-scale culture of live food organisms is more appropriate for small-scale culture of *Euprymna* in view of low operating costs at present. Development of artificial feed is necessary to reduce costs, but it could be postponed on a small scale. Artificial feed for cephalopods has not been commercially developed anywhere, but many studies are being completed, focusing on species that are aimed to be cultured as human food. Investigating various types of feed may give insight as to whether bobtail squids can also use artificial feed in such a manner.

E. hyllebergi and *E. tasmanica* can be cultured through multiple consecutive generations (3 generations for both *E. hyllebergi* and *E. tasmanica*) with similar growth rates (under similar conditions) without apparent effects of inbreeding on decreased growth (Nabhitabhata et al. 2005). Similar growth among generations enables a reliable supply of broodstocks for aquaculture and provides an alternative to continued fishing for wild-caught specimens, which can be time consuming and costly. However, the feasibility of inbreeding effects on decreasing of growth and fertility must be considered when producing future generations from the same broodstock. Broodstocks cannot rely solely on cultured batches, and wild broodstocks should be added intermittently to provide both genetic variation and possibly the induction of beneficial microbes that are necessary to keep squid healthy during their lifetime. Growth in captivity and culture methodology of both *E. hyllebergi* and *E. tasmanica* as well as their congeners should be further studied in views of maximising the aquaculture production and increasing our ability to provide a useful resource for a variety of research studies as well as the development of model aquaculture cephalopods.

15.8 Conclusions

The ability to maintain and grow small benthic squids such as *Euprymna* has opened up a new avenue for instigating the use of these animals as model systems in both bioengineering (adhesion) and biomedical (beneficial bacteria) research. The requirements for housing, maintaining and raising sepiolids is minimal and not as costly as other, more gregarious squid species, and this allows laboratories to set up facilities that may not necessarily be close to the ocean (such as NMSU). Presently, there are 14 laboratories in the USA alone that have culture facilities for raising

Euprymna; such facilities would not be feasible unless these animals were easy to transport long distances and maintained continually and without a nearby marine station or source of seawater. Additional research must be considered for the effects of inbreeding (when maintaining a constant broodstock) as well as comparing species for different traits that can be used for certain research foci. The inception of using cephalopods as research and not "feed" organisms is a new and exciting avenue that multiple areas of research can benefit from for furthering our knowledge in aquaculture, bioengineering, medicine, ecology and evolutionary biology.

References

Anderson RC, Mather JA (1996) Escape responses of *Euprymna scolopes* Berry, 1911 (Cephalopoda: Sepiolidae). J Mollusc Stud 62:543–545

Anderson RC, Mather JA, Steele CW (2002) The burying behavior of *Euprymna scolopes* Berry, 1913 (Cephalopoda: Sepiolidaae). West Soc Malacol Annu Rep 33:1–7

Aungtonya C, Nateewathana A, Tendal OS, Nabhitabhata J (2011) New records of the Bobtail Squid, *Euprymna hyllebergi* Nateewathana, 1997, with designation of a neotype. Phuket Mar Biol Cent Res Bull 70:15–27

Boletzky Sv (1998) Cephalopod eggs and egg masses. Oceanogr Mar Biol Ann Rev 36:341–371

Byern Jv, Grunwald I (eds) (2010) Biological adhesive systems from nature to technical and medical application. Springer-Verlag, Austria

Guerrero-Ferreira RC, Nishiguchi MK (2011) Bacterial biodiversity in natural environments. In: López-Pujol J (ed) Biodiversity/book 3, In Tech, Open access publisher. http://www.intechopen.com/articles/show/title/bacterial-biodiversity-in-natural-environments

Guerrero-Ferreira RC, Gorman C, Chavez AA, Willie S, Nishiguchi MK (2012) Characterization of the bacterial diversity in Indo-west Pacific loliginid and sepiolid squid light organs. Microb Ecol. doi: 10.1007/s00248–012-0099–6

Hanlon R (1990) Maintenance, rearing, and culture of teuthid and sepiolid squids. In: Gilbert DL, Adelman WJ, Arnold JM (eds) Squids as experimental animals. Plenum Press, New York

Hanlon RT, Claes MF, Ashcraft SE, Dunlap PV (1997) Laboratory culture of the sepiolid squid *Euprymna scolopes*: a model system for bacterial-animals symbiosis. Biol Bull 192:364–374

McFall-Ngai MJ, Heath-Heckman EAC, Gillette AA, Peyer SM, Harvie EA (2012) The secret languages of coevolved symbiosis: insights from the *Euprymna scolopes-Vibrio fischeri* symbiosis. Sem Immunol 24:3–8

Moltschaniwskyj NA, Carter CG (2010) Protein synthesis, degradation, and retention: mechanisms of indeterminate growth in cephalopods. Physiol Biochem Zool 83:997–1008

Moltschaniwskyj NA, Hall K, Lipinski MR, Marian JEAR, Nishiguchi MK, Sakai M, Shulman DJ, Sinclair B, Sinn DL, Staudinger M, Van Gelderen R, Villanueva R, Warnke K (2007) Ethical and welfare considerations when using cephalopods as experimental animals. Rev Fish Biol Fish 17:455–476

Moynihan M (1982) Notes on the behavior of *Euprymna scolopes* (Cephalopoda: Sepiolidae). Behav 85:25–41

Nabhitabhata J, Nilaphat P, Promboon P, Jaroongpattananon C (2005) Life cycle of cultured bobtail squid, *Euprymna hyllebergi* Nateewathana, 1997. Phuket Mar Biol Cent Res Bull 66:351–365

Nateewathana A (1997) The sepiolidae (Cephalopoda) of the Andaman Sea, Thailand, with description of *Euprymna hyllebergi* sp. nov. Phuket Mar Biol Cent Spec Publ 17:465–481

Nateewathana A, Nabhitabhata J, Nilphat P (2001) A new record of a bobtail squid, *Euprymna hyllebergi* Nateewathana, 1997, in the Gulf of Thailand. Phuket Mar Biol Cent Spec Publ 25:501–506

Norman MD, Lu CC (1997) Redescription of the southern dumpling squid *Euprymna tasmanica* and a revision of the genus *Euprymna* (Cephalopoda: Sepiolidae). J Mar Biol Assoc UK 77:1109–1137

Nyholm SV, Nishiguchi MK (2008) The evolutionary ecology of a sepiolid squid-vibrio association: from cell to environment. Vie Milieu 58:175–184

Reid AL, Jereb P (2005) Family Sepiolidae. In: Jereb P, Roper CFE (eds) Cephalopods of the world an annotated and illustrated catalogue of cephalopod species known to date vol. 1 chambered nautiluses and sepioids (Nautilidae, Sepiidae, Sepiolidae, Sepiadariidae, Idiosepiidae and Spirulidae) FAO Species catalogue for fishery purposes no. 4, vol. 1. FAO, Rome

Reid AL, Norman MD (1998) Sepiolidae bobtail squids. In: Carpenter KE, Niem VH (eds) The living marine resources of the western central Pacific vol 2 cephalopods, crustaceans, holothurians and sharks. FAO, Rome

Ruby EG (1999) The *Euprymna scolopes-Vibrio fischeri* symbiosis: a biomedical model for the study of bacterial colonization of animal tissue. J Molec Microbiol Biotech 1:13–21

Ruby EG, Lee K (1998) The *Vibrio fischeri-Euprymna scolopes* light organ association: current ecological paradigms. Appl Env Microbiol 64:805–812

Ruby EG, McFall-Ngai MJ (1992) A squid that glows in the night: development of an animal-bacterial mutualism. J Bacteriol 174:4865–4870

Shears J (1988) The use of a sand-coat in relation to feeding and diel activity in the sepiolid squid *Euprymna scolopes*. Malacologia 29:121–133

Sinn DL (2005) From individuals to populations: personality traits in southern dumpling squid (*Euprymna tasmanica* Pfeffer, 1884) and their life history correlates. Doctor of Philosophy Thesis, University of Tasmania

Sinn DL, Moltschaniwskyj NA (2005) Personality traits in dumpling squid (*Euprymna tasmanica*): context-specific traits and their correlation with biological characteristics. J Comp Psychol 119:99–110

Sinn DL, Gosling SD, Moltschaniwskyj NA (2008) Development of shy/bold behaviour in squid: context-specific phenotypes associated with developmental plasticity. Anim Behav 75:433–442

Steer MA, Moltschaniwskyj NA, Nichols DS, Miller M (2004) The role of temperature and maternal ration in embryo survival: using the dumpling squid *Euprymna tasmanica* as a model. J Exp Mar Biol Ecol 307:73–89

Summers WC (1985) Comparative life history adaptations of some myopsid and sepiolid squids. NAFO Sci Stud 9:139–142

Chapter 16
Loligo vulgaris and *Doryteuthis opalescens*

Érica A. G. Vidal and Sigurd von Boletzky

Abstract The medium-sized loliginid squids *Loligo vulgaris* and *Doryteuthis opalescens* have a long record as experimental models in cephalopod culture. The respective size of eggs (2.0–2.5 mm) and of hatchlings (2.5–3.5 mm mantle length) make them an interesting material for small-scale culture, but the high requirements for live food (copepods, live crustacean larvae and/or mysids at early juvenile stages), highly active swimming behaviour and the relatively modest adult size have prevented them from becoming target species in commercial culture projects. The inherent difficulties in providing large volume tanks to hold groups of schooling squid have an impact on profitability. However, the data obtained in culture experiments involving these species have been useful in a wider biological context. This chapter summarizes our present knowledge of their culture potential refraining from suggesting standardized methodologies.

Keywords *Doryteuthis opalescens* · *Loligo vulgaris* · Cephalopod · Culture · Paralarvae · Rearing · Squid

16.1 Importance of these Species in the Market

Loligo vulgaris (Lamarck 1798) and *Doryteuthis opalescens* (Berry 1911) are two medium-sized squids of the family Loliginidae and both are of great interest to fisheries.

The European squid *L. vulgaris* (Fig. 16.1) is distributed in the eastern Atlantic Ocean and North and Mediterranean Seas (approximately from 55°N to 20°S). For this species, Jereb and Roper (2010) stated the following: '*Loligo vulgaris* is taken throughout its distributional range all the year round, mainly as by catch of the

É. A. G. Vidal (✉)
Center for Marine Studies, University of Parana—UFPR,
Cx P. 61, Pontal do Paraná, PR 83255-976, Brazil
e-mail: ericavidal2000@yahoo.com.br; erica.vidal@ufpr.br

S. von Boletzky
CNRS, Laboratoire Arago, 66651 Banyuls-sur-Mer, France
e-mail: sigurd.boletzky@sfr.fr

J. Iglesias et al. (eds.), *Cephalopod Culture,* DOI 10.1007/978-94-017-8648-5_16,

Fig. 16.1 *Loligo vulgaris*. Adult individuals in a large outdoor tank, with an egg mass attached close to the *upper end* of the overflow pipe. On the *left side* is a mated pair, the female is turning to the *right* and beginning to approach the egg mass. (Original image)

multi-species bottom and pelagic trawl fisheries. Major fishing grounds are located off Portugal, on the West African Banks and in the western Mediterranean, where the species is caught in the international fisheries with otter trawls and purse seines in daytime and occasionally at night with light attraction'.

The California market squid (Fig. 16.2) was recently renamed from *Loligo opalescens* to *Doryteuthis opalescens*, based upon the molecular phylogeny (Anderson 2000). This key forage species lives in the nearshore pelagic environment and is endemic to the California Current region, being found from British Columbia, Canada to the southern tip of Baja California, Mexico (from 50°N to 25°N).

The fishery is focused entirely on adults aggregated for spawning over the shelf, which occurs from April to July in Monterey during the upwelling season and from November to March in southern California during the productive storm season. Fisheries shifted from Monterey to southern California (Channel Islands) from 1970 to 2000. Landings grew in the past 20 years to become the largest US squid fishery in terms of tonnage and ex-vessel value (Jereb and Roper 2010, p. 64). The fishery is managed with weekend closures and an annual limit of 118,000 t with yields of ~US$ 26 million. However, since 2003, a record year, landings in Monterey have dropped off to zero in 2008, and fishing in southern California has gone from a winter activity to a year-round venture (Jackson and Domeier 2003). 'They typically are harvested on shallow nearshore spawning grounds; specialized light boats shine high intensity lights on the water to attract and congregate the squids

Fig. 16.2 *Doryteuthis opalescens*. (Photo credits: Roger T. Hanlon)

near the surface, then seiner boats capture them with purse-seine nets' (Jereb and Roper 2010 p. 64).

One of the most popular capture methods in recreational and artisanal fishery is hand jigging. For *L. vulgaris* in the Mediterranean, there are at least two variants, one with bare jiggers (generally used along with a surface light) and the other using jiggers partly wrapped with fish fillets (supposed to act chemically as a 'tasteful lure', G. Celentano, personal communication). Hand jigging is minimally traumatic for specimens of medium to large size, and therefore can be recommended for field collections serving the establishment of a breeding stock.

16.2 State of the Art

Before approaching the questions of culture techniques and the aims of specific culture projects, it may be useful to survey the scientific literature on the two species considered, especially with regard to their biology.

Many of the early accounts of the fishery and biology of *L. vulgaris* were related to the activities of the Zoological Station at Naples (Italy), beginning with Jatta (1896) and Lo Bianco (1899), continuing with Bauer (1909), Naef (1921/1923, 1928), Sereni (1930) and with Grimpe (1928) who also worked at the Helgoland Station and studied the cephalopods of the North Sea, including *L. vulgaris* (Grimpe 1925). A highlight of cephalopod biology research at the Naples Station was the description of giant nerve fibres in *L. vulgaris* (Young 1938, 1939). A great part of the detailed studies of Bidder (1950) on digestive mechanisms in squid was achieved at the Naples Station. A major contribution from the Naples Station is a catalogue of body patterning in cephalopods (see Packard 1995), describing and illustrating also the chromatic components, body patterns and displays observed in *L. vulgaris* (Borrelli et al. 2006).

A survey of the biology of *L. vulgaris* in the western Mediterranean was performed by Mangold-Wirz (1963) giving information on sexual maturation, mating and spawning.

Statoliths studies revealed a lifespan around 15 months, with males maturing earlier and at smaller sizes than females (Moreno et al. 2007). However, longer lifespans were estimated in males from the Galician waters, but shorter (9 months) in squid from the Portuguese waters and the Sahara shelf. Maturity was ultimately found to be primarily dependent on size rather than age. Detailed information on life cycle, maturation and reproduction of *L. vulgaris* can be found in Moreno et al. (2013).

Following earlier reports by Collins (1892), Scofield (1924) and others, Fields (1950) surveyed the fishery and biology of *D. opalescens*, providing catch statistics from the Monterey area beginning in the 1930s, and offered detailed observations on maturation and spawning. McGowan (1954) summarized his biological study of *D. opalescens*, especially laboratory observations on animals collected by diving. A richly illustrated monograph covering fisheries, general biology and development of *D. opalescens* is due to Fields (1965) who also described feeding behaviour in the aquarium.

Hurley (1977, 1978) studied the mating behaviour and school structure of *D. opalescens* in the laboratory and in the field, using artificial eggs to study the visual stimulus of egg masses on spawners. Recksiek and Frey (1978) coordinated a publication uniting a dozen original articles covering the results of 3 years of research on *D. opalescens*. Starting from background information on market squid research, basic life history and the California squid fishery, presented by the two coordinators, there are studies of spermatogenesis (Grieb and Beeman 1978), oögenesis (Knipe and Beeman 1978), age and growth (Spratt 1978), feeding dynamics (Karpov and Caillet 1978), importance of *L. opalescens* in the marine vertebrate food web (Morejohn et al. 1978), morphological indicators of population structure (Kashiwada and Recksiek 1978), biochemical genetic population structure (Ally and Keck 1978), electrophoresis of select proteins (Christofferson et al. 1978), acoustic sensing (Vaughan and Recksiek 1978), target strength of individual squid (Vaughan 1978) and correlations between squid catches and oceanographic conditions (McInnis and Broenkow 1978).

More presentations on these loliginid species are available in volume I of *Cephalopod Life Cycles* (Boyle 1983), with individual chapters for *L. vulgaris* (Worms 1983) and *D. opalescens* (Hixon 1983) covering the 'Egg stage', the 'Juvenile stage', 'Growth', 'Maturation', 'Reproduction', 'Mortality' and 'Ecology'. In the abovementioned volume, Worms (1983) deals with characteristics of *L. vulgaris*, unfortunately introducing several errors. The correct graphs for embryonic development of *L. vulgaris* that should have been given by Worms (1983) in his Fig. 9.1 are recapitulated in volume II of *Cephalopod Life Cycles* (Boletzky 1987, Fig. 2.1b). For *D. opalescens*, recent in-depth information on life cycles can be found in Zeidberg (2013).

Loliginid females lay their egg capsules over other egg masses of the same species, forming egg masses that can vary greatly from small clusters of a few capsules to very large ones. In *L. vulgaris*, each capsule contains from 130 to 180 eggs (Mangold-Wirz 1963) and newly laid eggs measure 2.0–2.2 mm in length by 1.5–1.6 mm in width (Boletzky 1974a). The size of eggs are close to those of *D. opalescens* (2.0–2.5 mm in length by 1.3–1.6 mm in width), although in this species the number of eggs per egg capsule ranges from 90 to 150 as indicated by Hixon (1983). In summary, these two loliginid species form oocytes of roughly similar size, much larger than, e.g. *Doryteuthis pealeii* (Hanlon et al. 1979).

16.3 Broodstock Acclimatization to Captivity

Broodstock acclimatization is necessary when juvenile, subadult or adult individuals are captured from the sea and transferred to holding facilities where reproduction can be achieved.

In a brief section on ecological characteristics of *L. vulgaris*, Bauer (1909, p. 154) had already stated the following: '*Loligo* is pelagic like most of the decapods and in general lives in shoals. In the aquarium the animals are in continuous movement, swimming intermittently forward and backward without turning; indeed all the individuals in a swarm move simultaneously in the same direction'.

Tardent (1962) summarized earlier experience (e.g. Grimpe 1928) with squid maintenance at the Naples Aquarium (which used an open seawater circuit, functioning as a closed circuit during stormy periods with high water turbidity), where groups of individuals have been kept continuously for periods lasting up to 2 months, being fed mainly on live and dead fish. Freshly caught individuals were acclimatized in circular rubber tanks (inflatable splash pools; see Neill 1971) before being transferred to a rectangular show tank, where they formed well-organized 'schools'. Bentivegna (1987) indicated maximum survival time of 3 months for small individuals of *L. vulgaris* in the Naples Aquarium, under optimal feeding conditions (live prawn).

For highly active swimmers such as loliginid squids (Neumeister et al. 2000), which begin schooling at an early juvenile stage, sufficient tank space accommodating relatively large groups is a major requirement (Neill 1971; Neill and Cullen 1974). Hanlon and Messenger (1996, p. 151) drew attention to a distinction between shoaling and schooling: '*Shoaling* emphasises the social behaviours related to aggregating as distinct from the physical synchrony (velocity and direction) and polarization (parallel swimming) of individuals in the group, which are termed *schooling*'.

Visual landmarks in their artificial environment (e.g. striped tank walls or curtains, or entirely black-walled cylindrical tanks, as described by Mladineo et al. 2003) may help the animals to avoid contact with technically imposed barriers in their artificial environment (Hanlon 1978, p. 10). But even when a high water quality (cf. Gilly and Lucero 1992), a regular supply of appropriate prey (DeRusha et al.

1989) and a complementary assortment of sensory inputs are offered in captivity, the question remains whether these highly active, gregarious animals can be fully acclimatized to captivity without losing significant elements of their natural behaviours (cf. Zeidberg 2008). Only if ethological aspects are clearly outside the scope of the research field envisaged, can the question of possibly impoverished behavioural repertoires be ignored in a squid culture (cf. Yang et al. 1986). Little is known about how broodstock feeding affects eggs and paralarval quality in squid. Ideally, females should be maintained under controlled conditions that match those to which they have been exposed in the wild.

16.3.1 Food Supply

Considering wild food sources of squid (mainly living marine fish and shrimp), the question of food supply in captivity raises some practical and economic problems: Are the prey species known from wild squid stomach contents available, and if so, can their procurement be afforded by the squid culturist? What are the acceptable alternatives, e.g. other marine prey items or fresh water animals? Can live prey be replaced by dead animals or food preparations? These questions again relate to the problem of behavioural repertoires mentioned above (Boletzky 2004).

Tardent (1962) indicated for *L. vulgaris* kept in the Naples Aquarium that 'cut dead sardines and anchovies' and live prawns '*Leander* and *Lysmata* (are) given at least once a day'. In particular, this author observed that 'food is as readily picked up from the bottom or walls of the tank as from the water surface'. Bentivegna (1987) confirmed these observations.

Pierce et al. (1994) summarized information on diets of wild *L. vulgaris* from Scotland, northwestern Spain, western and southern Portugal and the western Mediterranean, and found fish, crustaceans, cephalopods and polychaetes in various combinations.

For *D. opalescens*, Fields (1965) summarized his observations and literature surveys: 'Thus *L. opalescens* probably eats two or more meals each day, foraging from the sea surface to its floor, and any available animals of appropriate size may be its prey. Pelagic crustaceans and fellow squid are important components of its diet; it feeds with, and probably upon, sardines (*Sardinops caeruleus*), herring (*Clupea pallasi*), mackerel (*Scomber diego*), sauries (*Cololabis saira*), and anchovies (*Engraulis mordax*)'. The observation that 'fellow squid' are regularly eaten raises the question whether cannibalism could be 'exploited' in squid cultures (e.g. in 'sacrificing' small, immature males).

16.3.2 Sex Ratio

Although the sex ratio appears to be on an average roughly 1:1 in loliginid squids, a special situation arises with 'sneaker mating', a phenomenon also observed in other cephalopods. Zeidberg (2008) reports that in *D. opalescens* mating groups involved 10–20 individuals and had a sex ratio of 1–2 males for each female. Although sneaker mating has been observed in other species of *Doryteuthis* and *Loligo*, so far no such reports exist for *L. vulgaris*. However, Zeidberg (2008) considers that multiple paternity in individual egg capsules may be common in most loliginid squids.

16.3.3 Physical and Chemical Parameters

Natural or synthetic seawater of high quality is indispensable for cephalopod maintenance, rearing and culture. A salinity between 34 and 38 psu is recommended, with pH around 8 (7.7–8.2) and O_2 levels close to saturation (5 mg L^{-1}), along with levels of ammonia and nitrite <0.1 mg L^{-1}, and <20 mg L^{-1} for nitrate are considered normal (Boletzky and Hanlon 1983). Water temperature should best be maintained between 12 and 22°C in culture systems for the two species considered here.

16.4 Spawning Process

The structure of the egg capsules in both species is essentially as described by Jecklin (1934), but the way these capsules are deposited is different. In *L. vulgaris*, the egg capsules are always suspended in dense clusters from overhanging substrates (Boletzky 1974a), whereas in *D. opalescens*, they are deposited on (generally soft) bottoms (Fields 1965). Spawning can be stimulated by the visualization of natural or artificial egg masses in the tanks (Arnold 1962).

An open question remains with regard to the actual making of an egg capsule by the spawning female: Is the spirally coiled string of eggs embedded in oviducal jelly released along with the outer capsule, the material of which is produced by the nidamental glands, or is the outer capsule released first and the egg string 'pumped' into it, as described by Zeidberg (2008)?

Ideally, females spawning in a tank should remain in contact with other adult individuals (Hanlon 1990), to be able to regularly feed, mate and lay eggs (Fig. 16.1). However, if mating males are generating too much stress for spawning females, they should be separated.

16.5 Embryonic Development

Egg care is one of the most important steps for rearing success. During embryogenesis, eggs should be closely monitored to evaluate progress of development and proximity of hatching. Naef (1928) and Arnold (1965) provided embryonic developmental scales for loliginids and the latter scale is used as reference in this chapter.

16.5.1 Egg Collection and Transportation

Egg masses can be obtained from females spawning in captivity and collected from the spawning grounds. In the second case, it entails careful selection of eggs in good condition, following proper transportation. Embryos and paralarvae can also be produced by in vitro fertilization methods as demonstrated by *D. pealeii* (Crawford 2002). Studies that focus on aspects of embryonic development should acquire eggs from captive females or in vitro fertilization, since fundamental information such as the exact incubation time and the physical conditions the eggs were exposed to can be obtained and controlled. When eggs are collected from the field, it is advisable to get hold of them around stages 16–22 (i.e. late gastrulation stages). Collection of eggs at early stages frequently results in death of the embryos if conditions during transport and maintenance are not ideal, for instance, major oscillations in water parameters (temperature, salinity, dissolved oxygen and pH) can lead to abnormalities and death of embryos (Marthy 1972). On the other hand, collection of eggs at late stages (25–28) entails other problems. Mechanical disturbance may trigger premature hatching and may cause mass hatching during transport. Additionally, late-stage embryos metabolize at high rates and require more water and oxygen than at early stages (Rosa et al. 2012). Thus, depletion of oxygen can occur rapidly causing mass premature hatching due to deterioration of water quality (Hanlon 1990). Prematurely hatched paralarvae have not yet fully absorbed the outer yolk sack, a crucial step prior to hatching; thus, the outcome from premature hatching is high mortality of paralarvae (Boletzky and Hanlon 1983, Hanlon 1990), thus jeopardizing rearing trials.

The effects of transportation conditions on loliginid eggs are largely undervalued; however, when proper conditions are endowed the eggs can withstand long transport times (~24 h, Yang et al. 1983a, b; Vidal et al. 2002a, 2005) at relatively small shipping costs compared to juveniles and adults (Hanlon 1990). During transport, eggs should be conditioned in double plastic bags containing from 30 to 50% of oxygenated seawater with the remaining space filled with pure oxygen. To avoid thermal shock, they should be slowly acclimatized to the temperature and salinity of the rearing tanks at destination (1 °C h^{-1}, Villanueva 2000b), especially if there is a difference in any aspect of water quality. Hanlon (1990) provided detailed information on egg transportation and handling.

Eggs produce large amounts of ammonia and require high levels of dissolved oxygen (Hanlon 1990), demanding pristine water quality and aeration. Excellent hatching success has been obtained by carefully tying single collective capsules (strands) or bunches of five to ten egg strands with nylon thread and suspending them in highly aerated water underneath the water inflow of the tanks and/or maintaining an upwelling flow of water through the suspended eggs (Boletzky and Hanlon 1983; Paulij et al. 1990; Villanueva 2000a; Vidal et al. 2002a). This simple procedure exposes the eggs to adequate levels of oxygen that in combination with high flow rates (Villanueva et al. 2003) and proper current speed inside the tanks (Vidal et al. 2002a) maximize timely hatching. The aim of keeping egg strands in

Fig. 16.3 *Doryteuthis opalescens*. Late-stage embryos (stage 28). (Arnold 1965; original image)

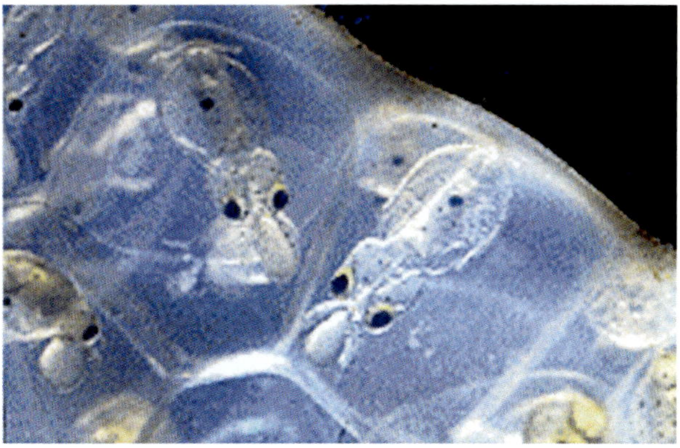

small bunches and in constant motion is to promote an even aeration between them. Steer and Moltschaniwskyj (2007) have shown that the relative position of the eggs within the egg mass of *Sepioteuthis australis* dictates their chances of survival. Eggs located near the attachment point of the egg strand or within the interior of the egg mass are subject to the highest mortality rates due to reduced levels of oxygen supply. This effect is amplified with increasing egg mass size. Furthermore, the development of the embryos within a single egg strand is asynchronous as they are exposed to differential oxygen levels. Embryos located at the periphery and exposed to higher oxygen supply will hatch first, as particularly in the case of those species which produce egg capsules containing dozens of eggs as both *L. vulgaris* and *D. opalescens* (Fig. 16.3). Through the hatching process the external egg envelope is punctured, triggering a reduction of the diffusive distance to the centrally located embryos, and allowing them to develop further, as demonstrated for *Sepia apama* (Cronin and Seymour 2000). Thus, during embryonic development oxygen levels should be maintained close to saturation, and flow rates and the current generated inside the tanks should be closely monitored to ensure optimal survival of the embryos. Low-speed currents reduce aeration between the egg strands, leading to death of the embryos, and high current speeds can provide mechanical stimuli that will cause premature hatching at late embryonic stages. Air bubbles must also be avoided because they adhere to either the surface of the egg strands or the hatchlings' skin, thus causing mortality (Boletzky and Hanlon 1983).

Loliginid eggs collected in the wild are sometimes infested with capitellid polychaete worms (McGowan 1954, Fields 1965; Boletzky and Dohle 1967; Yang et al. 1986; Vidal et al. 2002a; Zeidberg et al. 2011). Infestations by these worms have been related to deterioration of the external egg envelope, exposure of the chorion of the eggs, premature hatching and subsequent high mortality of paralarvae (Vidal et al. 2002a). An exception to this pattern, however, was found by Zeidberg et al. (2011), who observed that by perforating and feeding on the external egg envelope the worms slightly increased the hatch rate (3.1 %), suggesting the existence of a symbiotic relationship between the squid eggs and the worms. However, these authors did not supply information on the condition of the hatchlings (premature, normal or late; see below) or on their subsequent survival and performance to attest that infestation by the worms is indeed beneficial during embryonic development. In any case, infestation can be prevented or greatly reduced by removing the sand from the apical tip of the egg strands. Thus, the relationship between the squid eggs and these worms remains not fully explained, as well as other ecological relationships during the egg phase. It has been shown that the egg capsules of *D. opalescens* and *D. pealeii* hold a complex bacterial community via the accessory nidamental glands, recognized as the host organ and inocula for bacterial species (Barbieri et al. 2001). The precise function of the egg capsule-associated bacteria is unknown, but it was suggested by Biggs and Epel (1991) that the defence of the eggs against microscopic and macroscopic predators is a possible explanation, as demonstrated by the antifungal effects of *D. pealeii*-associated *Pseudoalteromonas strains* (Barbieri et al. 1997).

16.5.2 Incubation Conditions and Influence of Environmental Factors during Embryogenesis

Both natural and artificial seawater have been used to incubate *L. vulgaris* and *D. opalescens* eggs, although developmental problems have been observed mainly in artificial seawater. The perivitelline fluid inside the egg chorion is in osmotic equilibrium with seawater and will reflect changes in the concentration of major ions of seawater; therefore, these ions must be within the required concentration for normal egg development (i.e. calcium and potassium 9–15 mM, magnesium 48–68 mM and sulphate 15–37 mM) (D'Aniello et al. 1987). Even though late-stage eggs developed normally when incubated in artificial seawater containing these major ions in proper concentrations, early stage eggs did not, suggesting that other elements are required. This was later addressed by Hanlon et al. (1989), who demonstrated that the embryo statoliths do not develop normally when strontium levels are below 8 mg L^{-1}. Paralarvae hatching without or with abnormal statoliths were termed 'spinners', as they displayed erratic swimming patterns making continuous loops at the surface and being unable to descend in the water column. The importance of strontium in the biomineralization process during embryogenesis remains unknown and other factors must play a role in the formation of statoliths, since the spinners are occasionally found from natural seawater cultures. Natural seawater is therefore recommended for egg incubation.

Eggs of both species have been incubated in closed and open, flow-through culture systems (Tables 16.1 and 16.2) and exposed to both constant and fluctuating water temperatures and salinity conditions. Temperature strongly influences the rate of embryonic development, incubation time and hatching rate (Boletzky 1987) and there is some evidence that it might also have an effect on the hatching rhythm (Paulij et al. 1990). Sen (2005a) evaluated thermal tolerance limits of *L. vulgaris* intermediate stage eggs (11–13) from Izmir Bay, Turkey. When eggs were exposed to temperatures from 6 to 28 °C (at 37 psu), hatching rates (defined as number of viable

Table 16.1 Selected literature with information on type of culture, water system, embryonic development and survival, stocking density and growth rate during rearing of paralarvae and juveniles of *Loligo vulgaris*

Type of culture	Water system	Mean temperature (°C)	pH	Duration (days)		Paralarval survival rate (%)	Initial stocking density (paralarvae L^{-1})	Growth rate		Reference
				icp	dah			Weight (% ww day^{-1})	Size (mm ML day^{-1})	
ED	CL	13, 15, 17 and 19	8.1 ± 1	14 (13 °C), 27 (19 °C)		–		–		Rosa et al. (2012)
ED	OP	16.2–16.7								Villanueva et al. (2007)
ED										Sen (2004, 2005a, b)
ED										Villanueva et al. (2003)
ED and PR	OP	11, 19	–		62 and 50	16.8 (30 d); 0.2 (62 d, 11 °C); 10% (30 d); 1.4 (50 d, 19 °C)	25	3.6 (62 d, 11 °C); 7.8 (50 d, 19 °C)	1.3 (62 d, 11 °C); 2.6 (50 d, 19 °C)	Villanueva (1994, 2000a)
ED	Semi-CL									Villanueva (2000b)
PR	CL	15.6–18.5	7.9–8.3		54, 123 and 140	10 (10 dah) 6.3 (30 dah) 0.1 (60 dah)	~ 1.5		2.28 (123 d)	Turk et al. (1986)
PR	OP	18.5–22			75					Boletzky (1979)
ED and PR	OP				45					Boletzky (1974, 1971)

ED embryonic development, *PR* paralarval rearing, *CL* closed system, *OP* open system, *dah* days after hatching, *icp* incubation period, *d* day, *ww* wet weight, *ML* mantle length

Table 16.2 Selected literature with information on type of culture, water system, embryonic development and survival, stocking density and growth rate during rearing and culture of paralarvae and juveniles of *Doryteuthis opalescens*

Type of culture	Water system	Mean temperature (°C)	pH	Duration (days)		Survival rate (%)	Initial stocking density (paralarvae L^{-1})	Growth rate		Reference
				icp	dah			Weight (% ww day^{-1})	Size (mm ML day^{-1})	
ED	OP	9.2–24.4	–	93 (9.2°C) 19 (24.4°C)						Zeidberg et al. (2011)
PR	CL	16 ± 0.5	8.1–8.4		10	86 (14–24 dah)	2.3–3.1	5.4–8.4 (14–24 dah)		Vidal et al. (2006)
PR	CL	16 ± 0.5	8.1–8.4		45–60	36–60	6.8–13.6	6.4–7.4 (0–20 dah)	2.4 (0–20dah)	Vidal et al. (2002a)
PR	CL	12 ± 0.5 16 ± 0.5	8.1–8.4		20		1.4	3.3 (12°C) 7.4 (16°C)		Vidal et al. (2002b)
PR	OP	13–16	–		60					Preuss and Gilly (2000)
PR					60					Preuss et al. (1997)
PR	OP	17–18	–		52	0.13–16	~1.0			Chen et al. (1996)
ED	CL	18–20	7.8–8.4							D'Aniello et al. (1989)
CT	CL	15	7.8–8.2		235–248	10 (49 and 120 dah)	0.27–0.93	5.6–8.4 (0–60 dah), 1.6 (0–240 dah)	2.4 (0–56 dah), 3.2 (0–60 dah), 0.63 (0–240 dah)	Yang et al. (1986)
PLC	CL	15–17	7.8–8.2		233	25 (60 dah), 10 (76 dah), 0.23 (233 dah)	0.46		1.69	Yang et al. (1983a and b)
PR	CL									Hanlon et al. (1979)
PLC	CL	15–17			100	~2–12% (70 dah)	5.2		0.5–4.5	Hurley (1976)

ED embryonic development, *PR* paralarval rearing, *PLC* partial life cycle, *CT* culture, *CL* closed system, *OP* open system, *dah* days after hatching, *icp* incubation period, *ww* wet weight, *ML* mantle length

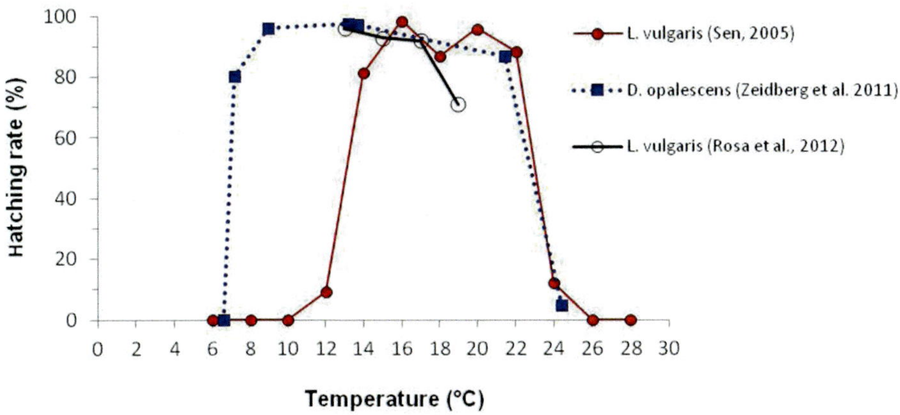

Fig. 16.4 Hatching rate (defined as number of viable hatchlings × 100/number of incubated eggs) versus temperature for *Loligo vulgaris* and *Doryteuthis opalescens*

paralarvae × 100/number of incubated eggs) above 80 % were found within the range of 14–22 °C (Fig. 16.4). Temperatures above and below this range resulted in low hatching rates, deformities and death of embryos. This is consistent with results of Villanueva et al. (2003), who observed irregular statolith growth and low hatching rates from *L. vulgaris* eggs incubated at 12 and 24.7 °C. More recently, Rosa et al. (2012) obtained similar results by incubating early stage eggs from the western coast of Portugal at similar temperatures (13, 15, 17 and 19 °C), but at 34 psu. These authors, however, found relatively lower hatching rates (71 %) and the highest percentage of premature hatchlings at the highest temperature (Fig. 16.4). They have also shown that late-stage embryos were more heat tolerant than hatchlings, obtaining CL50 (temperature at which 50 % of the initial population died) values of 34 and 35 °C from eggs incubated at 13 and 19 °C, respectively. This was partially attributed to the physical protection provided by the egg envelopes.

It is noteworthy that eggs at different developmental stages have different tolerance limits to stress factors, and this should be taken into consideration to avoid misinterpretations. Tolerances are also closely related to habitat, reproductive season and geographic range of species (Kinne 1971), and could uncover seasonal or geographical cohorts characteristics as a result of genetic variability. Off the south coast of Portugal, *L. vulgaris* eggs are found year-round within a temperature range of 13–19 °C (Villa et al. 1997).

D. opalescens eggs collected from southern Monterey Bay seem to be more tolerant than *L. vulgaris* to lower temperatures as hatching rates above 80 % were obtained from eggs incubated at 7.2 °C. In addition, hatching rates above 95 % were recorded at temperatures of 9–14 °C, but decrease sharply outside this range (Fig. 16.4) (Zeidberg et al. 2011). These authors suggested that this is the best range for egg incubation of *D. opalescens* as it coincides with the temperatures commonly experienced by the eggs in the spawning grounds, which are subjected to intermittent, wind-driven, coastal upwelling (Koslow and Allen 2011).

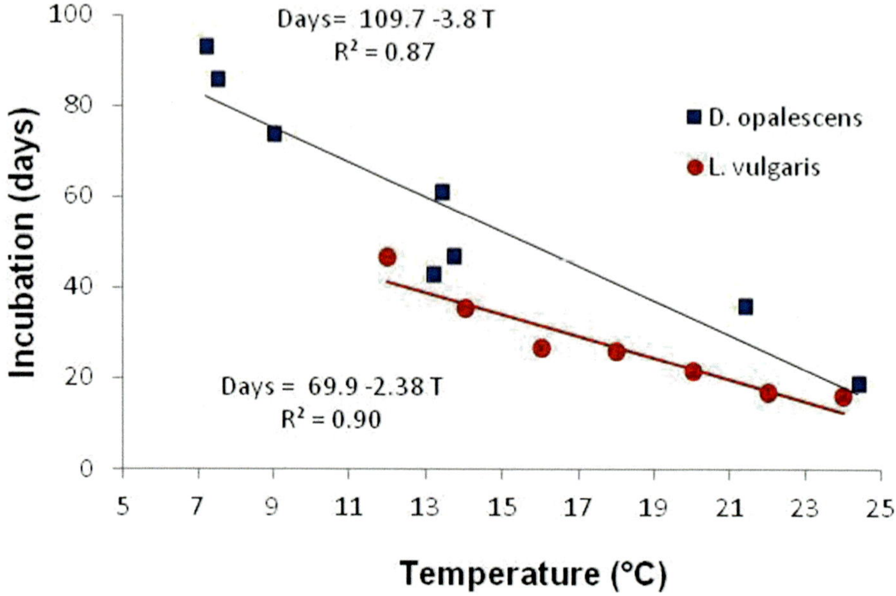

Fig. 16.5 Embryonic developmental time versus temperature for *Loligo vulgaris* and *Doryteuthis opalescens*

For *L. vulgaris*, incubation times at extreme temperatures were recorded for 70 days at 10 °C (Boletzky 1987) and between 13 and 19 days at 24 °C (Sen 2005a). While for *D. opalescens*, incubation times varied between 93 days at 7.2 °C and 19 days at 24.4 °C (Zeidberg et al. 2011). Incubation time is inversely proportional to temperature and egg size in cephalopods (Boletzky 1974a, 1987). Therefore, differences in incubation time suggest that more attention should be given to the exact range of variation in the respective sizes of ova in both species as well as to the exact relationship between the rates of embryonic development at different temperatures and the individual size of ova.

When the data available for both species are plotted to fit a linear relationship between temperature and developmental time, the predicted response shows that each 1 °C increase in temperature will reduce incubation time by approximately 3.8 days for *D. opalescens* and by 2.4 days for *L. vulgaris*, with incubation time turning out to be very similar for both species at the highest temperatures (Fig. 16.5). This information can be very useful to minimize collection efforts and allow that several experimental trials be undertaken from eggs maintained at different temperatures (Boletzky and Hanlon 1983). Although temperature exerts a major effect on incubation time, other factors such as dissolved oxygen (Rosa et al. 2012) and salinity (Sen 2005a) must also play an important role that is likely masked by the dominant effect of temperature.

Very few studies focused attention on the sensitivity of loliginid eggs to changes in salinity. When *L. vulgaris* eggs (stages 15–17), from the Bay of Naples, Italy, were transferred to salinities from 19 to 57 at 18–20 °C, normal development was only observed between 34 and 42 psu (D'Aniello et al. 1989). Sen (2004, 2005b) also evaluated the salinity tolerance of *L. vulgaris* eggs (stages 7–11) from the Izmir Bay, Turkey in two experiments. In the first, egg capsules were introduced at salinities of 0, 28, 31, 34 and 37 psu under a mean temperature of 12.2 °C and in the second, at 32–42 psu (mean temperature of 19.8 °C). The eggs developed normally only within the range of 34–38 psu. These results suggest that salinity tolerance of eggs decreased at lower temperatures, in accordance with the result obtained by Nabhitabhata et al. (2001) for *Sepioteuthis lessoniana*. Nonetheless, in the experiments of Sen (2004, 2005b) eggs were transferred to different salinities without acclimatization and no information on hatchings competence was provided. There is a good evidence that sudden temperature and salinity changes must be avoided, especially at early embryonic stages (D'Aniello et al. 1989; Boletzky and Hanlon 1983).

Changes in pH have the potential to cause a major impact on the physiological aerobic performance of cephalopods (Pörtner and Zielinski 1998), particularly during late embryonic stages. D'Aniello et al. (1989) exposed *L. vulgaris* intermediate stage eggs to pH levels between 7.1 and 9.1, observing no development at these extreme pH values. Eggs developed normally and hatching followed at pH 7.6 and 8.6, but the paralarvae did not survive. Normal egg development and paralarval survival was only found within a pH range of 7.8–8.4. Vidal et al. (2002a) observed that pH below 8.0 produced abnormal development and high mortality of the newly hatched *D. opalescens* paralarvae and recommended an even narrower pH range (8.1–8.4) for egg incubation. More recently, Lacoue-Labarthe et al. (2011) evaluated the effects of ocean acidification on accumulation of trace elements in embryos and hatchlings of *L. vulgaris*. It was shown that the combined effect of low pH on the adsorption and protective properties of the egg capsules during embryogenesis and of high CO_2 partial pressure on the metabolism of embryos and paralarvae resulted in changes to the bioaccumulation of metals. These physiological responses were believed to be a result of ionic and acid–base imbalance.

Eggs are incubated under constant low light intensities (1–6 Lx—Vidal et al. 2002a; 1.5μE m^{-2} s^{-1}—Villanueva 2000b) and also to a variety of manipulated (Paulij et al. 1990; Vidal et al. 2002a, b; Villanueva et al. 2007) and natural photoperiods (Boletzky 1979; D'Aniello et al. 1989). In any event, it is advisable to maintain light at low intensities. Paulij et al. (1990) demonstrated that photoperiodicity has a major influence on the timing of hatching of both *L. vulgaris* and *L. forbesii*. Independent of the timing and duration of the dark period, the transition from light to dark conditions function as a synchronizer of hatching. This synchronization occurs only if light and dark transitions are detected by the embryos, given that hatching could not be stimulated by twilight (<50 μE m^{-2} s^{-1}). Embryos exposed to constant light showed random hatching, but when exposed to a dark shock hatched shortly after the onset of darkness. It was also revealed that an endogenous hatching rhythm can only be preserved after stage 15. Synchronous hatching seems to be very important to increase the chances of hatchlings survival in the wild and could have a foundation on selective intrinsic factors, such as light (L)–dark (D) cycles, although there are no field studies to attest this argument. There seem to be no reports of the negative effect of light on the embryonic development of the loliginid squid. However, reliable evidence suggest that continuous illumination might lengthen the embryogenesis in *Sepia officinalis* by making the duration of the hatching period longer and hatching asynchronous (Paulij et al. 1991). In addition, it was also shown that *L. vulgaris* embryos exposed to constant light conditions produced slower statolith growth when compared to embryos exposed to manipulated summer (16L/8D) or winter (8L/16D) photoperiod (Villanueva et al. 2007). This is consistent with observations made for *D. opalescens* intermediate stage eggs exposed to continuous illumination and to natural photoperiod, but under the same water quality and temperature conditions. Hatchlings from eggs exposed to constant light were smaller and had consumed nearly all their internal yolk reserves, when compared with those under natural photoperiod, suggesting that light might also affect the efficiency of yolk utilization, and consequently, hatchling size and condition (E.A.G. Vidal, personal observation). This requires validation, but it seems imprudent to expose the eggs to constant illumination if the main purpose of the experiment is rearing of paralarvae. Hatchlings exposed to suboptimal or stressed conditions during embryogenesis may be weak, have poor functionality or be more susceptible to infection.

Viability and competence of paralarvae will, to a large extent, depend on the developmental history and environmental influences during egg development. Environmental conditions are considered to play important roles on eggs and larval quality of invertebrates (Benzie 1998). Nonetheless, little progress has been made in elucidating the connections between the influence of environmental factors during embryonic development and the production of high-quality and competent paralarvae for rearing. Studies that focus on evaluating tolerances and conditions during embryonic development should provide information on a fundamental aspect of egg quality and development: paralarval survival and competence. For instance, temperature has a dramatic impact on embryonic development of cephalopods (Boletzky 1987) and both the rate and the efficiency of yolk utilization are temperature dependent (Vidal et al. 2002b, 2005). Eggs of both *L. vulgaris* and *D. opalescens* incubated at lower temperature yield larger hatchlings than those incubated at higher temperatures (Villanueva 2000a; Vidal et al. 2002b). This shows that development at lower temperatures maximizes yolk conversion efficiencies, with the opposite effect observed at high temperatures (Vidal et al. 2002b). The high metabolic demands of late-stage embryos maintained at the highest temperature tolerance limits may impose a substantial constraint on the efficiency of conversion of yolk into tissue. This is coherent with the production of premature hatchlings of *L. vulgaris* when incubated at 24.7 °C (Villanueva et al. 2003) and 19 °C (Rosa et al. 2012). Observations under experimental conditions showed that the yolk reserve at hatching, although variable, is proportional to the body mass, representing around 33–65 % of the body dry weight (dw) of *D. opalescens* paralarvae (Vidal et al. 2002b). Most importantly, the yolk content at hatching is crucial for the survival of hatchlings during the transition from endogenous (yolk) to exogenous (prey) feeding, when prey capture skills are developing and the highest mortality rates are registered.

An essential fact here is the absence of a morphological or physiological 'hatching stage' in cephalopods (D'Aniello et al. 1989; Boletzky 2003). According to Boletzky (2004), hatchlings can be: (1) premature, without complete absorption of the outer yolk sac; (2) normal, after complete absorption of the outer yolk sac; and

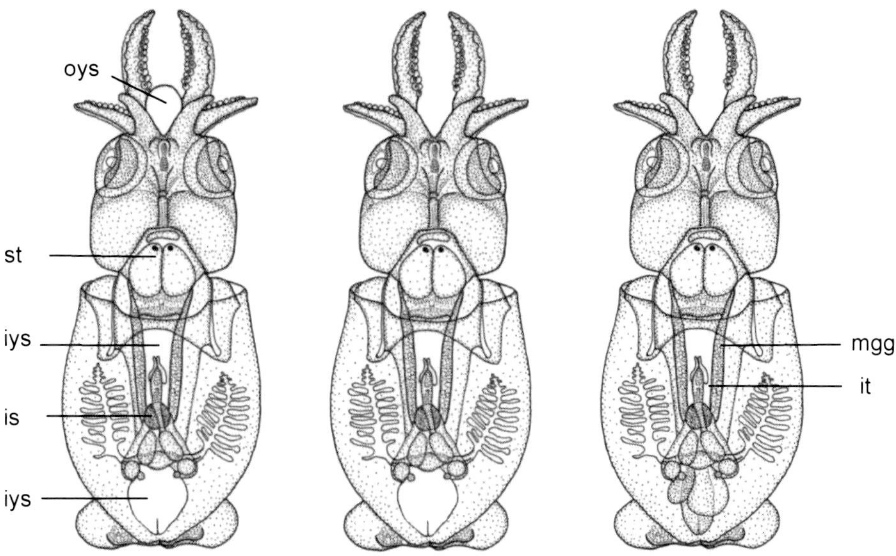

Fig. 16.6 Schematic drawing showing possible hatchling forms in ventral view: **a** Premature. **b** Normal. **c** Late. *oys* outer yolk sack, *mgg* mid gut gland, *iys* inner yolk sack, *st* statocysts, *in* intestine, *is* ink sac. (Adapted from Segawa et al. 1988)

(3) late, when most of the inner yolk sac was absorbed (Fig. 16.6). While development at lower temperatures tends to produce late, larger and likely more developed and competent hatchlings (Vidal et al. 2002b; Villanueva et al. 2003) adding a final time of embryonic 'perfection' (Boletzky 2004), these paralarvae lack the necessary energy reserves to make the transition to successful prey capture. The yolk reserve offers a 'safeguard of considerable adaptive value', allowing hatchlings to deal with an event of temporary food shortage (Boletzky 2010). Consequently, conditions during embryonic development should produce paralarvae with large internal yolk reserves, an advantage under any circumstances. Better knowledge must be achieved to understand the influence of environmental conditions not only on yolk utilization rates during embryonic development but also how these conditions can be manipulated to uphold the production of high-quality and competent paralarvae.

16.6 Paralarval Rearing

Paralarval rearing is the main bottleneck in squid culture. It is a demanding process for which the production of meaningful quantities of healthy specimens requires meeting all environmental conditions from embryo through the paralarval phase. A major breakthrough was the understanding that holding tanks with favourable circulation regimes enhance successful interactions with prey and minimize skin damage due to abrasion that occurs when paralarvae come in contact with tank surfaces (Vidal et al. 2002a). In addition, supplying a variety of live prey of high nutritional quality enhances feeding of the heterogeneously developing hatchlings by encouraging their natural hunting behaviour. The main requirements for paralarvae rearing are listed below.

16.6.1 Tank Design and Circulation Patterns

L. vulgaris and *D. opalescens* paralarvae have been reared in both open and closed seawater systems (Tables 16.1 and 16.2). Boletzky and Hanlon (1983) and Hanlon (1990) address the advantages and disadvantages of each system, with emphasis on filtration, a fundamental element of closed systems.

A tank must provide adequate space to reduce stress of paralarvae and allow them to express normal behaviour. Cylindrical tanks generally are preferable to rectangular tanks, as the corners of the latter create 'dead' zones of poor circulation that make it more difficult to establish suitable circulation for the rearing of paralarvae.

Early attempts to rear *L. vulgaris* hatchlings in a 'bell jar holding about 25 L' were unsuccessful, with a maximum survival of only 1 week (Portmann and Bidder 1928). Hatchlings of this species have been raised in 5-L glass cylinders (25 cm diameter with opaque walls), either in standing seawater, partially renewed twice daily or in slowly running seawater; and in circular 45-L polyvinyl chloride (PVC) tanks (40 cm diameter, 40 cm deep), in slowly running seawater (Boletzky 1974a). An experiment made under similar conditions but with continuous light and slowly decreasing temperatures (18.5–12 °C) developed a film of red and green algae on the inner tank surface. This presented a visual barrier that helped the animals avoid contact with the tank walls (Boletzky 1979), thus preventing injuries, especially abrasion of the outer rim of the fins. Reduced fin size due to abrasion prevents animals from stabilising the forward darting movement needed when attacking prey.

L. vulgaris was also reared by Turk et al. (1986) from eggs shipped in from the Mediterranean, in Galveston (USA) in both natural and artificial seawater in a closed system previously used by Yang et al. (1983a) for *D. opalescens*. To evaluate the effect of temperature on statolith growth, Villanueva (2000a) raised *L. vulgaris* paralarvae in an open system of 13-L cylindrical bags (20 cm diameter, 42 cm deep) with constant water flow (~35 L h^{-1}), using near-natural photoperiod with fluorescent lamps (10.3–3.3 μE m^{-2} s^{-1} luminosity from the top to the middle of the bags).

D. opalescens paralarvae first were reared by Hurley (1976) in a closed system of aerated black cylindrical tanks (48 L). Subsequently, pioneering experiments were conducted in Galveston (Hanlon et al. 1979; Yang et al. 1980, 1983a, b, 1986). The work of Yang et al. (1983a) became the foundation for experiments with loliginids in which paralarval rearing and growth was improved compared to previous trials. These authors used a closed system of two relatively large tanks (1,300 L, 1.8 m diameter, 0.75 m deep, total volume 2,600 L). One was the rearing unit and the other the biofilter. Photoperiod was set at 17L:7D and light intensity varied between 11 and 15 Lx at midwater. Water was processed through the biofilter, a 15-W ultraviolet (UV) sterilizer, a 5-μm polyester filter and activated carbon. Mortality still was high, but an improvement compared to earlier works. This was attributed to greater reduction of skin and fin damage owing to a lower wall-surface-to-volume ratio; improved prey capture with black sidewalls and adequate illumination; and higher water quality from the efficient filtration. The larger tanks, however, required higher current speeds for thorough water mixing. This was later found to decrease survival because paralarvae still were subject to fin and skin damage, as explained below.

For the first 2 months of life in *D. opalescens*, Vidal et al. (2002a) used natural seawater collected from offshore locations, which was maintained at about 16°C and between 30 and 35 psu, and thus obtained high hatchling survival rates. Light intensities were low (<6.0 Lx at the surface, Vidal et al. 2002a) and evenly distributed to avoid patchy distribution of the paralarvae and their prey. Sudden light changes should be avoided as they will set off escape reaction by backward jets making the squids hit the walls of the tanks and causing skin damage (Boletzky and Hanlon 1983). Thus, light conditions can greatly influence swimming and feeding behaviour and consequently promote stress. Homogeneous flat black colour tanks have improve feeding of paralarvae by reducing reflectance of light and increasing the visual contrast of prey organisms (Yang et al. 1983a, b; Vidal et al. 2002a).

The minimum water volume for rearing paralarvae or adults is not known, but a safe practice is to use a larger volume than required (Boletzky and Hanlon 1983). Rearing densities ranged substantially from 0.67 L^{-1} for *D. opalescens* paralarvae (closed system, Yang et al. 1986) to 25 L^{-1} for *L. vulgaris* paralarvae (open system, Villanueva 2000) (Tables 16.1 and 16.2). Best survival has been found from 6.8 to 13 hatchlings L^{-1} (Vidal et al. 2002a).

Vidal et al. (2002a) advanced rearing of *D. opalescens* by reproducing circulation patterns in tanks in which paralarvae had higher survival and lower incidence of fin damage. Tanks were of intermediate diameter (1 m) and smaller height (0.4 m). This provided adequate swimming space. Fine-scale adjustment of the position and intensity of water inflow allowed optimizing current speed to achieve circulation of low speeds (1.0–1.4 cm s^{-1}) and gentle mixing. Higher speeds (>2.0 cm s^{-1}) and greater turbulence were directly responsible for the lowest survival rates of 30-day-old paralarvae. Circulation thus is a crucial design consideration. Properly implemented, it noticeably reduces mortality caused by contact of the fragile paralarvae with the tank walls and also produces a uniform distribution of hatchlings and prey, and this enhances their interaction.

If circulation is too slow, the negatively buoyant paralarvae (Martins et al. 2010) tend to remain near the tank bottom while prey items, such as *Artemia nauplii* and zooplankton, have a patchy distribution near the surface. This segregation of paralarvae and their food source reduces feeding interactions and, as a consequence, both survival and growth. In contrast, the stronger circulation needed for adequate mixing in large-diameter tanks (Yang et al. 1980, 1983a) increases abrasive damage of fins, thus lowering survival.

Optimal circulation and mixing greatly reduce the mechanical stress on the delicate paralarvae and can be achieved in several ways, as for example: (1) placing spray bars near the surface at intermediate angles (40–50°) and (2) introducing water at the tank bottom to generate a gentle upwelling. In either case, one indication of ideal circulation is a relatively smooth water surface (see Vidal et al. 2002a) and long horizontal and vertical displacements of squids away from the walls. A tank-side window permits observation of distribution, feeding and swimming behaviour without disturbing the paralarvae.

The effect of small-scale turbulence on swimming performance and prey capture yet is an unexplored field of paralarvae research. As noted above, current speed and mixing directly affect survival rates (Vidal et al. 2002a). Further, studies of fish larvae demonstrate the negative effect of turbulence in removing prey from a larva's effective pursuit field, thereby reducing capture success (Mackenzie and Kiørboe 2000). As a consequence, the effect of turbulence on encounter rates has been suggested as an explanation for the lack of a straightforward relationship between larval feeding and prey density. Turbulence is also believed to be a critical stressor that severely increases energy expenditure, which also reduces survival. Understanding how small-scale turbulence affects swimming and feeding behaviour is a fundamental step towards improving paralarvae growth and survival. Future efforts should also focus on the influence of tank design on feeding behaviour.

16.6.2 Water Quality

Water of excellent quality is a primary requirement for successful rearing of loliginid paralarvae. Both natural and artificial seawater have been used, but more research is needed on the efficacy of artificial seawater in supplying essential elements for paralarval development (Boletzky and Hanlon 1983).

Water quality in open (or flow-through) systems is managed by continual replacement of culture water with fresh seawater. Though effective, this may afford little control over specific water-quality parameters; and this, in turn, can make it more difficult to resolve low quality, contamination and pollution problems that jeopardize rearing.

In closed (or recirculation) systems, water quality is managed by implementing mechanical, physical adsorption, disinfection and biological processes. The main parameters that require control are particulate organic matter (POM), nitrogenous wastes, dissolved gases (oxygen, carbon dioxide and nitrogen), pH, alkalinity and pathogens. Each is discussed briefly below. Detailed discussions on the theory and application of these processes are found in Spotte (1979) and Losordo et al. (2001).

16.6.2.1 Particulate Organic Matter

The POM load is expected to be very low for paralarvae rearing. Yang et al. (1983a) estimated POM at < 1.0 g m^{-3}, consisting mainly of dead paralarvae and prey. Denser POM sinks to the bottom of the rearing tanks and is removed easily by siphoning.

Suspended POM is removed by mechanical filtration, such as employing a fine-meshed (50 μm) screen in the biofilter inflow stream. Vidal et al. (2002a) found that this had a considerable impact on reducing pH and nitrogenous compounds variations and maintaining a stable and high-quality rearing environment for *D. opalescens* paralarvae. Dissolved organic matter (DOM) is not a problem when POM is removed quickly. If, however, DOM increases owing to some malfunction in the system, it can be reduced effectively by protein skimmers operated outside of the rearing tanks.

16.6.2.2 Nitrogenous Wastes

Potentially deleterious inorganic nitrogen compounds are un-ionized ammonia (NH_3-N), nitrite (NO_2-N), nitrate (NO_3-N) and dissolved nitrogen gas (N_2). Ammonia management deserves special attention in closed systems because its un-ionized form (NH_3-N) is particularly toxic to many marine species, even at very low concentrations. Together with its ionized form, ammonium (NH_4^+-N), the two are referred to as total ammonia nitrogen (TAN) and exist in an equilibrium determined by water temperature, salinity, pH and pressure. Experiments at high temperature and (especially) high pH must thus monitor NH_3-N closely. Ammonia is produced by heterotrophic excretion and so increases, although not linearly, with stock ingestion rate and feed protein content. Thus, at higher stocking densities, more feed will be ingested and more ammonia will be produced. Excess feed, faeces and dead animals not removed from the rearing unit will be mineralized by heterotopic microorganisms, which will add to the ammonia load.

Tolerance of loliginid embryos and paralarvae to nitrogen species is largely unknown, so current guidelines follow those for most marine invertebrates: Un-ionized ammonia should not exceed 0.1–0.05 mg L^{-1} (Timmons and Ebeling 2007); nitrite-N, <0.10 mg L^{-1}; and nitrate-N, <20 mg L^{-1} (Spotte 1979; Yang et al. 1989).

It is also very important to consider the system's filter capability, which should have a surface-area-to-volume ratio as large as possible to offset total animal biomass. Recommendations for loliginids suggest a ratio >10 for small biomass loads, as in the case of paralarvae (Yang et al. 1989). In a robust biofilter, ammonia is converted to NO_2-N that is oxidized to NO_3-N. This NO_3-N may accumulate to relatively high levels, but it eventually must be removed by water exchange. Care also must be taken with oxygen levels, as a low oxygen concentration leads to insufficient biological oxidation and subsequent build-up of toxic NO_2-N. Even with good filtration, water exchange must be done periodically to ensure proper types and concentrations of trace elements, as these have important metabolic functions in growth and survival of squid hatchlings (Boletzky and Hanlon 1983; Yang et al. 1983a).

16.6.2.3 Dissolved Gases

In loliginids, oxygen demand increases considerably after hatching because of the highly active paralarvae as shown for *L. vulgaris* (Rosa et al. 2012). Feeding, respiration and excretion of paralarvae and their prey all impact water quality. Dissolved oxygen levels thus should be maintained at saturation at all times (Boletzky and Hanlon 1983). Oxygen solubility depends on temperature, salinity and the partial pressure gradient across the air–water interface (Appleford et al. 2003). Saturation is lower at high temperature, exactly when oxygen demands by respiration, feeding and ammonia excretion increase. Therefore, when rearing at high temperatures, continual monitoring of oxygen levels is very important, as hatchlings must be able to extract enough oxygen to match their higher metabolic demands. Special attention must be paid at higher stocking densities.

Air bubbles (especially extra-fine bubbles) must be avoided in rearing tanks because they adhere to the skin and are trapped inside the mantle, causing entrapment of the hatchlings in the surface tension at the air–water interface. This impairs swimming and prey capture, which increases mortality of paralarvae (E.A.G. Vidal, personal observation). If artificial aeration is used, it should be in a separate treatment tank and not in the rearing tank (Boletzky and Hanlon 1983). Gas bubble disease can be a problem in some systems. It is caused by the supersaturation of total dissolved gasses: oxygen, carbon dioxide, nitrogen and argon. This can result when injecting high-pressure air at depth; when water near saturation is heated; or when an outdoor system near supersaturation is exposed to a low-pressure weather front. The gas bubbles so formed can adhere to hatchlings, hampering mobility in a way that can increase mortality. In general, total gas saturation should not exceed about 105 % and is managed by degassing.

16.6.2.4 pH

pH is a critical parameter that plays a role in determining the concentrations of other key parameters, such as the aforementioned toxic nitrogen species. As with eggs, paralarvae are very sensitive to pH changes and it was shown that low pH values might affect the accumulation of essential and nonessential metals in *L. vulgaris* paralarvae due to the disturbance of ionic and acid–base physiological processes (Lacoue-Labarthe et al. 2011). The ideal pH range is 8.1–8.4 (Vidal et al. 2002a), similar to the typical values in their natural oceanic habitat.

16.6.2.5 Disinfection and Pathogens

The seawater used for rearing should pass through UV sterilization to reduce levels of pathogenic bacteria. However, filtration to remove suspended particles should be done before the utilization of the UV sterilizer as POM is the major substrate for bacteria in recirculation systems (Appleford et al. 2003). UV sterilization has been useful in minimizing bacterial infection (Hanlon 1990; Vidal et al. 2002a) and should be employed during the rearing of paralarvae.

The surfaces of eggs and larvae are well known to be good substrates for bacterial colonization (Bergh et al. 1992) and paralarvae stressed by suboptimal conditions are more subject to skin damage and infection. A particularly promising line-of-research is the interaction between paralarvae and bacteria, particularly the ability of paralarvae to withstand pathogens.

16.6.2.6 Temperature

L. vulgaris paralarvae have been reared from 11 to 22°C (Boletzky 1974a; Villanueva 2000a) and *D. opalescens* from 12 to 18°C (Vidal et al. 2002b; Chen et al. 1996) (Tables 16.1 and 16.2). Interestingly, *L. vulgaris* hatchlings showed lower thermal tolerance than the embryos. This was attributed mainly to the greater oxygen demand of the highly active hatchlings. In addition, thermal tolerance of hatchlings increased with temperature, with LT50 (temperature at which 50% of hatchlings had died) values of 31 and 33°C from hatchlings incubated at 13 and 19°C, respectively (Rosa et al. 2012).

Despite our current deficiencies in establishing realistic criteria on tolerance and chronic toxicity limits for loliginid hatchlings, as indicated above, available information provides essential features for the main factors required in successful rearing.

16.6.3 Food and Feeding

In the first days after hatching (dah), paralarvae combine both endogenous (yolk) and exogenous (prey) energy sources (Boletzky 1975; Vidal et al. 2002b). The yolk content at the moment of hatching although highly variable seems to be proportional to the body mass, representing 33–63% of body dw and 10–18% of body wet weight (ww) of paralarvae. The time required to exhaust yolk reserves decreases exponentially with increasing temperature. For example, *D. opalescens* hatchlings can survive longer (6 days) on their yolk reserve at 12°C than at 16°C (4 days; with 80% mortality, Vidal et al. 2002b). Clearly, the yolk reserve must fuel metabolism while the prey capture skills of paralarvae are still developing. This implies that the 'window of opportunity' is very short and hatchlings must quickly capture sufficient food to meet their high metabolic demands (Vidal et al. 2002b; Rosa et al. 2012). Accordingly, it was shown that early *D. opalescens* paralarvae are unable to withstand even short periods of food deprivation, being extremely sensitive to starvation (Vidal et al. 2006). This is consistent with the high mortality rates observed during the first 10 dah for both *L. vulgaris* and *D. opalescens* that delimits the critical transition period from yolk absorption to successful prey capture (Vecchione 1987; Villanueva 2000a; Vidal et al. 2002b).

Extra care should be taken to provide food of adequate type and size, high nutritional quality and in sufficient quantities to the newly hatched paralarvae. Delaying first feeding will most certainly increase mortality due to energy deficit and starvation. It has been recognized that the first prey items a hatchling encounters are decisive for the immediate predatory success or failure and the subsequent neurophysiologic development and learning processes in the young predator (Chen et al. 1996; Preuss and Gilly 2000). The catchability of such prey may be defined in relation to its escape response, its defensive equipment (e.g. spiny integuments) acting after seizure, and ultimately its resistance to attempted ingestion (hard carapaces, etc.).

Preferably, paralarvae are fed on a variety of live foods (Vidal et al. 2002a). The prey types used with success as food for rearing *L. vulgaris* and *D. opalescens* paralarvae are *Artemia nauplii*, metanauplii and adults, decapod crab zoeae, copepods, mysid shrimp, fish larvae and other zooplankton organisms (Fig. 16.7) (Tables 16.3 and 16.4). Hatchlings of both *L. vulgaris* and *D. opalescens* perform external predigestion and ingest only the flesh of their crustacean prey (Boletzky 1974b; Franco-Santos and Vidal in press). Their beaks are denticulated (Boletzky 1971), colourless and show a conspicuous slit, features that seems to be an adaptation to ease the ingestion of predigested flesh.

One of the most demanding chores in rearing squid paralarvae is to provide food in the necessary quantity, quality and reliability. Cultured feed organisms such as *Artemia*, copepods, mysids and decapod crab zoeae (especially those of hermit crabs that can be kept in culture for long-term production of larvae; Villanueva 1994) have been demonstrated to represent a good standard food for many cephalopod hatchlings, including loliginid squids, and are very important to warrant a steady supply of food as natural zooplankton supply are unreliable.

Paralarvae can capture and ingest prey of a very wide size range including prey much larger than their own size (Fig. 16.7; Hanlon 1990; Yang et al. 1986). However, as size at hatching shows high individual variability, the ability of paralarvae to capture prey of different sizes and swimming capabilities also vary considerably (Vidal et al. 2002a).

D. opalescens hatchlings have limited swimming ability and prey capture skills (Chen et al. 1996; Preuss et al. 1997), and feed on relatively smaller prey, such as *Artemia nauplii*, zoeae and copepods. Although some paralarvae are also able to capture large prey, such as early juvenile mysids, it is critical to supply smaller and 'easy to catch' prey during the first 2 weeks after hatching (Tables 16.3 and 16.4). Later, feeding a variety of prey of different sizes and types are important so as to match the different sizes and hunting abilities of same-aged but heterogeneously developing squid. When paralarvae reach 40–50 days of age (6–10 mm mantle length, ML) and start to swim in schools, they require large and more energetically rewarding prey types, such as adult mysids, large zoeae, shrimp mysis and fish larvae (Yang et al. 1983a; Vidal et al. 2002a).

It is perhaps significant that feeding interactions are often observed between same-aged but different sized paralarvae during the first month after hatching and before formation of schools. As large paralarvae are better fit to capture larger prey, which cannot be enclosed within the arms, other paralarvae (usually smaller) can eventually attack and feed on the same prey item (E.A.G. Vidal, personal observation) (Fig. 16.8). There seems to be a preference for attacking large prey already subdued by another paralarvae instead of a free prey, even when prey are abundant.

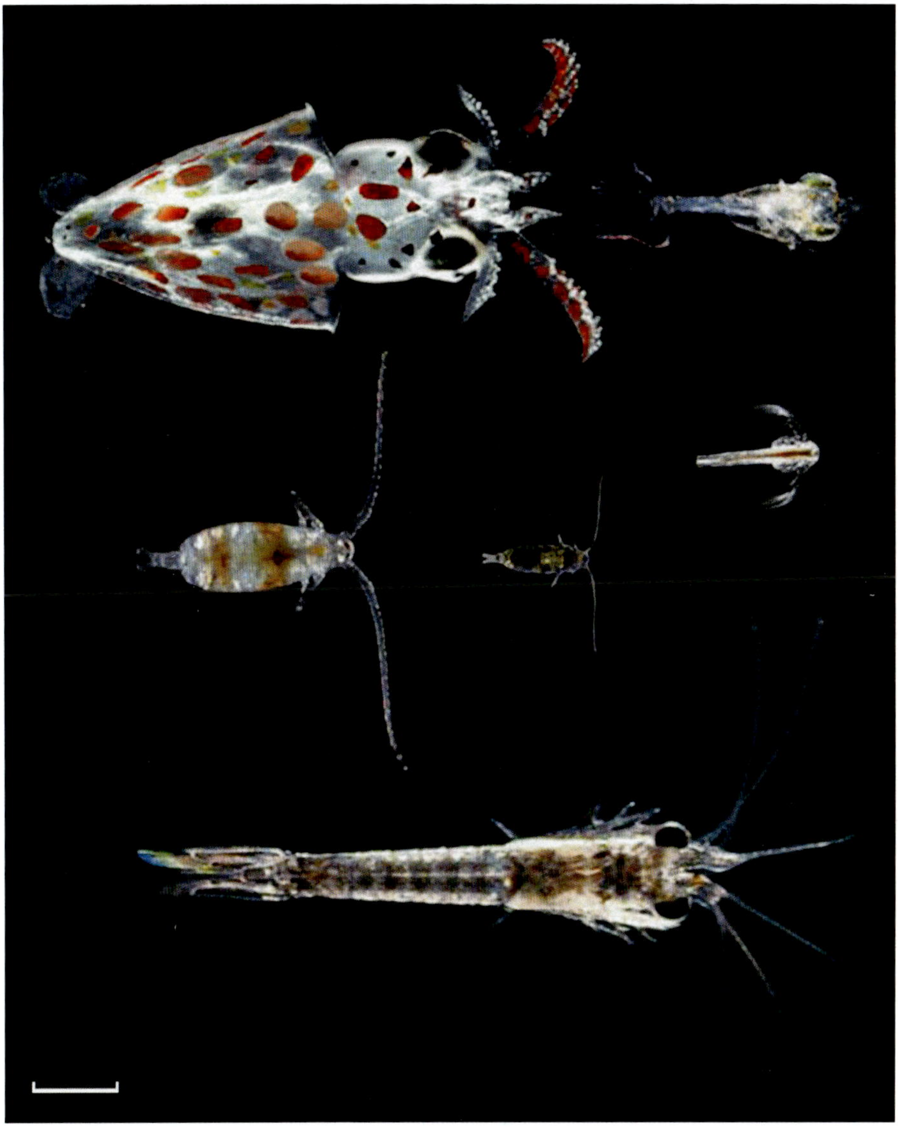

Fig. 16.7 Relative size of *Doryteuthis opalescens* hatchling and some prey types offered as food during rearing (Vidal et al. 2002a). From *top right*, decapod crab zoea (photo R Villanueva), *Artemia metanauplii*, *Acartia tonsa*, Pontelid copepod (*left*) and *Americamysis almyra*. Scale bar = 1 mm. (Original image)

This behaviour was first reported by Hurley (1976) and allows smaller hatchlings to feed on prey that they could not subdue alone and that a large squid could not possibly ingest fully. Whether this is a natural behaviour or is induced under rearing conditions due to high stocking densities needs to be evaluated.

Table 16.3 Live prey composition, size and developmental stage offered to *Loligo vulgaris* paralarvae and juveniles during rearing

Taxa	Species	Developmental stage	Size (mm)	System type	Reference
Phylum Arthropoda					
Class Crustacea					
Subclass Branchiopoda					
	Artemia spp.	Metanauplii	–	RT	Boletzky (1971, 1974, 1979)
Subclass Copepoda	*Acartia tonsa*	Adults	0.8–1.2	RT	Turk et al. (1986)
	Anomalocera ornata	Adults	2.0–3.0	RT	Turk et al. (1986)
	Centropages velificatus	Adults	1.4–1.8	RT	Turk et al. (1986)
	Labidocera aestiva	Adults	1.5–2.5	RT	Turk et al. (1986)
	Temora longicornis	Adults			Portmann and Bidder (1928)
Subclass Malacostraca					
Order Mysidacea					
	Americamysis almyra	Juveniles and adults	2.0–11.0	RT	Turk et al. (1986)
	Leptomysis mediterranea	Adults	5.0–15.0	RT	Boletzky (1979)
		–	–	RT	Villanueva (1994)
Order Decapoda					
	Pagurus prideaux	Zoea	2.8–3.3	RT	Villanueva (1994)
	Dardanus arrosor	Zoea	2.8–3.2	RT	Villanueva (1994)
	Palaemonetes spp.	Larvae, post-larvae	2.2–2.8	RT, RC	Turk et al. (1986)
Phylum Chordata					
Class Osteichthyes					
	Cyprinodon variegatus		10–28.0	RC	Turk et al. (1986)
	Fundulus variegatus		15–31.0	RC	Turk et al. (1986)
	Gambusia affinis		12–28.0	RC	Turk et al. (1986)
	Menidia beryllina		18–52.0	RC	Turk et al. (1986)
	Poecilia latipinna		22–41.0	RC	Turk et al. (1986)

RT rearing tanks, *RC* raceways

Table 16.4 Live prey composition, size and developmental stage offered to *Doryteuthis opalescens* paralarvae, juveniles and adults during rearing

Taxa	Species	Developmental stage	Size (mm)	System type	Reference
Phylum Arthropoda					
Class Crustacea					
Subclass Branchiopoda					
	Artemia spp.	Nauplii	~0.3–0.7	RT	Hurley (1976); Hanlon et al. (1979); Yang et al. (1983a, 1986); Vidal et al. (2002a)[a]
		Metanauplii and adults	2–5	RT	Hurley (1976); Hanlon et al. (1979)
Order Cladocera	*Evadne sp.*	Adults	0.6–1.0	RT	Vidal et al. (2002a)
Subclass Cirripedia		Nauplii	0.2–0.5	RT	Vidal et al. (2002a)
Subclass Copepoda	*Acartia tonsa*	Nauplii and copepodites	0.2–0.5	RT	Vidal et al. (2002a)
		Adults	0.8–1.2	RT	Yang et al. (1983a); Vidal et al. (2002a)
	Acartia lilljeborgi	Adults	1.0–1.3	RT	Vidal et al. (2002a)
	Anomalocera ornata	Adults	2.0–3.0	RT	Hanlon et al. (1979); Yang et al. (1983a); Vidal et al. (2002a)
	Calanopia sp.	Adults	1.2–1.5	RT	Vidal et al. (2002a)
	Centropages velificatus	Adults	1.4–1.8	RT	Vidal et al. (2002a)
	Corycaeus spp.	Adults	0.8–1.1	RT	Vidal et al. (2002a)
	Eucalanus hyalinus	Adults	6.0–6.5	RT	Hanlon et al. (1979)
	Euterpina acutifrons	Adults	0.5–0.7	RT	Vidal et al. (2002a)
	Labidocera aestiva	Copepodites	0.8–1.5	RT	Vidal et al. (2002a)
		Adults	1.5–3.5	RT	Hanlon et al. (1979); Yang et al. (1983a); Vidal et al. (2002a)
	Paracalanus spp.	Adults	0.5–1.0		Vidal et al. (2002a)
	Pontella spp.	Adults	3.3–4.1	RT	Hanlon et al. (1979); Vidal et al. (2002a)
	Subeucalanus pileatus	Copepodites	1.7–2.1	RT	Hanlon et al. (1979)
	Temora styilifera	Copepodites	0.4–0.6	RT	Vidal et al. (2002a)
		Adults	1.2–1.6		Vidal et al. (2002a)
	Temora turbinata	Nauplii and copepodites	0.3–0.7	RT	Vidal et al. (2002a)
		Adults	1.1–1.6	RT	Vidal et al. (2002a)
		Nauplii	0.3–0.7	RT	Vidal et al. (2002a)
Subclass Malacostraca					

Table 16.4 (continued)

Taxa	Species	Developmental stage	Size (mm)	System type	Reference
Order Mysidacea					
	Americamysis almyra	Juveniles and adults	2.0–11.0	RT	Hanlon et al. (1979); Yang et al. (1983a); Vidal et al. (2002a)
Order Decapoda					
		Zoeae	7.0–10.0	RT	Vidal et al. (2002a)
		Zoeae and myses	2.0–5.0	RT	Vidal et al. (2002a)
	Palaemonetes spp.		1.5–25.0	RT, RC	Yang et al. (1983a, 1986)
	Penaeus duorarum	Zyses and post-larvae	2.5 –6.0	RT, RC	Yang et al. (1983a, 1986)
Phylum Chaetognatha			9.0–13.0	RT	Vidal et al. (2002a)
Phylum Chordata					
Class Osteichthyes					
	Anchoa mitchilli		20–25.0	RC	Yang et al. (1986)
	Adinia xenica		–	RC	Yang et al. (1986)
	Brevoortia spp.		15–31.0	RC	Yang et al. (1986)
	Cyprinodon variegatus		10–28.0	RC	Yang et al. (1986)
	Eucinostomus gula		–	RC	Yang et al. (1986)
	Fundulus variegatus		15–31.0	RC	Yang et al. (1986)
	Gambusia affinis		12–28.0	RC	Yang et al. (1986)
	Hemicaranx amblyrhynchus			RC	Yang et al. (1986)
	Lagodon rhomboids		–	RC	Yang et al. (1986)
	Menidia beryllina		18–52.0	RC	Yang et al. (1986)
	Mugil spp.		18–38.0	RC	Yang et al. (1986)
	Poecilia latipinna		22–41.0	RC	Yang et al. (1986)
	Pogonias cromis		10–15.0	RC	Yang et al. (1986)
	Sciaenops ocellatus		1.5–14.5	RC	Yang et al. (1986)
	Scomber japonicus			RT	Hurley (1976)

RT rearing tanks, *RC* raceways

[a] Nauplii enriched with highly unsaturated fatty acids (HUFAs)

Fig. 16.8 *Doryteuthis opalescens* paralarvae. Thirty-day-old paralarvae feeding on a juvenile mysid (*Americamysis almyra*). The largest paralarva (*top right*) captured the mysid and the other two smaller paralarvae attacked the same prey after its capture. Scale bar = 2 mm. (Original image)

Indeed, the key mechanisms by which paralarvae select their prey persist as an area of much needed research. Prey type and swimming pattern play an important role in prey capture success (Hanlon 1990). Successful prey capture increases with age and experience (Chen et al. 1996), and experience is enhanced under high prey density conditions. However, little has been reported on the ideal prey densities for rearing loginid squid paralarvae.

Villanueva (1994) exposed *L. vulgaris* paralarvae to prey densities from 100 to 300 zoeae L^{-1} after addition of food and with two to six feeding per day. Yang et al. (1983a, 1986) provided information on the number of different prey organisms fed to *D. opalescens* paralarvac per litre per day and reported prey densities values of nearly 0–60 prey L^{-1}. Vidal et al. (2002a) recommended as minimum prey density as 50 prey L^{-1} and exposed paralarvae to prey densities of 50–150 L^{-1} in four to five feeding per day during the first 2 months after hatching. Increasing prey density was proven to increase encounters between squid and prey, improving feeding and survival.

Even when prey density is already high, the addition of food stimulates the predatory behaviour of paralarvae. Thus, several small feedings during the day at consistent time intervals can improve feeding rate when compared to a single feeding, even if the same amount of food is offered. When prey organisms are added to the tanks, they are patchily distributed, representing visual, probably chemical and mechanical stimuli that instigate attack responses (Vidal et al. 2002a).

Care has to be taken not to overfeed, as the addition of excess food can reduce subsequent feedings and uneaten prey remaining in the tanks will have reduced nutritional quality if no food is provided for them, thereby decreasing survival (Vidal et al. 2002a). This is especially true for the enriched *Artemia nauplii*, which will contain significantly less nutritional value if not eaten within a few hours of being enriched. Accumulation of low-quality prey items in the rearing tanks can be avoided by slightly adjusting prey density at every feeding. In addition, prey density should increase gradually as paralarvae grow at high temperatures as well, when feeding intake needs to be higher to support the metabolic demand. Hurley (1976) has estimated a feeding rate of 35–80% ww day^{-1} for *D. opalescens* paralarvae reared at 15–17°C.

While prey availability is essential, prey quality is just as important for paralarval growth. The vital importance of highly unsaturated fatty acids (HUFAs) in the nutrition of marine fish larvae and cephalopod paralarvae is well established (Sargent 1995; Navarro and Villanueva 2000, 2003). As in the case of fish larvae, squid paralarvae also require food rich in (n-3) HUFA (eicosapentaenoic acid—EPA and docosahexaenoic acid—DHA). The main lipid classes and the fatty acid composition of the newly hatched *L. vulgaris* were determined by Navarro and Villanueva (2000). These authors found that paralarvae are rich in (n-3) HUFA, but particularly low in (n-6) HUFA, having a very high n-3/n-6 HUFA ratio. They further concluded that hatchlings need food rich in (n-3) HUFA, phospholipids and cholesterol and with a reasonable content of neutral lipids.

Thus, the inclusion of prey with high level of (n-3) HUFA, mainly DHA and high ratios of DHA/EPA is very important in determining dietary value for paralarvae. Natural preys, such as mysids, decapod crustacean zoeae and marine copepods (Drillet et al. 2006), have a fatty acid prolife very rich in (n-3) HUFA. Their DHA and EPA content are also very high with a ratio of 1:1 (Navarro and Villanueva 2000). On the other hand, it is well known that *Artemia* is deficient in long chain (n-3) HUFA, suggesting that it can cause a nutritional imbalance in the fatty acid profile (i.e. DHA/EPA ratio) of paralarvae, affecting their survival and growth (Navarro et al. 1992, Navarro and Villanueva 2000). This implies that to provide better nourishment as live food for paralarvae, *Artemia* require HUFA enrichment. However, recent studies have shown that the supply of juvenile *Artemia* containing higher DHA levels was unsuccessful to promote higher survival and growth of *Octopus vulgaris* paralarvae during rearing (Seixas et al. 2010). These authors have found that the protein/lipid ratio, on the other hand, seemed to have a more important effect on the paralarval growth and survival. This seems meaningful if we consider that cephalopods metabolism is basically amino acid and protein driven (Lee 1994). Paralarvae need to synthesize protein at high rates to support their high growth rates (Vidal et al. 2006). In fact, high levels of the amino acids, lysine, leonine and argentine have been found in *L. vulgaris* paralarvae, suggesting that they could be limiting essential amino acids in their diets (Villanueva et al. 2004).

Other studies on the elemental composition of hatchlings and wild juveniles of *L. vulgaris*, and on both their natural and artificial prey strongly suggest that they must demand a food rich in cooper. This is likely related with the haemocyanin requirements for oxygen transport. In addition, S, Na, K, P and Mg were the main elements found in hatchlings and juveniles (Villanueva and Bustamante 2006). These studies underscore the importance of elucidating ontogenetic size-specific nutritional requirements of paralarvae to improve feeding and should be ranked as a high-priority research area.

16.6.4 Survival

L. vulgaris hatched from egg masses in the laboratory were raised by Boletzky (1974a) to a maximum of 45 days and subsequently to 70 days of age (Boletzky 1979) (Table 16.1). Turk et al. (1986) reared *L. vulgaris*, hatched from eggs shipped in from the Mediterranean, in Galveston, recording high mortality rates (90 %) in the first 10 dah, with the last surviving squid reaching 140 days of age (Table 16.1). To assess the effect of temperature on growth of paralarvae and their statoliths, Villanueva (2000a) reared this species up to 62 and 50 dah at a mean temperature of 11 and 19 °C, respectively. Survival rates ranged from 0.2 to 1.4 % and were likely affected by the tetracycline staining procedure used.

Hurley (1976) and Hanlon et al. (1979) were the first to rear *D. opalescens* to a maximum age of 100 and 79 days, respectively, with survival rates of 1–12 % at 70 dah (Table 16.2). Subsequently, best results were obtained by Yang et al. (1980, 1983a, b, 1986) with survival rates of 25 % at 60 dah. To study the development of circular mantle muscles and the giant axon system control over both prey capture and escape behaviour of paralarvae, Preuss et al. (1997) and Preuss and Gilly (2000) raised *D. opalescens* paralarvae up to 60 dah, but no information is given on the survival rates.

By refining rearing methodology, Vidal et al. (2002a) were able to obtain survival rates of 70–78 % at day 10 and of 42–60 % during the first 60 dah (Table 16.2). This was a step forward when compared with the best previous results of only 25 % during the same period, thus offering new possibilities for squid culture. The main factors that contribute to the high survival rates obtained were attributed mainly to: (1) stabilising water quality parameters; (2) offering a combination of live prey types of different sizes and including enriched *Artemia nauplii* to overcome problems of prey availability; (3) keeping prey density above 50 prey L^{-1} at all times; (4) feeding small amounts of food at regular intervals; and (5) reducing turbulence and maintaining current speed inside the tanks between 1.0 and 1.4 cm s^{-1}.

A common pattern emerges from the aforementioned experiments, showing that the highest mortality rates occur during the first 10–15 dah and are primarily a consequence of the critical transition between full yolk absorption and successful prey capture (Hanlon 1987; Villanueva 2000a; Vidal et al. 2002a, b). Survival level after this period and thereafter remains relatively steady until day 45–70 (Hanlon 1987;

Vidal et al. 2002a), when it again decreases. This second peak in mortality could be a consequence of inadequate diet due to the lack of appropriate sized food and a need for larger tanks. By this age, squid have a mean ML of about 6–11 mm and require higher daily food intake, larger prey types and more swimming space as they start swimming in schools.

A reproducible high rate of survival and growth of paralarvae is necessary to achieve successful culture. Reliable methods to assess paralarvae quality should be developed for use soon after hatching and allow comparison of results. A good measure of paralarvae competence seems to be their ability to withstand short periods of starvation. It has been shown that starvation elicits a strong selection for inherently faster growing and stronger paralarvae (Vidal et al. 2006). Behaviour may also be an effective tool for the evaluation of paralarval quality. Weak or abnormal hatchlings will display irregular behavioural patterns.

16.6.5 Growth

It has been shown experimentally that there are two distinct growth phases during the early development of loliginid squid: (1) the no net growth phase and (2) the exponential growth phase. The no net growth phase comprises a negative and a positive growth sub-phases. In the first, a decrease in body weight of hatchlings is observed (from 20 to 30 %; Vidal et al. 2002b) because the yolk is being consumed at exponential rates and there is little, if any, conversion of yolky matter into somatic tissue due to the high metabolic cost of swimming and maintenance (Vidal et al. 2002b; Rosa et al. 2012). In the positive sub-phase, the weight loss is regained over the next few days by prey capture, resulting in a very short phase where no net growth takes place. As a consequence, at the end of this no net growth phase, the body weight of paralarvae is the same as at hatching. The duration of this phase is temperature dependent and indicates the time required to recover the weight loss due to yolk utilization. For example, for *D. opalescens* the no net growth phase lasted 7 days at 16 °C and 15 days at 12 °C. This early growth pattern has also been observed in *L. vulgaris* (Villanueva 2000a) and *L. reynaudii* (Vidal et al. 2005) reared in the laboratory.

Interestingly, this short phase where no net body growth occur has a reflection in the growth of the statoliths as demonstrated by Villanueva et al. (2007). Statoliths growth rates of newly hatched *L. vulgaris* showed a conspicuous decrease of more than half in comparison with those of pre-hatching embryos, clearly indicating the negative growth sub-phase as a result of the exponential rate of yolk utilization.

This no net growth phase set the limits of a critical period in the early life history of squid that correspond to the transition from endogenous to exogenous feeding and is mainly caused by the constrains imposed by the high metabolic demands of the active swimming mode of hatchlings (Rosa et al. 2012) and their short-lasting energy reserves to overcome metabolic suppression and starvation (Vidal et al. 2002b, 2006). The high mortalities that occur during this critical period are due to the inability of hatchlings to withstand food deprivation and to capture prey in the required amounts for daily maintenance and growth. This reveals a short initial delay in growth of paralarvae, which are subjected to an intense and effective selection for successful feeders. The survivors shortly attain the exponential growth rates that are the fundamental hallmark of most cephalopod species (Forsythe and Van Heukelem 1987).

Temperature exerts an instantaneous and extreme effect on growth rate of cephalopods and this is particularly true during early life when the highest growth rates are registered (Tables 16.1 and 16.2). The influences of temperature on metabolic rates govern growth and food conversion efficiencies. Activity or stress may impact growth at different temperatures as a result of increased energy demand. Nevertheless, growth rates should increase up to an optimum with temperature, above which energy requirements will be too high to maintain feed conversion and other metabolic processes and growth rates will be reduced.

While it is useful to determine optimum temperatures for growth, it is also very important to know the tolerance limits. This is because the range of temperature (and salinity) tolerances for growth is usually narrower than that for survival. Salinity also influences thermal response: optimal temperature is lower at lower salinities, and this naturally affects development and growth rates, as observed for the hatching rates of *S. officinalis* (Palmegiano and D'Apote 1983) and *L. vulgaris* (Sen 2004, 2005a). It is important to emphasize that ideal temperatures for growth vary with life stage and many biological (e.g. diet, crowding) and abiotic factors (e.g. salinity, light), as well as, their many interactions.

Fed ad libitum in the laboratory, hatchlings of *D. opalescens* showed exponential growth rates during the first 20 dah that were more than twofold higher at 16°C (7.4% body ww day^{-1}) than at 12 °C (3.3% body ww day^{-1}; Vidal 2002b). As a consequence, small squid hatching at high temperatures (2.5 mm ML) soon attain larger sizes than larger squid (3.0 mm ML) hatching at lower temperatures. Growth in size during the same period was 2.4% ML day^{-1} (Tables 16.1 and 16.2).

By day 15, at the higher temperature, paralarvae were almost threefold heavier than the newly hatched squid and at day 60, squid were growing more rapidly in length and in mass, as paralarvae were transforming to the arrow-shaped adult form. Mean body dw reached 16.5 mg and mean ML 14 mm. Thus, during the first 60 days of life, *D. opalescens* hatchlings double their mean dw five times and mean ML twice, every 12 and 31 days, respectively (Vidal et al. 2002b; E.A.G.Vidal unpublished data). Hurley (1976) obtained highly variable growth rates (0.5–4.5 mm ML day^{-1}) rearing *D. opalescens* for the first 3 months after hatching on a diet composed of *Artemia nauplii* and adults, while Yang et al. (1983a, 1986) obtained exponential growth rates of 5.6 to 8.4 ww day^{-1} and 2.1 ML day^{-1} for *D. opalescens* during the first 2 months of life at 15 °C (Table 16.2).

Very similar results were obtained by Villanueva (2000a) for *L. vulgaris* reared at 11 and 19 °C for 62–50 dah. Growth rates were 3.6 ww day^{-1} and 1.3 mm ML day^{-1} at 11°C and 7.8 ww day^{-1} and 2.6 mm ML day^{-1} at 19 °C (Table 16.1). Paralarvae subjected to the higher temperature grew faster and showed final mean weight five times higher than those reared at the lower temperature. At the coldest temperature,

L. vulgaris paralarvae doubled their ww at every 19 days; as a result, after a period of 50 days they attained 32.7 mg and 6.3 mm ML. However, at the warmest temperature, they doubled their ww at every 9 days, attaining 157 mg and 11.7 mm ML at the same time period (Villanueva 2000a).

Recent studies on the effects of starvation and recovery on survival, growth and RNA/DNA ratio of *D. opalescens* paralarvae have demonstrated that food availability is an important regulator of growth and survival. Paralarvae showed substantial growth plasticity in response to food availability in a very short period of time (10 days). In addition, paralarvae subjected to 2–3 days of starvation were able to increase growth rates above that of the continuously fed squid, showing compensatory growth that mitigated the effects of starvation. Noticeably, starvation caused a strong selection for faster growing squid, suggesting that discontinuous feeding regimes would perhaps promote better growth to closer approximation of the natural environment (Vidal et al. 2006).

16.6.6 Social Behaviour

'Squids are schooling, social creatures with large brains and complex behaviour' (Hanlon 1990). During rearing of *D. opalescens* changes in social interactions between same-aged paralarvae can be often observed. Early paralarvae (0–15 days old) swim dispersed, holding a certain distance from each another and often display aggressive behaviour towards one another. About 30–40-day-old paralarvae usually swim at 45–65°- angles, less vertically oriented than the newly hatched squid and are able to swim against low current speeds (~ 1.4 cm s^{-1}), frequently copying the position of other squid swimming nearby (Vidal 2000). Schooling can be first noticed in 35–45-day-old paralarvae (6–8 mm ML) and seems to be dependent on size of squid and flow intensity (Vidal 2000). In fact, Yang et al. (1983a, 1986) first noticed the ability of *D. opalescens* to hold position at 40–45 days (10 mm ML), and to form schools between 60–80 dah (15 mm ML). These authors have used larger rearing tanks and current speeds.

16.7 Ongrowing

So far, no reports are known to us about large-scale ongrowing culture in the two species under consideration. Given the rather expensive food required (Tables 16.3 and 16.4), it is not surprising that no industrial squid culture using *L. vulgaris* and *D. opalescens* has ever been achieved. What potential market should be interested in these two species? The year-round production of squid that must be kept alive until consumption (e.g. as Sashimi) can be achieved much more easily with other species, especially *Sepioteuthis* (see Chap. 17). Only for *D. opalescens* the complete life cycle was close under laboratory conditions (Yang et al. 1986). For any attempt to set up a large-scale culture of *L. vulgaris* or *D. opalescens*, the following conditions would have to be fulfilled.

16.7.1 Tank and Sea Cage Conditions

Since fin abrasion is one of the most important dangers in loliginid squid culture (Boletzky 1979; Hanlon et al. 1979), any tank or sea cage would have to be designed to provide sufficient tank space to minimize contact of individuals with the tank or cage walls. Large circular tanks or long-oval raceway tanks have proven to be effective to reduce fin damage during the ongrowing of *D. opalescens* (Yang et al. 1983a, 1986), although they have high costs of maintenance. Probably sea cages would have to be equipped with light sources allowing the captive animals to stay clear of the surrounding mesh walls during darkness.

16.7.2 Density

Density as such (number of individuals per volume of tank water) is not a meaningful criterion as soon as young individuals start schooling. The question then is more about how many organized groups of individuals can be maintained in one tank or sea cage. The constant risk of cannibalism draws attention to the likely necessity of a rather strict separation of size classes in different holding facilities.

16.7.3 Food

Live fish and shrimps (Tables 16.3 and 16.4) or dead food items offered (e.g. pieces of fish or entirely artificial preparations) are seized and eaten by hungry squids, and any combination of these feeding regimes may be used to maintain a high degree of appetence (for food other than conspecifics!). What can be considered a hungry squid may be assessed by the actual feeding behaviour.

Fields (1965, p. 22) already described feeding behaviour in the aquarium: 'On one occasion, about 200 shrimp (*Spirontocaris* sp.) were added to a tank (about 12 by 4 by 2 feet deep) in which about 40 *D. opalescens* had been shuttling back and forth for 3 days. Some of the squid immediately showed excitement: clouds of colour raced over their translucent mantles, the respiratory rate increased, the animal's movements quickened and in each the sessile arms, hiding the tentacles, formed a sharp cone pointing directly at a swimming shrimp. A squid would make two or three short darts toward and from the prey, coming closer with each approach until, on the last, the tentacles shot out to seize the shrimp and the arms opened for its reception'.

Yang et al. (1986) offered a diet consisting of mysids, palaemonid shrimp and estuarine fishes to *D. opalescens* juveniles and adults (Table 16.4), obtaining mean daily group feeding rates of 14.9 and 18.0 % body ww.

16.7.4 Cleaning Techniques

Any action likely to disturb the animals should be avoided or at least minimized. Thus, cleaning techniques should be 'squid friendly'. Cleaning robots moving slowly over the walls and the bottom of tanks or sea cages may offer an optimal approach. For such robots, circular tanks are optimal.

16.7.5 Growth, Survival, Spawning and Sampling

The right food items offered at short intervals to achieve an ad libitum regime, optimal water quality and temperature should generate high growth rates. However, differences in individual growth rates are inevitable and may lead to increased risks of cannibalism (small individuals being attacked by large ones).

The whole life cycle of *D. opalescens* was covered by Yang et al. (1986) in less than 8 months, thus representing 'culture' in the strict sense of the term. Squid exhibited exponential growth for the first 2 months of life and thereafter showed slower logarithmic growth of 1.6 % ww day^{-1} and 0.63 % mm ML day^{-1} and mean growth of 16 mm per month at 240 dah (Table 16.2). At this period, they were doubling their weight and length at every 42 days and 109 days, respectively. Mortality slowed after 60–70 days post hatching, and again after 180 days due to spawning. Maximum lifespan was 235 and 248 days in the two experiments conducted, with squid attaining a maximum size of 116 mm ML. Maturity was observed at 60 mm ML in females and 70 mm ML in males. Spawning took place between 196 and 239 dah producing viable eggs. Fin or skin damage and senescence after reproduction accounted for late mortality. Thus, considering the aforementioned culture conditions, time to marketable size cannot be reduced below a minimum of 4–8 months.

16.8 Trends in Research and Industrial Level

Both loliginid species considered in this chapter have been used extensively as experimental models in several research areas. Nevertheless, as indicated above, there is no hint of a market pressure for industrial levels of culture in *L. vulgaris* and *D. opalescens*. When considering these two species as candidates for large-scale cultures in research, it would be interesting to assess their potential in relation to that of other loliginid species.

16.9 Conclusions

Considering loliginid squid hatchlings as miniature cephalopods that already have adult-like sensory organs and effector systems (Budelmann 1996), including muscular suckers allowing them to seize and hold live prey (Schmidtberg 1997), but also special structures such as tailored denticulated beaks (Boletzky 1971; Franco-Santos and Vidal in press), the complex requirements of their culture are not really surprising.

The production of high-quality and competent paralarvae is dependent on the developmental history and environmental influences during embryogenesis. Research progress has been slow in identifying the factors that endorse the production of functional and competent paralarvae with large internal yolk reserves for rearing. This is a precursory step to reduce mortality during the critical transition from endogenous (yolk) to exogenous feeding (prey capture) that takes place during the first 10 dah when the highest mortality rates are observed. In this context, many open questions remain about *how* the known physical and chemical parameters act via the physiological mechanisms to control biological performance (including predatory behaviour, digestion, growth) during the different phases of the life cycle.

So far, survival rates thus are the most reliable indicators for 'good' culture systems. A reproducible high rate of survival and growth of paralarvae is necessary to achieve successful culture. Tank circulation and turbulence have a direct impact on paralarval survival by increasing the incidence of skin and fin damage caused by the contact of the fragile paralarvae with tank walls. More attention should be given to the rearing systems as improvements on survival rates of paralarvae have been related to advances in water quality and the design of culture systems (Yang et al. 1983a; Vidal et al. 2002a). Nevertheless, large numbers of loliginid paralarvae can be reared in the laboratory with relatively good survival rates in recirculating systems if live food are provided.

Paralarvae require a variety of live prey of different sizes and types so as to match the different sizes and hunting abilities of same-aged but heterogeneously developing squid. This is still one of the main bottlenecks in culturing, and a steady supply of live, fresh, cultured food organisms (copepods, crab zoeae, mysids) can greatly contribute to excellent survival rates on experimental level. Paralarvae feeding behaviour is noticeably experience based (Chen et al. 1996).

While food availability is essential during rearing, food quality is more important for development and growth. When squid reach 40–50 days of age (6–10 mm ML) they require higher daily food intake, large and more energetically rewarding prey types and more swimming space as they start swimming in schools. If these requirements are not met, a second peak in mortality is often observed. Elucidating ontogenetic size-specific nutritional requirements of squid is essential to improve feeding and should be ranked as a high-priority research area.

For highly active swimmers such as loliginid squids, which begin schooling at an early juvenile stage, sufficient tank space to allow expression of normal swimming and feeding behaviour and to accommodate relatively large groups is a major requirement. Thus, intrinsic practical and economical difficulties of culturing loliginid squid lie mainly in avoiding skin and fin damage by offering large volume tanks or raceways and carrying their high cost of maintenance. In addition, production is constrained largely by the high protein demand of squid and the dilemma of feeding them on a commercial scale, all of which impact on profitability.

Finally, each new attempt to culture one or both of the loliginid squids described in this chapter should therefore focus on improving earlier methods rather than promote uncritical application of ostensibly reliable prescriptions.

Acknowledgement EAGV was supported by the Brazilian National Research Council—CNPq (Grant no. 307204/2011-1).

References

Ally JRR, Keck SA (1978) A biochemical-genetic population structure study of market squid, *Loligo opalescens*, along the California coast. In: Recksiek CW, Frey HW (eds) Biological, oceanographic, and acoustic aspects of the market squid, *Loligo opalescens* berry. Calif fish and game. Fish Bull 169:113–121

Anderson FE (2000) Phylogenetic relationships among loliginid squids (Cephalopoda: Myopsida) based on analyses of multiple data sets. Zool J Linn Soc 130:603–633

Appleford P, Lucas JS, Southgate PC (2003) General principles. In: Lucas JS, Southgate PC (eds) Aquaculture: farming aquatic animals and plants, pp 11–46

Arnold JM (1962) Mating behavior and social structure in *Loligo pealeii*. Biol Bull 123:53–57

Arnold JM (1965) Normal embryonic stages of the squid *Loligo pealeii* (Lesueur). Biol Bull 128:24–32

Barbieri E, Barry K, Child A, Wainwright N (1997) Antimicrobial activity in the microbial community of the accessory nidamental gland and egg cases of *Loligo pealeii* (Cephalopoda: Loliginidae). Biol Bull 193:275–276

Barbieri E, Paster BJ, Hughes D, Zurek L, Moser DP, Teske A, Sogin ML (2001) Phylogenetic characterization of epibiotic bacteria in the accessory nidamental gland and egg capsules of the squid *Loligo pealeii* (Cephalopoda: Loliginidae). Environ Microbiol 3(3):151–167

Bauer V (1909) Einführung in die Physiologie der Cephalopoden. Mitt Zool Station Neapel 19(2):149–268

Bentivegna F (1987) Des observations techniques et biologiques sur les Céphalopodes de l'aquarium de Naples. Vie Marine (Hors série no) 8:74–80

Benzie JAH (1998) A review of the effect of genetics and environment on the maturation and larval quality of the giant tiger prawn *Penaeus monodon*. Aquaculture 155:69–85

Bergh Ø, Hansen GH, Taxt RE (1992) Experimental infection of eggs and yolk sac larvae of halibut, *Hippoglossus hippoglossus* L. J Fish Diseases 15:379–391

Bidder AM (1950) The digestive mechanism of the European Squids *Loligo vulgaris*, *Loligo forbesi*, *Alloteuthis media* and *Alloteuthis subulata*. Quart J Micr Sci 91:1–43

Biggs J, Epel D (1991) Egg capsule sheath of *Loligo opalescens* Berry: structure and association with bacteria. J Exp Zool 259:263–267

Boletzky Sv (1971) Mandibules denticulées chez les larves des Teuthoidés et des Octopodes (Mollusca, Cephalopoda). C R Acad Sc Paris 272:2904–2906

Boletzky Sv (1974a) Elevage de Céphalopodes en aquarium. Vie Milieu 24(2)A:309–340

Boletzky Sv (1974b) The "larvae" of Cephalopoda: a review. Thalass Jugosl 10(1/2):45–76

Boletzky Sv (1975) A contribution to the study of yolk absorption in the Cephalopoda. Z Morphol Tiere 80:229–246

Boletzky Sv (1979) Observations on the early post-embryonic development of *Loligo vulgaris* (Mollusca, Cephalopoda). Rapp Comm Int Mer Médit 25/26(10):155–158

Boletzky Sv (1987) Embryonic phase. In: Boyle PR (ed) Cephalopod Life Cycles Vol II, Academic Press, London, pp 5–31

Boletzky Sv (2003) Biology of early life stages in cephalopod molluscs. Adv Mar Biol 44:143–204

Boletzky Sv (2004) A brief survey of cephalopod culture techniques. In: Öztürk B, Salman A (eds) 1st National Malacology Congress, Izmir—Turkey, Tr J Aquatic Life 2(2):229–240

Boletzky Sv (2010) The 'yolk organ' of cephalopod embryos: on transient functions from crawling substratum to provisional knapsack. Ferrantia 59:14–21

Boletzky Sv, Dohle W (1967) Observations sur un capitellidé (*Capitella hermaphrodita* sp. n.) et d'autres polychètes habitant la ponte de *Loligo vulgaris*. Vie Milieu 18:79–98

Boletzky Sv, Hanlon RT (1983) A review of the laboratory maintenance, rearing and culture of cephalopod molluscs. Mem Natl Mus Vic 44:147–187

Borrelli L, Gherardi F, Fiorito G (2006) A catalogue of body patterning in Cephalopoda. Stazione Zoologica A. Dohrn, Firenze University Press

Boyle PR (ed) (1983) Cephalopod life cycles vol. I. Academic Press, London

Budelmann BU (1996) Active marine predators: the sensory world of cephalopods. Mar Freshw Behav Physiol 27(2–3):59–75

Chen DS, VanDykhuizen G, Hodge J, Gilly WF (1996) Ontogeny of copepod predation in juvenile squid (*Loligo opalescens*). Biol Bull 190:69–81

Christofferson JP, Foss A, Lambert WE, Welge B (1978) An electrophoretic study of select proteins from the market squid, *Loligo opalescens* Berry. In: Recksiek CW, Frey HW (eds) Biological, oceanographic, and acoustic aspects of the market squid, *Loligo opalescens* Berry. Calif Fish and Game, Fish Bull 169:123–133

Collins JW (1892) Report on the fisheries of the Pacific Coast of the United States. US Fish Com, Rept For 1888:3–269

Crawford K (2002) Culture method for *in vitro* fertilization to hatching of the squid, *Loligo pealeii*. Biol Bull 203:216–217

Cronin E, Seymour R (2000) Respiration of the eggs of the giant cuttlefish *Sepia apama*. Mar Biol 136:863–870

D'Aniello A, D'Onofrio G, Pischetola M, Denucé JM (1987) Effect of the ionic concentration of calcium and potassium on hatching of the embryo of the squid, *Loligo vulgaris*. A preliminary study. Arch Int Phisiol Biochim 95, B66

D'Aniello A, D'Onofrio G, Pischetola M, Denucé JM (1989) Effect of pH, salinity and Ca^{2+}, Mg^{2+}, K^+ and SO_4^{2+} ions on hatching and viability of *Loligo vulgaris* embryo. Comp Biochem Physiol A 94:477–481

DeRusha RH, Forsythe JW, DiMarco FP, Hanlon RT (1989) Alternative diets for maintaining and rearing cephalopods in captivity. Lab Anim Sci 39:306–312

Drillet G, Jørgensen NOG, Sørensen TF, Ramløv H, Hansen BW (2006) Biochemical and technical observations supporting the use of copepods as live feed organisms in marine larviculture. Aquac Res 37:756–772

Fields WG (1950) A preliminary report on the fishery and on the biology of the squid *Loligo opalescens*. Calif Dept Fish Game 36:367–377

Fields WG (1965) The structure, development, food relation, reproduction and life history of the squid *Loligo opalescens* Berry. Dept Fish Game Calif, p 108

Forsythe JW, Van Heukelem WF (1987) Growth. In: Boyle PR (ed) Cephalopod life cycles, Vol II. Academic Press, London, pp 135–156

Franco-Santos RM, Vidal EAG (in press) Beak development of early squid paralarvae (Cephalopoda: Teuthoidea) may reflect an adaptation to a specialized feeding mode. Hydrobiol 721

Gilly WF, Lucero MT (1992) Behavioral responses to chemical stimulation of the olfactory organ in the squid *Loligo opalescens*. J Exp Biol 162:209–229

Grieb TM, Beeman RD (1978) A study of spermatogenesis in the spawning population of the squid, *Loligo opalescens*. In: Recksiek CW, Frey HW (eds) Biological, oceanographic, and acoustic aspects of the market squid, *Loligo opalescens* Berry. Calif Fish and Game. Fish Bull 169:11–21

Grimpe G (1925) Zur Kenntnis der Cephalopodenfaune der Nordsee. Wiss Meeresunters Abt Helgoland 16(3):1–124

Grimpe G (1928) Pflege, Behandlung und Zucht der Cephalopoden für zoologische und physiologische Zwecke. Abderhalden Handbuch der biologischen Arbeitsmethoden, Abt. IX. Teil 5:331–402

Hanlon RT (1978) Aspects of the biology of the squid *Loligo* (*Doryteuthis*) *plei* in captivity. Ph. D. Dissertation, University of Miami

Hanlon RT (1987) Mariculture. In: Boyle PR (ed) Cephalopod life cycles vol II, Academic Press, London, pp 291–305

Hanlon RT (1990) Maintenance, rearing and culture of teuthoid and sepioid squids. In: Gilbert DL, Adelman Jr WJ, Arnold JM (eds) Squid as experimental animals. Plenum Press, New York, pp 35–62

Hanlon RT, Messenger JB (1996) Cephalopod Behaviour. Cambridge University Press, London

Hanlon RT, Hixon RF, Hulet WH, Yang WT (1979) Rearing experiments on the California Market Squid *Loligo opalescens* Berry 1911. Veliger 21:428–431

Hanlon RT, Ridwell JP, Tait R (1989) Strontium is required for statolith development and thus normal swimming behaviour of hatchling cephalopods. J Exp Biol 141:187–195

Hixon RF 1983. *Loligo opalescens*. In: Boyle PR (ed) Cephalopod life cycles, vol. I, pp 95–114

Hurley AC (1976) Feeding behaviour, food consumption, growth, and respiration of the squid *Loligo opalescens* raised in the laboratory. Fish Bull 74:176–182

Hurley AC (1977) Mating behaviour of the squid *Loligo opalescens*. Mar Behav Physiol 4:195–203

Hurley AC (1978) School structure of the squid *Loligo opalescens*. Fish Bull 76:433–442

Jackson GD, Domeier ML (2003) The effects of an extraordinary El Niño/La Niña event on the size and growth of the squid *Loligo opalescens* off Southern California. Mar Biol 142:925–935

Jatta G (1896) I cephalopodi viventi nel Golfo di Napoli. Fauna Flora Golf Neapel Monograph 23

Jecklin L (1934) Beitrag zur Kenntnis der Laichgallerten und der Biologie der Embryonen decapoder Cephalopoden. Rev Suisse Zool 41(37):593–673

Jereb P, Roper CFE (2010) Cephalopods of the world, vol. 2. Myopsid and Oegopsid Squids. FAO Species Catalogue for Fishery Purposes No. 4

Karpov KA, Caillet GM (1978) Feeding dynamics of *Loligo opalescens*. In: Recksiek CW, Frey HW (eds) Biological, oceanographic, and acoustic aspects of the market squid, *Loligo opalescens* Berry. Calif Fish and Game. Fish Bull 169:45–65

Kashiwada J, Recksiek CW (1978) Possible morphological indicators of population structure in the market squid, *Loligo opalescens*. In: Recksiek CW, Frey HW (eds) Biological, oceanographic, and acoustic aspects of the market squid, *Loligo opalescens* Berry. Calif Fish and Game. Fish Bull 169:99–111

Kinne O (1971) Salinity: 3. Animals: 1. Invertebrates. In: Kinne O (ed). Marine ecology: a comprehensive, integrated treatise on life in oceans and coastal waters: 1. Environmental factors: 2, pp 821–995

Knipe JH, Beeman RD (1978) Histological observations on oögenesis in *Loligo opalescens*. In: Recksiek CW, Frey HW (eds) Biological, oceanographic, and acoustic aspects of the market squid, *Loligo opalescens* Berry. Calif Fish and Game. Fish Bull 169:23–33

Koslow JA, Allen C (2011) The influence of the ocean environment on the abundance of market squid, *Doryteuthis* (*Loligo*) *opalescens*, paralarvae in the southern California Bight. CalCofi Rep 52:205–213

Lacoue-Labarthe T, Réveillac E, Oberhänsli F, Teyssié JL, Jeffree R, Gattuso JP (2011) Effects of ocean acidification on trace element accumulation in the early-life stages of squid *Loligo vulgaris*. Aquat Toxicol 105(1–2):166–176

Lee PG (1994) Nutrition of cephalopods: fueling the system. Mar Freshw Behav Physiol 25:35–51

Lo Bianco S (1899) Notizie biologiche riguardanti specialmente il periodo di maturità sessuale degli animali del golfo di Napoli. Mitt Zool Station Neapel 13:448–573

Losordo TM, Masser MP, Rakocy J (2001) Recirculating aquaculture tank production systems. An overview of critical considerations. World Aquacul 32:18–22

MacKenzie BR, Kiørboe T (2000) Larval fish feeding and turbulence: a case for the downside. Limnol Oceanogr 45:1–10

Mangold-Wirz K (1963) Biologie des Céphalopodes benthiques et nectoniques de la Mer Catalane. Vie Milieu, suppl 13:1–285

Marthy HJ (1972) Sur la localisation et la stablité du plan d'ebauches d'organes chez l'embryon de *Loligo vulgaris*. C R Acad Sci Paris 275:1291–1293

Martins RS, Roberts MJ, Chang N, Verley P, Molooney CL, Vidal EAG (2010) Effect of yolk utilization on the specific gravity of chokka squid (*Loligo reynaudii*) paralarvae: implications for dispersal on the Agulhas Bank, South Africa. ICES J Mar Sci 67:174–183

McGowan JA (1954) Observations on the sexual behaviour and spawning of the squid, *Loligo opalescens*, at La Jolla, California. Calif Fish Game 40:47–54

McInnis RR, Broenkow WW (1978) Correlations between squid catches and oceanographic conditions in Monterey Bay, California. In: Recksiek CW, Frey HW (eds) Biological, oceanographic, and acoustic aspects of the market squid, *Loligo opalescens* Berry. Calif Fish and Game. Fish Bull 169:161–170

Mladineo I, Valic D, Jozic M (2003) Spawning and early development of *Loligo vulgaris* Lamarck 1798, under experimental conditions. Acta Adriat 44(1):77–83

Morejohn GV, Harvey JT, Krasnov LT (1978) The importance of *Loligo opalescens* in the food web of marine vertebrates in Monterey Bay, California. In: Recksiek CW, Frey HW (eds) Biological, oceanographic, and acoustic aspects of the market squid, *Loligo opalescens* Berry. Calif Fish and Game. Fish Bull 169:67–98

Moreno A, Azevedo M, Pereira J, Pierce GJ (2007) Growth strategies in the squid *Loligo vulgaris* from Portuguese waters. Mar Biol Res 3:49–59

Moreno A, Boavida-Portugal J, Pimentel M, Pereira J, Rosa R (2013) *Loligo vulgaris*, European Squid. In: Rosa R, O´Dor R, Pierce GJ (eds) Advances in Squid Biology, Ecology and Fisheries, Part I—Myopsid Squid. Nova Science Publishers, New York, pp 3–32

Nabhitabhata J, Asawangkune P, Amorujaruchit S, Promboon P (2001) Tolerance of eggs and hatchlings of neritic cephalopods to salinity changes. Phuket Mar Biol Center Spec Publ 25:91–99

Naef A (1921/1923) Die Cephalopoden (Systematik). Fauna Flora Golf Neapel 35A. Monogr:1–863

Naef A (1928) Die Cephalopoden (Embryologie). Fauna Flora Golf Neapel. 35B. Monogr:1–357

Navarro JC, Villanueva R (2000) Lipid and fatty acid composition of early stages of cephalopods: an approach to their lipid requirements. Aquaculture 183:161–177

Navarro JC, Villanueva R (2003) The fatty acid composition of *Octopus vulgaris* rearead with live and inert food: deviation from their natural fatty acid profile. Aquaculture 219:613–631

Navarro JC, Amat F, Sargent JR (1992) Fatty acid composition of coastal and inland *Artemia* sp. populations from Spain. Aquaculture 102:219–230

Neill SSJ (1971) Notes on squid and cuttlefish; keeping, handling and colour-patterns. Pubbl Staz Zool Napoli 39:64–69

Neill SRSJ, Cullen JM (1974) Experiments on whether schooling by their prey affects the hunting behaviour of cephalopods and fish predators. J Zool. Lond 172:549–569

Neumeister H, Ripley B, Preuss T, Gilly WF (2000) Effects of temperature on escape jetting in the squid *Loligo opalescens*. J Exp Biol 203:547–557

Packard A (1995) Discrimination of coupled muscle fibre populations in the skin of squids (*Loligo vulgaris* and *L. opalescens*) following unilateral nerve section. J Physiol 489P:134–135

Palmegiano GB, D'Apote MP (1983) Combined effects of temperature and salinity on cuttlefish (*Sepia officinalis* L.) hatching. Aquaculture 35:259–264

Paulij WP, Herman PMJ, Van Hannen EJ, Denucé JM (1990) The impact of photoperiodicity on hatching of *Loligo vulgaris* and *Loligo forbesi*. J Mar Biol Ass UK 70:597–610

Paulij WP, Herman PMJ, Roozen MEF, Denucé JM (1991) The influence of photoperiodicity on hatching of *Sepia officinalis*. J Mar Biol Ass UK 71:665–678

Pierce GJ, Boyle PR, Hastie LC, Santos MB (1994) Diets of squid *Loligo forbesi* and *Loligo vulgaris* in the northeast Atlantic. Fish Res 21(1–2):149–164

Portmann A, Bidder AM (1928) Yolk absorption in *Loligo* and the function of the embryonic liver and pancreas. Quart J Microsc Sci 72:301–324

Pörtner HO, Zielinski S (1998) Environmental constraints and the physiology of performance in squids. S Afr J Mar Sci 20:207–217

Preuss T, Gilly WF (2000) Role of prey-capture experience in the development of the escape response in the squid *Loligo opalescens*: a physiological correlate in an identified neuron. J Exp Biol 203:559–565

Preuss T, Lebaric ZN, Gilly WF (1997) Post-hatching development of circular mantle muscles in the squid *Loligo opalescens*. Biol Bull 192:375–387

Recksiek CW, Frey HW (eds) (1978) Biological, oceanographic, and acoustic aspects of the market squid, *Loligo opalescens* Berry. Calif Fish and Game. Fish Bull 169:1–189

Rosa R, Pimentel MS, Boavida-Portugal J, Teixeira T, Trübenbach K, Diniz M (2012) Ocean warming enhances malformations, premature hatching, metabolic suppression and oxidative stress in the early life stages of a keystone squid. PlosOne 7:e38282

Sargent J (1995). Origins and functions of egg lipids: Nutritional implications. In: Bromage NR, Roberts RJ (eds) Broodstock management and egg and larval quality. The University Press, Cambridge, pp 353–372

Schmidtberg H (1997) The structure of suckers of newly hatched *Sepia officinalis*, *Loligo vulgaris* and *Octotpus vulgaris*. Vie Milieu 47(2):155–159

Scofield WL (1924) Squid at Monterey. Calif Fish and Game 10:176–182

Seixas P, Rey-Méndez M, Valente LMP, Otero A (2010) High DHA content in Artemia is ineffective to improve *Octopus vulgaris* paralarvae rearing. Aquaculture 300:156–162

Segawa S, Yang WT, Marthy H-J, Hanlon HT (1988) Illustrated embryonic stages of the eastern Atlantic squid *Loligo forbesi*. Veliger 30(3):230–243

Sen H (2004) A preliminary study on the effects of salinity on egg development of European squid (*Loligo vulgaris* Lamarck 1798). Isr J Aquacult 56:93–99

Sen H (2005a) Temperature tolerance of loliginid squid (*Loligo vulgaris* Lamarck 1798) eggs in controlled conditions. Turk J Fish Aquat Sci 5:53–56

Sen H (2005b) Incubation of european squid (*Loligo vulgaris* Lamarck 1798) eggs at different salinities. Aquac Res 36:876–881

Sereni E (1930) The chromatophores of the cephalopods. Biol Bull 59(3):247–268

Spotte S (1979) Fish and invertebrate culture: water management in closed systems, 2nd edn. Wiley- Interscience, New York

Spratt JD (1978) Age and growth of the market squid, *Loligo opalescens* Berry, in Monterey Bay. In: Recksiek CW, Frey, HW (eds). Biological, oceanographic, and acoustic aspects of the market squid, *Loligo opalescens* Berry. Calif Fish and Game. Fish Bull 169:35–44

Steer MA, Moltschaniwskyj NA (2007) The effects of egg position, egg mass size, substrate and biofouling on embryo mortality in the squid *Sepiotheutis australis*. Rev Fish Biol Fish 17:173–182

Tardent P (1962) Keeping *Loligo vulgaris* L. in the Naples Aquarium. Bull Inst Océanogr Monaco, special No. A:41–46

Timmons B, Ebeling JM (2007) Recirculating aquaculture. Cayuga Aqua Ventures, LLC, New York

Turk PE, Hanlon RT, Bradford LA, Yang WT (1986) Aspects of feeding, growth and survival of the European squid *Loligo vulgaris* Lamarck, 1798, reared through the early growth stages. Vie Milieu 36(1):9–13

Vaughan DL (1978) The target strength of individual market squid, *Loligo opalescens*. In: Recksiek CW, Frey HW (eds) Biological, oceanographic, and acoustic aspects of the market squid, *Loligo opalescens* Berry. Calif Fish and Game. Fish Bull 169:149–159

Vaughan DL, Recksiek CW (1978) An acoustic investigation of market squid, *Loligo opalescens*. In: Recksiek CW, Frey HW (eds) Biological, oceanographic, and acoustic aspects of the market squid, *Loligo opalescens* Berry. Calif Fish and Game, Fish Bull 169:135–147

Vecchione M (1987) Juvenile ecology. In: Boyle PR (ed) Cephalopod life cycles, vol II, Comparative reviews. Academic Press, London, pp 61–84

Vidal EAG (2000) Optimizing survival, growth and feeding in hatchling squid (*Loligo opalescens*). Ph. D. Dissertation, Texas A&M University

Vidal EAG, DiMarco FP, Wormuth JH, Lee PG (2002a) Optimizing rearing conditions of hatchling loliginid squid. Mar Biol 140:117–127

Vidal EAG, DiMarco FP, Wormuth JH, Lee PG (2002b) Influence of temperature and food availability on survival, growth and yolk utilization in hatchling squid. Bull Mar Sci 71:915–931

Vidal EAG, Roberts MJ, Martins RS (2005) Yolk utilization, metabolism and growth in reared *Loligo vulgaris* reynaudii paralarvae. Aquat Living Resour 18:385–394

Vidal EAG, DiMarco P, Lee PG (2006) Effects of starvation and recovery on the survival, growth and RNA:DNA ratio in loliginid squid paralarvae. Aquaculture 260:94–105

Villa H, Quintela J, Coelho ML, Icely JD, Andrade JP (1997) Phytoplankton biomass and zooplankton abundance on the south coast of Portugal (Sagres), with special reference to spawning of *Loligo vulgaris*. Sci Mar 61:123–129

Villanueva R (1994) Decapod crab zoeae as food for rearing cephalopod paralarvae. Aquaculture 128:143–152

Villanueva R (2000a) Effect of temperature on statolith growth of the European squid *Loligo vulgaris* during early life. Mar Biol 136:449–460

Villanueva R (2000b) Differential increment-deposition rate in embryonic statoliths of the loliginid squid *Loligo vulgaris*. Mar Biol 137:161–168

Villanueva R, Bustamante P (2006) Composition in essential and non-essential elements of early stages of cephalopods and dietary effects on the elemental profiles of *Octopus vulgaris* paralarvae. Aquaculture 261:225–240

Villanueva R, Riba J, Ruíz-Capillas C, González AV, Baeta M (2004) Amino acid composition of early stages of cephalopods and effect of amino acid dietary treatments on *Octopus vulgaris* paralarvae. Aquaculture 242:455–478

Villanueva R, Arkhipkin A, Jereb P, Lefkaditou E, Lipinski MR, Perales-Raya C, Riba J, Rocha F (2003) Embryonic life of the loliginid squid *Loligo vulgaris*: comparison between statoliths of Atlantic and Mediterranean populations. Mar Ecol Prog Ser 253:197–208

Villanueva R, Moltschaniwskyj NA, Bozzano A (2007) Abiotic influences on embryo growth: statoliths as experimental tools in the squid early life history. Rev Fish Biol Fish 17:101–110

Worms J (1983) *Loligo vulgaris*. In: Boyle PR (ed) Cephalopod life cycles, vol. I, pp 143–157

Yang WT, Hanlon RT, Krejci ME, Hixon RF, Hulet WH (1980) Culture of California market squid from hatching – first rearing of *Loligo* to sub-adult stage (in Japanese with English abstract). Aquabiol 2:412–418

Yang WT, Hanlon RT, Krejci ME, Hixon RF, Hulet WH (1983a) Laboratory rearing of *Loligo opalescens*, the market squid of California. Aquaculture 31:77–88

Yang WT, Hixon RF, Turk PE, Krejci ME, Hanlon RT, Hulet WH (1983b) Culture of California market squid from hatching—completion of the rearing cycle to second generation hatchlings (in Japanese with English abstract). Aquabiol 5:328–339

Yang WT, Hixon RF, Turk PE, Krejci ME, Hulet WH, Hanlon RT (1986) Growth, behavior and sexual maturation of the market squid, *Loligo opalescens*, cultured through the life cycle. Fish Bull 84:771–798

Yang WT, Hanlon RT, Lee PG, Turk PE (1989) Design and function of closed seawater systems for culturing loliginid squids. Aquac Eng 8:47–65

Young JZ (1938) The functioning of the giant nerve fibres of the squid. J Exp Biol 15:170–185

Young JZ (1939) Fused neurons and synaptic contacts in the giant nerve fibres of cephalopods. Phil Trans B 229:465–505

Zeidberg LD (2008) First observations of sneaker mating in the California market squid, *Doryteuthis opalescens* (Cephalopoda: Myopsida). J Mar Biol Assoc 2 Biodiversity Records, 5998

Zeidberg LD (2013) *Doryteuthis opalescens*, Opalescent Inshore Squid. In: Rosa R, O'Dor R, Pierce GJ (eds) Advances in Squid Biology, Ecology and Fisheries, Part I – Myopsid Squid. Nova Science Publishers, New York, pp 159–204

Zeidberg LD, Isaac G, Widmer CL, Neumeister H, Gilly WF (2011) Egg capsule hatch rate and incubation duration of the California market squid, *Doryteuthis* (= *Loligo*) *opalescens*: insights from laboratory manipulations. Mar Ecol 32:468–479

Chapter 17
Sepioteuthis lessoniana

Jaruwat Nabhitabhata and Yuzuru Ikeda

Abstract *Sepioteuthis lessoniana* is a demersal neritic species that inhabits coral and rock reefs, seaweed, sea grass beds, and estuaries. Due to its wide distribution range in the Indo-Pacific region, *S. lessoniana* is an economically important resource of many countries. *S. lessoniana* has been successfully cultured through multiple generations since the 1960s in both open and closed seawater systems in Thailand, Japan, and the USA. The objectives of aquaculture studies are the production of human food in tropical countries and experimental animals in temperate countries. *S. lessoniana* hatchlings are larger than other loliginid squids, which enables good adaptation to culture conditions and a very high growth rate through the entire life cycle. In tropical waters, individuals can grow to 500 g in less than 150 days. This rapid growth results from a high feeding rate and requires a massive supply of live feed organisms during the early phase of life. The grow-out phase begins after *S. lessoniana* can accept dead feed. Further studies of artificial feed or mass production of live feed is required in order to make aquaculture of *S. lessoniana* economically viable on a large scale. The method and studies of *S. lessoniana* culture in tropical and temperate waters are reviewed.

Keywords *Sepioteuthis lessoniana* · Tropical waters · Temperate waters · Open seawater system · Closed seawater system

17.1 Importance of this Species in the Market

The fins of *Sepioteuthis lessoniana* are very large and broad compared to those of other loliginid squids. This has led to its common names, the bigfin reef squid, the bigfin squid, and the oval squid. The fin length (FL) of this squid is 90–100% of

J. Nabhitabhata (✉)
Excellence Centre for Biodiversity of Peninsular Thailand (CBIPT), Faculty of Science, Prince of Songkla University, Hatyai, Songkhla 90112, Thailand
e-mail: jaruwat.n5@gmail.com; jaruwat.n@psu.ac.th

Y. Ikeda
Department of Chemistry, Biology and Marine Sciences, Faculty of Science, University of the Ryukus, Nishihara, Japan
e-mail: ikeda@sci.u-ryukyu.ac.jp

J. Iglesias et al. (eds.), *Cephalopod Culture*, DOI 10.1007/978-94-017-8648-5_17,

the mantle length (ML), and the fin width is more than 75% of the ML (Jereb et al. 2010). The bigfin squid is a demersal neritic species and is found between the surface and depths of about 100 m. Its habitats include coral and rock reefs, seaweed and sea grass beds, and estuaries. Its geographical distribution covers the Indo-Pacific region, 20°E–120°W, 40°N–40°S (Dunning 1998; Jereb et al. 2010; Okutani 2005). The maximum body weight (BW) of *S. lessoniana* is about 2 kg, while the maximum ML is 422 mm for males and 382 mm for females (Jereb et al. 2010). In the shallow waters of the Gulf of Thailand, the maximum ML of these squid is around 305 mm, and squid that are 70–260 mm in length are commonly caught (Chotiyaputta 1995). The first maturation size of female squid is 130–140 mm and their fecundity is approximately 1,000 (400–3,500) eggs (Chotiyaputta 1984, 1988, 1995; Nabhitabhata 1996; Roongratri 1997; Segawa 1987; Segawa et al. 1993b).

S. lessoniana is of commercial and artisanal interest throughout the Indo-West Pacific region. It constitutes about 7% of Indian east coast cephalopod landings and 5% of Thai landings (Jereb et al. 2010; Supongpan et al. 1988; Supongpan 1995). It is captured using a variety of fishing gear and comprises about 10% of the catch by offshore trawlers and purse seine. In the Southeast Asian region where *S. lessoniana* is the most abundant, i.e. Thailand and Indonesia, squid traps are popular inshore gear for capturing this species. Squid trap fisheries in Thailand annually yield about 7,000 t, and 90–95% of the catch is comprised of *S. lessoniana,* the main target species (>6,000 t). The remaining 5–10% is comprised of by-catch pharaoh cuttlefish, *Sepia pharaonis* (Supongpan et al. 1988; Chotiyaputta and Yamrungreung 1998).

The trap used to capture these squid is artisanal, can be made from local materials, and reflects a good understanding of squid behaviour by local fishermen. In Thailand, two types of traps are used: the semicylindrical type (Fig. 17.2; for details, see also Chap. 7, 'Aquaculture to Restocking') and the rectangular box type. Both types are 500–800 × 800–1,200 × 500–650 mm, and most of them are 600 × 1,000 × 500 mm (Boongerd and Rachaniyom 1990; Chotiyaputta and Yamrungreung 1998). Trap frames are made of locally available wood and are covered by a polyethylene fishing net that has a 25-mm mesh size. In fishing operations, the trap is suspended in the water column 2–3 m above the bottom using rock sinkers with Styrofoam and bamboo floats. The fishing depth is approximately 6–40 m, depending on the local fishermen's experience regarding the optimal trap depth. Coconut fronds are used to cover the trap, making it look like a shelter. *S. lessoniana* egg clusters are hung inside the trap to persuade squid to enter. When egg clusters are scarce, stripped plastic shopping bags are sometimes used as a substitute. Female *S. lessoniana,* escorted by mated males, enter the trap in search of appropriate substrates for spawning. After entering the trap, a female attaches her eggs to the net covering. Chotiyaputta (1984, 1988) and Roongratri (1997) reported that the sex ratio of squid captured in traps in Thai waters is close to 1:1.

The squid trap is an appropriate fishing gear in view of natural resource conservation, as only full-grown stocks are exploited. Chotiyaputta (1988) reported that 80% of the squid captured in traps are larger than 140 mm and are therefore fully mature. The fishing period consists of 20 days per month for about 9 months each year. The remaining periods are reserved for natural stock recruitment.

Aquaculture of *S. lessoniana* has been successfully performed in various places and times. Descriptions of aquaculture efforts include those by: Choe and Ohshima (1961, 1963), Ohshima and Choe (1961), and Choe (1966a, b) of efforts in Korea–Japan; SEAFDEC (1975) of efforts in the Philippines; Nabhitabhata (1978, 1996), Nabhitabhata and Kbinrum (1981), and Nabhitabhata et al. (1984, 1992b, 1996) of efforts in Thailand; Saso (1979), Tsuchiya (1982), Segawa (1987, 1990, 1993), and Ikeda et al. (2003, 2009a) of efforts in Japan; Sivalingam et al. (1993) of efforts in India; Lee et al. (1994, 1998a, b) of efforts in the USA; and Ahmad and Usman (1997) of efforts in Indonesia. All of these studies agree that the very high growth rate and large final size of *S. lessoniana* make them a promising candidate for commercial aquaculture.

In Japan, *S. lessoniana* rearing experiments date from the early 1960s when Ohshima and Choe (1961) reared hatchlings for up to 45 days in an open seawater system. Similar rearing experiments were also carried out by regional fisheries experimental stations. In the 1980s, Tsuchiya (1982) succeeded in rearing *S. lessoniana* in an open seawater system from hatchlings to maturity in 306 days. Later, Segawa (1987, 1993) reported the embryonic development and feeding characteristics of juvenile squid using observations made during rearing experiments. Since then, *S. lessoniana* has been recognized as an advantageous squid for aquaculture because this species can survive for long periods in captivity, as can cuttlefishes. Lee et al. (1994) succeeded in culturing *S. lessoniana* for three generations in a closed water system and suggested this squid might be used as a model for neuroscientific and behavioural studies. Advanced trials were carried out by Walsh et al. (2002) who cultured *S. lessoniana* for seven generations. Ikeda et al. (2003) also established culture traits of *S. lessoniana* in a closed seawater system, with the goal of using this species to study brain science.

17.2 State of the Art

The life cycle of *S. lessoniana* can be completed under aquaculture conditions. Broodstock are either collected from the wild, as they will spawn in captivity, or cultured from egg capsules collected as a by-product of fishing. This is the initial phase of culture. Hatchlings are fed for up to 30 days on live feed (i.e. mysid shrimps or shrimp postlarvae), collected from the wild or produced by hatcheries. Similar facilities and management can be used in both closed and open systems for the nursing of eggs and the nursing of young squid after hatching. Training the young to feed on dead feed is the initial and critical period of the ongrowing phase. A very high growth rate is a hallmark of the aquaculture of this species. Feeding efficiency is high from hatching to maturity. Cultured squid can reach 200 mm length and 500 g weight in 120 days. Culture of this species in floating net cages and other in situ facilities can reduce the cost of production. Future studies should focus on live and artificial feed as well as taking advantage of size dimorphism, including monosex culture and induction of sex reversion from females into larger males and differences in the growth rate of different morphs.

17.3 Culture Methodology

Culture methodology is expressed in the Sects. 17.3.1 ‘Open Seawater System in Tropical Countries’ and 17.3.2 ‘Open and Closed Seawater System in Temperate Countries’ because of two reasons. First, the methodology developed in different countries serves different main purposes: production of human food in tropical countries and production of experimental animals in temperate countries. Second, some aspects of biological information obtained from studies in different climate zones are contradictory, which are considered to be the results of different natural environment as well as different conditions and management in different culture systems.

17.3.1 Open Seawater System in Tropical Countries

17.3.1.1 Background

Countries in tropical zones are located near the equator, where temperatures are higher (20–30 °C), and fluctuate less, than those in temperate zones. The high temperature enables an impressively high *S. lessoniana* growth rate in culture. On the other hand, the high growth rate is accompanied by shorter life span and smaller final size. Studies of cephalopod culture in tropical countries are performed with the goal of providing food for human consumption. The information in this section comes mainly from Thailand and other Southeast Asian countries such as Indonesia, the Philippines, and India. Because these countries are developing countries, open seawater systems are suitable and typically used in studies of *S. lessoniana* culture. Open seawater systems require low investment, have low operational costs, and consume less energy than closed seawater systems. In order to reduce costs, researches in these countries are studying *S. lessoniana* aquaculture using in situ facilities, i.e. net cages or pens.

17.3.1.2 Broodstock Maintenance

S. lessoniana broodstock can be collected using hand jigs, squid traps, purse seine, or lift nets operating near the shore. Squid-trapping fishermen in Thailand maintain live squid in a ‘live chamber’ in their boats. The live chamber is a partition of the boat that water can flow through and is similar to that used for *Sepia pharaonis* (see Chap. 12). Jigs and lift nets are also proper gear for collection of broodstock. At landing, squid can be transferred to 500-L plastic tanks for transport to hatcheries. However, bigfin squid have a high metabolic rate and must be placed in oxygen-rich seawater immediately after landing. A change of water or a flow-through system is necessary after inking by the squid that is caused by handling stress. The high dissolved oxygen content of the water should also be maintained during transportation. In hatcheries, broodstock are maintained in 2 m^3 concrete tanks at a male to female ratio of 1:1 (Nabhitabhata et al. 2005). Pieces of fishing net with sinkers are placed in the tanks as artificial substrates for the laying of eggs.

S. lessoniana reproduce in tanks, but the acclimatization period depends on the condition of the animals. A healthy pair will mate within a few hours after release into a tank. Consequent spawning takes place soon after (Nabhitabhata et al. 2005). The female swims around touching substrates with her arms in order to examine the surface and shape of materials before attaching her egg capsules. Each female spawns one to three egg masses. Each egg mass contains 100–400 egg capsules that are attached to each other at the capsule stalks. Spawned egg capsules are sometimes removed from their substrates by other males, although the male mate will attempt to defend them. The spawning period may last for as long as 10 days (Nabhitabhata 1983, 1996). After the last spawning of the female mate, the healthy male turns to another female. If another female is already accompanied by her male mate, agonistic contests between two males will occur. The larger male (regardless of the prior mating status of the male) generally prevails, provided he is healthy (normal performance of routine and ritual behaviour is observed).

In vitro fertilization between mature oocytes from the female oviducts and sperm from the spermatophores in the male Needham's sac has been studied. About 50 % of the oocytes were successfully fertilized (Nabhitabhata et al. 2001a). The oviducal and nidamental gland jellies are not applied to the eggs. The eggs develop for about 4–10 h to the blastula stage before ceasing.

17.3.1.3 Nursing of Egg Capsules

Egg-Capsule Characteristics

Egg capsules are white, opaque, slender, and cylindrical in the first stage of development. Between 2 and 11 eggs in separate chambers aggregate into a one-string capsule. About five to six eggs per capsule are regularly observed in tropical waters with an average temperature of 28 °C. Larger females tend to lay longer capsules with larger numbers of eggs than smaller females (Nabhitabhata 1983). Egg capsules are attached in clusters of 100–400 capsules. Egg capsules gradually absorb water and increase in size along with the size of the embryos. The shape of each egg capsule changes to a tamarind- or finger-like shape, and the capsule wall becomes more transparent as development proceeds. Eggs can better tolerate transportation and handling during early embryonic developmental stages than during later stages.

Egg capsules collected from the wild vary in size (Fig. 17.1) and embryonic stages. Therefore, the egg capsules must be graded. Newly spawned capsules are at least 50 % smaller than those almost ready to hatch. Egg capsules are graded into four stages according to their size and shape. Newly spawned egg capsules are in the first stage and have a cylindrical shape. Egg capsules in the second stage are similar in shape to those in the first stage, but have comparatively larger egg chambers. In the third stage, the capsules have a finger-like shape, and each chamber of the egg capsule is globular in shape and has become more transparent. Each chamber of the egg capsule becomes cube shaped in the fourth stage.

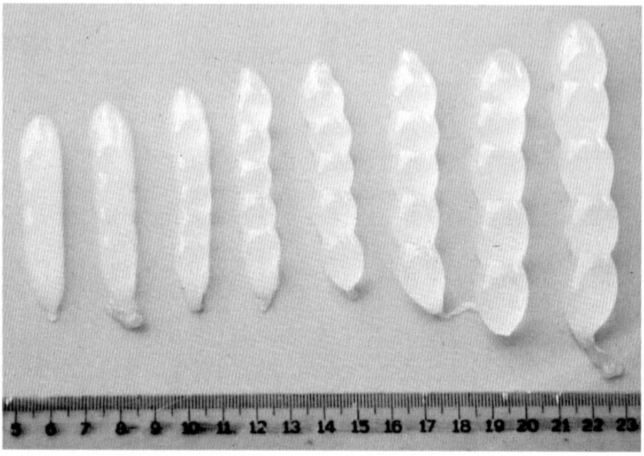

Fig. 17.1 *Sepioteuthis lessoniana* egg capsules in sequential stages of development (*left* to *right*). (Photograph of J. Nabhitabhata)

Water Quality and System Requirements

Egg capsules that contain many eggs require efficient aeration. There must be enough space between egg capsules to allow a sufficient water current to carry enough oxygen to the capsules. Also, with regard to sanitation, elimination of unfertilized eggs and empty egg capsules is conveniently carried out in plastic baskets that float in concrete tanks until the eggs hatch (Nabhitabhata 1978, 1996; Nabhitabhata and Kbinrum 1981; Chindamaikul et al. 1994b). Before spreading in plastic baskets, the capsules must be separated from their clusters.

Nursing of egg capsules can continue under controlled conditions in the tanks previously used for acclimatization in hatcheries. In an open system, changes in physical and chemical parameters should be minimized. Egg capsules are visually checked every day, and those that die, containing abnormally developed embryos, or are empty are discarded in order to prevent infection of others or decomposition and reduction of water quality. Mechanical stimuli and brief changes in temperature or salinity can cause premature hatching. A water flow-through method is used to exchange tank water and to minimize temperature fluctuation in open systems, and flow is maintained at a rate of 1 L min^{-1}. Fresh seawater is pumped through a carbon filter and stored in a concrete tank before it is gravity-fed to the hatchery. Using this system, the water in nursing tanks can be maintained at an average temperature of 28 °C, salinity of 30–33 psu, and pH of 6.0–8.0 (Nabhitabhata et al. 2005). The salinity required to achieve a hatching rate of at least 50 % is 22–37 psu (Nabhitabhata et al. 1991b, 2001c).

Other important physical conditions include turbidity, or suspended solids, and light. Turbidity should be kept as low as possible (Nabhitabhata 1993) through filtration and/or prior sedimentation, particularly in open systems. High levels of suspended solids or high turbidity are critical detrimental parameters. The use of lighting and unnatural daily light–dark periods should be avoided. The growth of algae and fungi on the surface of the egg capsules due to excess light reduces the hatching rate by blocking oxygen supply to the embryos. The attachment of algae can also initiate fungal growth on the capsule membrane and lead to infection of the embryo. The most convenient way to reduce incident light is by curtaining the

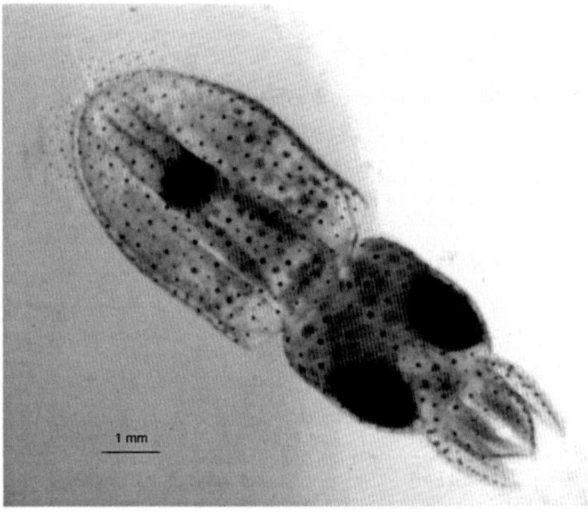

Fig. 17.2 Planktonic hatchling of *Sepioteuthis lessoniana*. (Photograph of J. Nabhitabhata)

hatchery with a camouflage net that reduces light efficiency by 80 % (Nabhitabhata et al. 2005).

Hatching Event

The hatching rate depends on the biotic and abiotic history during incubation, and embryonic development. The average embryonic period is 20 days (a range of 17–23 days) at about 28 °C (Nabhitabhata 1978, 1996; Nabhitabhata and Kbinrum 1981) and is longer at lower temperatures. The hatching period, from the hatching of the first egg to the hatching of the last eggs in the same cluster of the same spawning, is 3–7 days. More than 50 % of the hatching occurs at night after 2400 h on the second and third day from the first hatching. The hatching rate is generally more than 90 % but may vary depending on the embryonic stage at the time of collection from the wild. Hatching rates of eggs spawned in captivity are low due to a larger percentage of unfertilized eggs and abnormally developed embryos than are found in eggs collected from the wild. The hatchlings pass through the basket mesh into the nursing tank. When the optimum density of hatchlings in the tank is achieved, the egg baskets are transferred to another tank containing (previously prepared) water of similar quality.

Egg baskets are transferred between tanks to control the hatchling density in tanks and to grade the size of hatchlings. Egg-capsule-nursing tanks are then used to nurse the young. Hatchlings may be handled if extreme caution is used to avoid skin damage that could subsequently cause mortality. Transfer of hatchlings is better avoided, and it is much more appropriate to transfer egg capsules sometime before hatching.

17.3.1.4 Nursing of Young

Characters and Habits of Young

S. lessoniana hatchlings are planktonic (Fig. 17.2) and, compared to other loliginids, are much larger with mantles that are 3.5–6.4 mm, or approximately 5.4 mm, in length (Lee et al. 1994; Nabhitabhata 1978, 1996), and with BWs of approximately 0.04 g. Hatchlings in good health swim with their heads down and hover at a 45–60° angle at approximately 75% of the total water depth above the bottom of the tank. They begin to feed immediately after hatching. Hatchlings are photopositive and prefer low-intensity light (Nabhitabhata 1978; Nabhitabhata and Kbinrum 1981). Hatchlings are dark brown in the light and are pale or transparent in dim light or dark areas. The arms are still comparatively short, but they are strong enough to catch prey that are equal to, or sometimes larger than, their own ML. When feeding on small crustaceans, hatchlings seize prey using their arms and the prey are bitten at the posterior part of the carapace. The tail parts of the prey are eaten first, and then the flesh is gradually consumed. The hard structures of the prey are left uneaten.

Water Quality and System Requirements

The planktonic habit of hatchlings requires a directed current that allows them to hover in the water column. Aeration has to be managed to generate a directional flow in the tanks. Adjusting the current velocity to an optimum velocity can reduce stress and the energy consumption required for countercurrent swimming. The optimum current velocity can be observed from the angle of hovering. The hovering angle should be 45–60° to the tank bottom at all times. A simple air-lift system is used to generate the appropriate horizontal flow in the cylindrical tank. In a circular tank that is 2–3 m in diameter with a water depth of 0.6 m, such a system can be constructed from a PVC pipe that is 50 mm in diameter and 800 mm in length and that has been cut longitudinally in half (for details of a similar system on a smaller scale, see Chap. 15 '*Euprymna hyllebergi*'; Fig. 15.3). Each half is drilled at one end where an air pipe and air stone are inserted into the interior face. Both prepared half-pipes are placed facing the same direction at a 45° angle with the air stone at the bottom of the tank. The number of these flow-generating pipe sets used in one tank depends on the size and shape of the tank. At least two sets are required in each circular tank and at least four sets are needed in each rectangular tank, one set at each corner. The current velocity can be adjusted by adjusting the aeration rate.

The maintenance of water quality and tank management in the nursing phase may need to be adjusted from those of the egg-nursing phase. Hatchlings can tolerate a range of salinity from 20 to 40 psu for at least 24 h. The salinity required for more than 50% survival is 23–36 psu and the pH required for more than 50% survival is 6.3–8.4 (Nabhitabhata et al. 1991a, 1992a, 2001c; Nabhitabhata and Kbinrum 1984). *S. lessoniana* hatchlings can tolerate ammonia concentrations (NH_4-N) up to 0.06 $mg \cdot L^{-1}$, nitrite (NO_2-N) up to 0.08 $mg \cdot L^{-1}$, and phosphate (PO_4-P) up to 6.0 $mg \cdot L^{-1}$ (Chindamaikul et al. 2001; Chainak et al. 2003). The routine management of nursing tanks follows directly from the egg-nursing phase.

In larger tanks, the water level in the nursery tanks is adjusted to accommodate the number and size of live feed organisms. An appropriate level of the feed and cephalopod biomass in the tank reduces excess food-hunting activity. The initial water depth is 500 mm at hatching and the density of hatchlings is 5–10 individuals·L^{-1}. As the squid grow, the water depth is increased by 50 mm every day or every second day. Squid are graded by size every 10 days, and squid of similar size are grouped together for subsequent nursing. The density of squid is decreased by about 20–25 % in each group after grading.

Feeding

S. lessoniana hatchlings begin to feed immediately after hatching and must be provided with live feed of appropriate size, which is 50–200 % of their ML. Wild-caught live mysids (*Mesopodopsis orientalis*) that are about 6–7 mm in total length are key to the successful nursing of hatchlings (Nabhitabhata 1978). The live feed is stocked and fed to the squid ad libitum. Pygmy goby fry (*Stigmatogobius romeri*) are also a good feed organism (Chankaew et al. 2003a). Other wild zooplankton and fish larvae are suitable food, but as the supply of these organisms is not consistent, they are not reliable enough for the mass production of squid.

Feed organisms that are commercially produced can ensure more reliable and consistent success of squid hatchery operations. Postlarvae of the penaeid shrimp (*Penaeus merguiensis*) that are 5–10 mm in total length and sea bass fry (*Lates calcarifer*) that are 5 mm in length can be used as the primary foodstock. About 7 (1–22) postlarvae or fish are consumed daily by each squid hatchling (Chankaew et al. 2003b; Nabhitabhata et al. 1996). Other crustaceans that have been tried did not result in good squid production. The crab mysis *Portunus pelagicus* can initiate seizures in squid hatchlings if fed in the first 2–3 days after hatching. Young squid fed freshwater prawn mysis (*Macrobrachium rosenbergii*) each consumed about 140 individuals per day, but only 8 % of the hatchlings survived longer than 10 days (Nabhitabhata et al. 1992b). One reason for this low rate of hatchling survival was that the prawn mysis could not survive in 30 psu seawater long enough to provide the required quantity of feed for the squid hatchlings.

The survival rate of young squid fed shrimp postlarvae (*P. merguiensis*) is approximately 80 % (22–100 %) from hatching to 10 days of age (Chindamaikul et al. 1994a; Nabhitabhata 1978; Nabhitabhata and Kbinrum 1981; Nabhitabhata et al. 1996). The survival rate during this period decreases if the supply of feed organisms is insufficient (Nabhitabhata 1978, 1996; Nabhitabhata and Kbinrum 1981).

The daily feeding rate of squid hatchlings is about 28 % of their BW and the average conversion rate is more than 50 % (28–95 %; Nabhitabhata et al. 1992b, 1996). The unit cost of production in Thailand has been estimated to be about US$ 10 for 100 hatchlings with a feed cost of about US$ 9 in 1996 (Nabhitabhata et al. 1996), when US$ 1 is approximately equal to Thai Baht 25 (US$ 1 equals Thai Baht 30 in 2013). At 30 days after hatching, survival is about 60 % (10–74 %; Nabhitabhata 1978; Nabhitabhata and Kbinrum 1981; Nabhitabhata et al. 1996; Table 17.1).

Young squid can alternatively be nursed in a floating net cage. Ahmad and Usman (1997) nursed their *S. lessoniana* from hatchlings in a net cage that was 1.5 × 1.5 × 1.5 m and provided *Mesopodopsis* mysids as feed organisms. The growth was rapid with an increase of 2.9–4 g after 30 days, as compared to about 2 g in

Table 17.1 Comparison on culture conditions and production of *Sepioteuthis lessoniana* from various studies

Aspect	Ohsima and Choe (1961); Choe and Ohshima (1961, 1963); Choe (1966a, b)	SEAFDEC (1975)	Nabhitabhata (1978, 1996)	Tsuchiya (1982)	Segawa (1987)	Sivalingam et al. (1993)	Lee et al. (1994)
Tank system	Open	Open	Open	Open	Open	Open	Closed
Egg-capsule incubation (days)	25–28	14–21	20.3 (17–23)	23.2 (21–25)	24–27, 19–23	–	–
Temperature (°C)	23.5–24		28	24	25, 30		
Hatching period (days)	–	–	–	6–10	–	–	–
Hatchling size (mm, g)	5.6, 0.04	6.5, 0.03	5.4, 0.04	5.8, 0.04	5.0, 0.06	5.2, 0.014	5.3, 0.08
Initial density		m^2		500 m^3			
Age at maturity (first mating: days)	90	–	60	–	–	–	130–160
First spawn size at maturity (mm, g)	–	113	112 (90–166)	136	113	–	148–230
Daily growth (%: ML, W)	4.6, 9.6 (45)	2.8, 8.12	2.8, 6.4	3.6, 2.6	2.3, 4.4	4.4, 10.5 (57)	–, 3.9–5.3
Mean final size (mm, g)	–	143, 158	213, 497	–	–	–	214–260, 682–1,073
Max size (mm, g)	–	368, –	252, 605	260, 810	–	–	350, 2,210
Mean span (days)	–	–	137	136	–	–	–
Max span (days)	–	115	176	306	179	–	333

ML mantle length, *W* weight

concrete tanks. The cost of production using this method may be lower than that in tanks because the cost of the aeration and water quality maintenance required for both open (water flow-through) and closed systems is eliminated.

Growth

Beginning 10 days after hatching, the habit of the young *S. lessoniana* obviously changes from planktonic to pelagic or nektonic. The young lower their level of hovered swimming from near the surface level to a depth of about 50% of the total water depth. They begin to gather in schools of similar body size and respond to stimuli in a synchronized manner. The ritual colour pattern on the dorsum changes from dark brown to green and golden brown with white spots. The FL increases from approximately 33% of the ML at hatching to 74% of the ML at 30 days of age (Nabhitabhata 1978, 1983, 1996; Nabhitabhata and Kbinrum 1981).

The rate of growth is highest over the first 10-day period. In open systems, *S. lessoniana* grow to 11 mm and 0.55 g by the tenth day after hatching, with a daily growth rate of 6.7% for ML and 17.3% for weight (Fig. 17.3). During the first month (0–30 days), the daily growth rate for ML is 2.8–6.7% and the daily growth rate for weight is 6.4–17.3%. The relationship between ML (mm) and weight (W; g; Fig. 17.4) in the first (nursing) phase from hatching to 30 days of age at 28 °C in open seawater systems (Nabhitabhata 2002) is given by:

$$W = 1.250 \times 10^{-3}\ ML^{1.508} \tag{17.1}$$

Growth in terms of the ML–age (T; d) relationship (Fig. 17.5) in the first (nursing) phase at 28 °C is given by:

$$ML = 0.590\ e^{0.0496T} \tag{17.2}$$

17.3.1.5 Ongrowing

Water Quality and System Requirements

The ongrowing tanks and water management systems for open systems are similar to those for nursing of the young. Open systems operate at 24–30 °C, a higher temperature range than closed systems (20–24 °C). This results in a shorter life span and smaller final size.

Feeding

Training *S. lessoniana* to feed on dead feed is a critical process which affects the subsequent success of the ongrowing phase. At 20 days after hatching, young

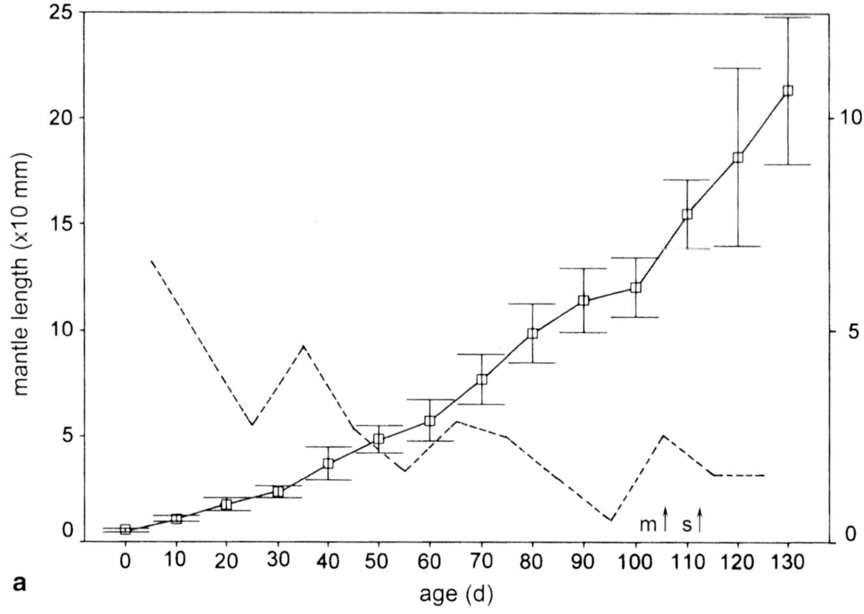

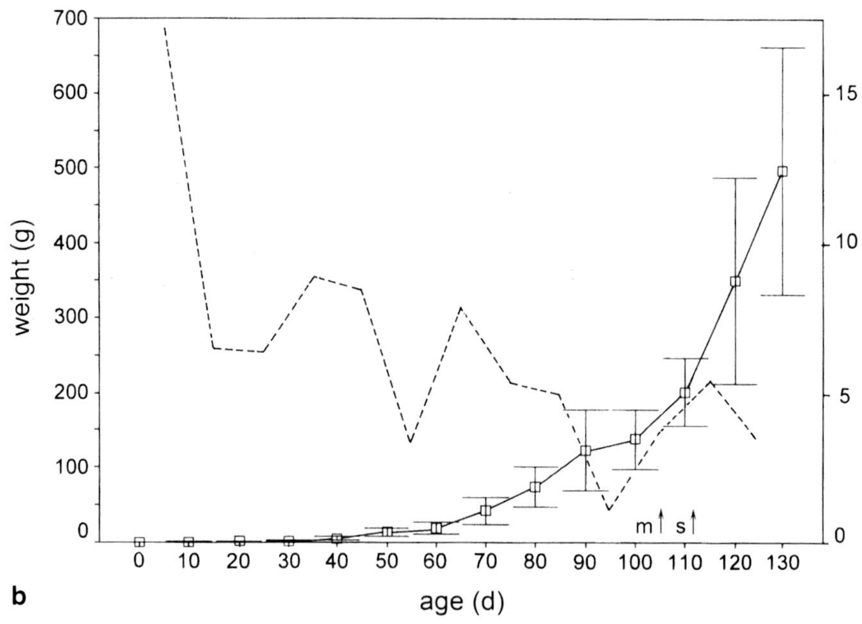

Fig. 17.3 Growth in terms of the relationships between mantle length (mm, *above*), weight (g, *below*), age (d, *solid line*), and daily growth rate (%, *broken lines*) in length and weight in an open seawater system. *Arrows* indicate mating (*m*) and spawning (*s*). (From Nabhitabhata 2002)

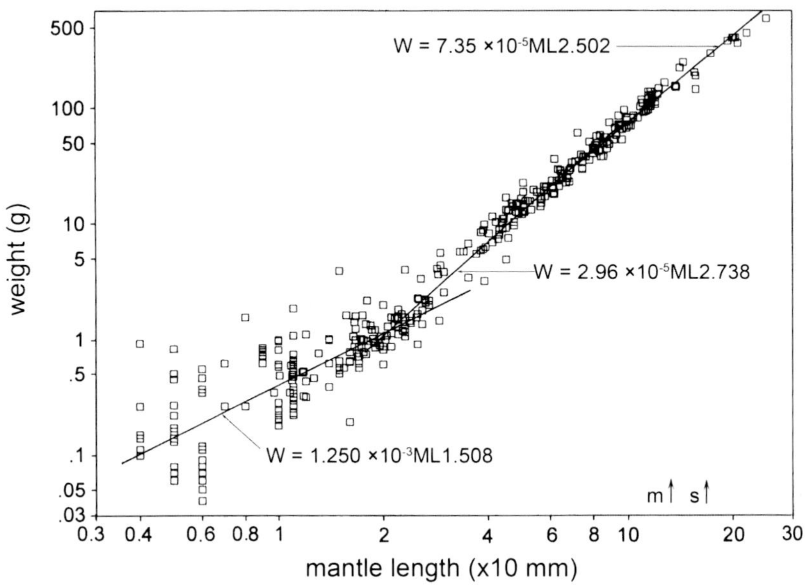

Fig. 17.4 Growth in terms of the relationship between mantle length (mm) and weight (g) in an open seawater system. *Arrows* indicate mating (*m*) and spawning (*s*). (From Nabhitabhata 2002)

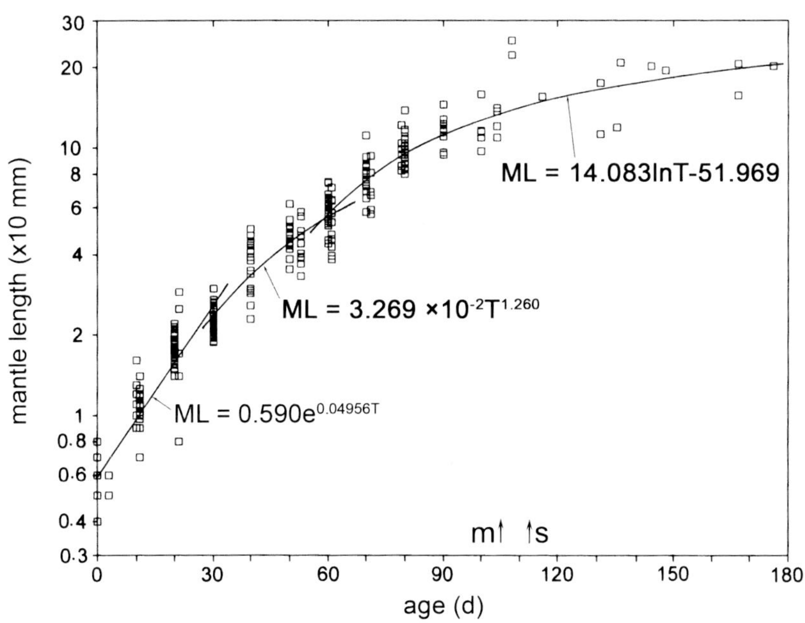

Fig. 17.5 Growth in terms of the relationship between mantle length (mm) and age (d) in an open seawater system. *Arrows* indicate mating (*m*) and spawning (*s*). (From Nabhitabhata 2002)

S. lessoniana are trained to feed on dead meat. Squid are fed on sliced fish meat to satiation twice daily at 0800 and 1600 h. There has been no sign of specific preferences for the species of fish consumed. The sizes of the fish chips correspond to the size of the squid, and are one to two times the ML. Squid seize their food in the water column and eat it while hovering. When squid feed on live prey, the prey is seized with the tentacles and retained using the arms. When the squid are fed dead feed, they change their feeding behaviour; they use only their arms to seize the food and do not perform a positioning step. This change indicates the success of the training period and that the squid are now accustomed to the dead feed and culture conditions (human appearance with feed and the environment of the culture tank). One reason for the change in behaviour is the lower motion of the dead feed compared to active live feed. This kind of behavioural change is similar to that observed for cultured cuttlefish, *Sepia pharaonis* and *Sepiella inermis* (see Chaps. 12 and 13). Aggressive behaviour and cannibalism become prominent at this age whenever food is in short supply and/or there are squid with a wide range of sizes in the same tank. Larger starved squid that have not learnt to consume dead feed may attack and feed on smaller squid in the same tank. Human trainers must ensure sufficient feeding and that every squid is fed.

Growth

Growth is very rapid although the rate of increase in size and weight decreases as maturity approaches. In open seawater systems, the mantle growth rate decreases to less than 3 % daily after 40 days of age and the rate of weight gain decreases to less than 5 % after 50 days of age. For *S. lessoniana* reared at 28 °C, the overall average daily growth rate from hatching to 130 days of age is 2.8 % for length and 6 % for weight (Nabhitabhata 1996). As in the first 30 days after hatching, the relationship between ML and W (Fig. 17.4) is exponential in the second phase (30–60 days after hatching) and the third phase (60–176 days after hatching; Nabhitabhata 2002). The 'b' values from aquaculture batches, 2.738 and 2.502, are higher than those from wild stocks, 2.218–2.477 (Rattana-anant 1978, 1979, 1980; Roogratri 1997; Sivashanthini et al. 2009), revealing higher growth rates under culture conditions than those that occur in the wild. Growth models for cultured batches at 28 °C in open seawater systems (Nabhitabhata 2002) are:

$$W = 2.96 \times 10^{-5}\, ML^{2.738} \tag{17.3}$$

$$W = 7.35 \times 10^{-5}\, ML^{2.502} \tag{17.4}$$

Growth, in terms of the ML–age relationship (Fig. 17.5), in the second phase and the third phase can be modelled as:

$$ML = 3.269 \times 10^{-2}\, T^{1.260} \tag{17.5}$$

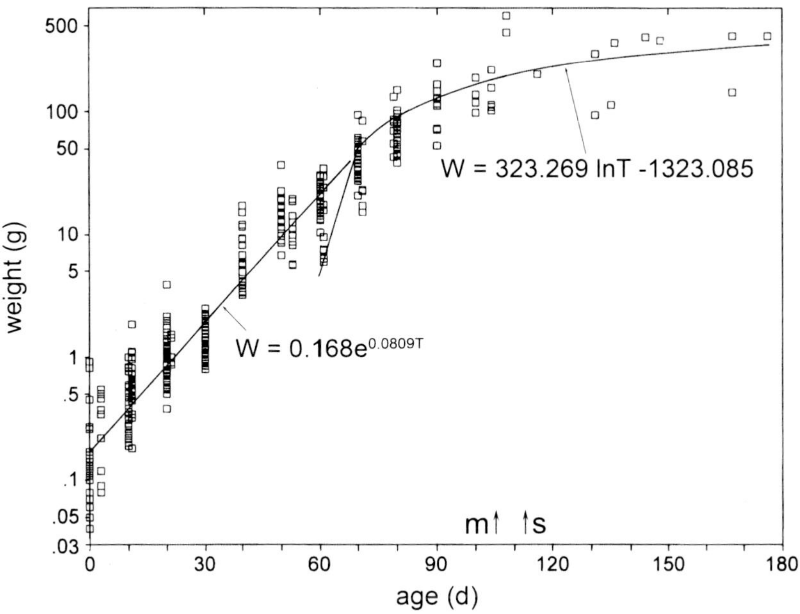

Fig. 17.6 Growth in terms of the relationship between weight (g) and age (d) in an open seawater system. *Arrows* indicate mating (*m*) and spawning (*s*). (From Nabhitabhata 2002)

$$ML = 14.083 \ln T - 51.969 \tag{17.6}$$

Models that fit plots of the weight–age relationship (Fig. 17.6) from hatching to 60 days of age (early phase) and from 60 to 176 days of age (ongrowing phase) are:

$$W = 0.168\, e^{0.0809T} \tag{17.7}$$

$$W = 323.269 \ln T - 1{,}323.085 \tag{17.8}$$

In open seawater systems at 28 °C, mating occurs at about 90 days of age and spawning is 1–3 weeks later, at about 110 days of age. The average life span of cultured batches is about 130–140 days, or within 1 month after spawning. The final size of cultured squid is about 250 mm in length and 600 g in weight (Nabhitabhata 1978, 1983, 1996; Nabhitabhata and Kbinrum 1981; SEAFDEC 1975; Tables 17.1 and 17.2).

Alternative Culture Facilities

Ongrowing in net cages that float in natural waters is another option for *S. lessoniana* culture. Nabhitabhata et al. (1984) suggested that the most appropriate structure was a box-type cage, hanging from a floating buoy or a bamboo raft with sinkers

Table 17.2 Expected growth, survival, and managed density of *Sepioteuthis lessoniana* under culture conditions (24–28 °C). Estimation from Choe (1966a, b); Choe and Ohshima (1961); Lee et al. (1994); Nabhitabhata (1978, 1997); Nabhitabhata and Kbinrum (1981); Ohshima and Choe (1961); SEAFDEC (1975); Segawa (1987); Sivalingam et al. (1993), and Tsuchiya (1982)

Culture period (days)	ML (mm)	DGRL (%)	W (g)	DGRW (%)	Density (ind.m^{-3})	Water depth (mm)	Survival (%)
0	5.5	–	0.04	–	500	50	–
30	25	4.5	2	9.5	150	60	60
60	50	2.5	15	6.0	90	100	50
90	95	2.0	80	5.0	40	100	40
120	150	1.5	200	2.5	20	100	30
130	160	–	220	–	–	–	–

ML mantle length, *DGRL* daily growth rate in mantle length, *DGRW* daily growth rate in body weight, *W* weight

at the corners. The cages are made from a monofilament nylon net with a 13-mm mesh size, corresponding to the initial size of squid. Underwater hard frames for cages are avoided in order to prevent collisions or rubbing of squid that cause skin damage and consequent infectious disease. In order to fit into the existing net-cage facilities, the cage dimensions are 2×2.5×2 m and are submerged to a depth of 1.5 m. Cages therefore contain 7.5 m^3 seawater and have a 5-m^2 surface area. The optimum squid density is reported to be about 5–10 individuals·m^{-2} (3–7 ind·m^{-3}) through a 60-day period for squid with an initial ML of about 50 mm length and weight of 16 g (40 days; Nabhitabhata 1978, 1996; Nabhitabhata and Kbinrum 1981; Nabhitabhata et al. 1996). Because the nets are made from monofilament, cages should be left at the ongrowing sites until fouling organisms are observed growing on the net. The fouled net has enhanced visibility to the squid and reduces their collisions with the net.

17.3.2 Open and Closed Seawater System in Temperate Countries

Methods of aquaculture for *S. lessoniana* in tropical countries and temperate countries can share many things. In this section, the methodology specifically applied to closed water systems and open water systems in temperate countries will be described for comparison.

17.3.2.1 Broodstock Maintenance

In Japan, jigging and set net are the main fishery tools for collecting subadult and adult *S. lessoniana*. Munekiyo and Kawagishi (1993) and Ueta (2000a) found that *S. lessoniana* caught by set net follow a lunar cycle; the catch amount increases at the full moon and decreases at the new moon. The sex ratio of *S. lessoniana* caught by set net is estimated to be 1:1 (Ueta and Jo 1989). Precise statistical data for the total landing of *S. lessoniana* from all Japanese waters are not available. Ueta (1992) estimated the annual *S. lessoniana* catch per year in the coastal waters of Tokushima Prefecture, where *S. lessoniana* is abundant, to be 37–57 t by set net and 8–24 t by jigging (resource from 1985 to 1988). Based on 13 years of survey data (resource from 1984 to 1996), Ueta (2000a) estimated the annual catch per year for this squid in Tokushima Prefecture to be 79.1 ± 44.1 t.

Because squid caught by set nets can be kept alive and in good condition in net cages, set nets are suitable for supplying live specimens, at a relatively high price, to traditional Japanese restaurants and sushi bars. Squid caught by set net are also suitable for broodstock maintenance and aquaculture. Ikeda et al. (2004) transported young and subadult *S. lessoniana* from set nets and reared them as the origin for culture trials. The approximately 740-km transfer took 22–23 h, and both water quality and temperature had decreased by arrival (pH < 7.0, NH_4-N 5.0 mg·L^{-1}, temperature > 4 °C). Squid survival is high when the BW to seawater volume ratio is below 30 at shipping. Survival is also relatively high in large containers, such as Styrofoam boxes with large (> 1,620 cm^2) bottom areas. An acclimation period of suitable duration is necessary when transported squid are moved to new aquaria. Squid with a short acclimation (< 1 h) period die soon after being transferred to the aquaria, but those with a longer acclimation (> 2 h) can survive (e.g. shipping container water differs from aquarium water by 5 °C and 1.2 pH units).

Set nets also act as a spawning bed for *S. lessoniana*. Therefore, it is easy to get freshly laid egg capsules for use in aquaculture. Three types of artificial *S. lessoniana* spawning beds were examined in the spawning grounds in the coastal waters of Tokushima in Japan (Ueta et al. 1995; Ueta and Kitakado 1996). These were fibreglass-reinforced plastic (FRP) type, mid-water cage type, and steel type. The FRP type consisted of 9- or 12-mm-diameter FRP that were spaced 5, 10, 15, or 20 cm apart in 1 × 1 × 0.5 m concrete blocks. The mid-water cage type consisted of 0.5-, 1-, or 2-m-high cylindrical cages with a 9-mm rope that was stretched between the walls, which were linked to three other cages, a float, and a concrete block. The steel type consisted of 49 vertical and 0–6 horizontal 19-mm steel poles fixed to a 1 × 1 × 0.5 m concrete block (Ueta et al. 1995; Ueta and Kitakado 1996). Egg capsules were laid by *S. lessoniana* on these artificial spawning beds with the number of egg capsules laid on FRP type > mid-water cage type > steel type. However, the differences between the numbers of eggs laid on each spawning bed type were small. When using the FRP-type spawning bed, female squid spawn many more egg capsules on poles set close together than on poles that are widely spaced. Furthermore, female squid prefer spawning on these artificial spawning beds to natural spawning beds composed of *Sargassum* and *Zostera*.

Late-stage egg capsules can be collected in order to decrease the necessary incubation time and to avoid cessation of embryonic development at earlier stages in captivity. However, special care must be taken when late-stage egg capsules are collected because accidental pre-hatching often occurs due to physical shock during transportation.

The availability of artificially fertilized eggs would be convenient for both culture programs and experimental studies. Techniques for artificial insemination

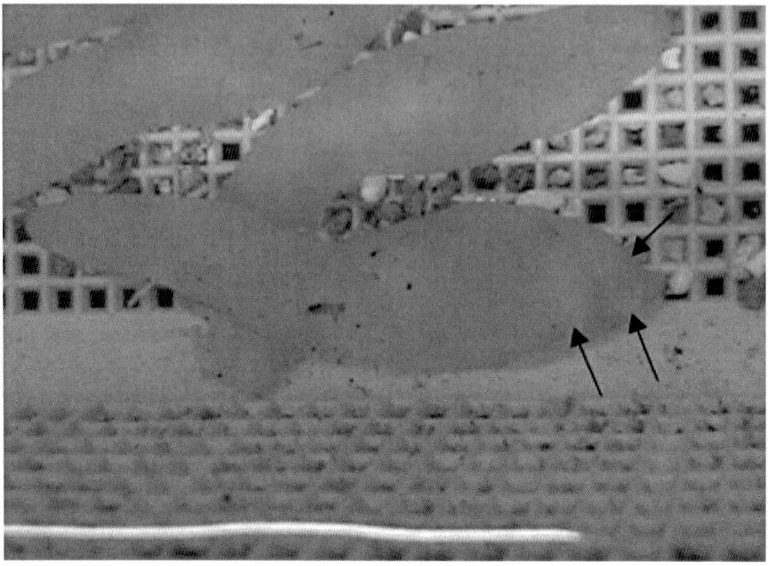

Fig. 17.7 Egg capsules spawned by cultured *Sepioteuthis lessoniana*, with *arrows* indicating ova. Note that one end of the egg capsule is widened due to the abnormal distribution of ova. (From Ikeda et al. 2009a)

have not yet been established for loliginid squid. These squid have relatively large ova that are wrapped in a thick gelatinous layer (i.e. egg capsule) during embryonic development. Artificial insemination of the ommastrephid squid *Todarodes pacificus* is successfully performed by the addition of oviducal gland jelly at, and after, insemination of ova from the oviduct with spermatozoa from the seminal receptacle (Ikeda et al. 1993). The gelatinous substance of oviducal gland origin is necessary for formation of the perivitelline space that allows embryonic development to proceed normally. Oviducal gland jelly would act similarly in *S. lessoniana*.

17.3.2.2 Nursing of Egg Capsules

Egg-Capsule Characteristics

Structural abnormality of spawned egg capsules occurs in later generations of cultured *S. lessoniana* (Ikeda et al. 2009a). The outer gelatinous coat appears to be normal but the arrangement of eggs in the capsules is abnormal. The ova connect at one end of the egg capsule (Fig. 17.7). Most of the abnormally arranged eggs do not develop, which results in a low rate of hatching. Walsh et al. (2002) also reported a decrease in egg viability in their cultured squid. Abnormalities, such as the abnormal arrangement of eggs in egg capsules, may be a consequence of inbreeding.

The number of ova contained in a single *S. lessoniana* egg capsule varies in temperate waters. For example, egg capsules are reported to contain the following number of ova: 2–8 (average 3.0 and 5.8) at Iriomote Island of the Ryukyu Archipelago (Tsuchiya 1981), 2 and 4–8 (mode 6) at Ishigaki Island of the Ryukyu Archipelago (Segawa et al. 1993b), 0–9 (mean 3.5–6.4) at Kominato of Chiba Prefecture (Segawa 1987), and 0–9 (mean 5.5) at Kii Channel in Tokushima Prefecture (Ueta et al. 1992). Ueta et al. (1992) provided the following equation for a natural population of *S. lessoniana* that describes the relationship between the average number of ova in a single egg capsule (X) and the number of ova per egg mass (Y):

$$Y = 214.1X - 621.9\ (r = 0.631,\ p < 0.01) \tag{17.9}$$

Water Quality and System Requirements

If large-capacity tanks that have enough space for floating several baskets cannot be provided, egg capsules can be separated and suspended from Styrofoam pillars that are 2 cm in diameter. Each egg capsule then freely floats on the surface layer of a smaller circular tank (30 cm in diameter, 27 cm in depth, 20 L; Ikeda et al. 2005) or a large circular tank (70 cm in diameter, 34 cm in depth, 120 L; Ikeda and Sugimoto 2013; Fig. 17.8). With this method, every surface of the egg capsules is completely exposed to water. This allows for sufficient oxygen exchange with the eggs, and gentle flow continuously cleans the capsule surface.

Lee et al. (1994); Forsythe et al. (2001), and Walsh et al. (2002; all at the Natural Resource Center for Cephalopods, Marine Biomedical Institute, University of Texas Medical Branch, USA) used 1.8-m-diameter cylindrical tanks for egg nursing in a closed water system. The tanks were supplied by fresh natural seawater drawn through an under-gravel filter and pumped back to the tanks after passing through a particle filter, an activated carbon filter, and two 30-W ultraviolet (UV) sterilizers. The concentration of NH_4-N, NO_2-N, and NO_3-N was maintained under 0.10, 0.10, and 50 mg·L^{-1}, respectively (Walsh et al. 2002). Ikeda et al. (2003, 2009a) circulated the water from 20- and 50-L tanks through coral gravel filters and UV sterilizers. Another option involves pumping water into eight cubic tanks used for growing macro-algae before gravity-feeding the water back to the main tank.

Hatching Event

Variation in the size of hatchlings occurs due to different hatching times. Those that hatch early tend to be smaller than those that hatch later (Ikeda et al. 1999). Smaller, early hatchlings are less able to move in order to feed or escape than their later-hatching counterparts. Special care must be taken for these early-hatched small juveniles.

Fig. 17.8 Tank systems for *Sepioteuthis lessoniana* culture. **a** 50-L cylindrical tank (Multi-hydense® AQUA INC; from Ikeda et al. 2005). **b** 10,000-L cylindrical tank connected to an algae tank. (From Ikeda et al. 2003)

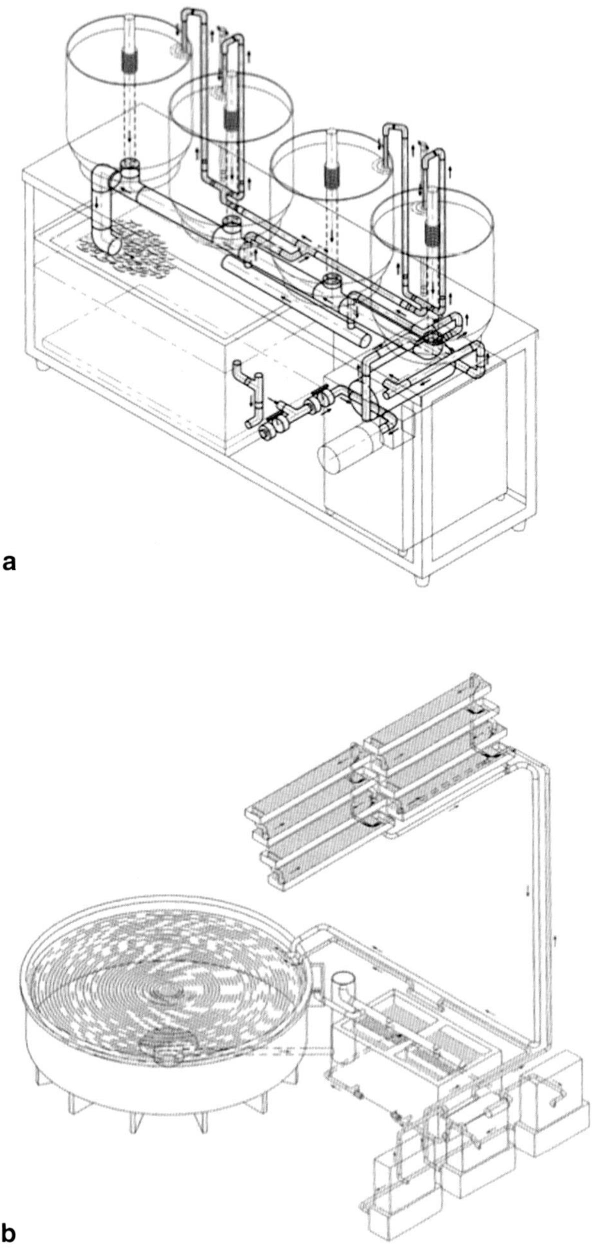

a

b

17.3.2.3 Nursing of Young

Characters and Habit of Young

Compared to other loliginids, *S. lessoniana* hatchlings are much larger, with a 3.5–6.4-mm ML (Lee et al. 1994). Hatching usually occurs at midnight and seldom occurs during daylight hours. After hatching, hatchlings (of ML 5–7 mm) swim gently to maintain their position in the middle to upper water layer, keeping their mantle upward and head obliquely downward (Segawa 1987). Hatchlings also assume an upward or downward V-shaped position, while body colour is kept either all dark or transparent. In the upward V-shaped position, the arms are divided into two rows, like the letter 'V', while in the downward V-shaped position, all arms droop downward (Ikeda, unpublished observation). Hatchlings show positive phototaxis to a beam of light at night (Segawa 1987). Because a similar behaviour is also observed for hatchlings of *Heterololigo bleekeri,* another loliginid squid (Ikeda et al. 2005), this may be a shared characteristic of loliginids at hatching. This characteristic likely contributes to keeping hatchlings at the surface layer where they are most likely to encounter prey organisms and/or avoid predators.

The young begin to form schools in which individuals are positioned parallel to their neighbours, and schools are completely formed within 2 months of hatching (Sugimoto and Ikeda 2012).

Water Quality and System Requirements

In closed water systems, nursing of the young proceeds in the 20- or 50-L cylindrical egg-nursing tanks at a temperature of about 23–24 °C (Ikeda et al. 2003, 2009a; Lee et al. 1994, 1998a, b). Such small tanks are advantageous in that prey organisms aggregate in them and they are convenient for aquaculturists to manage. In addition, less seawater is required for closed systems than for open systems and they are therefore economical. A 120-L circular tank can also be used to rear juveniles and the young in a closed system; this relatively large tank enables the observation of squid behaviours that require space (e.g. schooling and social interaction).

Feeding

Various species of mysids can be successfully provided as prey, e.g. the brackish water crustaceans *Neomysis japonica* (Choe and Ohshima 1961; Ohshima and Choe 1961; Choe (1966a, b), *Siriella longipes* (Segawa 1990), and *Mysidopsis almyra* (Lee et al. 1994). Other potential live feed organisms include small-sized or freshwater fish fry, e.g. guppy, *Poecilia reticulata,* or medaka, *Oryzias latipes* (Lee et al. 1994; Ikeda et al. 2003, 2009a).

Juvenile squid older than 1 month of age can adapt to feeding on dead organisms that are thrown by humans into the water column. At first, squid neglect the dead diet when caretakers throw it into the tank, but after repeated exposure, they begin to pay attention to it and catch it immediately. Juvenile squid will also seize dead diet that is provided by caretakers with forceps. This method ensures that all squid are fed, even when individuals do not hunt food that is thrown in the water column. Providing the diet ad libitum can avoid starvation and cannibalism and is important for culturing squid successfully.

Growth

A large decrease in survivorship occurs in cultures within 30 days after hatching (Lee et al. 1994; Ikeda et al. 2003, 2009a). The most likely reason for the high death rate of *S. lessoniana,* as for other squids, during the early phase of life is starvation as a result of insufficient feeding. Hatchlings of *S. lessoniana* consume both inner yolk and live dietary specimens for up to 1 week (Shimazoe 2010). Hatchlings may learn by trial and error how to catch food during this phase, as do other loliginid squid (Chen et al. 1996).

17.3.2.4 Ongrowing

Feeding

Because a live diet of marine organisms (e.g. zooplankton and fish larvae) is only obtainable by laboratories situated near oceans, dead diets provide a suitable and economical alternative for inland squid cultures. Young *S. lessoniana* are fed on a mixture of live and dead frozen prey items, including Japanese sardine *Sardinops* sp. and Japanese anchovy *Engraulis* sp. (Ikeda et al. 2003, 2009a). Subadult and adult squid can be fed on a variety of prey items, either alive or dead. Lee et al. (1994) estimated the feeding rate of captive *S. lessoniana* (60–300 days of age) in a closed water system to be 20–35 % BW·day^{-1}.

Growth and Survival

Fins of *S. lessoniana* hatchlings are subterminal, small, and round. However, they elongate with growth. Segawa (1987) used the following equation to describe the relationship between FL and ML for squid with ML 2.6–206 mm:

$$FL = 0.975\ ML - 4.34\ (r = 0.9993) \tag{17.10}$$

Kanamaru and Itoh (1996) reported that the ratio of FL to the length of the mantle margin reaches an adult-like value when *S. lessoniana* grow to 50–60 mm in ML.

Segawa (1987) used the following equation to describe the relationship between BW and ML for *S. lessoniana* with ML 5.7–128.2 mm cultured in open water systems (>20 °C):

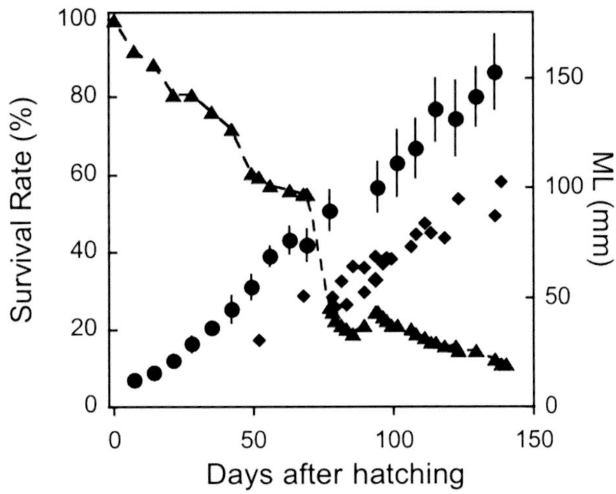

Fig. 17.9 Survivorship of *Sepioteuthis lessoniana* in culture conditions. *Triangles* indicate survival rate; *diamonds* indicate dead individuals; *circles* and *bars* indicate the average and standard deviation of dorsal mantle length (*ML*). (Sugimoto and Ikeda 2013)

$$\text{Squid} < 25\ \text{mm ML, BW} = 0.000398\ \text{ML}^{2.481} (r = 0.9734) \tag{17.11}$$

$$\text{Squid} > 25\ \text{mm ML, BW} = 0.000287\ \text{ML}^{2.680} (r = 0.9914) \tag{17.12}$$

Lee et al. (1994) estimated the instantaneous growth rate of cultured *S. lessoniana* (temperature of 23.2 ± 0.6 °C) to be 3.9–5.3 BW·day^{-1} for the entire life span and 8.2–12.0 BW·day^{-1} for the first 100 days after hatching in closed water systems. Forsythe et al. (2001) examined the effect of temperature on the growth of cultured *S. lessoniana* in a closed water system. They estimated that squid grown at approximately 27 °C attained a size of 10 g in as little as 45 days at a sustained growth rate of 12.2 % BW·day^{-1}, while squid cultured at 20 °C required almost 100 days to attain the same size at a rate of 5.7 % BW·day^{-1}. Kanamaru and Itoh (1996) did not find any difference in the growth of females and males with ML 45–160 mm. On the other hand, Ueta (2000a) used the following equation to describe the relationship between BW and ML for a natural population of *S. lessoniana* with ML 7–470 mm in Japanese waters:

$$\text{Female, BW} = 0.183\ \text{ML}^{2.581} (r = 0.996) \tag{17.13}$$

$$\text{Male, BW} = 0.186\ \text{ML}^{2.562} (r = 0.997) \tag{17.14}$$

Survival of squid older than 30 days does not often change drastically, but intermittent death from unknown causes sometimes occurs in closed water systems. Between 70 and 140 days after hatching, only 1–2 of 25 squid die daily, and those that die are always the smallest in the rearing population (Fig. 17.9; Sugimoto and Ikeda 2013). In addition to the size variation that occurs at hatching, much greater size variation occurs in culture tanks 2 months after hatching (Sugimoto and Ikeda

Fig. 17.10 Size variation of *Sepioteuthis lessoniana* during the post-hatching phase in culture conditions. *Circles* and *bars* indicate the average and standard deviation of dorsal mantle length (*ML*), respectively. (Sugimoto and Ikeda 2013)

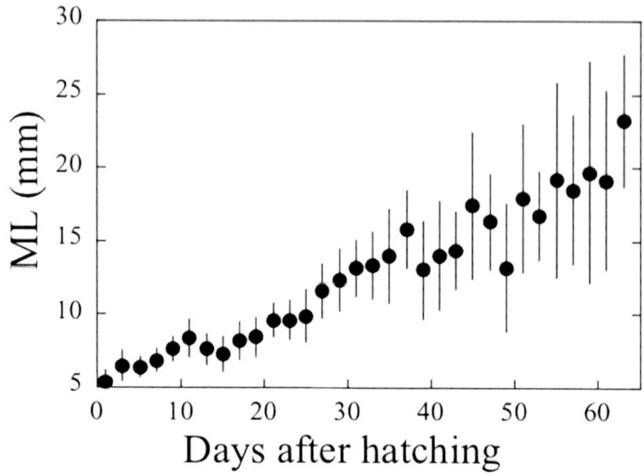

2013; Fig. 17.10). At 60 days post hatching, the largest squid may be more than three times the size of the smallest squid. Whether this difference is genetic or environmental is unknown, and size differences may become much larger in later phases of life under both cultured and wild conditions (Fig. 17.10; Ikeda et al. 2009a). Because smaller squid often become the target of cannibalism (Segawa 1993), selective ongrowing of squid with similar-sized squid in the same tank may contribute to successful culture.

Kanamaru and Itoh (1996) found that it is possible to completely determine the sex of *S. lessoniana* based on characteristic colour patterns on the dorsal mantle (for squid with ML > 165 mm) and the hectocotylized fourth left arm of males (for squid with ML > 85 mm). These criteria can be used to regulate the number of female and male squid in a single tank. Because male squid often become aggressive and engage in agonistic contests after they reach maturity, maintaining a suitable balance of males and females (e.g. male < female) leads to successful reproduction in the tank.

S. lessoniana are estimated to live for about 1 year in Japanese waters (Ueta and Jo 1989; Ueta 2000b). In general, cephalopods reach maturity earlier in captive conditions than in the wild (Mangold 1987). This is also true for *S. lessoniana*. Tsuchiya (1982), who first cultured *S. lessoniana* in an open water system (temperature of 20.81–29.89 °C), observed spawning at 136 days post hatching. Ikeda et al. (2003, 2009a) also observed precocious maturation of cultured squid (temperature of 21.1–23.8 °C; Fig. 17.11). They observed that males reached maturity at 140 days of age, 3–8 months earlier than wild-caught squid of the same generation, and initiated copulation repeatedly. This is also similar for females. Cultured squid begin to mature at least 2 months earlier than wild-caught squid. The constant, relatively high temperature in captivity may accelerate maturation. Female squid spawn 160–196 days post hatching. Sometimes, a single female will spawn more than twice within 40–48 days. Intermittent spawning by a single female was also observed by Wada and Kobayashi (1995) and Ueta (2000b) in an open water system. Both males and females die restively after repeated copulation and spawning. The life span of cultured *S. lessoniana* in closed water systems is reported to be 4–6 months

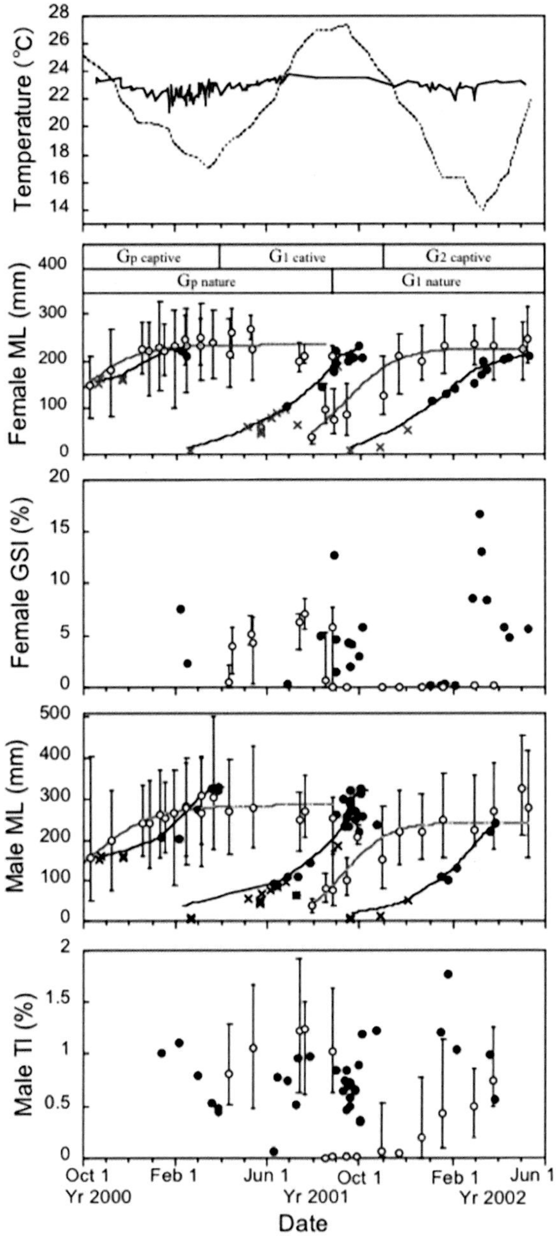

Fig. 17.11 Temperature regime for *Sepioteuthis lessoniana* measured in captivity (*solid line*) and estimated in nature (*dotted line*), the progress of dorsal mantle length (*ML*), female gonad somatic index (*GSI*), and male testis index (*TI*) of cultured and wild-caught squid by date (*X*, cultured squid of unknown sex; +, cultured squid; *W*, wild-caught squid). Data for wild-caught squid are expressed as means with maximum and minimum values. *Lines* in the ML graph show hypothesized growth lines. (From Ikeda et al. 2009a)

(Lee et al. 1994), 220–300 days (Forsythe et al. 2001), 169–262 days (Walsh et al. 2002), and 161–315 days (Ikeda et al. 2009a). Because cephalopod life spans may be controlled by programmed death (Wodinsky 1977), and cephalopods are terminal spawners, the early maturation of cultured squid must also induce early death (i.e. short life span).

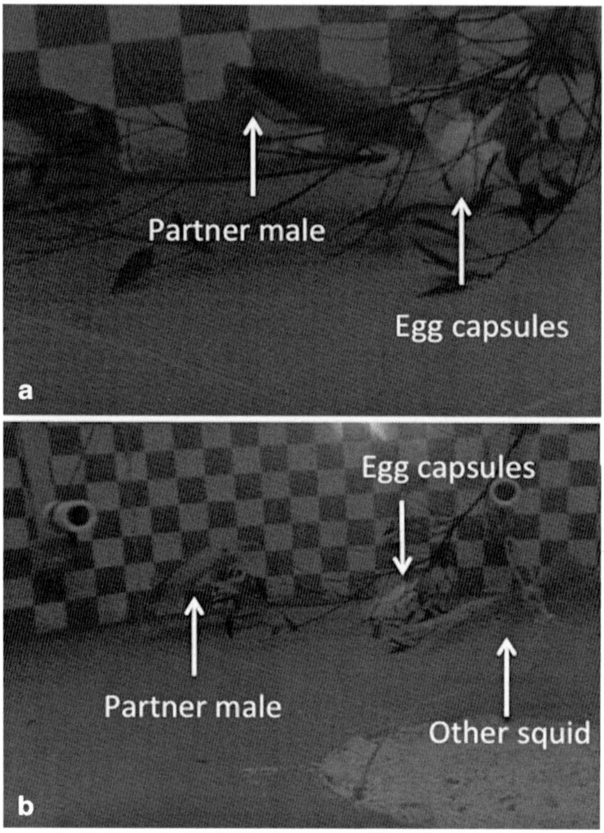

Fig. 17.12 Egg-touching behaviour by cultured *Sepioteuthis lessoniana*. **a** Male partner touches egg capsules. **b** Male partner stays near egg capsules to guard against other squid. (Ikeda, original observation)

Mating pairs touch their newly laid egg capsules and stay near them to guard them from other pairs (Fig. 17.12). On average, mating pairs touch their eggs for approximately 6 s and guard them for 45 s. Similar behaviour is observed in both cultured (Lee et al. 1994) and wild squid (Segawa 1987).

Cultured female squid have larger genital organs than wild-caught squid (Ikeda et al. 2009a). This is because captive females store a larger number of ripe ova in their oviducts, presumably due to inhibition of spawning. Spawning may be inhibited by particular environmental factors, such as strong lighting (Fig. 17.11). Wild females store 55–303 ripe ova, whereas cultured females store 4–2,096 ripe ova (Ikeda et al. 2009a). Mature females may be motivated to spawn, even without mating and without stored spermatangia and spermatozoa (Ikeda et al. 2009b).

Water Quality and System Requirements

Closed systems used for the ongrowing phase have filtration systems similar to those used for nursing the young, but use tanks of larger size. Lee et al. (1994, 1998a, b) and Walsh et al. (2002) used a 15,000-L raceway (2.4 × 6.1 × 0.9 m) and a 50,000-L circular tank (6.5 m circular line, 1.75 m depth) filled with artificial seawater. Juvenile squid of 30–40 mm ML were transferred from their nursing tanks to these ongrowing tanks. The ongrowing density of this system was 5.6 individuals·m^{-3}. An average water temperature of 23 °C is a major factor that can extend the life span and the final size of cultured squid. After about 300 days, squid are 214–260 mm in length and 682–1,073 g in weight. Six generations have been cultured using this system. Ikeda et al. (2003, 2009a) used a large circular tank with a 10,000-L capacity (4 m diameter, 1 m water depth) connected to an algae tank (Fig. 17.8) and smaller 20-, 50-, and 1,700-L-capacity tanks. The squid grew to 268 mm in length 243 days after hatching (20–23 °C). Three consecutive generations were cultured using this system. Lee et al. (1998b) suggested that, in a closed system, removal of ink (ejected by frightened squid) was an important consideration peculiar to filtration systems.

Closed water systems maintained at 21–26 °C have been used to successfully culture three to seven consecutive generations of *S. lessoniana* (Lee et al. 1994; Walsh et al. 2002; Ikeda et al. 2009a; Fig. 17.11). Ueta (2000b) reported that the mortality rate in a long-term rearing experiment of *S. lessoniana* using an open water system increased when the temperature fell below 20 °C. In the wild, lower temperatures (15 °C) affect the movement and distribution of *S. lessoniana* (Ueta and Jo 1990). Therefore, temperatures should be maintained between 20 and 30 °C for *S. lessoniana* culture.

Both natural light and artificial light are used for the culture of *S. lessoniana* in open and closed water systems (Tsuchiya 1982; Lee et al. 1994; Walsh et al. 2002; Ikeda et al. 2003, 2009a). In closed water systems, fluorescent and metal halide bulbs are used for illumination. Young *S. lessoniana* generally avoid both the brightest part (85 Lx) and darkest part (5 Lx) of the tank (Lee et al. 1994), whereas adult squid avoid the brightest areas of tanks (Ikeda, unpublished observation).

The required physical and chemical parameters of the seawater used for squid culture are, on average, pH 7.9–8.2, NH_4-N 0.009–0.044 mg·L^{-1}, NO_2-N 0.023–0.069 mg·L^{-1}, and NO_3-N 9.6–36.0 mg·L^{-1} (Walsh et al. 2002) or pH 7.8 ± 0.2, ammonia 4.7 ± 11.7 mg·L^{-1}, nitrite 0.3 mg·L^{-1}, nitrate 114 ± 73 mg·L^{-1}, and salinity 35.8 ± 1.2 psu (Ikeda et al. 2009a). As with other loliginids, ammonia concentrations that are too high cause sudden death of *S. lessoniana*. In addition, drastic changes in water quality and temperature can also cause immediate squid death.

17.4 Trends in Research and Industrial Level

One of the main focuses of future research should include the development of feed for large-scale culture, with live and artificial feed that are similar to that used for the cuttlefish, *Sepia pharaonis* and *Sepiella inermis* (see also Chaps. 12 and 13). Problems such as the cost and supply of live feed and the availability of an acceptable form of artificial feed are still encountered (Nabhitabhata et al. 2001b; Sangpradab et al. 1984).

Nursing and ongrowing in net cages and pens in situ should be studied as a means of reducing the cost of *S. lessoniana* culture. At selected sites, such facilities could substitute for land-based hatcheries. In tropical countries, such facilities must be protected from strong sunlight, as excess lighting causes stress to cultured squid, fouling of egg capsules, and fouling of cage cover nets.

Monosex culture may be a method for postponing reproduction and prolonging squid growth. To accomplish this, squid must be separated by sex before proceeding to, or as early as possible in, the ongrowing phase. The sex of *S. lessoniana* can be determined from their colour patterns 60 days after hatching (Nabhitabhata 1978, 1983).

Manipulation of the sex ratio or artificial sex induction for *S. lessoniana* can increase production. The degree of sexual size dimorphism is high. *S. lessoniana* males are much larger than females. Batch cultures with many large males will produce a higher yield than cultures with few large males. Manipulation of the development of the male external characters through the application of sex hormones, e.g. testosterone, is feasible. The methodology for this type of manipulation can be adapted from that which is already used in the finfish culture industry for tilapia (*Oreochromis niloticus*) and groupers (*Epinephelus* spp.). For example, developing *S. lessoniana* egg capsules can be immersed in water that contains a particular hormone and/or the young can be fed hormone-added feed. These methods need further study.

S. lessoniana is a complex species (Dunning 1998) comprised of at least three morphs with different final sizes. Segawa et al. (1993a, b) reported morphological, behavioural, and ecological differences among three morphs of *S. lessoniana* with the local names *shiro-ika* (white squid), *aaka-ika* (red squid), and *kuwa-ika* (small squid) from Ishigaki Island of the Ryukyu Archipelago, Japan. The three morphs have different final sizes, chromatophore arrangements, egg-capsule characteristics, spawning sites, and seasons (Izuka et al. 1996b; Segawa et al. 1993a, b). Molecular evidence indicates genetic differences between the three morphs (Izuka et al. 1994, 1996a; Yokokawa and Ueta 2000; Triantafillos and Adams 2005; Aoki et al. 2008). Further investigation and taxonomic description are required to determine whether these differences are at the specific or subspecific level. Differences in the growth rate of different morphs should be a very interesting research topic for the maximization of aquaculture production.

Although *S. lessoniana* is widely distributed in the Indo-Pacific region, collection sites for materials (i.e. eggs) used for previous culture or long-term rearing are predominantly located in Asian waters. For example, all of the *S. lessoniana* eggs used by Lee et al. (1994); Forsythe et al. (2001), and Walsh et al. (2002) were collected from Asian waters, i.e. Thailand, Mainland Japan, and Okinawajima Island of Japan. These Asian countries have major fisheries for this species (Ueta et al. 1992, 2000a; Chotiyaputta et al. 2002; Lu 2002). These countries are ideal locations for studying the culture of *S. lessoniana*. Aquaculture facilities for cephalopods and other marine organisms are already established in Asian countries. Squid cultured in such facilities can be used as scientific experimental models, species for restocking, and as human food. The network of *S. lessoniana* culture to be established in Asian countries is expected to form the future cephalopod culture centre of the world.

17.5 Conclusions

Sepioteuthis lessoniana is highly adaptable to captive conditions and can be cultured for multiple generations in either open or closed seawater systems. This success is mainly due to the relatively large hatchling size of *S. lessoniana* and behavioural characteristics that are advantageous for limited tank space. This is an advantage not only for industrial culturing per se but also for other purposes, such as providing this squid for restocking and as an experimental animal for ethology, physiology, and neuroscience for which live material is usually necessary. On the other hand, similar to other cephalopods, the life history of *S. lessoniana* has not been entirely documented in either tropical or temperate waters. This includes the reproductive process, survival after hatching, and population genetics. Modern techniques, such as bio-logging and DNA markers, are expected to provide new information about these unresolved questions. In addition, because the fisheries' demand for *S. lessoniana* is high, policies for the protection of this squid must be established.

The process of *S. lessoniana* culture includes broodstock collection, nursing of egg capsules, nursing of the young in the hatchery, and the grow-out phase. Water quality control procedures are different in closed and open seawater systems and are fully controlled in the former and partially controlled in the latter. The young are fed from hatching with live prey organisms, particularly mysids. The size of prey has to be similar to that of the squid. Sufficient supply of live feed ensures good survival. The young can be trained to accept dead feed, starting in the grow-out phase. The growth rate, final size, and longevity of these squid can vary, particularly with temperature and seawater system. Research and development of artificial feed is urgently required in order to resolve the live feed bottleneck and to reduce the cost of production.

References

Ahmad T, Usman (1997) Bigfin squid culture: the Indonesian experience. Phuket Mar Biol Cent Spec Pub 17:285–287

Aoki M, Imai H, Naruse T, Ikeda Y (2008) Low genetic diversity of oval squid *Sepioteuthis lessoniana* (Cephalopoda: Loliginidae), in Japanese waters inferred from a mitochondrial DNA non-coding region. Pac Sci 62:403–411

Boongerd S, Rachaniyom S (1990) Squid trap fishing. Tech Pap 1/1990, Fish Tech Subdiv, Mar Fish Div, Dept Fish

Chainak P, Chindamaikul T, Nabhitabhata J, Kwandee V (2003) Effect of phosphate concentration on survival of hatchlings of bigfin squid *Sepioteuthis lessoniana* Lesson, 1830. In:

Nabhitabhata J (ed) Report on cephalopod culture research and development programme, 2002–2004. Rayong Coast Fish Res Dev Cent Contrib 27

Chankaew K, Nabhitabhata J, Sanein N, Pattanakul S (2003a) Life history and management of pygmy goby as live feed for nursing of young bigfin squid. In: Nabhitabhata J (ed) Report on cephalopod culture research and development programme, 2002–2004. Rayong Coast Fish Res Dev Cent Contrib 27

Chankaew K, Nabhitabhata J, Sanein N, Pattanakul S (2003b) Management of sea bass fries as live feed of hatchlings of bigfin squid and pharaoh cuttlefish. In: Nabhitabhata J (ed) Report on cephalopod culture research and development programme, 2002–2004. Rayong Coast Fish Res Dev Cent Contrib 27

Chen DS, Dykhuizen GV, Hodge J, Gilly WF (1996) Ontogeny of copepod predation in juvenile squid (*Loligo opalescens*). Biol Bull 190:69–81

Chindamaikul T, Nabhitabhata J, Krainara T, Somsueb S (1994a) Survival of bigfin squid, *Sepioteuthis lessoniana* Lesson, in different stocking density. Tech Pap 30/1994, Phuket Coast Aquacult Dev Cent, Coast Aquacult Div, Dept Fish

Chindamaikul T, Nabhitabhata J, Somsueb S, Krainara T (1994b) Comparison on different methods for hatching of bigfin squid, *Sepioteuthis lessoniana*. Tech Pap 7/1994, Phuket Coast Aquacult Dev Cent, Coast Aquacult Div, Dept Fish

Chindamaikul T, Kaoian K, Chainak P, Nabhitabhata J, Samrarnrat N (2001) Effect of nitrite-nitrogen and ammonia-nitrogen on survival of hatchlings of bigfin squid, *Sepioteuthis lessoniana* Lesson, 1830. In: Nabhitabhata J (ed) Report on cephalopod culture research and development programme, 2002–2004. Rayong Coast Fish Res Dev Cent Contrib 27

Choe S (1966a) On the egg, rearing, habits of the fry, and growth of some cephalopoda. Bull Mar Sci 16:330–348

Choe S (1966b) On the growth, feeding rates and the efficiency of food conversion for cuttlefishes and squids. Korea J Zool 9:12–20

Choe S, Ohshima Y (1961) On the embryonal development and the growth of the squid, *Sepioteuthis lessoniana* Lesson. Venus 21:462–476

Choe S, Ohshima Y (1963) Rearing of cuttlefishes and squids. Nature 197:307

Chotiyaputta C (1984) Spawning season of *Sepioteuthis lessoniana*. Ann Rep 1984, Invertebr Sect, Mar Fish Div, Dept Fish

Chotiyaputta C (1988) Reproductive biology of bigfin reef squid from squid trap. Tech Pap 5, Mar Life Hist Sect, Mar fish Div, Dept Fish

Chotiyaputta C (1995) Biology of cephalopods. In: Nabhitabhata J (ed) Biology and culture of cephalopods. Rayong Coast Aquacult St, Rayong

Chotiyaputta C, Yamrungreung A (1998) Trap fisheries for squid and their impact on spawning. S Afr J Mar Sci 20:285–291

Chotiyaputta C, Nootmorn P, Jirapunpipat K (2002) Review of cephalopod fishery production and long term changes in fish communities in the Gulf of Thailand. Bull Mar Sci 71:223–238

Dunning MC (1998) Loliginidae inshore squids, pencil squids. In: Carpenter KE, Niem VH (eds) The living marine resources of the western central Pacific vol 2 cephalopods, crustaceans, holothurians and sharks. Food and Agriculture Organization of the United Nations, Rome

Forsythe JW, Walsh LS, Turk PE, Lee PG (2001) Impact of temperature on juvenile growth and age at first egg-laying of the Pacific reef squid *Sepioteuthis lessoniana* reared in captivity. Mar Biol 138:103–112

Ikeda Y, Sakurai Y, Shimazaki K (1993) Fertilizing capacity of squid (*Todarodes pacificus*) spermatozoa collected from various sperm storage sites, with special reference to the role of gelatinous substance from oviducal gland in fertilization and embryonic development. Invertebr Reprod Dev 23:39–44

Ikeda Y, Wada Y, Arai N, Sakamoto W (1999) Note on size variation of body and statolith in the oval squid *Sepioteuthis lessoniana* hatchlings. J Mar Biol Assoc UK 79:757–759

Ikeda Y, Sakurazawa I, Sakurai Y, Matsumoto G (2003) Initials trials of squid rearing, maintenance and culture at the Brain Science Institute of RIKEN. Suisanzoshoku 51:391–400

Ikeda Y, Ueta Y, Sakurazawa I, Matsumoto G (2004) Transport of the oval squid *Sepioteuthis lessoniana* Férussac, 1831 in Lesson 1830–1831 (Cephalopoda: Loliginidae) for up to 24 h and subsequent transfer to an aquarium. Fish Sci 70:21–27

Ikeda Y, Sakurazawa I, Ito K, Sakurai Y, Matsumoto G (2005) Rearing of squid hatchlings, *Heterololigo bleekeri* (Keferstein 1866) up to 2 months in a closed seawater system. Aquac Res 36:409–412

Ikeda Y, Ueta Y, Anderson FE, Matsumoto G (2009a) Reproduction and life span of the oval squid *Sepioteuthis lessoniana* (Cephalopoda: Loliginidae): comparison between laboratory-cultured and wild-caught squid. Mar Biodiversity Rec 2:e50. doi:10.1017/S175526720900061X

Ikeda Y, Oshima Y, Sugimoto C, Imai H (2009b) Multiple spawning by uncopulated oval squid (*Sepioteuthis lessoniana*) in captivity. Aquac Sci 57:39–43

Izuka T, Segawa S, Okutani T, Numachi K (1994) Evidence on the existence of three species in the oval squid, *Sepioteuthis lessoniana* complex in Ishigaki Island, Okinawa, South-western Japan, by isozyme analyses. Venus 53:217–228

Izuka T, Segawa S, Okutani T (1996a) Biochemical study of the population heterogeneity and distribution of the oval Squid *Sepioteuthis lessoniana* complex in southwestern Japan. Amer Malac Bull 12:129–135

Izuka T, Segawa S, Okutani T (1996b) Identification of three species in oval squid, *Sepioteuthis lessoniana* complex by chromatophore arrangements on the funnel. Venus 55:139–142

Jereb P, Vecchione M, Roper CFE (2010) Family Loliginidae. In: Jereb P, Roper CFE (eds) Cephalopods of the world an annotated and illustrated catalogue of cephalopod species known to date vol. 2 myopsid and oegopsid squids FAO Species Catalogue for Fishery Purposes No. 4, Vol. 2. FAO, Rome, pp 38–117

Kanamaru H, Itoh Y (1996) Development of fin base and appearance of secondary sexual characters in juvenile and young of the oval squid *Sepioteuthis lessonina*. Saibai Giken 25:15–20

Lee PG, Turk PE, Yang WT, Hanlon RT (1994) Biological characteristics and biomedical applications of the squid *Sepioteuthis lessoniana* cultured through multiple generations. Biol Bull 186:328–341

Lee PG, Turk PE, Forsythe JW, DiMarco FP (1998a) Cephalopod culture: physiological, behavioral and environmental requirements. Suisanzoshoku 46:417–422

Lee PG, Walsh L, Turk PE, DiMarco FP (1998b) Large-scale culture of the loliginid squid, *Sepioteuthis lessoniana*. In: Tiensongrasmee B, Tedengren M, Jarayabhand P, Tandavanitj N, Popongviwat A, Nujsawat A, Soisodsri S (eds) Mollusk research in Asia. Thailand Research Fund, Bangkok

Lu CC (2002) Cephalopod production from the waters around Taiwan. Bull Mar Sci 71:465–471

Mangold K (1987) Reproduction. In: Boyle PR (ed) Cephalopod life cycles vol. 2. Academic Press, London

Munekiyo M, Kawagishi M (1993) Diurnal behaviors of the oval squid, *Sepioteuthis lessoniana*, and fishing strategy of a small-sized set net (preliminary report). In: Okutani T, O'Dor RK, Kubodera T (eds) Recent advances in cephalopod fisheries biology. Tokai University Press, Tokyo

Nabhitabhata J (1978) Rearing experiment on economic cephalopod -I: long-finned squid *Sepioteuthis lessoniana* Lesson. Tech Pap 1978, Rayong Brackish Fish Sect, Rayong Fish Stn, BrackishFish Div, Dept Fish

Nabhitabhata J (1983) External sex characteristics and sexual activities of long-finned squid, *Sepioteuthis lessoniana* Lesson. Tech Pap 1/1983, Rayong Brackish Fish Stn, Brackish Fish Div, Dept Fish

Nabhitabhata J (1993) Concept, materials and method of cephalopod culture. Contrib 6, Rayong Coast Aquacult Stn, Coast Aquacult Div, Dept Fish

Nabhitabhata J (1996) Life cycle of cultured bigfin squid, *Sepioteuthis lessoniana* Lesson. Phuket Mar Biol Cent Spec Pub 16:83–95

Nabhitabhata J (2002) Life history and behaviour of reared neritic cephalopods from Thai Waters. PhD Thesis, Tokyo University of Fisheries

Nabhitabhata J, Kbinrum S (1981) The culture of long-finned squid, *Sepioteuthis lessoniana* Lesson. Rayong Brackish Fish Stn, Brackish Fish Div, Dept Fish, Ann Rep 1981:74–115

Nabhitabhata J, Kbinrum S (1984) Biological studies on economic cephalopods-III: note on survival of young cephalopods during continuous change of salinity. Tech Pap 6/1984, Rayong Brackish Fish Stn, Brackish Fish Div, Dept Fish

Nabhitabhata J, Kbinrum S, Kumpang P (1984) On floating net cage culture of bigfin squid, *Sepioteuthis lessoniana* Lesson, at different stocking densities. Tech Pap 7/1985, Rayong Brackish Fish Stn, Brackish Fish Div, Dept Fish

Nabhitabhata J, Kbinrum S, Soodmee Y (1991a) Effect of pH on survival rate of cephalopods hatchlings. Tech Pap 17/1991, Rayong Coast Aquacult Stn, Coast Aquacult Div, Dept Fish

Nabhitabhata J, Kbinrum S, Soodmee Y (1991b) Effect of salinity on hatching rate of *Sepioteuthis lessoniana* Lesson. Tech Pap 16/1991, Rayong Coast Aquacult Stn, Coast Aquacult Div, Dept Fish

Nabhitabhata J, Soodmee Y, Petrasatian K (1992a) Effect of salinity on survival rate of bigfin squid hatchlings, *Sepioteuthis lessoniana* Lesson. Tech Pap 3/1992, Rayong Coast Aquacult Stn, Coast Aquacult Div, Dept Fish

Nabhitabhata J, Jaroenvittayakul Y, Kakhai N, Soodmee Y, Nilaphat P (1992b) On live feed supply management for seed production of bigfin squid, *Sepioteuthis lessoniana* Lesson, and cost of production. Tech Pap 13/1992, Rayong Coast Aquacult Stn, Coast Aquacult Div, Dept Fish

Nabhitabhata J, Kakhai N, Suriyawarakul J, Nilaphat P (1996) Feeding and cost of seed production of cephalopods, *Sepioteuthis lessoniana* Lesson; *Sepia pharaonis* Ehrenberg; Sepiella inermis Fer. & d'Orb. Tech Pap 1/1996, Rayong Coast Aquacult Stn, Coast Aquacult Div, Dept Fish

Nabhitabhata J, Nilaphat P, Promboon P (2001a) Artificial fertilization and embryonic development of bigfin squid, *Sepioteuthis lessoniana* Lesson, 1830. In: Nabhitabhata J (ed) Report of cephalopod culture research and development programme, 2002–2004. Rayong Coast Fish Res Dev Cent Contrib 27

Nabhitabhata J, Nilaphat P, Promboon P (2001b) Behavioural approaches on the acceptable form of artificial feed of bigfin squid, *Sepioteuthis lessoniana* Lesson, 1830. In: Nabhitabhata J (ed) Report of cephalopod culture research and development programme, 2002–2004. Rayong Coast Fish Res Dev Cent Contrib 27

Nabhitabhata J, Asawangkune P, Amornjaruchit S, Promboon P (2001c) Tolerance of eggs and hatchlings of neritic cephalopods to salinity changes. Phuket Mar Biol Cent Spec Pub 25:91–99

Nabhitabhata J, Nilaphat P, Promboon P, Jaroongpattananon C, Nilaphat G, Reunreng A (2005) Performance of simple large-scale cephalopod culture system in Thailand. Phuket Mar Biol Cent Res Bull 66:337–350

Ohshima Y, Choe S (1961) On the rearing of young cuttlefish and squid. Bull Japan Soc Sci Fish 27:978–986

Okutani T (2005) Cuttlefish and squid of the world. Seizando-Shoten Publishing Co., Ltd., Tokyo

Rattana-anant T (1978) Biological studies on *Sepioteuthis lessoniana* Lesson (Cephalopoda; Loliginidae) in the Gulf of Thailand. Invertebr Sect, Mar Fish Div, Dept Fish

Rattana-anant T (1979) Biological studies on *Sepioteuthis lessoniana* Lesson (Cephalopoda; Loliginidae) in the Gulf of Thailand. Invertebr Sect, Mar Fish Div, Dept Fish

Rattana-anant T (1980) Biological studies on *Sepioteuthis lessoniana* Lesson (Cephalopoda; Loliginidae) in the Gulf of Thailand. Invertebr Sect, Mar Fish Div, Dept Fish

Roongratri M (1997) Biology of bigfin reef squid (*Sepioteuthis lessoniana*) in the eastern coast of the Gulf of Thailand. Tech Pap 65, East Mar Fish Dev Cent, Mar Fish Div, Dept Fish

Sangpradab S, Veschasatra V, Kbinrum S (1984) Experiments on feeding of long-finned squid (*Sepioteuthis lessoniana* Lesson) with three different kinds of feed. Tech Pap 1/1984, Rayong Brackish Fish Stn, Brackish Fish Div, Dept Fish

Saso N (1979) On the culture of aori-ika, *Sepioteuthis lessoniana*. In: Japan Association of Fisheries Cooperatives (ed) Proc 25 Japan Nat Fish Activ Convn

SEAFDEC (1975) Observations on the culture and life history of the broad finned squid *Sepioteuthis lessoniana* Fer. and Orb. SEAFDEC-MSU-Inst Fish Res Develop Ann Rep 1975:115–119

Segawa S (1987) Life history of the oval squid, *Sepiotuethis lessoniana* in Kominato and adjacent waters Central Honshu, Japan. J Tokyo Univ Fish 74:67–105

Segawa S (1990) Food consumption, food conversion and growth rates of the oval squid *Sepioteuthis lessoniana* by laboratory experiments. Nippon Suisan Gakk 56:217–222

Segawa S (1993) Field and laboratory feeding studies of the Japanese oval squid, *Sepioteuthis lessoniana*. In: Okutani T, O'Dor RK, Kubodera T (eds) Recent advances in cephalopod fisheries biology. Tokai University Press, Tokyo

Segawa S, Hirayama S, Okutani. T (1993a) Is *Sepioteuthis lessoniana* in Okinawa a single species? In: Okutani T, O'Dor RK, Kubodera T (eds) Recent advances in cephalopod fisheries biology. Tokai University Press, Tokyo

Segawa S, Izuka T, Tamashiro T, Okutani T (1993b) A Note on mating and egg deposition by *Sepioteuthis lessoniana* in Ishigaki Island, Okinawa, southwestern Japan. Venus 52:101–108

Shimazoe K (2010) Post hatching cephalopod nutrition and growth in tropical waters. Master degree thesis, University of the Ryukyus

Sivalingam D, Ramadoss K, Gandhi AD, Enose M (1993) Hatchery rearing of the squid, *Sepioteuthis lessoniana* and the cuttlefish, *Sepia pharaonis*. Mar Fish Info Serv 122:12–14

Sivashanthini K, Charles GA, Thulasitha WS (2009) Length-weight relationship and growth pattern of *Sepioteuthis lessoniana* Lesson 1830 (Cephalopoda: Teuthida) from the Jaffna Lagoon, Sri Lanka. J Biol Sci 9:357–361

Sugimoto C, Ikeda Y (2012) Ontogeny of schooling behavior in the oval squid *Sepioteuthis lessoniana*. Fish Sci 78:287–294

Sugimoto C, Ikeda Y (2013) Comparison of the ontogeny of hunting behavior in pharaoh cuttlefish (*Sepia pharaonis*) and oval squid (*Sepioteuthis lessoniana*). Biol Bull 225(1):50–59

Supongpan M (1995) Cephalopod resources in the Gulf of Thailand. In: Nabhitabhata J (ed) Biology and culture of cephalopods. Rayong Coast Aquacult Stn, Rayong

Supongpan M, Kongmuag K, Jittrapong S (1988) Bigfin reef squid fisheries in the Gulf of Thailand. Tech Pap 1/1988, Mar Fish Div, Dept Fish

Triantafillos L, Adams. M (2005) Genetic evidence that the northern calamary, *Sepioteuthis lessoniana*, is a species complex in Australian Waters. ICES J Mar Sci 62:1665–1670

Tsuchiya M (1981) On the spawning of the squid, *Sepioteuthis lessoniana* Lesson at Amitori Bay, Iriomote Island, Okinawa. Inst Ocean Res Develop. Tokai Univ Note 3:53–75

Tsuchiya M (1982) On the culture of the squid *Sepioteuthis lessoniana* Lesson (Loliginidae), from hatching to the whole life in Iriomote Island, Okinawa. Inst Ocean Res Develop. Tokai Univ Note 4:49–70

Ueta Y (1992) Distribution on the oval squid, *Sepioteuthis lessoniana* and its fisheries around Tokushima Prefecture. Fish Biol Oceanogr SW Water Jpn 8:61–69

Ueta Y (2000a) Fishery biological studies of the oval squid, *Sepioteuthis lessoniana* around Tokushima Prefecture. Bull Tokushima Pref Fish Exp Stn 1:1–79

Ueta Y (2000b) Feeding and growth of the oval squid, Sepioteuthis lessoniana in captivity. Tokushima Suishi Jiho 125–132

Ueta Y, Jo Y (1989) Notes on ecology of the oval squid *Sepioteuthis lessoniana* in outer waters adjacent to the Kii Channel. Nippon Suisan Gakk 55:1699–1702

Ueta Y, Jo Y (1990) Migration of the oval squid, *Sepioteuthis lessoniana* around Tokushima Prefecture. Suisanzoshoku 38:221–226

Ueta Y, Kitagado I (1996) Experimental studies on the form of spawning bed by oval squid, *Sepioteuthis lessoniana*. Suisanzoshoku 44:67–72

Ueta Y, Segawa S, Tenjin M, Jo Y, Kitakado I, Fukunaga M, Kangawa Y (1992) Occurrence and character of egg masses and estimation of egg-laying and hatching period on oval squid, *Sepioteuthis lessoniana* in outer waters adjacent to the Kii Channel. Suisanzoshoku 40:469–474

Ueta Y, Kitakado I, Segawa S, Tenjin M, Jo Y, Fukunaga M, Kanagawa Y (1995) Selection of spawning ground and spawning bed in oval squid *Sepioteuthis lessoniana*. Fish Eng 31:189–194

Wada Y, Kobayashi T (1995) On an iteroparity of the oval squid *Sepioteuthis lessoniana*. Nippon Suisan Gakk 61:151–158

Walsh LS, Turk PE, Forsythe JW, Lee PG (2002) Mariculture of the loliginid squid *Sepioteuthis lessoniana* through seven successive generations. Aquaculture 212:245–262

Wodinsky J (1977) Hormonal inhibition of feeding and death in *Octopus*: control by optic gland secretion. Science 198:948–951

Yokokawa K, Ueta Y (2000) Genetic analysis of oval squid (*Sepioteuthis lessoniana*) around Japan. Venus 59:45–55

Chapter 18
Amphioctopus aegina

Jaruwat Nabhitabhata

Abstract The marbled octopus, *Amphioctopus aegina*, is a moderately sized benthic octopus, inhabiting muddy substrates in the coastal zone of the Indian and Western Pacific Ocean. Capture fishing of octopus in these regions includes this species. The aquaculture methodology comprises collection of broodstocks from the wild, nursing of the eggs and brooding females, nursing of the planktonic young and growout of the benthic young. All phases of the culture process prior to growout can be performed on a small scale with a 50-L tank. The general task for feeding is to feed planktonic food to the planktonic young for about 30 days, and to then provide benthic food after the settling stage. Culturing of *Amphioctopus aegina* as a model for scientific experiments and home aquaria is the future commercial task.

Keywords *Amphioctopus aegina* · Benthic octopus · Planktonic young · Nursing · Growout · Small scale

18.1 Importance of this Species in the Market

Amphioctopus aegina, the marbled octopus, is a moderately sized benthic octopus with a mantle length (ML) of 40–100 mm. This species is quoted in fishery statistics under the name *Octopus dollfusi*, its junior synonym (Nateewathana 1997; Norman and Hochberg 2005). Its distribution is wide in the coastal zone of the Indo-West Pacific region, from 30° to 140°E and from 40°N down to 20°S (Roper et al. 1984; Norman 1998). This species is common in this region and inhabits sand and muddy sand areas (Nateewathana 1997). The marbled octopus is solitary with crepuscular behaviour. They live in holes and defend their home territories (Promboon et al. 2011). *Amphioctopus aegina* reproduces all year-round, at least in the Gulf of Thailand, the South China Sea, Pacific Ocean. The highest reproduction peak is during March to May, i.e. summer to the early rainy season (Phanichpong 1985).

J. Nabhitabhata (✉)
Excellence Centre for Biodiversity of Peninsular Thailand (CBIPT),
Faculty of Science, Prince of Songkla University, Hatyai, Songkhla 90112, Thailand
e-mail: jaruwat.n5@gmail.com, jaruwat.n@psu.ac.th

J. Iglesias et al. (eds.), *Cephalopod Culture*, DOI 10.1007/978-94-017-8648-5_18,

About 15,000 t (12% of the total cephalopod catch) of octopus including *Amphioctopus aegina* are caught annually from Thai waters, the Gulf of Thailand and the southern Andaman Sea (Supongpan 1995; Department of Fisheries 2012). The major fishing methods involve several types of trawlers. Considering the homing behaviour of this octopus, the most recently emerging gear in the Southeast Asian region is the artisanal octopus trap derived from the Japanese octopus pots. Fishermen have substituted the original clay pots with local and lower cost materials, e.g. gastropod shell, *Cymbiola nobilis*, concrete blocks, soft-drink bottles. Srikum and Somchanakit (2011) reported that one fisherman owns 200–5,000 trap shells and obtains a monthly average yield of 10 g per trap shell from the Gulf of Thailand. The octopus fishing grounds vary in their locations but mostly, it is between 200 and 8,000 m from the shoreline, at a water depth of 2–18 m. The octopus trap is an appropriate gear for fishing in view of natural resource conservation for two reasons. The first one is that the fishing period is 7–20 days per month, so the remaining period of every month is reserved for natural stock recruitment. The second reason is that most of the *Amphioctopus aegina* captured by traps are mature, of a total length of 190–220 mm (Srikum and Somchanakit 2011), and therefore only the full-grown stock is exploited.

18.2 State of the Art

As mentioned in previous chapters, the culture of octopus species that involve planktonic hatchlings, like *Amphioctopus aegina,* is more difficult compared to those that have benthic hatchlings. The outstanding problem is that these hatchlings are plankton feeders and require live food organisms of a specific size range and characteristics (Hanlon 1987; Boyle and Rodhouse 2005; Villanueva and Norman 2008). The subsequent transitional period before their complete settlement also involves a change to a new characteristic type and size of food organisms.

The marbled octopus, *Amphioctopus aegina,* has a moderate size and as an adult presents a benthic habit. This is an advantage since it is possible to provide culture conditions on a small scale. Broodstock can be collected from the wild and consecutive generations can spawn in captivity, which can ensure a continuous supply of hatchlings for aquaculture. Small-scale facilities can be used for culture of the species through its entire life cycle. *Amphioctopus aegina* is the second benthic octopus species with planktonic hatchlings, after *Octopus vulgaris* (Iglesias et al. 2004), that can also be cultured throughout its entire life cycle. Such success in aquaculture requires the problems associated with the rearing of the planktonic hatchlings and also to overcome the transitional phase prior to settlement. To resolve these difficulties, hatchlings are commonly fed for the first month with several types of wild-collected live zooplanktonic species. After complete settlement, they can be then trained to accept dead feed or inert diets. The daily growth rate is 1.9% in length and 6.4% in weight throughout the culture period of 140 days (Promboon et al. 2011). A growing demand for using this octopus as a neurophysiologic experimental model as well as an ornamental animal cannot be overlooked.

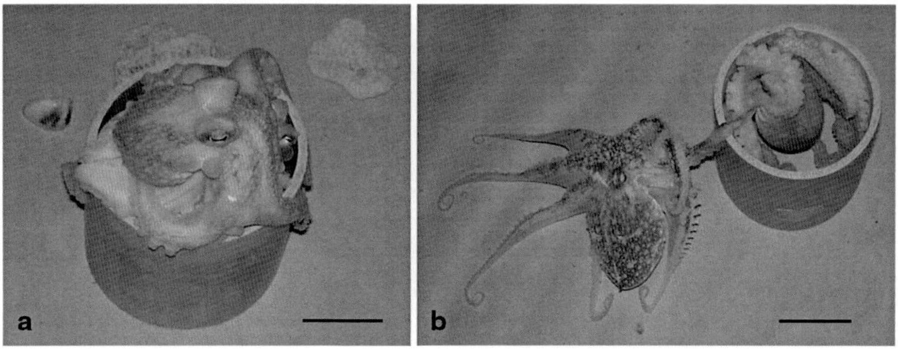

Fig. 18.1 The mating behavioural pattern of *Amphioctopus aegina,* in the culture tanks; tactile contact or enveloping pattern (**a,** male on *top*) and a remote copulation pattern (**b,** male on *left*); *scale bars* 50 mm. (Adapted from Promboon et al. 2011)

18.3 Broodstock Maintenance and Spawning

The broodstock for the marbled octopus, *Amphioctopus aegina,* are collected from otter board and beam trawlers or by using octopus traps, operating along the eastern part of the Gulf of Thailand, South China Sea and the Pacific Ocean. Onboard, the octopuses are maintained in 50-L-capacity cylindrical fibreglass tanks containing 30 L of fresh sea water with aeration and then, upon landing, are transported to the cephalopod hatchery. The octopuses are separated by sex and maintained for a 7-day acclimatization period in an open system of circular concrete tanks of 1,800-mm diameter, with a capacity of 2 m^3 and a 30-cm water depth. In order to minimize any change in water quality, particularly temperature, tanks are equipped with aeration devices and supplied with running filtered sea water that flows through a central drainage (the sea-water supply system in this chapter is also described in Chap. 7: 'Aquaculture to Restocking'). Cylindrical polyvinyl chloride (PVC) pieces (50 mm diameter, 150 mm length) are previously placed on the tank bottom as dens or shelters. Octopuses are fed twice daily at 0800 and 1700 h with chopped fish meat, *Selaroides leptolepis*. Tanks are cleaned daily by siphoning out the water containing debris and replacing about 50 % of their volume.

After 7 days, the octopuses are transferred to make a 1 : 1 pair in one tank (two individuals per tank) under similar conditions. Under these conditions, octopuses mate and spawn spontaneously in the tanks. The mating behavioural sequence can follow two different patterns: by a tactile contact and subsequent enveloping pattern (Fig. 18.1a) or by a remote copulation pattern (Fig. 18.1b). The copulation in the first pattern takes the male 7–10 min to mount and spread his arm-web umbrella to envelop the female. In the second pattern, the male spends a shorter period of 3–5 min, to extend its hectocotylus and insert it into the female mantle cavity without any other contact between both sexes (Promboon et al. 2011).

The interval between mating and spawning is approximately 32 days (range of 20–40 days). After spawning, the male is excluded and removed from the tank. The female broods her eggs by regular cleaning and cradling them in her arm web behind the mantle. The spawning female stops feeding until the hatching of her eggs and onward to her subsequent senescent death. The number of eggs laid by a single octopus female is on average 6,900 (5,500–8,500).

18.4 Nursing of Eggs and Brooding Females

Since the eggs are brooded by the female, there is no direct management for the incubation of the egg masses. The open system and facilities used for broodstocks can be also used in this phase. The management focuses on not disturbing the females and reducing stress from any external stimuli, i.e. sudden change of physical, particularly light, and chemical (water quality) parameters. Eggs during the early stages of a disturbed female are rejected, and those in the later development stages hatch prematurely. Near to hatching, around day 13 (at 28 °C), the brooding female with her shelter is carefully transferred to 50-L-capacity cylindrical PVC tanks containing 30 L of fresh sea water with aeration. Tanks are cleaned by siphoning and the water is changed daily to about 40 % of their volume. Any temperature change is minimized by means of an outside running water bath around the tank base (Nabhitabhata et al. 2005; Promboon et al. 2011).

The eggs are small (as categorized by Boletzky 1977), approximately 4.4 % of the adult ML, and each egg produces a planktonic hatchling. Each egg has a size of about 3.2 × 1.25 mm (Ignatius and Srinivasan 2006; Nabhitabhata et al. 2010; Promboon et al. 2011). The egg is telolecithal, with a rice-grain shape and a fine peduncle attached to the festoon (Fig. 18.2). The newly laid egg is opaque white and gradually turns more transparent after 5 days at 30 °C. Organogenesis and the subsequent first inversion occur simultaneously at this time. Genesis of the chromatophores, the black retina and the ink sac happens from day 9 onwards. The second reversion is from day 13 after the laying of the egg. After this embryonic stage, the female is transferred to another tank. The first hatching occurs on day 16 (Fig. 18.3) with an average embryonic period of 18.3 days (16–22) at 28 °C. Eggs hatching from the same cluster of a single female continue to do so for a period of 3–8 days. The hatching rate is higher than 90 %.

18.5 Nursing of the Young

18.5.1 Feed and Tank Management

After hatching, the female is removed from the tank, and all hatchlings are retained for nursing. The tank previously used for maintenance of the brooding female is

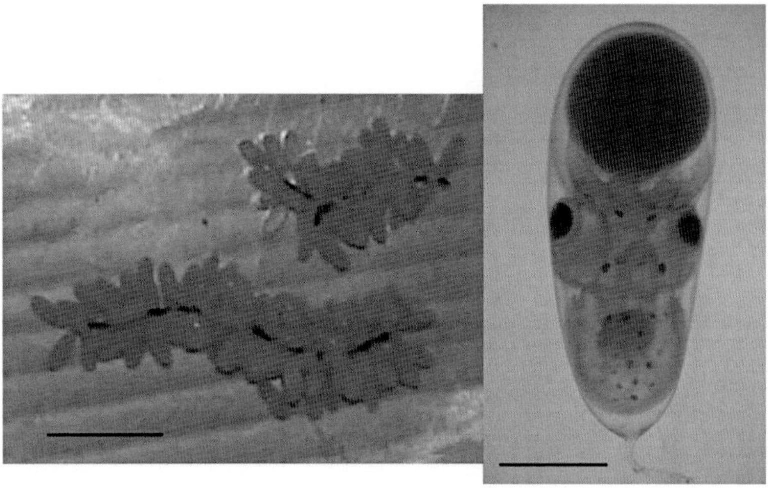

Fig. 18.2 Egg string (*left*) and egg (*right*) of *Amphioctopus aegina,* with embryo on day 11 (28 °C, *scale bars* 10 and 1 mm, respectively). (Adapted from Promboon et al. 2011)

Fig. 18.3 Dorsal view of marbled octopus paralarvae, *Amphioctopus aegina,* on day 16 (28 °C, *scale bar* 1 mm). (Adapted from Promboon et al. 2011)

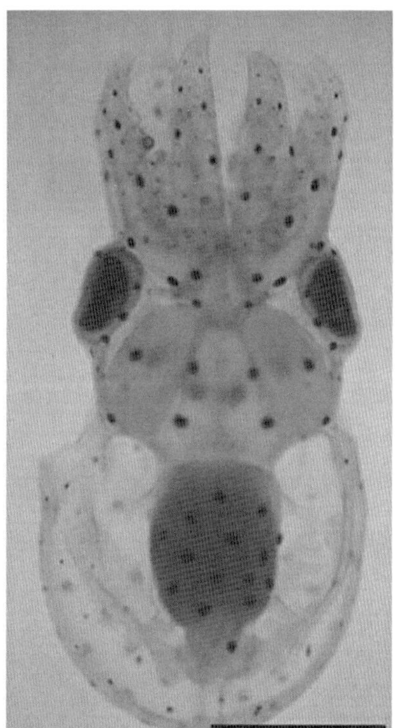

then used for the nursing of the young. The nursing tank has dark-blue walls and is filled with filtered sea water to a water depth of 150 mm. In order to generate and direct an artificial current, the tank is equipped with air stones attached to two pieces of longitudinal PVC pipes (25 mm diameter, 400 mm length). Tanks are cleaned by siphoning and about 80 % of the volume is replaced daily. The management method is similar to that for the previous phase and is also used for *Euprymna hyllebergi* (see also Chap. 15, Fig. 15.3; Nabhitabhata et al. 2005; Promboon et al. 2011). Temperature changes are avoided using a water bath outside the rearing tank. The average temperature should be maintained at approximately 30°C, a pH of 8 and salinity of 30 psu. (Promboon et al. 2011)

Amphioctopus aegina has planktonic hatchlings, and the pelagic phase lasts for around 20–25 days at 30°C, before they subsequently complete their settling stage about 1 week later, at day 30. From day 15 after hatching, the early settlers temporarily attach themselves to the tank wall and gradually increase their period of attachment and their proximity of attachment to the tank bottom, their final habitat. During the transitional settlement period, the depth of the attachment is gradually increased by approximately 50 mm day^{-1}. During this period, the young still attack their prey in the water column but swim down to eat them at the tank bottom.

The general feeding schedule is to feed hatchlings planktonic food when they are pelagic and column feeders, and provide them with benthic food after the settling stage, when they become benthic feeders. Hatchlings are fed twice daily with live prawn zoea, *Macrobrachium sintangense* (available from commercial hatcheries) of 1–2 mm total length for 30 days after hatching. Live amphipods, *Unciolella* sp., wild palaemonid shrimps, *Exopalaemon styliferus* and dwarf gobies, *Redigobius tambujon*, collected from the wild, are used to feed the octopus young from their transitional stage to their early post-settlement, days 25–50 (Fig. 18.4). Feeding has to be ad libitum to generate a similar growth rate, hence there being less difference in size or individual production, and to avoid subsequent cannibalism. Cannibalism has occasionally been observed when size difference is more than 50 %, from hatching to about 70 days.

The initial stocking density for hatchlings is one individual per litre at 150 mm water depth. Such a depth is required to aggregate the food organisms and reduce excessive activity of the octopus. Water depth is gradually increased by about 50 mm every day to attain a final depth of 300 mm at 30 days. The increasing water volume corresponds to the increasing size of the octopus. Survival rate in this nursing phase is only about 10 % because the live food might be not enough due to interruptions to the daily supply. The requirement for live prawns at an exact life stage (zoea), although commercially produced, makes daily and consistent supply difficult.

18.5.2 Growth

Marbled octopus hatchlings grow from 2.7 mm average ML and 0.003 g mean wet weight (WW) at hatching to 6.3 mm ML and 0.068 g WW at the settling stage (30 days after hatching). The daily growth rate in terms of instantaneous relative

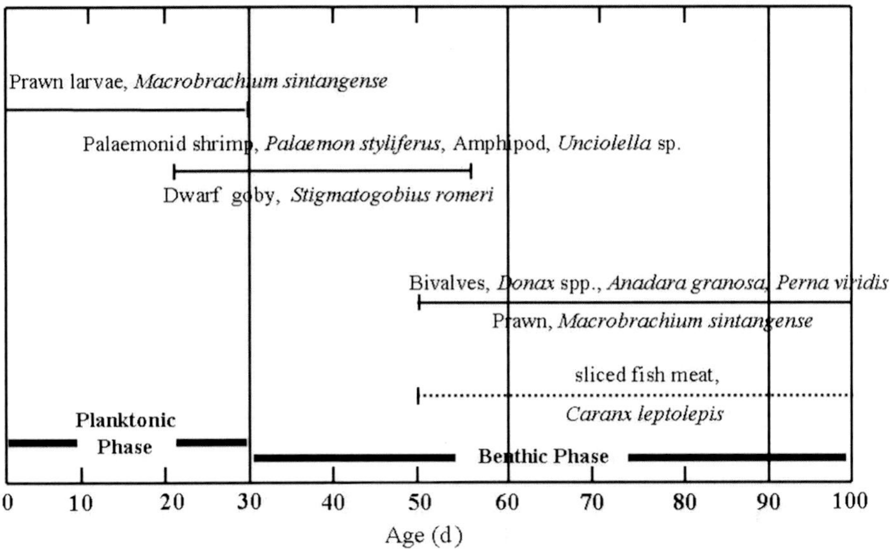

Fig. 18.4 Diagram of the feeding of *Amphioctopus aegina* on various live (*full line*) and non-live (*dashed line*) food corresponding to the age (days) after hatching. (Adapted from Promboon et al. 2011)

growth rate (IGR) during this period is about 4.1 % in length and 12.9 % in weight. After the settlement, the feeding rate peaks at about 240 % after 30–40 days (Fig. 18.5), and the growth rate increases to 5.1 % in ML (Fig. 18.6) and 17.5 % in weight (Fig. 18.7). Food conversion efficiency is about 6–11 % from the planktonic stage to settlement (Fig. 18.8), and this reveals a high energetic cost for the planktonic mode of the young (Promboon et al. 2011).

In the early period, from the planktonic to settling stages, the relationships between ML (mm) and age (days; from 0 to 40 days) and between WW (g) and age (days; from 0 to 30 days) can be described by exponential models (Figs. 18.9 and 18.10):

$$ML = 2.397e^{0.034A} \tag{18.1}$$

$$W = 2.559 \times 10^{-3} e^{0.103A} \tag{18.2}$$

18.6 Ongrowing

18.6.1 Feed, System Requirements and Management

The settling young (30–50 days) are transferred to 1,800 mm diameter circular concrete grey-walled tanks and maintained at a density of one individual per litre. Water depth is initially decreased to 150 mm and later on is gradually increased

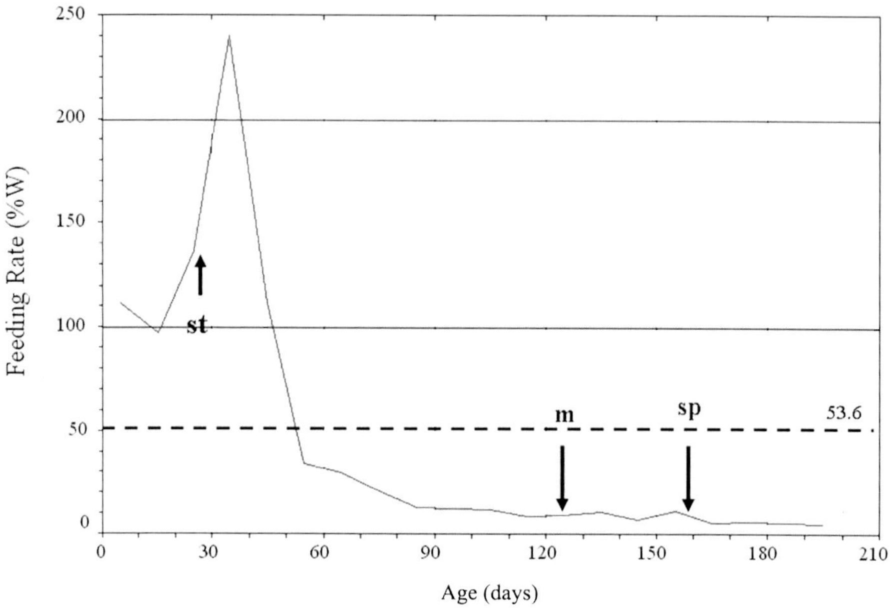

Fig. 18.5 Feeding rate (*z(%W)*) according to the age (days) after hatching. *Arrows* indicate settlement (*st*), mating (*m*) and spawning (*sp*). Individuals were reared at 28 °C. The *horizontal dashed line* with a figure indicates the average value. (Adapted from Promboon et al. 2011)

(50 mm daily) to 400 mm depth. Coral pebbles are supplied as shelters from 30 to 50 days and after 50 days, they are replaced by PVC pipe pieces (50 mm diameter, 150 mm length) for more effective tank cleaning. Other culture details are similar to those for the previous nursing phase (Nabhitabhata et al. 2005; Promboon et al. 2011).

After 50 days, the octopuses are able to feed on dead fish meat, *S. leptolepis*. Live bivalves, *Donax* spp., *Tegillarca granosa* and *Perna viridis* and prawns (*M. sintangense*) are occasionally used as supplementary feed. The feeding schedule of dead food is twice daily, at 0800 and 1700 h.

18.6.2 Growth

The peak conversion efficiency, 32–35 %, is at 50–70 days, when animals are entering maturity and sustain an above average value until they mate at 125 days. The average efficiency from hatching to 140 days is approximately 16 % (6–35 %). The daily growth rate rapidly decreases to achieve relative stability, about 1 %, after 80 days. The average daily growth from hatching to 140 days of age is about 1.9 % in ML and 6.4 % in weight (Promboon et al. 2011).

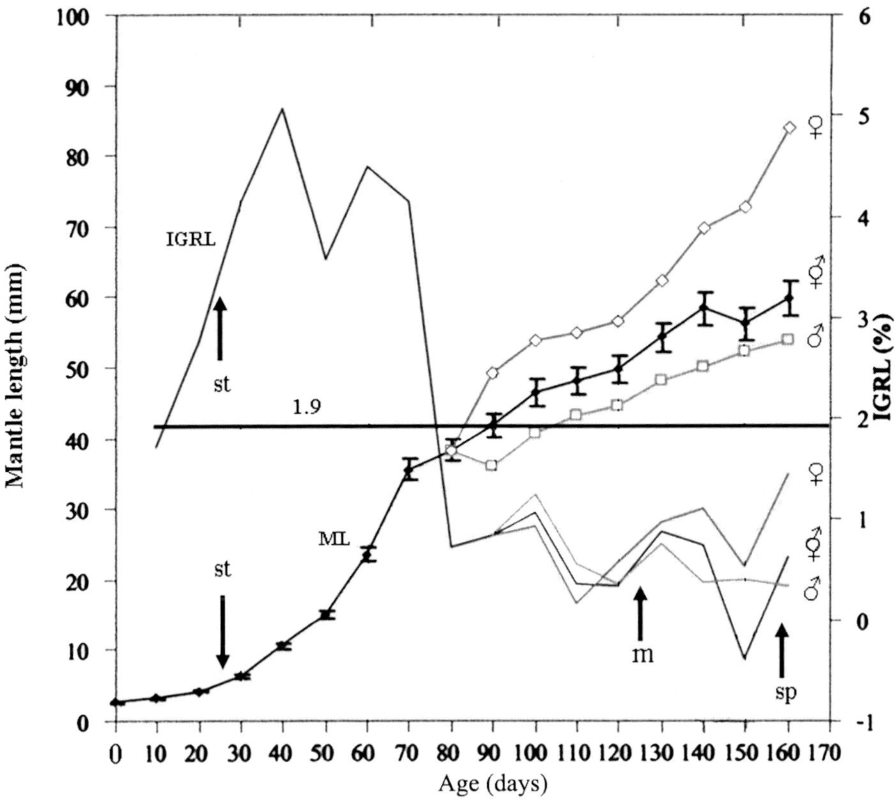

Fig. 18.6 Growth in terms of mantle length (mm), instantaneous relative growth rate (*IGRL(%)*) and age (days) after hatching. Individuals were reared at 28 °C. The *horizontal line* indicates the average value for *IGRL*. *Arrows* indicate settlement (*st*), mating (*m*) and spawning (*sp*). (Adapted from Promboon et al. 2011)

The relationships between the ML (mm) and WW (g) from hatching to 210 days can be expressed as a power regression (Fig. 18.11):

$$W = 1.75 \times 10^{-4} ML^{3.227} \quad (18.3)$$

The inflection point (Fig. 18.9) of the relationship between ML (mm) and age (days) is at about day 40, in the early benthic stage. In the second phase (40–160 days), the relationship can be described by a cubic regression model (Fig. 18.9):

$$ML = 1.342A - 6.685 \times 10^{-3} A^2 + 1.232 \times 10^{-5} A^3 - 34.451 \quad (18.4)$$

The inflection points of the relationship between weight (g) and age (days) are at about 30 (at the settling stage) and 70 days, where growth can be separated into two phases. The phase after the settlement can be expressed by a power regression

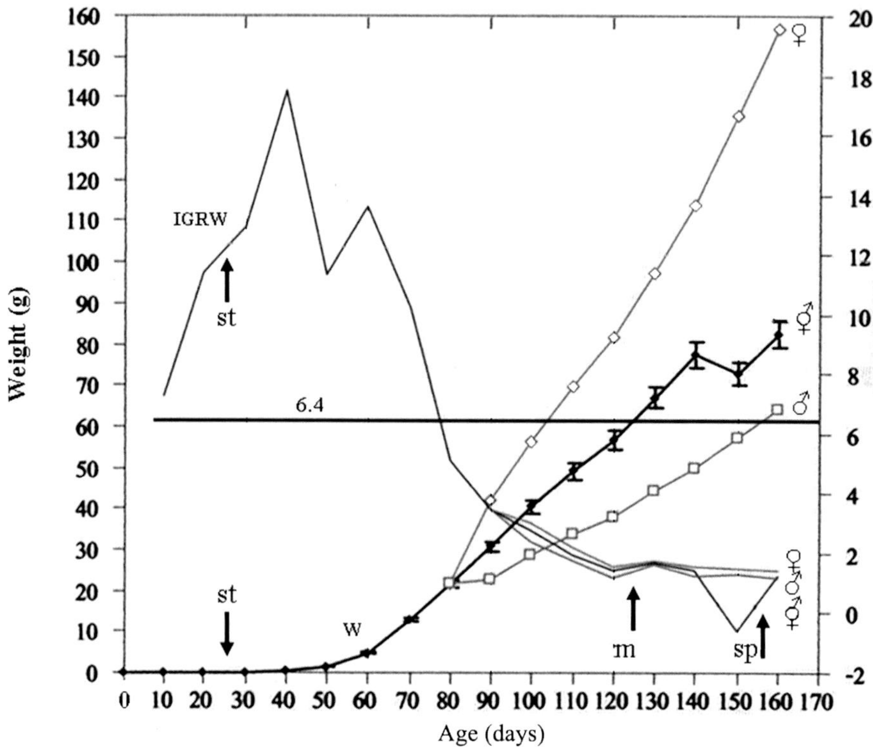

Fig. 18.7 Growth in terms of weight (g), instantaneous relative growth rate (*IGRW(%)*) and age (days) after hatching. Individuals were reared at 28 °C. The *horizontal line* indicates the average value for *IGRW*. *Arrows* indicate settlement (*st*), mating (*m*) and spawning (*sp*). (Adapted from Promboon et al. 2011)

model (30–70 days) and the following phase, from maturity onwards, by a quadratic regression model (70–160 days, Fig. 18.10):

$$W = 4.325 \times 10^{-11} A^{6.190} \tag{18.5}$$

and

$$W = 1.546A + 3.249 \times 10^{-3} A^2 - 80.765 \tag{18.6}$$

The conversion efficiency peaks for a period of about 2 months from 60 to 130 days, because of an energy storage process prior to reproduction, since mating is around day 125. Survival from hatching to mating is 1 %, and from settlement to mating is about 10 %. Spawning is at about 1 month after mating, at an age of around 150 days. Spawning females cease food intake and die within about 17 days after paralarval hatching. The lifespan lasts for approximately 200 days or about 6–8

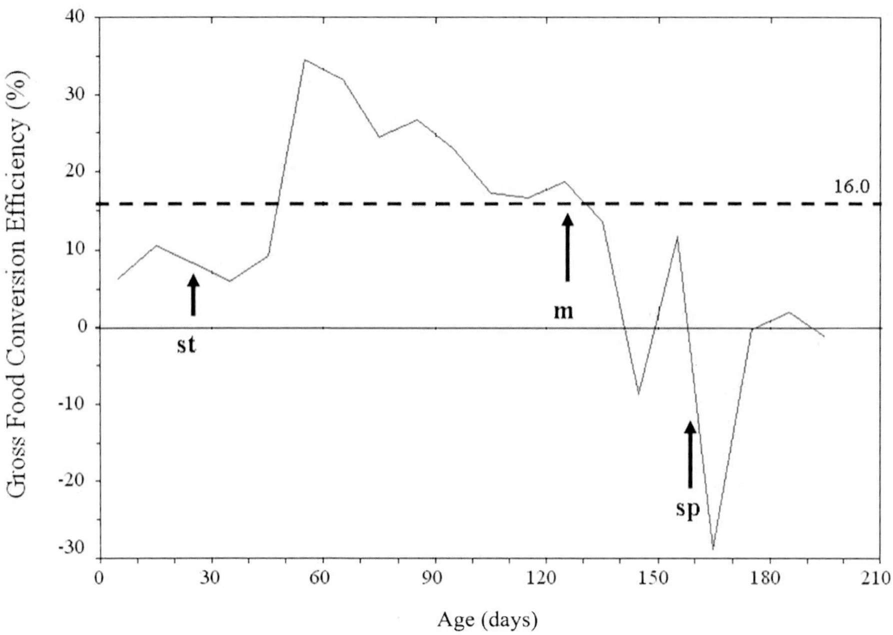

Fig. 18.8 The gross food conversion efficiency (*Gross Food Conversion Efficiency*(%)) during growth at age (days) after hatching. Individuals were reared at 28 °C. *Arrows* indicate settlement (*st*), mating (*m*) and spawning (*sp*). The *horizontal dashed line* with a figure indicates the average value. (Adapted from Promboon et al. 2011)

months. The subsequent reproductive phase lasts for 34 % of the species lifespan (Promboon et al. 2011). Since females cease food intake and subsequently lose their weight at spawning, harvest for a maximum aquaculture yield is recommended to be prior to spawning.

18.7 Trends in Research and Industrial Level

The purpose of culturing *Amphioctopus aegina* for human consumption has declined at present due to its high unit cost. Future research should focus on the reduction of unit costs through increasing the survival, hence increasing production. Such development encounters similar bottleneck obstacles to that of the other cephalopods, the requirement and availability for live food during the planktonic stage, as well as the efficient supply of such food. In the short term, other species used as food organisms should be studied as options, particularly the species that are already commercially produced. In the long term, development of artificial feed is necessary in order to replace the live food and eliminate these obstacles.

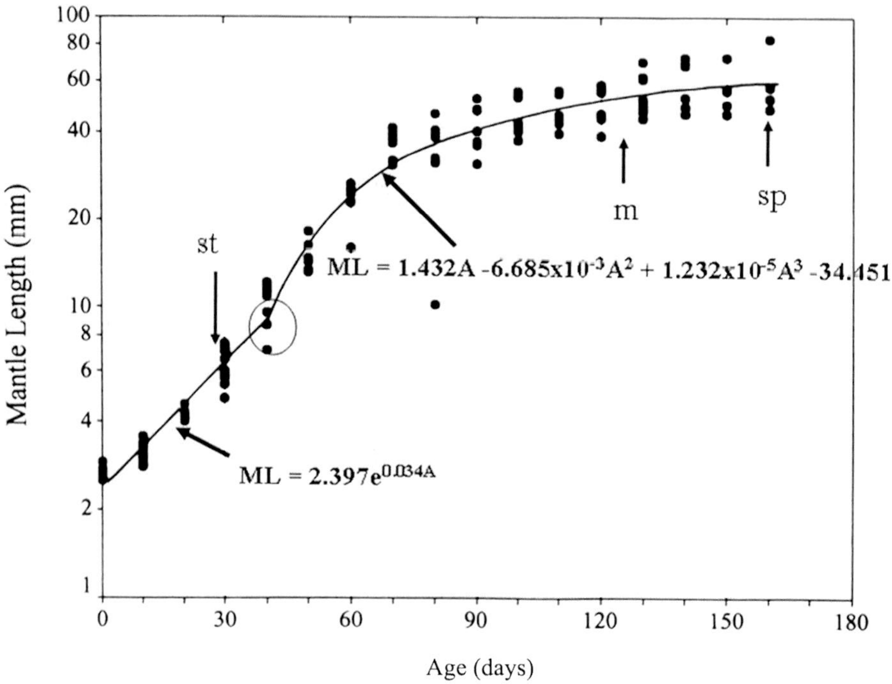

Fig. 18.9 Relationships between the mantle length (mm) and age (days) after hatching of *Amphioctopus aegina*. Individuals were reared at 28 °C. *Arrows* indicate settlement (*st*), mating (*m*) and spawning (*sp*). The *circle* indicates the inflection point. (Adapted from Promboon et al. 2011)

The small-scale facilities are appropriate for culturing a small-sized cephalopod like *Amphioctopus aegina.* Such facilities require less space and maintenance cost compared to the larger species. However, the differences in methodology between the culture of this species in an open system and in a closed system should be studied in order to maximise production and reduce costs.

The moderate size, benthic habit, good adaptability and tolerance to different culture conditions of this species are prominent characteristics that favour its aquaculture as an ornamental animal. Since *Amphioctopus aegina* is a 'newcomer' species, this demand needs to be initiated through commercial channels as a new 'pet'.

With those mentioned prominent characteristics, there is already a demand for other octopus species as experimental models for neurophysiology and ethology (i.e. *O. vulgaris*); this demand could include *Amphioctopus aegina* as an option and another model organism.

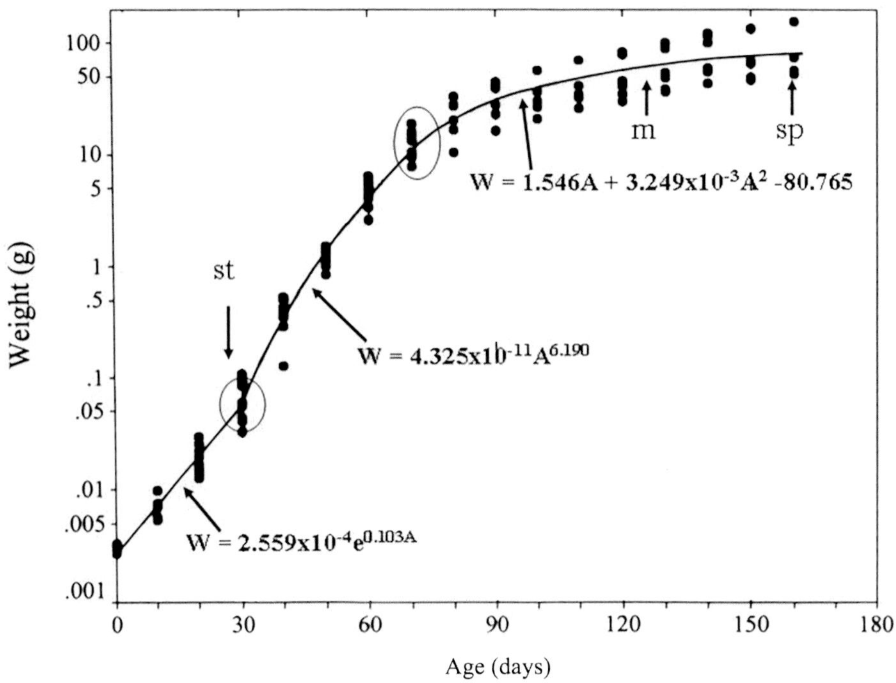

Fig. 18.10 Relationships between the weight (g) and age (days) after hatching of *Amphioctopus aegina*. Individuals were reared at 28 °C. *Arrows* indicate settlement (*st*), mating (*m*) and spawning (*sp*). The *circles* indicate the inflection points. (Adapted from Promboon et al. 2011)

18.8 Conclusions

The marbled octopus, *Amphioctopus aegina,* can be cultured on a small scale because of its small size and benthic habit. The nursing of brooding females is similar to the nursing of the eggs since the hatching rate is high only because of brooding by the female. Feeding of planktonic young in the first month during the nursing phase with live prey is a bottleneck similar to that for other cephalopods. However, live feed for *Amphioctopus aegina* is collected from the wild (mysids) as well as those produced in hatcheries (palaemonid shrimp larvae) can be supplied. The planktonic phase includes a transitional settling stage for about 10 days before the settlement is completed. After the settlement, the young can learn to feed on dead prey (chopped fish meat). The octopus reaches maturity in about 120 days and their lifespan is about 210 days. Survival is low and requires further improvements. *Amphioctopus aegina* is a new octopod species in aquaculture and the second benthic octopus species with planktonic hatchlings that can be cultured. The market demand for live octopus of moderate size should be initiated.

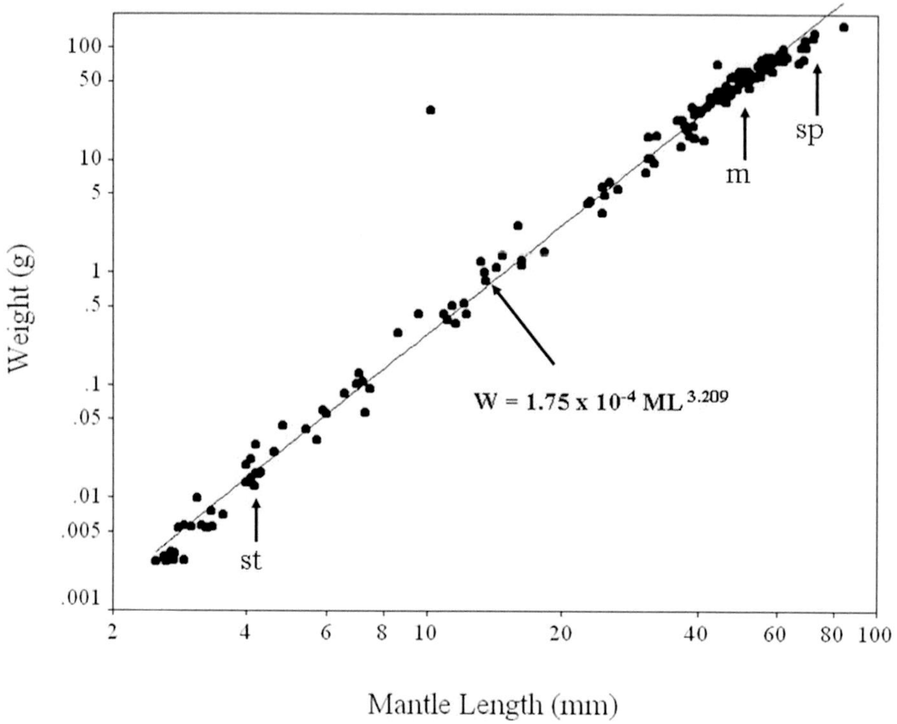

Fig. 18.11 Relationship between the mantle length (mm) and weight (g) of *Amphioctopus aegina*. Individuals were reared at 28 °C. *Arrows* indicate settlement (*st*), mating (*m*) and spawning (*sp*). (Adapted from Promboon et al. 2011)

References

Boletzky Sv (1977) Post-hatching behaviour and mode of life in cephalopods. Symp Zool Soc Lond 38:557–567

Boyle P, Rodhouse P (2005) Cephalopods ecology and fisheries. Blackwell Science, Oxford

Department of Fisheries (2012) Fisheries statistics 2009. www.fisheries.go.th/it-stat/. Accessed 15 Feb 2012

Hanlon RT (1987) Mariculture. In: Boyle PR (ed) Comparative reviews. Cephalopod life cycles, vol II. Academic Press, London, PP 291–305

Iglesias J, Otero JJ, Moxica C, Fuentes L, Sanchez FJ (2004) The completed life cycle of the octopus (*Octopus vulgaris* Cuvier) under culture conditions: paralarval rearing using *Artemia* and zoeae, and first data on juvenile growth up to 8 months of age. Aquacult Int 12:481–487

Ignatius B, Srinivasan M (2006) Embryonic development in *Octopus aegina* Gray, 1849. Curr Sci 19:1089–1092

Nabhitabhata J, Nilaphat P, Promboon P, Jaroongpattananon C (2005) Life cycle of cultured bobtail squid, *Euprymna hyllebergi* Nateewathana, 1997. Phuket Mar Biol Cent Res Bull 66:351–365

Nabhitabhata J, Promboon P, Duengdee T (2010) Embryonic and post-hatching development of marbled octopus, *Amphioctopus aegina* (Gray, 1849) (Cephalopoda: Octopodidae). Trop Nat Hist Suppl 3:300

Nateewathana A (1997) The octopod fauna (Cephalopoda: Octopoda) of the Andaman Sea, Thailand. Phuket Mar Biol Cent Spec Pub 17:407–452

Norman MD (1998) Octopodidae benthic octopuses. In: Carpenter KE, Niem VH (eds) Cephalopods, crustaceans, holothurians and sharks. The living marine resources of the western central Pacific, vol 2. Food and Agriculture Organization of the United Nations, Rome, pp 800–809

Norman MD, Hochberg FG (2005) The current state of octopus taxonomy. Phuket Mar Biol Cent Res Bull 66:127–154

Phanichpong W (1985) Studies on reproductive biology and some behaviors of the devil fish, *Octopus dollfusi* Robson, 1928. MSc Thesis, Kasetsart University

Promboon P, Nabhitabhata J, Duengdee T (2011) Life cycle of marbled octopus, *Amphioctopus aegina* (Gray, 1849) (Cephalopoda: Octopodidae) reared in laboratory. Sci Mar 75:811–821

Roper CFE, Sweeney MJ, Naun CE (1984) Cephalopods of the world: an annotated and illustrated catalogue of species of interest to fisheries. FAO fisheries synopsis no 125, vol 3. Food and Agriculture Organization of the United Nations, Rome

Srikum T, Somchanakit H (2011) Octopus fishery by noble volute shells in the Gulf of Thailand. Tech Pap 14/2011, Eastern Marine Fisheries Research and Development Center, Marine Fisheries Research and Technology Development Institute, Marine Fisheries Research and Development Bureau, Department of Fisheries.

Supongpan M (1995) Cephalopod resource in the Gulf of Thailand. In: Nabhitabhata J (ed) Biology and culture of cephalopods. Rayong Coast Fish Res Dev Cent Contrib 18, Rayong, pp 191–220

Villanueva R, Norman MD (2008) Biology of the planktonic stages of benthic octopuses. Oceanogr Mar Biol Annu Rev 46:105–202

Chapter 19
Enteroctopus megalocyathus

Íker Uriarte and Ana Farías

Abstract Advances in controlled cultivation of Patagonian red octopus *Enteroctopus megalocyathus* (Gould 1852) have fostered biological, physiological and nutritional studies of the species. Research on cultivation of *E. megalocyathus* is focused on obtaining juveniles under controlled cultivation conditions because it would be the only way to obtain a sustainable octopus production.

Experimental evidence indicates that it is possible to produce *E. megalocyathus* juveniles to supply an ongrowing production because broodstock can be reproductively conditioned in tanks, the controlled fertilization allows to produce developing embryos, incubation of eggs throughout the embryonic period allows the production of viable paralarvae and the rearing of paralarvae allows to successfully obtain completely settled benthic juveniles.

Currently, further investigations are needed on the physiology and growth of this species to determine the time and conditions required to obtain juvenile octopuses of 50 or 100 g to start the ongrowing, and to complete the total life cycle of the species.

Keywords *Enteroctopus megalocyathus* · Paralarval culture · Embryo development · Egg fertilization · Juvenile ongrowing

19.1 Introduction

19.1.1 State of the Art

In Chile, there are two species of octopus fit for the international market which are similar to the common octopus *Octopus vulgaris:* Changos octopus (*O. mimus*) and Patagonian red octopus (*E. megalocyathus*). Currently, fisheries of both species

Í. Uriarte (✉) · A. Farías
Instituto de Acuicultura, Sede Puerto Montt,
Universidad Austral de Chile, P.O. Box 1327, Puerto Montt, Chile
e-mail: iuriarte@uach.cl

A. Farías
e-mail: afarias@uach.cl

J. Iglesias et al. (eds.), *Cephalopod Culture*, DOI 10.1007/978-94-017-8648-5_19,

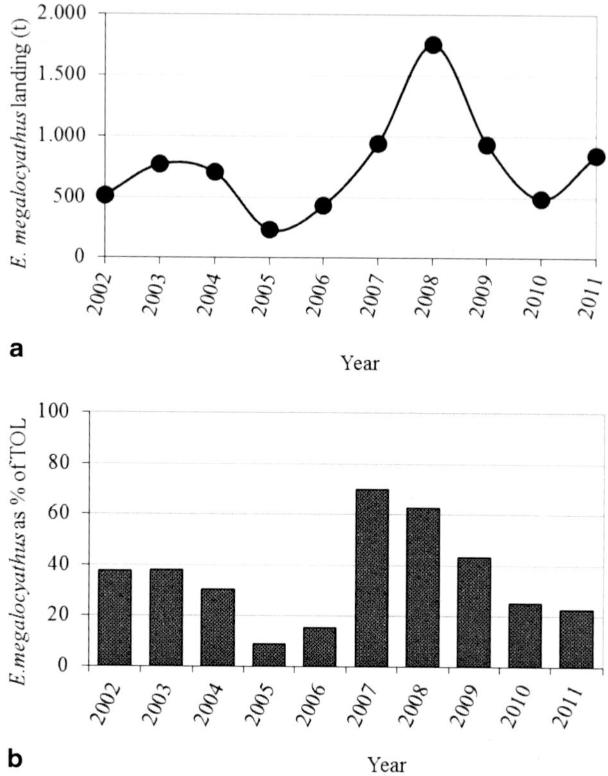

Fig. 19.1 *Enteroctopus megalocyathus*. **a** Landings (t) in Chile between the years 2002 to 2011. **b** Landings as percentage of total octopus landings (*TOL*) in Chile during 2002–2011. (Graphic made by the authors with data of Servicio Nacional de Pesca (2011) and taking in account the distribution of this species cited by Rocha and Vega (2003))

sustain processing plants. Patagonian red octopus (*E. megalocyathus*) inhabits the southern tip of South America, in the Pacific and Atlantic Oceans. In Chile, Patagonian red octopus has been fished for decades under the wrong denomination of 'common octopus', and for that reason it is indistinguishable from *O. mimus* in fishery statistics.

Rocha and Vega (2003) reported that the octopus fished down south of the VIII region of Chile (36°00 lat S) is the species *E. megalocyathus*. Under this framework, the recalculation of the offloading during the decade 2002–2011 (Fig. 19.1a) showed that Patagonian red octopus had relevant values of landings, reaching 1,700 t landed in 2008 (SERNAP 2011). In 2007, Patagonian red octopus accounted for 70 % of the octopuses landed in Chile (Fig. 19.1b). Consequently, a ban was established on the commercial fishing of this species between 2009 and 2011, with a Scientific Permit for Research being required to capture this native species. In this context, developing the cultivation of *E. megalocyathus* is important to support the octopus production and to reduce the pressure that fisheries might put on this resource in the future. For this reason, numerous studies have been carried out in the decade 2002–2011 on the ongrowing and controlled cultivation of juveniles.

Studies on Patagonian red octopus reported difficulties to obtain viable fertilizations, viable egg incubation and viable paralarvae cultivation (Ortiz et al. 2006; Farías et al. 2011; Uriarte et al. 2011a). For these reasons, multidisciplinary studies are focused on rearing the species from broodstock conditioning to production of juveniles under controlled cultivation. Once achieved, it would allow the sustainable ongrowing of the octopus. On the other hand, there are projects linked to artisanal fishermen focused only on the ongrowing of Patagonian octopus in the X region of Chile (Medrano and Godoy 2011).

19.1.2 Geographic Distribution

E. megalocyathus is geographically distributed around the southern tip of the South American continent, in both the Atlantic and Pacific Oceans (Ibáñez and Chong 2008). According to Ibáñez et al. (2009), Chilean cephalopods are distributed along the entire Pacific Coast of Chile in three biogeographic zones that are differentiated according to average temperature: north of 30°S, south of 42°S and one central zone between 30°S and 40°S. In Chile, *E. megalocyathus* is commonly found between 42°S (X region) and 56°S (XII region), and it has even been reported in the VIII region between 1991 and 1994 (Rocha and Vega 2003). Along the Argentine coast, *E. megalocyathus* is distributed from the Golfo San Matías (41°S) to the Strait of Magellan and Beagle Channel, Malvinas Islands and Burdwood Bank (Ré 1998; Ibáñez and Chong 2008).

E. megalocyathus is found in the Pacific coast at a depth of 15 to 30 m (Chong et al. 2001), while on the Atlantic coast it is found at depths of less than 20 m when the water temperature oscillates between 12 and 13 °C. When the water temperature rises above 13 °C, the octopus can be found at 170 m depth (Ortiz et al. 2011).

19.1.3 Methods of Capture

In Chile, this species is commercially fished through diving to insert a hook between rock crevices, the main octopus habitat (Chong et al. 2001). The minimum catch size is 1 kg, but a wide range in body size is displayed by the species, which greatly affects yields. Therefore, producers, processors and exporters specialized in Chilean octopuses are interested in the ongrowing process, in order to obtain larger specimens (≥4 kg). The largest species are strongly demanded in Asia and Europe, and the Network of Artisan Fishermen for Sustainable Development (RECOPADES 2009) even applied for a grant to ongrow octopuses of up to 4 kg from octopuses of 1 kg approximately.

The extraction of *E. megalocyathus* in the Argentine coast is carried out through the artisan fishing method of using hooks in the intertidal zone, while scuba diving using hooks and harpoons are utilized in the subtidal zones up to 25 m. In the Nuevo Gulf and the San Jose Gulf, the extraction is carried out from March to November–December. In these areas, the fishery operates in the tidal and intertidal strips, whereas in the Northern Patagonian gulfs, the fishery primarily operates in subtidal zones (Ré and Ortiz 2007).

Table 19.1 *Enteroctopus megalocyathus*. Main advantages and disadvantages of the different techniques of extraction of Patagonian red octopus, including traps, that were used in the project 'Network of Artisanal Fishermen for Sustainable Development'. (With permission of RECOPADES 2009)

Method of capture	Advantages	Disadvantages
Hooks and harpoons in the subtidal zones	There is great experience of artisanal divers. Gets a considerable number of Patagonian red octopus in a few hours	Very aggressive method for Patagonian red octopus. It is not recommended to obtain live octopuses or to select sizes or avoid females with egg layings. High percentage of loss or poor quality of raw material
Galician-style trap	Avoids sacrificing octopuses under the minimum catch size or under the commercial weight. Avoids females incubating eggs into dens. Captures at greater depths than the diving method	High investment and less catch than diving method. Low certainty in the success of the catches and difficult stability of supply of raw material. Risk of theft or loss of the fishing gear
Manual extraction	Live octopuses can be obtained. Selective. It prevents the capture of females with egg layings	Lower capture. Used as sport fishing

The best method for catching octopuses is the Galician-style trap because it does not damage the females laying eggs, and also allows to sort out size and gender after capture (Table 19.1). Furthermore, this method avoids damaging the females that are in the dens incubating eggs. Unfortunately, this method is not used commercially.

19.1.4 Morphological Characteristics

Gari and Ré (2002) gave a very detailed description of the morphological characteristics of *E. megalocyathus* based on the following structures: mandible morphology, with regard to the number and distribution of the teeth in the radula; and the digestive system regarding the location and morphometrics of the jaw, oesophagus, stomach, caecum, intestines, glands and rectum. In summary, the main differentiating characteristics of Patagonian red octopus are: oval head; head shorter than mantle; unequal, moderately long arms; biseriate suckers which are notably greater in the first proximal third of the arms; noticeable hectocotylus in males; and medium-developed interbranchial membrane, with ink sac and large ligula. Each external hemibranch has 11–13 lamellae.

19.1.5 Reproductive Cycle

The reproductive cycle of *E. megalocyathus* in Chilean Patagonian waters starts in winter reaching the maximum maturity in spring, and it continues during the summer with reproductive peaks in December and January (Ibáñez and Chong 2008), while in the Patagonian waters of Argentina there is evidence of two annual laying periods, one in winter and another starting in November–December (Ré 1980). Chong et al. (2001) estimated the size of sexual maturity of females and males at 71.7 and 69.9 cm total length ($LT_{50\%}$), respectively, and 14.9 cm of mantle length ($ML_{50\%}$) for both sexes. In the Patagonian coastal zone of Argentina, mature males are found during long periods of time (April–November). Males show an average of seven spermatophores in the Needham's sac. As early as July, spermatangia are found in females. The sex ratio remains 1:1 in breeding grounds, and in July the number of males increases due to the displacement of fertilized females to deeper waters. A study on the reproductive biology of *E. megalocyathus* in the Nuevo Gulf (42°49′58″S, 64°50′10″W) reported the presence of a pseudohermaphroditic individual that could be classified externally as male, but internally had female and male sexual structures (Ortiz and Ré 2006).

19.1.6 Natural Feeding

Most of the food items hunted by *E. megalocyathus* are brachiura and anomura crustaceans, followed by fish and other octopuses (Ibáñez and Chong 2008). The variability in the diet depends mainly on the size of Patagonian red octopus; therefore, small-sized octopuses feed on small crustaceans whereas larger specimens prefer large crabs and octopuses of the same species, either juveniles or eggs. The variation in the items available for hunting depends mainly on the habitat of the octopus. Considering that they can reach a large size, and the abundance and type of prey that is being hunted, it has been suggested that *E. megalocyathus* has a significant impact on the structure of the subtidal communities around the southern tip of South America (Ibáñez and Chong 2008).

19.2 Broodstock Conditioning

Iglesias et al. (2000) concluded that reproduction is not a limiting factor on the feasibility of cultivating the species *O. vulgaris,* and the diet during reproductive conditioning should include crustaceans to ensure good quality of eggs. Similar recommendations are given by Rosas et al. (2006) for the conditioning of the holobenthic species *O. maya*.

After courtship, in both species the mating occurs at 'distance' when the male inseminates the female without embracing but introducing his hectocotylus into the mantle of the female to pass one or more spermatophores (Rosas et al. 2006). However, *E. megalocyathus* is a merobenthic species with a complex reproductive behaviour, with courtship preceding mating where the male mounts the female (Gutiérrez et al. 2012). In this sense, Patagonian red octopus exhibits a reproductive behaviour similar to *E. dofleini.* Furthermore, the courtship and mating periods depend on the size relation between the males and females (Gutiérrez et al. 2012).

Studies carried out during the broodstock conditioning period of *E. megalocyathus* demonstrate that fecundity in these animals can be reduced while maintaining the biochemical composition of the eggs, when food is scarce or when changes in diet occur (Farías et al. 2011). These studies also show that Patagonian red octopus can be conditioned without a crustacean diet and still maintain fecundity.

After laying the eggs, the females stop feeding and it has been suggested that this behaviour generates population differences between genders since feeding habits affect the age when they reach senescence and death (Ibáñez and Chong 2008).

The reproduction area at the Hatchery of Marine Invertebrate of Universidad Austral de Chile (HIM-UACh) consists of an isolated room of 5 × 6 m, with controlled temperature and light. For the reproductive broodstock, 700 L (1.75 × 1.1 × 0.35 m) tanks with sea water filtered at 1 µm and ultraviolet (UV) sterilized were used. Sea water had the following characteristics: 29–31 psu, 11 ± 1 °C, 6–8 mg L^{-1} of dissolved O_2 and lower than 1 mg L^{-1} total ammonia. Light brightness was maintained at levels below 50 Lx cm^{-2}, while the temperature was controlled with a chiller equipment.

The sea water was recirculated through deep-bed filters with anthracite as filter material. Anthracite also has the property of retaining the dissolved ammonium. Also, the water was subjected to a sterilization by UV lamps in order to reduce the bacterial load.

The individuals came from the natural environment of Hueihue (Ancud, Chiloé, X region, Chile). They were captured using a traditional method where divers trap and place them in a floating fibreglass pond of 500 L which is tied to the boat. Then they are landed and individually placed in mesh bags into 500-L tanks to prevent their escape and damage between them, while being transported to the laboratory at 12 °C and with oxygen injection throughout the entire journey of 2.5 h from the capture area to the laboratory.

In the laboratory quarantine room, females are placed in acclimation ponds where they are maintained for 48 h until recovery. Before placing the octopuses in breeding tanks, they are sexed, measured and weighed. The animals are prepared for weighing by preparing a bath of ice-cold water at 2 °C for 3–5 min, which works as an anaesthetic that allows to measure and weigh the animals without harming them. Then, they are placed on a scale to record their weight and, finally, total length is recorded.

19.3 Spawning Process

Egg layings collected from the wild (Ortiz et al. 2006, 2011), as well as layings obtained by broodstock conditioning (Fig. 19.2) in the laboratory (Farías et al. 2011; Uriarte et al. 2011a, 2013; Gutiérrez 2012), confirm that this species presents absolute fecundity between 1,000 and 5,000 eggs per female (Table 19.2). On the other hand, relative fecundity under controlled conditions is 1,101 ± 159 eggs kg^{-1} female (Gutiérrez 2012). Leporati et al. (2008) showed that other factors besides the size of the female, such as seasonal changes in temperature, substrate availability for laying,

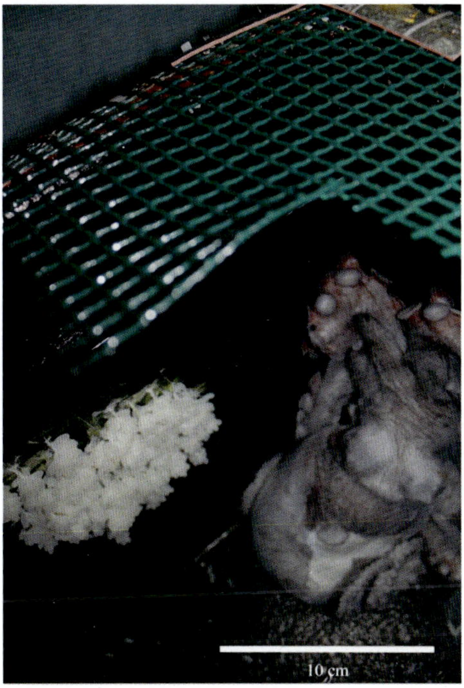

Fig. 19.2 *Enteroctopus megalocyathus*. Female octopus with clutches obtained at Hatchery of Marine Invertebrates of the Universidad Austral de Chile (HIM-UACh). The measuring slide indicates 10 cm. (Photo of the authors)

Table 19.2 *Enteroctopus megalocyathus*. Absolute values of fecundity and other reproductive characteristics

Fecundity per female	Method to estimate fecundity of females	Egg total length (mm)	Author
1,000–20,000	Number of oocytes hydrated from subsamples of the gonad		Chong et al. (2001)
1,469	Samplings on natural egg layings	10.7	Ortiz et al. (2006)
2,500–3,600	Not indicated	14.5	Ré (2007)
2,129 ± 1,182	Samplings of egg layings obtained from the broodstock conditioning of octopus in laboratory	7–11	Farías et al. (2011)
2,052 ± 368	Samplings of egg layings obtained from the broodstock conditioning of octopus in laboratory	8.5 ± 0.1 (5.8 % FML)[a]	Gutiérrez (2012)
2,321 ± 536	Samplings of egg layings obtained from the broodstock conditioning of octopus in laboratory	9.3 ± 0.1 (6.4 % FML)[a]	Uriarte et al. (2013)

FML female mantle length

[a] Calculated from the results of Gutiérrez (2012) and Uriarte et al. (2013)

den quality and the reabsorption rate of the eggs in the ovary, can lead to varying fecundity in females. In accordance with Chong et al. (2001) and based on the ovocytes contained in the ovary, the expected fecundity of *E. megalocyathus* could

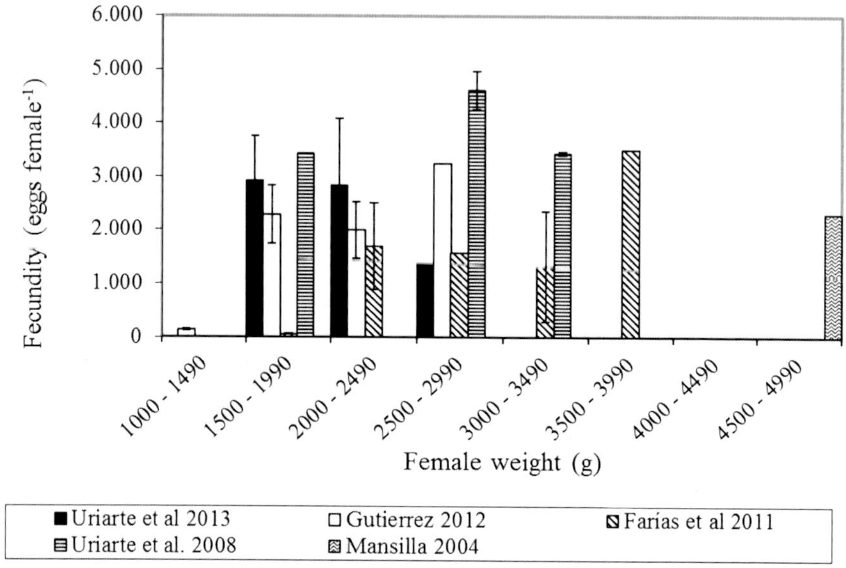

Fig. 19.3 *Enteroctopus megalocyathus*. Variation of fecundity under controlled culture conditions in relation to the female weights. Taken from different studies carried out in the Hatchery of Marine Invertebrates of the Universidad Austral de Chile (HIM-UACh). (Data of the authors and their thesis students)

reach up to 20,000 eggs per female, while Ortiz et al. (2011) reported an expected fecundity of 3,729 ± 1,443 oocytes; these authors found significant correlations between expected fecundity of males and females and body weight. Under laboratory conditions, the observed fecundity ranged from 63 to 5,307 eggs (Fig. 19.3), being affected by quantity or quality of the feed (Uriarte et al. 2008; Farías et al. 2011; Gutiérrez 2012) but without significant effect of the female weight (Farías et al. 2011; Uriarte et al. 2013). Therefore, it is possible that environmental conditions moderate the reabsorption rate of eggs in order to maintain the females during the period previous to laying and during incubation.

If the quantity of egg layings of Patagonian red octopus is reduced due to poor nutrition, the quality of the eggs in terms of their proximal contents and fatty acids is maintained, which is essential for the conservation and long-term success of incubation and hatching (Farías et al. 2011).

The broodstock conditioning period for *E. megalocyathus* is 4 months at 12 °C (Uriarte et al. 2011a; Farías et al. 2011). During broodstock conditioning, the weight of the females increases exponentially, regardless of the original size. It is suggested that during broodstock conditioning, the female *E. megalocyathus* invests a large part of its food energy in the production of eggs, leading to an increase of weight that is registered as growth rate. On the other hand, the specific growth rate (SGR) of males is reduced as the size of each octopus increases (Uriarte et al. 2013). When the males increase in size, the courting period is shortened (Gutiérrez et al. 2012), which could be an advantage for mating many times during the reproductive period.

At HIM-UACh, male and female octopuses are placed in the conditioning tanks where 101.6-mm polyvinyl chloride (PVC) elbows with perforated lids on one end are offered as dens. For the females, this lair serves as nest for egg laying. During the maturation period, the females are fed crabs only at a rate of 5–15 % of their wet body weight (Table 19.3). The crabs are offered alive without chelae, to prevent damage when defending from the octopus. The ration of crab every other day is maintained until spawning, after they are fed at demand since females require low amounts of food at this stage.

19.4 Egg Incubation

The length of octopus eggs is linked to lifestyle after hatching so that a species laying small eggs (under 10 % of the adult ML) generates a planktonic paralarvae; therefore, it is a merobenthic species. However, species that lay larger eggs (more than 10 % of the adult ML) originate benthic juveniles that hatch directly from the egg; therefore, it is a hobenthic species (Boletzky 2003; Ignatius and Srinivasan 2006). These strategies are also linked to fecundity, so that species laying smaller eggs are more fertile and originate many planktonic paralarvae that ensure their survival when they are dispersed and colonize new habitats. In the case of *E. megalocyathus,* a merobenthic species whose eggs measure between 5.8 and 6.4 % of the ML (Table 19.3), an intermediate strategy with medium-sized eggs and hatchlings that are planktonic during more than 60 days is observed (Uriarte et al. 2011b).

Currently, the length of the embryonic period of *E. megalocyathus* is reported for temperatures between 10 and 12 °C with normal embryo development (Uriarte et al. 2011a, 2013).

The embryos of Patagonian red octopus observed at HIM-UACh grow exponentially in total length, arm length and ML, and also in eye diameter. This occurs in other embryonic octopods such as *Robsonella fontaniana* but only for total length and arm length (Uriarte et al. 2009). The exponential growth rate observed in Patagonian red octopus embryos at HIM-UACh is 2 % day^{-1}, a value below the exponential growth rate observed in *R. fontaniana* which is 3 % day^{-1}, and paralarvae take 70 days to hatch at 12 °C (Uriarte et al. 2009). In contrast, *O. maya* embryos reach 182 % day^{-1} and take 30 days at 25 °C hatching directly to a benthic juvenile (Rosas et al. 2006). These different strategies could be due to the interaction between egg size, temperature at which embryos are developed and quantity of yolk available after hatching.

Eggs incubated under controlled conditions show a complete embryonic development when the fertilization of the females is assured (Uriarte et al. 2011a, 2013). For *E. megalocyathus,* embryonic development under laboratory conditions at 12 °C takes between 150 and 180 days (Table 19.3), being similar to *E. dofleini* as described by Villanueva and Norman (2008). During this long period, the eggs are exposed to bacterial contamination and detachment from the walls of the tanks, usually caused by the same female reproducers (Uriarte et al. 2011a). Therefore,

Table 19.3 *Enteroctopus megalocyathus*. Culture conditions cited by different authors. (Original results are referenced as part of the current work of the authors at HIM-UACh)

Stage of culture	Density ($kg\ m^{-3}$)	Length complete period	Temperature	Salinity	Culture system	Tanks	Sterilization	Feeding		References
		(days)	(°C)	(psu)				(preys)	(ration)	
Broodstock conditioning	1.4–2.1	120	12	30	Recirculation	1000 L, conic	Ozone	100% fish or 3:1 fish:crab	10% BW day^{-1}	Farías et al. (2011)
		120–150	11		Recirculation	500 L, conic	UV	70–100% crab + fish supplementation		Uriarte et al. (2008, 2011a, b)
	3.3	106±40	12±1	30±1	Recirculation	250 L, rectangular	UV	100% crab, 100% fish, 100% squid, mix	5% BW day^{-1}	Gutiérrez (2012)
		97–171	12	30				100% crab	5% BW day-1	Uriarte et al. (2013)
Embryo incubation		150–176	12		Closed system with laminar flow or air lift	3 L, 25 L	UV			Uriarte et al. (2011a, b)
		180			Recirculation	50 L, rectangular	UV			Uriarte et al. (2013)
Paralarval culture[a]	2	90–114	12	30	Recirculation	25 L rectangular	UV	Artemia nauplii		HIM-UACh
Ongrowing culture of wild juveniles	5–10		Environmental		Deep culture on the bottom	Cages				Medrano and Godoy (2011)

UV ultraviolet

[a] Density extrapolated to $kg\ m^{-3}$ calculated from 10 paralarvae L^{-1} with an average weight of 0.2 g for a paralarvae of 50 days after hatching. Original data of the authors

new strategies and incubation systems are required to ensure the paralarval hatching. The first would be to separate the eggs from females as soon as they have been released, because eggs could be incubated without parental control at HIM-UACh.

At HIM-UACh, after the eggs are laid they are separated from females and their clusters are suspended in incubation tanks with recirculating system, UV sterilization, temperature control, tank water exchange of 70 % day^{-1} and concentration of O_2 over 85 % saturation.

The eggs are incubated until hatching, a period of 4–5 months for this species. The percentage of egg fertilization reaches 75 % success, and is detected by the appearance of the eye pigmentation which occurs at stage IX described by Naef (1928).

19.5 Paralarval Growth

At the moment, the production of a merobenthic octopus species is a challenge. So far, juveniles from five octopus species are reported to have been obtained under controlled conditions: *E. dofleini* (Okubo 1980), *O. vulgaris* (Villanueva et al. 1994, 1996; Villanueva 1995; Iglesias et al. 2004; Carrasco et al. 2006), *O. joubini* (Forsythe 1984), *R. fontaniana* (Uriarte et al. 2010) and *Amphioctopus aegina* (Promboon et al. 2011).

The growth of paralarvae would allow juvenile cultivation under controlled conditions in adequate and stable quantities to sustain industrial ongrowing. This could mitigate the overfishing impact on this valuable resource. However, until now the technology to cultivate paralarvae either is incomplete or results in low survival rates (Villanueva and Norman 2008; Iglesias et al. 2007; Berger 2010; Uriarte et al. 2011a), limiting commercial culture development.

Ortiz et al. (2006) observed paralarvae of *E. megalocyathus* of up to 5 days of post-hatch age with suprabenthic behaviour, so these paralarvae share both paralarvae and benthic juvenile characteristics. Uriarte et al. (2011a) described that the length of the paralarval period was too extended, reporting it to be more than 60 days, ending with a den-searching behaviour. Currently, at HIM-UACh at 12 °C the paralarval period lasts between 90 and 114 days, similar to *E. dofleini* paralarvae as described by Villanueva and Norman (2008). At 114 days post hatching, paralarvae settle definitely and exhibit a juvenile morphological character and behaviour using dens permanently (Fig. 19.4).

In general, the feeding habit of octopus paralarvae is one of the most complex and limiting aspects for its cultivation. The scarce nutritional knowledge and the lack of an appropriate prey to satisfy their food requirements (Iglesias et al. 2007; Berger 2010; Uriarte et al. 2011a) are the most critical aspects for rearing paralarvae during more than 90 days, as in the case of Patagonian red octopus.

Another critical issue with paralarval cultivation is cannibalism in *E. megalocyathus*. Uriarte et al. (2013) observed that the cannibalism in well-fed paralarvae accounts for 4 % of total mortality, while poorly fed paralarvae cannibalism accounts for up to 13 % mortality. This means that another of the greatest challenges in the cultivation of these paralarvae is the reduction of cannibalistic behaviour.

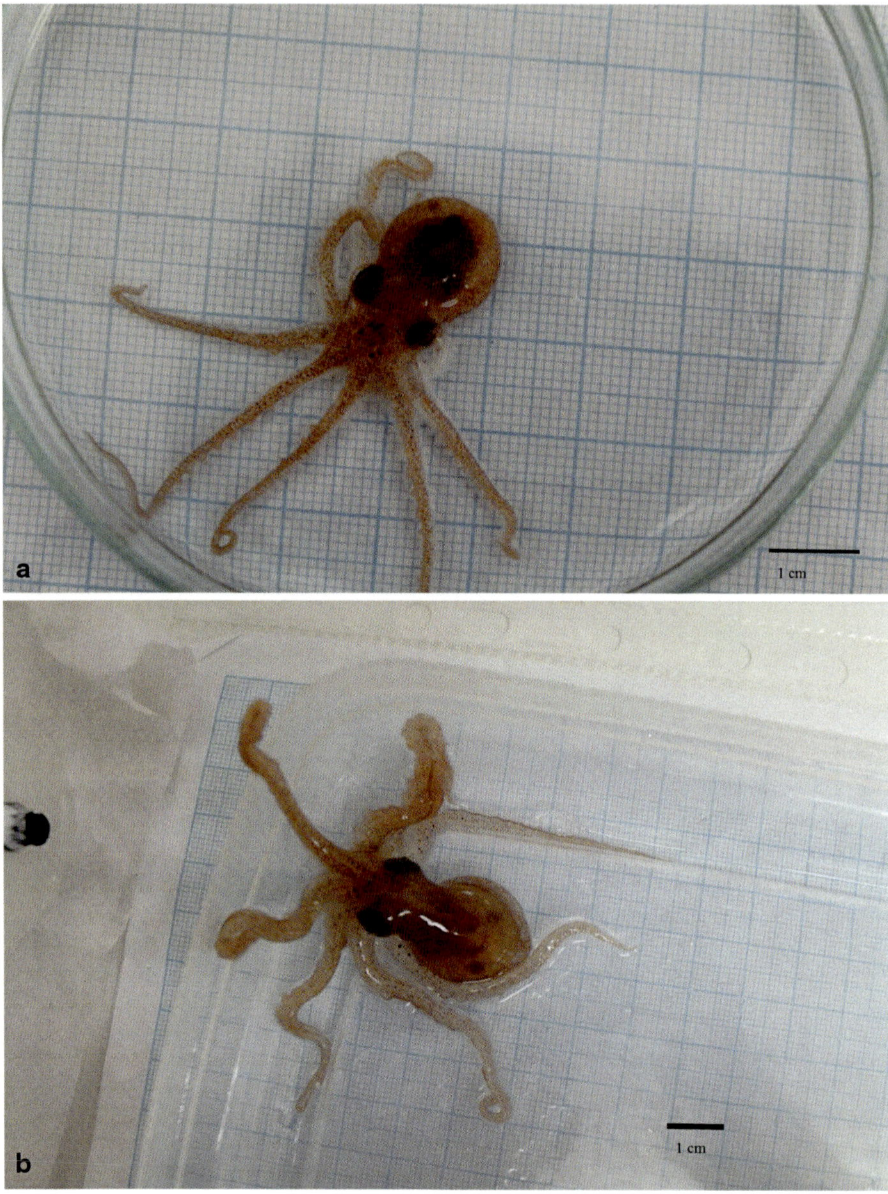

Fig. 19.4 *Enteroctopus megalocyathus.* Juveniles obtained after the broodstock conditioning, embryo incubation and paralarval rearing under laboratory conditions of Hatchery of Marine Invertebrates of the Universidad Austral de Chile (HIM-UACh). **a** Juvenile of 1.7 g. **b** Juvenile of 4.5 g. The measuring slide indicates 1 cm. (Original photos of the authors for the first juveniles produced under laboratory conditions)

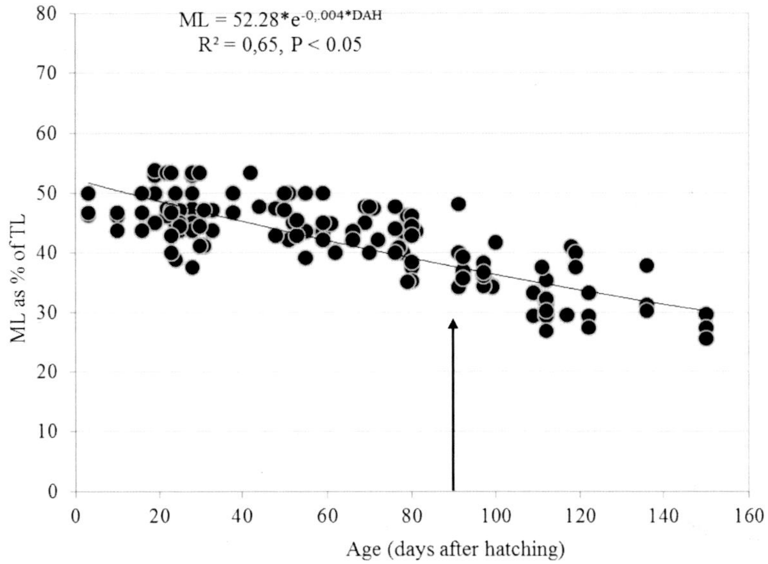

Fig. 19.5 *Enteroctopus megalocyathus*. Relationship observed between relative decrease in mantle length (ML) as percentage of total length (TL) and hatchling age. *Arrow* shows the initial settlement period. R^2 is the coefficient of determination and *P* indicates the significance of the coefficient of regression. (Original data of the authors for the first juveniles produced under laboratory conditions)

At HIM-UACh, the egg incubation period ends when the paralarvae start to hatch at regular intervals during 5–10 days. *Artemia* nauplii is supplied for feeding in the incubation tanks, so they can feed after hatching. Early paralarvae are transferred to rectangular ponds of 28 L, at a density of 10 paralarvae L^{-1} (Table 19.3). The density of *Artemia* nauplii is maintained at 2 nauplii paralarvae^{-1} day^{-1}. Paralarvae cultivation is carried out with a recirculated sea-water system, sterilized with UV, with temperature control, water replacement of 70 % of the tank day^{-1} and O_2 concentration above 85 % saturation. The unconsumed and dead *Artemia* nauplii waste is removed daily throughout the culture in order to maintain water quality at appropriate levels of oxygen and ammonia. The paralarval development period takes 4–5 months before reaching settlement. Presettlement reflexes of *E. megalocyathus* paralarvae start when ML reaches 37 % or less of the total length (Fig. 19.5). This relationship is lower than values of *O. vulgaris* (~50 %) and *E. dofleini* (45–55 %) reported by Villanueva and Norman (2008). It is interesting to note that the total length of *E. megalocyathus* paralarvae at hatching (13 mm) is similar to that of *E. dofleini* paralarvae at hatching (10 mm) and, under our laboratory conditions, paralarval Patagonian red octopus settle at total lengths between 2.5 and 3.5 cm (Fig 19.5), very similar to the 30 mm cited for *E. dofleini*.

At the HIM-UACh, the newly settled juveniles of 0.4 g are fed adult *Artemia*, in tanks of 28 L with PVC dens of 20 mm diameter. The diet consists of small adult *Artemia* and small crabs of the genus *Petrolistes* or *Acanthocyclus*. The water of recirculated system is maintained with UV sterilization, temperature control, replacement of sea water of 70 % tank day^{-1} and O_2 concentration above 85 % saturation. After reaching a weight of 1–2 g (Fig. 19.4a), juvenile octopuses are transferred to individual tanks of 27 L (Fig. 19.4b) and fed a fresh crab-based diet until they reach 20 g.

19.6 Ongrowing, Tank and Sea Cage Conditions

So far, the results in ongrowing of *E. megalocyathus* have not been published, but there are studies of juveniles captured in the wild and brought to the laboratory for ongrowing. González et al. (2008) reported for *E. megalocyathus* under individual cultivation, from 300 to 1,500 g, a high dispersion in the growth rates related to the type of feed used for ongrowing. The authors found that Patagonian red octopus did not grow with a diet of bivalve molluscs, but did show significant growth when fed crustaceans such as crabs, and also fish.

According to Farías et al. (2010), individuals that are cultivated from 300 to 1,500 g with formulated diets (pellet) do not show a good intake (0.3–0.7 % body weight day^{-1}), even though the digestibility was acceptable (47.1–61.8 %). The low intake or rejection of the artificial diets explains the absence of growth, and it probably indicates that better attractants are needed. However, the best formulated diet showed similar results to those tested with live crab, and for that reason it is assumed that it is possible to use a formulated diet, although it remains a challenge.

A comparative study on the energy physiology of *O. maya* and *E. megalocyathus* carried out by Farías et al. (2009) showed that Patagonian red octopus has a high growth rate that is similar to *O. maya*, despite the large difference in temperatures of their habitats: 10 and 25 °C, respectively. This similarity in growth is due to a better energy efficiency of *E. megalocyathus*, despite being a cold-water species. This better energy efficiency involves a higher food intake, a lower metabolic investment (lower respiratory rate) and a more efficient use of protein towards energy. According to the authors, once the octopus is acclimated to its environmental conditions, it can optimize its growth rate.

According to the results of Uriarte et al. (2011a), the exponential growth rates of *E. megalocyathus* at 18 °C with a recirculation system varied between 0.33 and 1.25 % day^{-1} for specimens between 10 and 500 g. Gutiérrez et al. (2013) show that the exponential growth rate for juveniles fed natural diets is between 0.77 and 1.01 % day^{-1} in octopuses weighing 297 g at temperatures of 12.9 °C with flow-through systems.

The field results of ongrowing in a deep farming system have been reported by Medrano and Godoy (2011). Their results showed SGR of 0.52 and 0.77 % day^{-1} for females and males at a density of 5 kg m^{-3} and a reduction to 0.40 and 0.24 % day^{-1}, respectively, when density increases to 10 kg m^{-3} (Table 19.3).

19.7 Trends in Research and Culture at Industrial Level

There is an important, unsatisfied demand for octopus, mainly in European countries such as Spain, as well as in Japan and other Asian countries (González et al. 2008). This demand, added to the need to recover the fisheries, generates a unique opportunity for the development of farming systems for this species.

The octopus is an organism that represents many challenges because of its highly developed nervous system and behavioural factors that can be complex; therefore, the prevalence of cannibalism must be reduced in order to achieve an effective cultivation of paralarvae and juveniles. For an efficient ongrowing, it is necessary to lower the competition for space, food and reproduction. Furthermore, behavioural factors make it difficult for the octopus to accept artificial or formulated food; however, the same capacity of learning should allow not only the medium-term development of balanced formulated feeds but also the use of automatic feeders that would allow the octopus to feed on demand.

The defence generated by the innate immune system of cephalopods against diseases and infections is relevant for the development of commercial production systems, especially when a high density of specimens is necessary for a profitable cultivation. The characterization of the innate immune system of *E. megalocyathus* demonstrates that the lysozyme activity and the activation of respiratory burst (ROS) are independent of the size of the octopus, but it could be dependent on diet (Silva et al. 2013). It has also been demonstrated that in the innate immune system of *E. megalocyathus*, the haemocyanin is a humoral component with special characteristics and high antibacterial activity (Uriarte et al. 2012).

Future tendencies are determined by the market, but presently the experimental results on Patagonian red octopus indicate that juvenile production is possible and it could involve:

- A larviculture process including broodstock conditioning, egg laying, embryo incubation and paralarval rearing until early juveniles that could last for about 12 months.
- An ongrowning process of juveniles of 100 g aiming to reach a commercial size of 2.7 kg that could last for about 9 months.

It is necessary to determine the growth rate from an early juvenile of 0.4 g to a juvenile octopus of 100 g to establish the total time needed under cultivation until reaching commercial size. Once these aspects are covered, the production cycle of Patagonian red octopus would be closed.

19.8 Conclusions

The main conclusions of the chapter are related to the potential success to obtain early juvenile or 'seed' for the ongrowing process. However, more information is required about the results of ongrowing in sea cages because there are relevant questions in need of an answer, for example: How long does it take to obtain octopus of 4 kg when having subadults of 500 g? How to delay the mating processes when environmental or feeding conditions affect the males or females by triggering them? How is the feed efficiency behaving throughout the ongrowing?

Acknowledgements We are very grateful for the support of FONDECYT 1131094.

References

Berger E (2010) Aquaculture of *Octopus* species: present status, problems and perspectives. Plymouth Stud Scient 4:384–399

Boletzky SV (2003) Biology of early life stages in cephalopod molluscs. Adv Mar Biol 44:143–203

Carrasco JF, Arronte C, Rodríguez C (2006) Paralarval rearing of the common octopus *Octopus vulgaris* (Cuvier). Aquac Res 37:1601–1605

Chong J, Cortés N, Galleguillos R, Oyarzún C (2001) http://www.fip.cl/resultadosProyectos.aspx?sub=LTE&an=MTk5OQ&rec=cHVscG8&tit=cHVscG8 Estudio biológico pesquero del recurso pulpo en las X and XI regiones. (Biological and fishery study of octopus at X and XI regions). Technical Report FIP, FIP/IT/99–20

Farías A, Uriarte I, Hernández J, Pino S, Pascual C, Caamal C, Domíngues P, Rosas C (2009) How size relates to oxygen consumption, ammonia excretion, and ingestion rates in cold (*Enteroctopus megalocyathus*) and tropical (*Octopus maya*) octopus species. Mar Biol 156:1547–1558

Farías A, Pereda S, Uriarte I, Dörner J, Cuzon G, Rosas C (2010) Evaluating the effects of formulated moist diets on juveniles of Patagonian octopus *Enteroctopus megalocyathus* (Gould 1852). J Shellfish Res 29:793–798

Farías A, Navarro JC, Cerna V, Pino S, Uriarte I (2011) Effect of broodstock diet on the fecundity and biochemical composition of eggs of Patagonian red octopus (*Enteroctopus megalocyathus* Gould 1852). Ciencias Marinas 37(1):11–21

Forsythe JW (1984) Octopus joubini (Mollusca Cephalopoda): a detailed study of growth through the full life cycle in a closed sea water system. J Zool (London) 202:393–417

Gari R, Ré M (2002) Morfología del aparato digestivo de Enteroctopus megalocyathus y Loligo sapaulensis (Mollusca, Cephalopoda). Iheringia Sér Zool 92(2):81–91

González ML, Arriagada S, López DA, Pérez MC (2008) Reproductive aspects, eggs and paralarvae of *Robsonella fontanianus* (d'Orbigny, 1834). Aquac Res 39:1569–1573

Gutiérrez R (2012). Reproducción y crecimiento en pulpo rojo patagónico *Enteroctopus megalocyathus*. Ph. D. Thesis, Programa Cooperativo Universidad de Chile, Universidad Católica del Norte y Universidad Católica de Valparaíso

Gutiérrez R, Farías A, Yany G, Uriarte I (2012) Interacciones macho-hembra del pulpo rojo patagónico *Enteroctopus megalocyathus* (Cephalopoda: Octopodidae) durante el comportamiento de apareamiento. Lat Am J Aquat Res 40(3):808–812

Gutiérrez R, Uriarte I, Yany G, Farías A (2013) Efecto de la relación proteína/energía de dietas frescas sobre el desempeño de juveniles del pulpo rojo patagónico *Enteroctopus megalocyathus* (Cephalopoda: Octopodidae). In: Iriarte JL, Vargas-Chacoff L, Rosas C, Farías A (eds) Ciencia y tecnología para una Acuicultura sostenible. Proceedings of IV Chilean Congress of Aquaculture, Puerto Montt

Ibáñez CM, Chong JV (2008) Feeding ecology of *Enteroctopus megalocyathus* (Cephalopoda: Octopodidae) in southern Chile. J Mar Biol Ass UK 88:793–798

Ibáñez CM, Camus PA, Rocha FJ (2009) Diversity and distribution of cephalopod species off the coast of Chile. Mar Biol Res 5:374–384. doi:10.1080/17451000802534873

Iglesias J, Sánchez FJ, Otero JJ, Moxica C (2000) Culture of octopus (Octopus vulgaris, Cuvier): present knowledge, problems and perspectives. Recent advances in Mediterranean marine aquaculture finfish species diversification. Cah Options Méditerr 47:313–322

Iglesias J, Otero JJ, Moxica C, Fuentes L, Sánchez FJ (2004) The completed life cycle of octopus (*Octopus vulgaris*, Cuvier) under culture conditions: paralarval rearing using *Artemia*, and zoeae, and first data on juvenile growth up to 8 months of age. Aquac Int 12:481–487

Iglesias J, Sánchez F, Bersano J, Carrasco J, Dhont J, Fuentes L, Linares F, Muñoz J, Okumura S, Roo J, Van der Meeren T, Vidal E, Villanueva R (2007) Rearing of *Octopus vulgaris* paralarvae: present status, bottlenecks and trends. Aquaculture 266:1–15

Ignatius B, Srinivasan M (2006) Embryonic development in *Octopus aegina* Gray, 1849. Curr Sci 91:1089–1092

Leporati SC, Pecl GT, Semmens JM (2008) Reproductive status of *Octopus pallidus*, and its relationship to age and size. Mar Biol 155:375–385

Mansilla E (2004) Bases sobre el manejo y mantención del pulpo del sur (*Enteroctopus megalocyathus*): Crecimiento y reproducción (Bases for maintenance of southern octopus *Enteroctopus megalocyathus*). Universidad Austral de Chile, Valdivia

Medrano S, Godoy J (2011) Informe Término Proyecto FONDEF D06I1051 (Technical report of Project "Commercial ongrowing of Southern Octopus (*Enteroctopus megalocyathus*) in a Deep Farming System"). Instituto de Ciencia y Tecnología, Universidad Arturo Prat, Puerto Montt

Naef A (1928) Cephalopoda Embriology. In: Fauna and flora of the Bay of Naples. Part I, vol II. Final part of Monograph n° 35. Translated from German, Smithsonian Institution Libraries, Washington, D.C., 2000. pp 470

Okubo S (1980) Culture of the paralarvae and juveniles of *Octopus dofleini* in an aquarium for 1 year and 2 months. Shima Marineland Quarterly 25:4–5

Ortíz N, Ré ME (2006) First report of pseudohermaphroditism in cephalopods. J Mollus Stud 72:321–323

Ortiz N, Ré ME, Márquez F (2006) First description of eggs, hatchlings and hatchling behaviour *of Enteroctopus megalocyathus* (Cephalopoda, Octopodidae). J Plankton Res 28:881–890

Ortiz N, Ré ME, Márquez F, Glembocki NG (2011) The reproductive cycle of the red octopus *Enteroctopus megalocyathus* in fishing areas of Northern Patagonian coast. Fish Res 10:217–223

Promboon P, Nabhitabhata J, Duengdee T (2011) Life cycle of the marbled octopus, *Amphioctopus aegina* (Gray) (Cephalopoda: Octopodidae) reared in the laboratory. Sci Mar 75(4):811–821

Ré ME (1980) Estudio taxonómico de *Enteroctopus megalocyathus* (Gould 1852) (Cephalopoda, Octopoda) con notas sobre su biología y pesca. (Taxonomic study of *Enteroctopus megalocyathus* with notes about its biology and fishery). Centro Nacional Patagonico, Puerto Madryn

Ré ME (1998) Pulpos octopodidos (Cephalopoda: Octopodidae). In: Boschi EE (ed) El Mar Argentino y sus Recursos Pesqueros. Tomo 2: Los moluscos de interés pesquero. Cultivos y estrategias reproductivas de bivalvos y equinoideos. Instituto Nacional de Investigación y Desarrollo Pesquero, Mar del Plata, pp 69–98

RECOPADES (2009) Report of project "Artes de pesca sustentables. Ancud: acciones concretas para la conservación de la biodiversidad". Fondo de Protección Ambiental Ministerio de Medio Ambiente, Ancud, pp 1–3

Rocha F, Vega M (2003) Overview of cephalopod fisheries in Chilean waters. Fish Res 60:151–159

Rosas C, Caamal C, Cázares R, Rodríguez D, Romero M, Chay D (2006) Manual preliminar para el cultivo del pulpo *Octopus maya*. UNAM, Yucatán

SERNAP (2011) Anuario Estadístico de Pesca (Annual fishing report). http://www.sernapesca.cl/index.php?option=com_remository&Itemid=246&func=select&id=728

Silva H, Uriarte I, Farías A, Amthauer R, Enríquez R, Romero A (2013) Análisis preliminar de la respuesta inmune innata de pulpo rojo patagónico *Enteroctopus megalocyathus*. In: Iriarte JL, Vargas-Chacoff L, Rosas C, Farías A (eds) Ciencia y tecnología para una Acuicultura sostenible. Proceedings of IV Chilean congress of aquaculture, Puerto Montt

Uriarte I, Farías A, Paschcke K, Ausburger A, Romo C (2008) Informe Término Proyecto FONDEF D04 I1401 (Technical report of project "Desarrollo científico-tecnológico de la producción de juveniles de cefalópodos"). Universidad Austral de Chile, Puerto Montt

Uriarte I, Zúñiga O, Olivares A, Espinoza V, Cerna V, Farías A, Rosas C (2009) Morphometric changes and growth during embryonic development of *Robsonella fontaniana*. Vie Milieu 59:315–323

Uriarte I, Hernández J, Dörner J, Paschke K, Farías A, Crovetto E, Rosas C (2010) Rearing and growth of the octopus *Robsonella fontaniana* (Cephalopoda: Octopodidae) from planktonic hatchlings to benthic juveniles. Biol Bull 218:200–210

Uriarte I, Iglesias J, Rosas C, Viana MT, Navarro JC, Domíngues P, Seixas P, Vidal E, Ausgburger A, Pereda S, Godoy F, Paschke K, Farías A (2011a) Current status, potential and prospects of cephalopod larviculture in Latin America. J World Aquac Soc 42:735–752

Uriarte I, Espinoza V, Herrera M, Gutiérrez R, Pino S, Carbonell P, Dörner J, Farías A (2011b) Larvicultura del pulpo rojo patagónico, *Enteroctopus megalocyathus*: una perspectiva tecnológica con futuro. In: Abstracts of VIII Latinoamerican Congress of Malacology, Puerto Madryn, Argentina, 12 a 17 junio 2011

Uriarte I, Amthauer R, Romero A, Enríquez R, Rosas C, Farías A (2012) Informe Término Proyecto FONDEF D09I1153 (Technical report of Project "Desarrollo biotecnológico para la producción sustentable de juveniles de pulpo patagónico (*Enteroctopus megalocyathus*"). Universidad Austral de Chile, Puerto Montt

Uriarte I, Espinoza V, Dorner J, Rosas C, Farías A (2013) Avances en el acondicionamiento reproductivo y en la producción de huevos y paralarvas del pulpo patagónico (*Enteroctopus megalocyathus*) en el Hatchery de Invertebrados Marinos de la UACh, Puerto Montt. In: Iriarte JL, Vargas-Chacoff L, Rosas C, Farías A (eds) Ciencia y tecnología para una Acuicultura sostenible. Proceedings of IV Chilean Congress of Aquaculture, Puerto Montt, pp 247–251

Villanueva R (1995) Experimental rearing and growth of planktonic *Octopus vulgaris* from hatching to settlement. Can J Fish Aquat Sci 52:2639–2650

Villanueva R, Norman MD (2008) Biology of the planktonic stages of benthic octopuses. Oceanogr Mar Biol Annu Rev 46:105–202

Villanueva R, Nozais C, Boletzky SV (1994) The planctonic life of octopuses. Nature 377:107

Villanueva R, Nozais C, Boletzky SV (1996) Swimming behaviour and food searching in planktonic *Octopus vulgaris* Cuvier from hatching to settlement. J Exp Mar Biol Ecol 208:169–184

Chapter 20
Octopus maya

Carlos Rosas, Pedro Gallardo, Maite Mascaró, Claudia Caamal-Monsreal and Cristina Pascual

Abstract *Octopus maya* culture is being developed at the Universidad Nacional Autónoma de México at pilot scale. A closed culture system of *O. maya* is operating at out of Yucatán facilities, in Sisal. Adult *O. maya* were obtained after 7 months of culture. *O. maya* broodstock is renewed every year with adults from the wild population. Spawns are artificially incubated at 25 °C during 45–50 days. Hatchlings pass through a post-embryonic phase that lasts approximately 10 days until they reach the juvenile stage. During the first 15 days after hatch, territorialism and cannibalism are absent and animals practically do not grow. After that age, increments on ingested food are observed together with an exponential growth rate. The culture system has been organized in two steps: (1) ongrowing of hatchlings and (2) ongrowing of juveniles. In the first step, hatchlings are cultured in indoor 7.5 m^2 tanks at a density of 50–60 individuals m^{-2} for 60 days. During this time, every 20 days animals are weighed and separated according to size to avoid cannibalism. At this phase, survival varies between 75 and 80 %. The juvenile ongrowing phase occurs in outdoor 6 m diameter tanks where juveniles are maintained at a density of 10–25 individuals m^{-2}, depending on whether a semi-intensive or intensive culture is applied. Juveniles are cropped when they reach between 80 and 150 g in wet weight (WW), and are sold to the gourmet market. Detailed seawater characteristics and other useful data are included in the present chapter in an attempt to offer an overview of *O. maya* culture.

C. Rosas (✉) · P. Gallardo · M. Mascaró · C. Caamal-Monsreal · C. Pascual
Unidad Multidisciplinaria de Docencia e Investigación, Facultad de Ciencias, Universidad Nacional Autónoma de México (UNAM),
Puerto de abrigo s/n Sisal, Yucatán, México
e-mail: crv@ciencias.unam.mx

P. Gallardo
e-mail: pedrogallardo2@gmail.com

M. Mascaró
e-mail: mmm@ciencias.unam.mx

C. Caamal-Monsreal
e-mail: caamons@hotmail.com

C. Pascual
e-mail: pascual.cristina@gmail.com

J. Iglesias et al. (eds.), *Cephalopod Culture*, DOI 10.1007/978-94-017-8648-5_20,

Keywords *Octopus maya* · Culture conditions · Broodstock · Hatchling production · Embryonic development

20.1 Introduction

20.1.1 Importance of this Species in the Market

The Mexican four-eyed octopus, *Octopus maya* (Voss and Solís-Ramírez 1966), is endemic to the Yucatán Peninsula. The species occurs in the states of Campeche, Yucatán and Quintana Roo (ranging from the waters off Ciudad del Carmen to those beyond Isla Mujeres). The *O. maya* fishery generates 15,000 jobs and is worth more than 27 million dollars annually (Jurado-Molina 2010), making it one the most socioeconomically important cephalopod species in America. Presently, the yearly fisheries production varies between 10,000 and 20,000 t, only in the Yucatán Peninsula, México. Eighty percent of the catch is exported to Europe and Asia, generating a major economy for the Yucatán region. Voss and Solis (1966) found that *O. maya,* after *O. hummelincki,* is the second described ocellated species of the tropical western Atlantic with large eggs, with embryos that hatch as holobenthic juveniles.

The *O. maya* fishery was established in 1949 and has since expanded across the distributional range of the species. On a regional scale, *O. maya* is the main component of octopus catches, representing 63 % of the landings in 2004 (Pérez et al. 2006). The increasing economic importance of this fishery has resulted in a steady increase of people joining the fishery each year, with spillover effects to coastal communities of the peninsula. Increasing demand for this resource has also resulted in social conflict among fishing communities throughout the peninsula. Some advocate local-scale fishing (i.e. fishing in areas close to their communities), while others defend "chasing" the resource wherever it is found (personal communications from fishermen). In the apparent absence of any physical or biological barriers, the species is currently regarded as a single population and is managed as a single stock (Begg et al. 1999).

20.1.2 O. maya *Culture at the Hawaii Institute of Marine Biology and Marine Biomedical Institute of University of Texas*

Since the 1960s, *O. maya* has been the focus of research attention from researchers in Mexico and the USA. Following Voss and Solis' (1966) work on the taxonomic description of the species, Walker et al. (1970) reported the first study on *O. maya* on the improvement of culture conditions, maintaining and training octopus in laboratory conditions. In this study, Walker et al. (1970) found that *O. maya* does not escape from an uncovered tank after being in the laboratory for a few hours, except when it is hungry or when it is being studied in an aversive experimental situation (involving oxygen deprivation or shock) or when the seawater is foul. They also found that a mean number of four crabs (mean weight: 4.1 g per crab) were required to satiate 400-g octopuses deprived of food for 1–4 days. Afterwards, Van Heukelem (1976, 1977) transported several egg strings from Campeche Bay to the Hawaii Institute of Marine Biology, where the eggs were artificially incubated. In that laboratory, five generations of *O. maya* were maintained to study if *O. maya* could be used as a model for neurobiology, behaviour, immunology, endocrinology and ageing studies. During his studies, Van Heukelem defined some maintenance conditions that were the base for the later studies on *O. maya* culture and pointed several knowledge gaps in the biological aspects of rearing octopuses in captivity. Van Heukelem found that *O. maya* eggs can be artificially incubated with a 100% success rate for fertilized eggs developing into hatchlings with incubation periods of about 45 days at 25 °C. To incubate the eggs, Van Heukelem (1977) used a glass funnel with filtered seawater supplied to the stems. In that funnel, seawater flow was adjusted "so that the eggs strings tumbled slowly and rubbed against one another, keeping the surface of the eggs clean, free of fungus, and well aerated". Van Heukelem also reported that air bubbles in the incubation system were harmful to embryos interfering with yolk epithelium development. Cannibalism was recognized as the major source of death of juveniles when animals where underfed or fed with dried foods. Frozen crab and shrimp produced good growth. A detailed study using 27 isolated *O. maya* juveniles showed that, from hatching (at a weight of 0.1 g) to 90 days old, animals grew exponentially, doubling their weight every 11–12 days at 25 °C, reaching 3,231 g after 8.5 months of culture. During that study, this author reported that one of the bottlenecks of the *O. maya* culture was the lack of a dried and/or storable food for massive maintenance of octopus in laboratory conditions.

Other studies reported during the 1980s were mainly dedicated to laboratory rearing conditions, pathology, mineral deficiencies and nutrition of several cephalopods species, including *O. maya* (Hanlon 1987; Hanlon et al. 1989; Hanlon and Forsythe 1985; Hanlon and Forsythe 1990; Hanlon et al. 1984; Lee 1994; Lee et al. 1991). From those studies, infection by *Vibrio carcharie* was identified to affect *O. maya,* and treatment with intramuscular injections of chloramphenicol was recommended. However, nitrofurazone and furazolidone bath applications were used as a first protocol at the Marine Biomedical Institute of University of Texas at Galveston when *Vibrio* spp. infections were presented (Forsythe et al. 1990). A review of cephalopod mariculture was published by Hanlon (1987), showing that at that time there were 20 species cultured, only six being from the Octopodidae family. Among those, *O. maya* and *O. bimaculoides* were considered to be the best candidates for aquaculture (Hanlon 1987).

20.1.3 **O. maya** *Culture at Universidad Nacional Autónoma de México*

The first assays to cultivate *O. maya* at the facilities of Universidad Nacional Autónoma de México (UNAM) started in 2004 when some wild juveniles were placed in 5-m-diameter ponds to determine the growth rate and survival of animals fed heads of fish and frozen crabs. Three trials were conducted between 23 and

32 days. Octopuses were held in three outdoor tanks with 19.6 m^2 of bottom area and 0.5 m water depth, with aerated seawater and water flow allowing 10% of water exchange per day. Initial density was between 2.9 and 3.8 kg m^{-3} with different initial mean weight of 542.3±18.8, 493±11.9 and 321±7.8 g, for trials 1, 2 and 3, respectively. Specific growth rate varied between 1.8 and 2.7% body weight (BW) day^{-1} with no apparent relationship with the culture temperature. Results showed that the tanks used and water quality at the UNAM facilities were adequate for the growing of *O. maya* juveniles, with commercial size being attained in a few weeks after the wild juveniles were captured (Domingues et al. 2012). This led to a formal research programme being developed to culture *O. maya* at the UNAM. In this programme, the spouses of the fishermen were included. Integrating the fishing community into the aquaculture operation was considered to be a key element of technological transference to the social sector of Yucatán, México. Considering that nowadays in Yucatán the octopus fisheries is around 10,000 t $year^{-1}$ of animals larger than 450 g, we designed the culture technology to obtain animals with smaller weights than those obtained from the fisheries. To date, the UNAM facility is producing animals lower than 250 g in living weight (LW), in an attempt to cover the gourmet octopus market that cannot be supplied by the fisheries. The present chapter presents a review of the actual status and bottlenecks of *O. maya* culture.

20.2 Conditions for Maturation, Spawn and Embryo Development

Only functionally mature females are selected to spawn under controlled conditions. Functional maturation is recognized when the oocytes are located in the reproductive coelom of the female and can be seen as small eggs. This is the phase of vitellogenesis before spawning (Arkhipkin 1992). The gonadosomatic index of vitellogenic females is >2% (Rosas et al. 2012). One week before being transferred to the functional maturation area, wild or cultured females are placed with males in outdoor tanks just to be sure that they will have enough spermatic material to ensure maximum fecundity of oocytes. The functional maturation facility was designed to maintain a 10:14 h red light–dark photoperiod. A light intensity of 30 Lx cm^{-2} is maintained at the surface of the tank to reduce female stress. Tanks are connected to a recirculation seawater system, in which seawater is filtered through an anthracite earth filter, UV and protein skimmer. Seawater temperature was maintained at 24–26 °C, since we observed higher temperatures can inhibit vitellogenesis.

O. maya females >500 g wet weight (WW) are individually placed in a 320-L dark tank, where a fibreglass box is offered as a refuge and posterior spawning nest (Rosas et al. 2008a; Fig. 20.1). Other refuges or nests can also be used to obtain spawns of *O. maya,* such as polyvinyl chloride (PVC) tubes (100 mm diameter), and different types of pots and cages. Different food types were used to feed females during functional maturity. At the beginning of the programme when only wild females where used for hatchling production, animals were fed frozen

Fig. 20.1 Nest and incubator for *Octopus maya*. *Top left to right:* Nest with an *Octopus maya* female and incubator device to maintain the eggs during embryonic development. *Lower left and right:* details of *O. maya* eggs in the incubator. (Rosas et al. 2014)

crabs (*Callinectes* spp.; 70 %) and mussels (*Mytilus* spp.; 30 %) at 15 % WW ration. Presently, a mixed diet composed of fish, squid and shrimp is used as a broodstock diet for *O. maya*-cultivated females. Wild and cultivated females normally spawn 37±23 days after entrance into the maturation facilities. Spawning takes at least 5 days to complete, during which the females can eat. Hence, it is important to feed females with half of the daily ration during this period. There is no relationship between hatchlings per female production and LW of *O. maya* females suggesting that egg production depends on each female characteristic, independently of the age and/or LW (Fig. 20.2).

O. maya is a semelparous species; females reproduce once and then die (Van Heukelem 1976). Considering this biological characteristic, a massive hatchling production system for *O. maya* was designed, permitting large-scale management of spawns, ensuring viability of embryos and reducing the space required for management. The incubation process starts when females deposit eggs in the cap of the nest (Fig. 20.1). Such a cover is then placed in the incubation system with seawater recirculation under controlled conditions to provide an adequate environment for embryonic development.

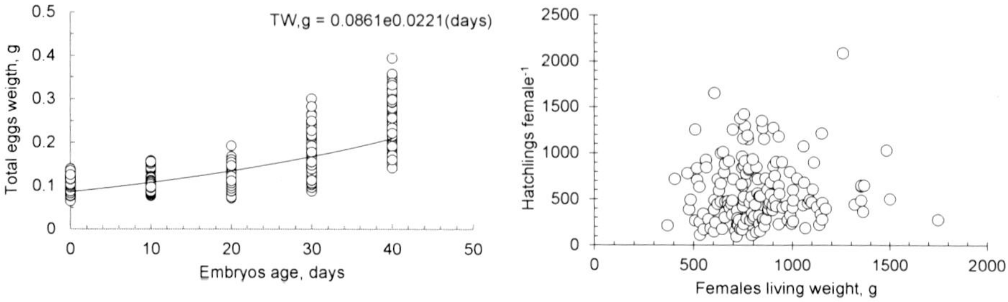

Fig. 20.2 Eggs' and embryos' total weight changes with age and hatchlings obtained per female of wild *Octopus maya* between 375 and 1,750 g living weight (LW) at 25±1 °C. *N*=213. (Rosas et al. 2014)

Laboratory trials have demonstrated that this system is as effective as when females care for the eggs, with the advantage that the artificial incubation system involves less space, less water and allows removing the females and selling them as a product of the cultivation system. At 26 °C, the total embryonic development lasts between 40 and 50 days, depending on the temperature stability. At that stage, eggs reach a mean value of 0.25±0.05 g total WW (Fig. 20.2).

20.3 Hatchlings Growing out

Newly hatched octopuses weigh approximately 0.146±0.02 g (*N*=1,565) and resemble an adult. However, recent studies have demonstrated that hatchlings pass through several changes before they reach the real juvenile stage (Moguel et al. 2010; Martínez et al. 2011). In those studies, changes in the histology, physiology and enzymatic activity of the digestive gland of *O. maya* during rearing—as well as changes in morphology and feeding behaviour—were examined to define the phases characterizing post-hatching development. Morphometric changes showed that newly born *O. maya* exhibited a non-growth phase during the first 15 days post hatching (DPH). A histological analysis revealed that the digestive gland morphology changed with age, from a simple tubular gland in octopuses 2 DPH to a tubulo-acinar and vacuolar structure with digestive cells characterized by vacuoles in octopuses 45 DPH. Digestive enzyme activities were erratic until 14 DPH, stabilizing afterwards. *O. maya* at 2 and 3 DPH rarely presented attack responses to either visual or both visual and chemical stimuli from prey. In contrast, at 4 DPH, octopuses responded to visual stimuli from crabs and palaemonids, but did not display preference in attacking either prey type. Based on our results, we have defined for the first time two phases in the early life history of *O. maya:* post-hatching (1–14 DPH) and juvenile (14 DPH) (Moguel et al. 2010). These findings explain why in all experiments with *O. maya* juvenile culture showed that hatchlings during the first 30 days of culture do not have cannibalistic behaviour and can be maintained in relatively high densities (Van Heukelem 1977; Moguel et al. 2010). *O. maya* hatchlings have an internal yolk sack that is used to fuel the animals during the post-hatching stage. For that reason, the animals do not ingest food during the first 5–7 days (Martínez et al. 2011). When the internal yolk is consumed and the digestive gland is ready to digest external food, the raptorial behaviour of hatchlings initiates. Despite that, under culture conditions hatchlings are fed from day 1, because hatchlings coming from one spawn will be between 1 and 5 days old, since embryos did not hatch at the same day.

Van Heukelem (1977) recommended feeding hatchlings with live food, mainly palaemonids, amphipods or small crabs. Feeding *O. maya* hatchlings in massive cultures with live feed is frequently impractical, although we also found that marine gammarids certainly improve the growth rate and survival between day 1 and 15 DPH (6.9±0.2% day^{-1}; 92% survival) when compared with animals fed *Artemia* spp. adults (4.8±0.2% day^{-1}, 71% survival) or freshwater gammarids (5.0±0.3% day^{-1}, 41% survival) (Baeza-Rojano et al. 2012). In an attempt to replace live prey, several sources of protein were tested. A semi-humid squid paste-bound gelatin has recently been developed and can be used as a main food of *O. maya* hatchlings and juveniles (Rosas et al. 2008b; Rosas et al. 2012).

When an elaborated diet is used, a ration should be established. At the experimental level with individualized animals, an adequate ration of 30% of WW day^{-1} was reported for this species (Quintana et al. 2011). However, in massive culture, this type of ration is not precise, since it is not possible to assure that every animal obtains the same amount of food. At present, hatchlings and juveniles are fed twice a day (around 50% WW each time) to avoid competition between them.

The type of tank and culture density for *O. maya* hatchlings were defined recently. At our facility, hatchling culture is done in 7.5 m^2 dark tanks (5.0×1.5×0.4 m) at a density of 50 individuals m^{-2}. Tanks are connected to a recirculatory seawater (25±1 °C) and aerated system that maintains oxygen, ammonia and nitrate levels at 5.5±0.5, 0.2±0.05 and 0.1±0.01 mg L^{-1}, respectively. In such conditions, octopuses are fed twice a day and maintained for 60 days until they reach around 2 g WW. In these tanks, survival can be variable and strongly dependent of the feeding rate. Less than 2 rations day^{-1} per animal will produce high cannibalism and in consequence low survival. We have registered survivals lower than 10% when not enough food was offered to hatchlings. Refuges are also important to avoid cannibalism and stress. Both for hatchlings and juvenile (until 10 g WW), *Strombus pugilis* conchs are used as refuges at a proportion of three conchs per octopus.

20.4 Juveniles Ongrowing

At 60 days old, juveniles are transferred to outdoor 6 m diameter tanks (Domingues et al. 2012). The octopuses are graded by size, reducing the high variability characteristic of *O. maya* hatchlings (Briceño-Jacques et al. 2010). Outdoor ponds are conditioned with greenhouse meshes that reduce the direct sunlight by 70% (Fig. 20.3).

The tanks are connected to a recirculation seawater system coupled to protein skimmers and 50 μm bag filters (one per tank). Squid paste is also used as food during this octopus ongrowing stage, offering a mean of 3 rations per animal day^{-1}. The culture density in outdoor tanks is 25 juveniles m^{-2} (700 juveniles per tank).

Fig. 20.3 Outdoor 6 m diameter tanks used for *Octopus maya* juvenile ongrowing at the Universidad Nacional Autónoma de México (UNAM) facilities in Yucatán, México. (Rosas et al. 2014)

When octopuses reach 10 g WW, PVC pipes are offered as a refuge in a proportion of three per animal. Results of massive juvenile growth suggest that an exponential growth curve and survival around 60 % can be obtained. Weights of 100–150 g at day 100–120 at 29 ± 2 °C have been obtained (Fig. 20.4).

Monthly grading in the tanks is necessary to avoid cannibalism and can provide survival rates of 60–70 %. Preliminary trials of *O. maya* growth indicate that a broodstock bank could be obtained within 6–7 months of culture, with animals > 500 g WW. Although at present there is no clear indication of what triggers functional maturity in Octopodidae species, temperature, changes in light intensity and photoperiod, as well as copulation could play an important role at the start of the vitellogenesis process in *O. maya*. Currently, broodstock of *O. maya* is obtained by separating young females (around 30 g WW) from males, just to avoid early spawns of animals lower than 60 g, since they produce few eggs.

20.5 Water Quality for *O. maya* Culture

20.5.1 Temperature

Water quality can be defined as a function of each phase of octopus culture. For females under functional maturation, levels of oxygen dissolved > 5 mg L^{-1}, ammonia < 1 mg L^{-1}, salinity > 32 psu and pH > 8 should be maintained to ensure spawning, with temperatures > 28 °C inhibiting female maturation. Embryos require

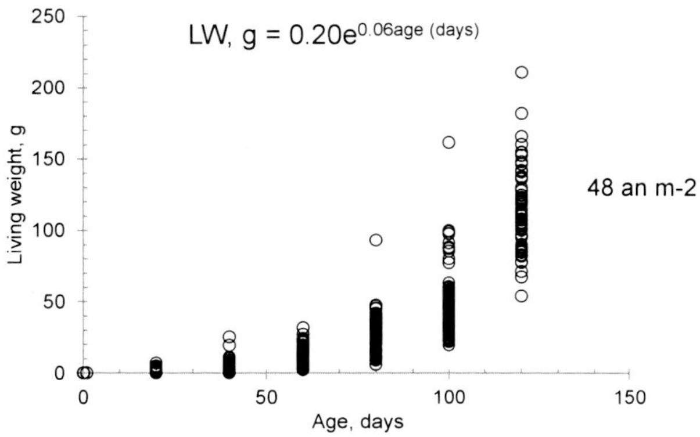

Fig. 20.4 Changes in living weight (LW, g) with age of *Octopus maya* juveniles maintained in outdoor tanks at a density of 50 animals m^{-2} and 28±2 °C. N=4 trials. (Rosas et al. 2014)

similar conditions to females, but are restricted by variations in physical and chemical parameters. Embryos (with or without female) should be maintained in a very stable environment where the temperature is 25±1 °C and oxygen dissolved levels >5 mgL^{-1}. We observed that at 30 °C, the embryo metabolic rate is 25 and 50% lower than at 26 and 22 °C, respectively. Relatively low pH can indirectly affect the embryo development via bacterial infection.

While hatchlings should be more tolerant that embryos, at present there is little information on the effect of environmental factors on early juvenile performance. From hatching (at a weight of 0.1 g WW) to 90 days old, animals grew exponentially, doubling their weight every 11–12 days at 25 °C (Van Heukelem 1977). More recent studies also demonstrated that *O. maya* juvenile growth is depressed at 30 °C, being 35% lower than registered at between 24 and 26 °C (Noyola et al. 2013a). In fact, animals acclimated at 30 °C during 40 days selected 26 °C when placed in a temperature gradient, suggesting that 30 °C is not adequate to *O. maya* early juveniles (Noyola et al. 2013b). In another study, Domingues et al. (2012) demonstrated that temperatures between 26 and 29 °C favour the growth rate of late juveniles and adults of *O. maya* via an increment on ingestion rate.

20.5.2 Dissolved Oxygen

For cephalopods, dissolved oxygen (DO) is essential because it is a modulator of metabolism and hence of the organism's capability to process ingested food and convert it into biomass. Oxygen consumption (R) has a power relationship with BW (W, g) (Farias et al. 2009; Briceño-Jacques et al. 2010):

$$R = 0.93W^{0.69} \tag{20.1}$$

With this equation, the oxygen consumption of animals between 0.13 and 3,000 g WW can be obtained and used to calculate the oxygen demands related to biomass in the culture tank. To maintain adequate DO in *O. maya* tanks, constant aeration

Fig. 20.5 Aeration system for *Octopus maya* ongrowing ponds based on airlift pumps. (Rosas et al. 2014)

should be considered. An aeration system based on airlift pump supply was designed to maintain oxygen dissolved levels higher than 5 mg L^{-1} (Fig. 20.5). With this system, it is possible to avoid damage produced by bubbles, which can eventually be trapped by octopuses' mantle if aeration was provided directly with a conventional stone.

20.5.3 Total Ammonia

Total ammonia (NH_3) results from protein catabolism produced by protein metabolism and bacterial activity. It is decomposed into nitrite and nitrate, depending on the nitrifying bacteria and pH. Although at present there are no formal studies that indicate the tolerance levels of total ammonia ($NH_3 \leftrightarrow NH_4^+$) or nitrite ($NO_2^-$) for *O. maya* juveniles in culture, in our culture system we maintain total ammonia levels below 1.4 mg L^{-1}. Considering that octopus have a protein metabolism, ammonia could be a problem if it is not eliminated from the system. Organic material, coming from food remains and faeces, in octopus outdoor ponds is controlled using a protein skimmer that is connected to a recirculation seawater system (Fig. 20.6).

20.5.4 Diseases

Minimal information is available on diseases for *O. maya* in captivity. However, Hanlon and Forsythe (1990) performed a review of diseases produced by

Fig. 20.6 Protein skimmer that eliminates organic material of recirculatory seawater system in *Octopus maya* culture. (Rosas et al. 2014)

microorganisms in cephalopods. In this review, they reported several bacteria that affected *O. maya,* including some from *Vibrio* spp. In culture conditions, high mortality could be observed when low-quality food was offered. Therefore, food supplied to *O. maya* juveniles must be clean and fresh. Other environmental factors can provoke stress and in consequence make octopus more vulnerable to bacterial infection. In outdoor ponds, temperatures around 15 °C provoked stress in 1 g WW juveniles and at the same time *V. alginoliticus* infection. To avoid that, we used an immersion heater to increase the temperature of the pond to 18 °C. Occasionally, infections with *Vibrio* spp. are observed, mainly when the marine food is contaminated. In such situations, oxytetracycline was used; a bath of 2 mg L^{-1} for 30 min during 7 days was effective against this type of infections.

20.6 Conclusions

Fig. 20.7 Women from Sisal village elaborating food, feeding animals and cleaning outdoor ponds of *Octopus maya* culture system at Universidad Nacional Autónoma de México (UNAM) facilities located in Sisal, Yucatán, México. (Rosas et al. 2014)

Presently, *O. maya* culture is closed; broodstock animals are grown from hatchlings cultured in outdoor ponds. However, studies on other cephalopod species have demonstrated that cephalopod culture in captivity reduces animal sizes after several generations. Therefore, the broodstock and the *O. maya* genetic bank in culture conditions needs to be refreshed every year, when the fisheries season is opened (August–December). So, every year broodstock is complemented with animals obtained in culture conditions.

A system that includes broodstock area, incubation system, pre-fattening ponds and fattening outdoor ponds works as a demonstrative unit just to show the technological process of *O. maya* culture. At this moment, the system has been transferred to the social sector through the "Moluscos del Mayab" cooperative that is performing the octopus culture at the UNAM facilities. This culture is conducted by women (fishermen spouses) of the Sisal village at Yucatán. They elaborate octopus food, feed animals and maintain the pilot-scale culture system (Fig. 20.7).

Although at present the bio-economic studies on *O. maya* production have only been done on a semi-pilot scale at the UNAM facilities at Sisal, results of this study indicate that, if the octopus price is maintained as a commodity (mean value of US$ 5 per animal of 80–200 g WW) of the gourmet market, aquaculture production can be profitable, independently of the massive fisheries production that is obtained in the Yucatán Peninsula, and in the rest of the world.

Acknowledgements This project was partially financed by the PAPIIT programme at UNAM through the project IN 212012. Also, thanks are given to CONACYT through the project CB-2010–01, No. 150810.

References

Arkhipkin A (1992) Reproductive system structure, development and function in cephalopods with a new general scale for maturity stages. J Northw Atl Fish Sci 12:63–74

Baeza-Rojano E, Domingues P, Guerra-García JM, Capella S, Noreña-Barroso E, Caamal-Monsreal C, Rosas C (2012) Marine gammarids (Crustacea: Amphipoda): a new live prey to culture *Octopus maya* hatchlings. Aquac Res. 44:1602–1612

Begg G, Friedland K, Pearce J (1999) Stock identification and its role in stock assessment and fisheries management: an overview. Fish Res 43:1–8

Briceño-Jacques F, Mascaró M, Rosas C (2010) GLMM-based modelling of growth in juvenile *Octopus maya* siblings: does growth depend on initial size? ICES J Mar Sci 67:1509–1516

Domingues P, López N, Rosas C (2012) Preliminary trials on the use of large outdoor tanks for the ongrowing of *Octopus maya* juveniles. Aquac Res 43:26–31

Farias A, Uriarte I, Hernández J, Pino S, Pascual C, Caamal C, Domingues P, Rosas C (2009) How size relates to oxygen consumption, ammonia excretion, and ingestion rates in cold (*Enteroctopus megalocyathus*) and tropical (*Octopus maya*) octopus species. Mar Biol 156:1547–1558

Forsythe JW, Hanlon RT, Lee PG (1990) A formulary for treating cephalopod mollusc diseases. In: Perkins FO, Cheng TC (eds) Pathology in Marine science. Academic Press, New York

Hanlon RT (1987) Mariculture. In: Boyle PR (ed) Cephalopod life cycles: comparative reviews. Academic Press, London

Hanlon RT, Forsythe JW (1985) Advances in the laboratory culture of Octopuses for biomedical research. Lab Anim Sci 35:33–40

Hanlon RT, Forsythe JW (1990) Diseases of Mollusc: cephalopod; 1.1. Diseases caused by micro-organisms; 1.3 structural abnormalities and neoplasia. In: Kinne O (ed). Diseases of marine animals. Vol. III: Cephalopoda to Urochordata. Biologische Anstalt Helgoland, Hamburg

Hanlon RT, Forsythe JW, Cooper KM (1984) Fatal penetrating skin ulcers in laboratory-reared octopuses. J Invert Path 44:67–83

Hanlon RT, Bidwell JP, Tait R (1989) Strontium is required for statolith development and thus normal swimming behaviour of hatchling cephalopods. J Exp Biol 141:187–195

Jurado-Molina JA (2010) Bayesian framework with implementation error to improve the management of the red octopus (*Octopus maya*) fishery off the Yucatán Peninsula. Cienc Mar 36:1–14

Lee PG (1994) Nutrition of cephalopods: fueling the system. Mar Freshw Behav Physiol 25:35–51

Lee PG, Forsythe JW, DiMarco FP, DeRusha RH, Hanlon RT (1991) Initial palatability and growth trials on pelleted diets for cephalopods. Bull Mar Sci 49:362–372

Martínez R, López-Ripoll E, Avila-Poveda OH, Santos-Ricalde R, Mascaró M, Rosas C (2011) Cytological ontogeny of the digestive gland in post-hatching *Octopus maya* (Mollusca: Cephalopoda) and cytological background of digestion in juveniles. Aquat Biol 11:249–261

Moguel C, Mascaró M, Avila-Poveda OH, Caamal-Monsreal C, Sanchez A, Pascual C, Rosas C (2010) Morphological, physiological and behavioral changes during post-hatching development of *Octopus maya* (Mollusca: Cephalopoda) with special focus on the digestive system. Aquat Biol 9:35–48

Noyola J, Caamal-Monsreal C, Díaz F, Re AD, Sánchez A, Rosas C (2013a) Thermal preference, tolerance and metabolic rate of early juveniles of *Octopus maya* exposed to different acclimation temperatures. J Therm Biol 38:14–19

Noyola J, Mascaró M, Caamal-Monsreal C, Noreña-Barroso E, Díaz F, Re AD, Sánchez A, Rosas C (2013b) Effect of temperature on energetic balance and fatty acid composition of early juveniles of *Octopus maya*. J Exp Mar Biol Ecol. doi: 10.1016/j.jembe.2013.1004.1008

Pérez MC, López DA, Aguila K, González ML (2006) Feeding and growth in captivity of the octopus *Enteroctopus megalocyathus*. Aquac Res 37:550–555

Quintana D, Rosas C, Moreno-Villegas E (2011) Relationship between nutritional and rearing parameters of *Octopus maya* juveniles fed different rations of crab paste. Aquac Nutr 17e:379–388

Rosas C, Caamal-Monsreal C, Cázares R (2008a) Proceso de incubación para pulpos e incubadora. Organización Mundial de la Propiedad Intelectual. Vol. MX/a/2008/011627

Rosas C, Caamal C, Cázares R (2010) Proceso de incubación para pulpos e incubadora, World Intelectual Protection Organization, WO2010/030155 A030151: 1-10

Rosas C, Tut J, Baeza J, Sánchez A, Sosa V, Pascual C, Arena L, Domingues P, Cuzon G (2008b) Effect of type of binder on growth, digestibility, and energetic balance of *Octopus maya*. Aquaculture 275:291–297

Rosas C, Valero A, Caamal-Monsreal C, Uriarte I, Farias A, Gallardo P, Sánchez A, Domingues P (2012) Effects of dietary protein sources on growth, survival and digestive capacity of Octopus maya juveniles (Mollusca: Cephalopoda). Aquaculture Research 44: 1029–1044

Van Heukelem WF (1976) Growth, bioenergetics and life span of *Octopus cyanea* and *Octopus maya*. PhD Thesis University of Hawaii

Van Heukelem WF (1977) Laboratory maintenance, breeding, rearing and biomedical research potential of the Yucatan octopus (*Octopus maya*). Lab Anim Sci 27:852–859

Voss GL, Solis-Ramirez MJ (1966) *Octopus maya*, a new species from the Bay of Campeche, Mexico. Bull Mar Sci 16:615–625

Walker JJ, Longo N, Bitterman ME (1970) The octopus in the laboratory. Handling, maintenance, training. Behav Res Meth Instrum 2:15–18

Chapter 21
Octopus mimus

Óscar Zúñiga, Alberto Olivares and Carlos Rosas

Abstract *Octopus mimus* (Gould 1852) is a member of the Octopodidae family. Its distribution includes intertidal and subtidal rocky shore habitats along the Southeast Pacific from northern Perú to Central Chile. The commercial supply of octopus has been diminished due to overfishing in many key fisheries. This chapter presents the results of the research on the reproduction of *O. mimus* in natural and captive conditions and the application of this knowledge for the potential of aquaculture at a commercial level. This species is particularly suitable for aquaculture; however, commercial cultures cannot be conducted until sound knowledge about cultivation technology and the physiology of the feeding processes in paralarvae are available. On the other hand, our studies have demonstrated that adequate growth rates of *O. mimus* reared in captivity can be obtained by feeding animals using a diet based on a mix of fish, mollusc and crustacean blended in sausages.

Keywords *Octopus mimus* · Aquaculture · Reproduction · Paralarval rearing · Ongrowing in captivity

21.1 Importance of this Species in the Market

Octopus mimus Gould 1839, 'pulpo común' (common octopus) or 'de los changos' from South America is distributed from northern Perú (Tumbes) to Central Chile (Bahía de San Vicente) inhabiting the littoral zone down to 30 m depth, although there are no precise limits regarding its latitudinal and bathymetric distri-

Ó. Zúñiga (✉) · A. Olivares
Facultad de Ciencias del Mar y de Recursos Biológicos,
Universidad de Antofagasta, 170 Antofagasta, Chile
e-mail: ozr@vtr.net

A. Olivares
e-mail: aolivares@uantof.cl

C. Rosas
Unidad Multidisciplinaria de Docencia e Investigación, Facultad de Ciencias,Universidad Nacional Autónoma de México (UNAM), Puerto de Abrigo s/n Sisal, Yucatán, México
e-mail: crv@ciencias.unam.mx

J. Iglesias et al. (eds.), *Cephalopod Culture*, DOI 10.1007/978-94-017-8648-5_21,

bution (Osorio et al. 1979; Guerra et al. 1999). This species is a representative of the Peruvian faunistic province and its distribution is restricted to temperate waters of the north. Due to its thermal sensitivity and endemic character, this species is very sensible to commercial exploitation (Ibáñez et al. 2009). This risk is enhanced as this octopus is one of the principal benthic resources of the artisanal fishery in northern Chile (Rocha and Vega 2003) and the fishery has been increasing over time due to a strong demand for cephalopods in international markets (Hunsicker and Essington 2010). Also, this species is easily captured using very basic gear such as hooks (in the intertidal) and semi-autonomous diving (in the subtidal zone), which makes this octopus very vulnerable to increments in fishing effort (Defeo and Castilla 1998).

Over decades, *Octopus mimus* was identified as *O. vulgaris* Cuvier 1797, being redescribed as *O. mimus* based on anatomic, morphometric and meristic characterization using adult individuals. This distinction between species was also verified in pelagic larval stages (Castro et al. 2002) and by ontogenetic studies (Warnke 1999). Molecular analysis showed a 12.7% of nucleotide divergence of the mitochondrial cytochrome oxidase III gen (Warnke et al. 2000; Söller et al. 2000). Randomly amplified polymorphic DNA (RAPD) showed different electrophoretic patterns in total DNA (Warnke et al. 2004). The genetic study by Pérez-Losada et al. (2002) suggests that *O. mimus* and *O. maya* belong to a different co-family distinct to that of *O. vulgaris*, as a result of allopatric speciation associated to the geomorphologic lifting of Central America. Apparently, the formation of the Panama Isthmus during the late Miocene and early Pliocene interrupted the genetic flow between Pacific and Atlantic populations (Porta 2003). Recently, Galleguillos et al. (2011) showed a high polymorphism in nine different loci of *O. mimus* using microsatellite markers. These results can help to address population genetic studies and ecological, conservation and fishery management issues.

Due to the high fishery and demand for this species, its aquaculture has been considered as a technological opportunity to satisfy market demands and to reduce fishing pressure. According to the Chilean National Fishery Service (SERNAPESCA), between 2001 and 2010 an average of 2,078t of octopus were caught mainly in the Antofagasta region. The average value of exports of this species reached ca. $ 10 million USD in the past 5-year period.

21.2 State of the Culture

O. mimus culture is in an early experimental stage and this research is carried out by Chilean and Peruvian research centres and universities (Uriarte et al. 2011). So far, octopus rearing in captivity (in tanks or floating cages) is the most promising advance on culture. Animals are fed with live and/or frozen food based on paste preparations using mollusc, crustaceans and fish (Olivares et al. 1996; Cortez et al. 1999; Baltazar et al. 2000; Carrasco and Guisado 2010; Zúñiga et al. 2011). However, one problem of these techniques is the mortality produced during fights among animals for territory or refuge and the high incidence of cannibalism. This is common even in animals kept at low densities (5 kg m^{-3}) with an ad libitum food supply. Also, a well-equilibrated diet needs to be formulated that ensures that all nutritional requirements are covered and that such a diet enhances the conversion index so far observed. That means that the animals reared in captivity must be larger and heavier than those captured by the fishery. In terms of larval culture, the current results are not very promising due to the high mortality occurring just before settlement of paralarvae on the substratum, which is attributable to the lack of adequate food.

In our experiments, different stages of Artemia nauplii enriched with microalgae and nutritional mixes, zoeas of decapods, rotifers and mashed crustacean eggs have been used as food in paralarvae culture. The best paralarval survival rates have been recorded when using densities of 25–50 paralarvae L^{-1}. Although tested, survival rates were not enhanced using water filtered at 1 μm, ultraviolet (UV) light, open flow, different light intensities and tank colours. Also, survivals rates were not enhanced when paralarvae were fed with food stained with vegetal dye which seems to facilitate the visualization and ingestion of food (unpublished data).

21.3 Broodstock and Spawning Process

21.3.1 Feeding

O. mimus is an opportunistic predator with a wide spectrum of prey. Studies using animals from the Tarapaca (Chile), Ilo and Callao (Perú) regions found that this species can consume 25 different prey species. Predator preferences vary depending on sex, maturity stage and seasonality, although a main order of preference consisted of decapods crustacean, bivalves, echinoderms and teleost fishes (Cortez et al. 1995a, 1999; Cardoso et al. 2004).

Based on prey information, different laboratory and field studies have been carried out in order to assess the feasibility of *O. mimus* culture using fresh and frozen food (Baltazar et al. 2000; Carrasco and Guisado 2010; Zúñiga et al. 2011). For instance, Cortez et al. (1999) fed octopuses just with fresh molluscs obtaining on an average very low growth rates (5.33 %/day in 40 days). Carrasco and Guisado (2010) reported a specific growth rate (SGR) of 0.26 % body weight/day in subadult octopus fed with the razor clam (*Tagelus dombeii*) in laboratory conditions. Lower rates were obtained when octopus were fed with the crabs *Mursia gaudichaudi*. Thus, these results showed that *O. mimus* growth may depend on specific food items since not all crustaceans and molluscs can satisfy the nutritional requirements. In order to optimize *O. mimus* culture, Zúniga et al. (2011) used a diet consisting of sausages made of fish and molluscs and obtained an absolute growth rate (AGR) of 7 g day^{-1}, SGR of 0.7 % body weight day^{-1} and survival close to 100 % on individuals kept in tanks with seawater flow of 60 L h^{-1}.

It has been observed that maturity stages are influenced by the diet of *O. mimus*, suggesting that energy and nutritional requirements may depend on maturity and even sex. The diet of *O. mimus* changes seasonally and this may be related to changes in prey availability. Mature males ingest a higher amount of prey in spring and summer compared to the consumption in fall and winter (Cortez et al. 1995b). Thus, feeding and predation patterns seem to be related to physiological changes occurring as the animal matures together with environmental variability. These constitute important aspects that must be considered during *O. mimus* aquaculture. *O. mimus* feeding seems to be affected by a combination of physiological and environmental factors, being considered as an opportunistic species like other octopods. This feature may be a consequence of adaptive mechanisms related to the wide environmental and ecological variability that the habitat of this species is exposed to, e.g. the frequent impacts of El Niño Southern Oscillation (ENSO) events.

21.3.2 Gonadal Development and Maturation on Males and Females

O. mimus from northern Chile has high growth rates that depend on water temperature and develop rapidly during oceanographic warming events like ENSO. The reproduction of this species occurs year-round with two maximum peaks occurring in summer and another, but minor, during fall (Wolff and Pérez 1992; Olivares et al. 1994; Cortez et al. 1995b; Olivares et al. 1996). Most of the octopus reproduction activity occurs at the rocky coastal platform, in shallow warm waters less than 10 m in depth. During reproduction periods, the animals inhabit rocky burrows surrounded by sandy bottoms. Females spawn and take care of the eggs during the embryonic development (Olivares et al. 1996). This period can last between 45 and 90 days depending on the season and sea temperature. Variations on temperature, light and food availability are the main factors that modulate the reproductive cycle in cephalopods (Mangold 1987).

Successful reproduction of *O. mimus* seems to occur in semi-darkness conditions while full illumination inhibits sexual maturation. It has been observed that low intensity of light (30 Lx cm^{-1} approximately) stimulates spawning, while higher levels inhibit it (Zúñiga et al. 1995). The role of this environmental factor in sexual maturity of cephalopods is regulated by the optical gland through the secretion of the octopus gonadotrophin-releasing hormone (Oct-GnRH) that promotes the synthesis and the release of sexual steroids that leads to ovarian maturation and yolk synthesis carried by follicular cells (Wodinsky 1977; Minakata et al. 2009). In *O. vulgaris* reared in dark conditions, the principal cells of the optical gland concentrate ARN associated with the synthesis and secretion of a protein with apparent gonadotropic effects (Wells and Wells 1977). The gland has the highest levels of synthesis when the exposure to light decreases, demonstrating that the aqueous extracts stimulate glandular tritiated thymidine incorporation and subsequent proliferation of germ cells and follicular cells under in vitro conditions (Koueta et al. 1995).

In the coast of northern Chile, the sex ratio of *O. mimus* fluctuates from 1:2 to 1:7 males per female from spring to fall, depending on the variation of the number of

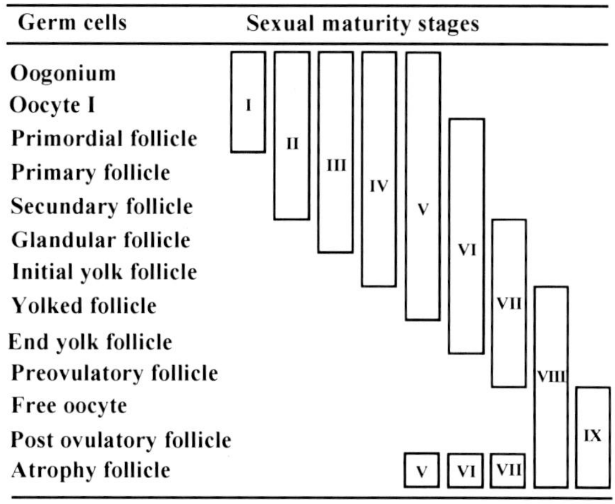

Fig. 21.1 Stages of ovarian sexual maturity of *Octopus mimus* in relation to the variation in germs cells. I: Proliferation, II: Genesis follicle, III: Gland, IV: Initial yolk, V: Yolked, VI: End yolk, VII: Preovulatory, VIII: Ovulation, IX: Post-spawning

females present in refuges from the spawning to the end of the parental care of the embryos. Females weighing more than 890 g are found on different stages of sexual maturity, which is determinate by the accumulation of yolk, follicle development and ovulation (Olivares et al. 1996).

Thirteen stages have been identified during oogenesis, without an absolute synchrony of the interfollicular development and nine stages of ovarian maturation which, used in complement to the different macroscopical levels of sexual maturation, allows the prediction of the reproductive cycle of this species. At the end of morphological maturation, oocytes on metaphase I stage are liberated to the ovarian lumen and the remaining follicular structures develop the post-ovulatory follicle (Olivares et al. 2001). The oocyte maturity is asynchronous, so females can spawn during several days. The ovary forms oocytes just once due to the nonrenewal of the epithelium. The oogonia do not proliferate and degenerate during the ovarian stage of vitellogenesis (Fig. 21.1).

After ovulation, the ovarian's trabecular tissue is completely disorganized and females unquestionably lose their reproductive function (Olivares et al. 2001). However, Ishiyama et al. (1999) determined eight development stages of the oocytes, considering variations of the size of the gonads, cytomorphic changes of the oocytes and macroscopic characteristics of the oviducal glands during the reproductive cycle.

Sexual maturity in females of *O. mimus* is associated with body growth, so normally at the end of the maturity processes and prior to spawning, the animal's body and the ovary reach their maximum size, with the ovary constituting on an average 11 % of the body weight (Fig. 21.2).

Studies about maturity and gonadosomatics (Cortez et al. 1995b; Olivares et al. 1996) conducted on different sizes of wild females of *O. mimus* determined that the ovogenesis is associated to the growth and follicular development. Neuroendocrine regulation modulates the endocrine and proliferative function of the ovary where

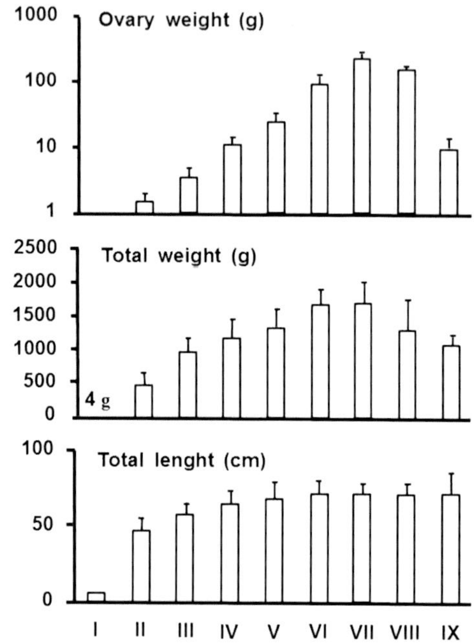

Fig. 21.2 *Octopus mimus*. Body growth and ovary development at the different sexual maturity stages. I: Proliferation. II: Genesis follicle. III: Gland. IV: Initial yolk. V: Yolked. VI: End yolk. VII: Preovulatory. VIII: Ovulation. IX: Post-spawning

steroidogenic function promotes anabolism and body growth. Consequently, females in pre-spawning have the highest body weight reaching the maximum growth of the ovary (Villegas and Tafur 2000).

O. mimus has a semelparous reproductive mode. Days before spawning females stop feeding and their metabolism is sustained by the stored energy in the body. Females die after hatching or sometimes just before hatching. A histological and biochemical study found changes in muscular tissue, digestive gland and ovary associated with the unique reproductive event of the female (Zamora and Olivares 2004). After spawning, the ovary does not show germinal cells that would allow the development of a new reproductive cycle. The digestive gland and muscular tissue show cellular atrophy and a large increment of connective tissue, especially on collagen fibres.

Protein, lipids and carbohydrate concentrations from the muscular tissue and digestive gland decrease during the time interval between spawning and paralarval hatching. Drastic biochemical alterations and irreversible structural deterioration of the muscular tissue and digestive gland interacting with degenerative changes reduce the life expectancy of females after reproduction (Zamora and Olivares 2004).

In pre-ovulatory conditions, the absorption surface on the digestive gland cells in *O. mimus* females are well developed, as evidenced by the presence of abundant microvilli that, during cellular functioning, are relatively longer in digestive cells of animals with high digestive content compared to females in reproductive fast. Although the appearance of cells' microvilli has not been evaluated in association with digestive absorption function as described in *Octopus spp.* (Bidder 1957), it is as-

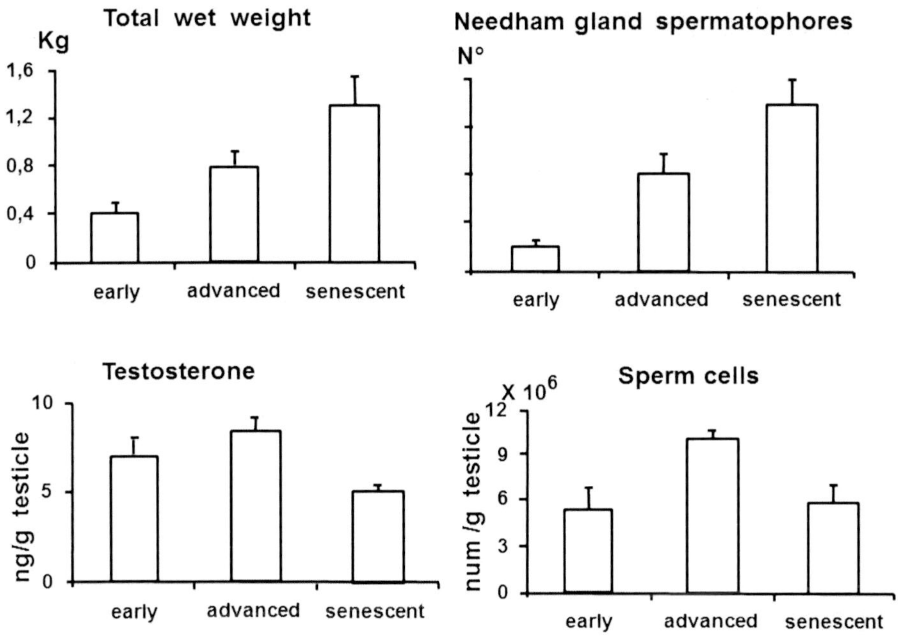

Fig. 21.3 Reproductive potential according to gametogenic and androgenic testicular function of *Octopus mimus* maturity

sumed that food intake in females induces hypertrophy and atrophy of the microvilli due to fasting conditions. No cilia have been seen in digestive cells of *O. mimus* as described on *Pteroctopus tetracirrhus* (Morales 1973; Zamora and Olivares 2004). The energetic costs for growth and development of the ovary imply the storage of 282.5 kcal g^{-1} of mostly proteins (47.6 %), carbohydrates (45.1 %) and few lipids (4 %). Approximately 90 % of the energy is deposited in the eggs and derived to sustain the development of the embryos (Zamora and Olivares 2004).

O. mimus fecundity is estimated between 60,000 and 400,000 eggs (Warnke 1999; Castro et al. 2002), similarly to other octopus species with planktonic larvae (Joll 1983; van Heukelem 1983), higher than species with larger eggs and direct development (Forsythe et al. 1983; Boletzky 1975; Solís-Ramírez 1967) and lower than squids with no parental care that destine more energy to produce a major quantity of eggs for ensuring reproductive success (Hixon 1983; Ikeda et al. 1993).

Mature males of *O. mimus* are present all year long (Olivares et al. 1996). Studies regarding the reproductive condition suggest that males have no seasonal reproductive cycle and they have the ability to transfer spermatophores once recovered from the last ejaculation (Olivares et al. 1996; Olivares et al. 2003). Nevertheless, further studies are needed to quantify the amount of spermatophores that can be transferred to the female during copulation and post-copulation recovery time in order to establish reproductive management in culture conditions (Uriarte et al. 2011). Males start sexual maturation at 75 g and conclude it at 200 g. The histological evaluation of testis gametic function and the endocrine evaluation by means of progesterone and testosterone quantification allow us to identify three categories of mature males: early, advanced and senescent (Olivares et al. 2003; Fig. 21.3).

Larger animals are mature and senescent, but their higher number of spermatophores enhances their fertilizing potential compared to males in the other two categories. It remains unclear if they do not have the ability to copulate due to structural and/or functional changes of the erectile tissue of the ligula in nature (Thompson and Voight 2003) associated with age, or there might be a reduced acceptability by females as a result of the decreased production of testosterone or if they have reached a size too large to be accepted by females. Advanced mature males of *O. mimus* have the best morphological and physiological parameters and they are the most suitable for mating. This assumption is useful in terms of management in captivity, especially in the event of establishing an integrated culture of the species (Olivares et al. 2003).

21.3.3 Reproduction in Captivity

O. mimus is easily acclimated to captivity conditions. Covering the tanks using a polyethylene mesh is useful in preventing the escape of octopus. In high density, the animals show strong aggression which can be reduced by keeping low densities close to 5 individuals m^{-3} (4.0–6.0 kg m^{-3}). It is also helpful to have similar initial sizes during rearing to reduce aggressive behaviour. Fibreglass and cement tanks of different sizes as well as high-density polyethylene (HDPE) tanks have been used for *O. mimus's* captivity with success. In addition, providing different types of refuges on the bottom of the tank such as polyvinylchloride (PVC) tubes with different diameters and length (depending on individual sizes) results in excellent survival rates. Rearing conditions used open flow and filtered water (5 % volume exchange h^{-1}) and constant aeration. Also, the daily removal of food debris using syphon suction prevents contamination caused by excessive ammonia.

Octopuses preferably consume live prey such as clams (*Protothaca thaca, Gari solida*), mussels (*Choromytilus chorus, Aulacomya ater*) and crabs such as *Cancer setosus, Cancer porteri*. Also, fresh or frozen fish can be provided. Live prey must be supplied ad libitum and frozen food should be 5 % of the body weight. Sex ratio in tanks can vary from 1:1 to 4:1 (female to male). All wild females with more than 500 g body weight are inseminated thus eliminating the need to keep males and females together in order to obtain fertilized eggs. In the absence of males, females in captivity spawn and are able to complete the embryonic development (by maternal care) with a 95 % hatching success of paralarvae. It is sound to use mature females with more than 800 g of body weight just before laying eggs. To expedite the maturity of the females, it is recommended to maintain the animals in semi-darkness conditions (Zúñiga et al. 1995).

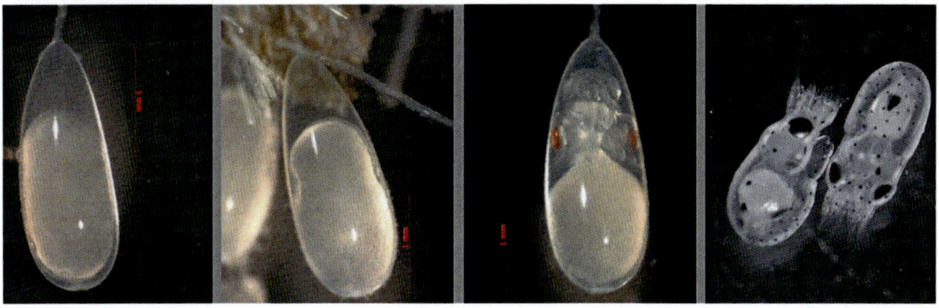

Fig. 21.4 View of some stages of the embryonic development of *Octopus mimus* in captivity, from egg to paralarvae (from left to rigth Naef's stages I, IV, XIII and paralarvae)

Table 21.1 Dimensions, weight and water percentage of *Octopus mimus* eggs, with embryos at different stages (Castro et al. 2002), in correspondence with Naef's stages (in brackets). Values are average ± standard deviation

	Fecundation (I)	First reversion (VIII)	Organogenesis (XIII)	Second reversion (XIX)
Length (mm)	2.14 ± 0.03	2.15 ± 0.02	2.16 ± 0.05	2.18 ± 0.10
Wide (mm)	0.79 ± 0.01	0.81 ± 0.02	0.85 ± 0.02	1.06 ± 0.06
Total wet weight (mg)	1.13 ± 0.10	1.13 ± 0.10	1.23 ± 0.20	2.71 ± 0.20
Total dry weight (mg)	0.28 ± 0.10	0.38 ± 0.10	0.36 ± 0.20	0.54 ± 0.20
Humidity (%)	77.0 ± 5.00	77.0 ± 6.00	71.0 ± 6.00	80.0 ± 5.00

21.3.4 Embryonic Development

The embryonic development of *O. mimus* (Fig. 21.4) in captivity, from fertilized egg to paralarvae, is similar to that described for *O. vulgaris* and other Octopodidae species (Naef 1928; Castro et al. 2002). The duration of the embryonic development extends from the initial release of the capsules until the hatching of paralarvae and this is inversely correlated to water temperature. From fertilization until the end of embryonic development in *O. mimus*, the following stages can be identified: meiosis, cleavage, gastrulation and displacement of the blastoderm, first reversion, organogenesis, second reversion and hatching. The presence of these identifiable stages concurs with the early classification proposed by Naef (1928). There are variations in size and weight during these stages, some of them detailed in Table 21.1. Distinctive patterns of paralarvae chromatophores in *O. mimus* are established during stage XIX (Castro et al. 2002). It is important to note that females of *O. mimus* show the same maternal behaviour described for other species (Mangold and Boletzky 1971; Overath and Boletzky 1974), although nursing behaviour was observed in females who had lost their spawns but were capable of taking care of those of other dead females. This is a conduct never observed before. Parental care of the spawn ensures survival of the embryos, without care, the eggs can be infected by microorganisms and/or consumed by predators.

The effects of temperature on embryonic development of *O. mimus* were first addressed by Warnke (1999). In that study, the time of embryo development was inversely related to water temperature. At 17±0.2 °C, embryonic development was between 61 and 66 days; at 18.5±0.1 °C, it lasted between 55 and 57 days, while at 19.7±0.1 °C the time interval ranged from 39 to 41 days. In a recent study, Uriarte et al. (2012) created a predictive model for temperature effects (12, 15, 18 and 21 °C) on embryonic development of *O. mimus*. This study also evaluated the temperature effects on the embryo's physiology and growth at the time that they reach XV stage, a moment when cardiac activity starts and eye development is defined. A third objective of the study was to determine if temperature affects the time interval for reaching XX stage. They used the physiological time expressed in degree-days (DD). As observed previously (Warnke 1999), increasing temperature reduces the time of embryonic development. The time interval for reaching XV stage at 21 °C was 24, 58 and 75% shorter than observed on eggs kept at 18, 15 and 12 °C, respectively. Logarithmic models that describe the relationships between development and time (in days; Naef Stage = b ln Age (d)−a marked the way temperature modulated the embryos' development speed (Uriarte et al. 2012), obtaining the following equations:

$$12^{\circ}\text{C Naef Stage} = 7.6 \ln \text{Age}(\text{days}) - 18.5$$

$$15^{\circ}\text{C Naef Stage} = 12.9 \ln \text{Age}(\text{days}) - 35.9$$

$$18^{\circ}\text{C Naef Stage} = 18.1 \ln \text{Age}(\text{days}) - 53.3$$

$$21^{\circ}\text{C Naef Stage} = 23.3 \ln \text{Age}(\text{days}) - 69.9$$

It is interesting to note that embryos kept at 18 and 21 °C grow at a similar or faster rate than those kept at 12 and 15 °C. Also, an inverse relationship between temperature and the use of the yolk was observed. Calculations obtained from DD necessary to reach XX stage suggest that *O. mimus* embryos require accumulating between 128 and 133 DD for a range of 15 and 21 °C. An apparent limit was detected for embryos kept at 12 °C. In these conditions, the embryonic development stopped almost completely by the time it reached XV stage, suggesting that this temperature value could be out of the species biokinetic range. According to our estimations, at 12 °C, embryos require 160 days to reach XX stage, 90 more days than those at 15 °C. These results suggest that this temperature could be very low for *O. mimus* embryos since embryos could remain in a dormancy state in these conditions. In natural conditions, if embryos are exposed to a low temperature, then the time interval for embryonic development would be longer (i.e. 80–90 days) than the survival time of the post-hatching females, leaving the embryos vulnerable to predators.

Table 21.2 A comparison of the morphological characteristics of the hatchling paralarvae of three *Octopus*'s species (MA: mantle length/arm length)

	O. mimus (Castro et al. 2002)	*O. vulgaris* (Boletzky 1975)	*O. maya* (Moguel et al. 2010)
Mantle length (mm)	2.21 ± 0.07	1.97 ± 0.06	6.58 ± 0.9
Arm length (mm)	0.64 ± 0.07	0.72 ± 0.02	6.93 ± 1.45
Index MA (Hochberg et al. 1992)	343	285	90
Sucker arm number	3	3	26
Koelliker organ	Present	Present	Present
Hatchling habitat	Planktonic	Planktonic	Benthic

Like in other invertebrates, oxygen consumption rates indicate that *O. mimus* embryos have a high sensitivity to temperature. The Arrhenius plot, showing the relationship between ln of oxygen consumption and temperature inverse in °K (1/kt), is a good description of the connection between metabolism and temperature. This allows comparisons between species or development stages in the same species or family (Clarke 2004). Uriarte et al. (2012) showed a−14.4 1/kt slope in the Arrhenius plot for *O. mimus* embryos reared at 12 and 21 °C. This slope is greater than that estimated in the analysis of the relationship between oxygen consumption and temperature in 13 octopus species (Farías et al. 2009; −5.8 1/kt), suggesting that *O. mimus* embryos are more sensitive to temperature than adults of the family Octopodidae. Further studies must be conducted for establishing a general slope on the Arrhenius relationship for octopus' embryos, like those proposed for fish larvae or adult octopuses (Clarke and Johnston 1999; Farías et al. 2009).

21.4 Paralarvae Culture

21.4.1 Characteristics of O. mimus *Paralarvae*

Similar to other experiences in octopus aquaculture, the lack of proper procedures for successfully rearing *O. mimus* paralarvae has limited the production of juveniles. At present, studies have reported some features on post-hatched paralarvae but no success-producing benthic juveniles. *O. mimus* paralarvae are planktonic (Villanueva and Norman 2008), swimming actively in the opposite direction of propulsion generated by the funnel. *O. mimus* hatchling paralarvae have eight arms with three suckers in each arm. The mean diameter of each sucker is 0.14 mm. The eye diameter is 0.5 mm and the funnel length is 0.7 mm. *O. mimus* paralarvae's morphological features (Table 21.2) are similar to *O. vulgaris*, both species with planktonic early stage after hatching.

Paralarvae present a constant number of chromatophores evenly distributed in the mantle and arms with a brown-reddish pigmentation, which can expand and

Table 21.3 Chromatophore pattern of *Octopus mimus* and *O. vulgaris* paralarvae of one day. *O. mimus* paralarvae from Southeastern Pacific (Antofagasta, Chile) and *O. vulgaris* paralarvae from Northeastern Atlantic (Galicia Spain) regions

	O. mimus (Castro et al. 2002)	*O. vulgaris* (Vidal et al. 2010)
Dorsal view		
Mantle	3–7	8–10
Mantle edge	3–4	2–3
Mantle total	6–11	10–13
Visceral	6–8	6–7
Head	2+4+4	2+2+2+2
Arm	3–4	3–4
Eyes	–	1
Ventral view		
Mantle	24–31	16–19
Funnel	4+2	2+2
Head	2	2–3
Arm	3–4	2–3
Eyes	–	1

shrink in response to environmental stimuli such as light, sound, tank bottom colour, turbulence and temperature. The number and distribution of chromatophores are morphological features important in octopus taxonomy (Villanueva and Norman 2008). The pattern of chromatophores' distribution in *O. mimus* was first described by Warnke (1999). Further, Castro et al. (2002) found different characteristics compared to those described for *O. vulgaris*. *O. mimus* paralarvae show a higher number of chromatophores at the ventral mantle surface compared to *O. vulgaris* (see Table 21.3) in addition to several other morphological characteristics.

21.4.2 Culture Conditions

The growth of *O. mimus* paralarvae was studied in conic tanks of 50 and 100 L following the adaptations and modifications proposed in Villanueva (1995), as well as for the maintenance of living food. Paralarvae feeding preference was evaluated using an open flow system with a supply of live food. Post-hatch paralarvae have an energy reserve in the endogenous yolk which is absorbed completely within 4 days after hatching (DAH) at 20 °C. Once the yolk is totally absorbed, paralarvae start feeding on food suspended in the water. This is a critical stage since high mortalities are observed between 4 and 6 DAH at 20±2 °C. Artemia metanauplii, zoea of decapods (*Leptograpsus variegatus, Cancer setosus*) and rotifers have been used as live diet (Zúñiga et al. 1997). It was evident that paralarvae capture and feed on zoea of decapods. In a density of 25 paralarvae L^{-1}, animals gained weight from 0.6 to 2.6 mg in 31 days at 20±2 °C. The maximum time of survival also occurred at this time.

Paralarval culture has become one of the most important challenges in species with planktonic stage. A fundamental aspect is feeding paralarvae and considering all the adaptations for living in the marine environment. At present, it is known that paralarvae from diverse species have an internal yolk which allows them to complete their development independently of external food (Villanueva and Norman 2008). Zúñiga et al. (2012) studied the mechanisms explaining how yolk reserve modulates the adaptive capacity of *O. mimus* paralarvae, evaluating the thermoregulatory capacity, thermal selection and oxygen consumption in individuals acclimated at 20 °C. The study suggests that post-hatched paralarvae (1 DAH) are less tolerant to extreme temperature, compared to 2–3 DAH individuals at the moment when paralarvae are fully developed and the yolk is totally absorbed. In addition, the paralarval thermal preference does not change with age. The temperature selected for paralarvae rearing was 22.8 ± 1.3 °C when they were exposed in a horizontal gradient of temperature. The paralarvae age and their nutritional condition affected the thermal tolerance range (TTR = Critical Maximum Temperature—Critical Minimum Temperature), with higher values for two and three DAH individuals (25 and 26 °C) than for 1 DAH (23.1 °C) and 4 DAH (22.7 °C).

Besides the adaptive implications and potential effects of the relationship between a lack of food and an increment of temperature on *O. mimus* paralarval survival in its natural environment, the results of Zúñiga et al. (2012) allowed the definition of a critical period, starting once yolk reserve is absorbed at 20 °C and after 4 DAH.

21.5 Ongrowing

In Antofagasta (northern Chile), the authors of this chapter have carried out different experiments focused on growth by feeding *O. mimus* using artificial food prepared with triturated fish flesh (54 %), fish meal (10 %) and clams compacted with 6 % non-flavour jelly all embedded in sausages (lamb tripe).

21.5.1 Tank Culture (Inland Facilites)

Octopuses with body weight less than 800 g were deployed at different densities (5, 10, 15 individuals m^{-3}) in metallic cages inside of raceway-type tanks made of HDPE of 2 x 6 x 1.5 m with continuous water flow. Water was changed two to three times, salinity was kept at 36 psu, ammonia at < 10 mg L^{-1} and dissolved oxygen at upper 4 mg L^{-1} level. The latter was also enhanced using constant aeration. Lower mortalities and higher growth were obtained at densities lower than 10 individuals m^{-3}. Octopuses preferred either fresh or frozen food, although they accepted the artificial diet previously described (sausage-type food) which facilitates the management of octopus growth. Mortality was high in this culture system. The survival average only reached 50 %. Growth was low and animals only attained a 70 % of the initial weight in 60 days of culture.

21.5.2 Floating Cage Culture

Growth experiments were carried out by using six floating cages made of HDPE ($1\ m^3$) installed in a raft. The experimental conditions were similar to those of the inland experiments. In these cages, the growth of the octopuses was similar to that of inland, but the survival rate was more than 70 %, and the growth was duplicated in 60 days of culture.

Results from *O. mimus* culture experiences show that this species can be managed in captivity at densities lower than 10 individuals m^{-3}, although we needed to improve the composition of the diet that enhances growth. Survival can be increased by separating the animals according to sex, which adds an extra cost to the operation. Growing female octopuses have the disadvantage of the high energy cost of the physiological and ethological reproductive processes for the reason that females lose weight during reproduction. However, the sexual maturity process can be inhibited using light which also favours growth.

It is necessary to address the problem of our inability for mass production of octopus paralarval and juvenile (seeds) in hatchery. Otherwise, the culture will always depend on juveniles and subadults captured in the wild by divers, which elevates the cost and reduces the profitability of the culture.

21.6 Trends in Research and Industrial Level

Violating legal regulations of the octopus fishery may lead to a collapse in some *O. mimus* populations that are favoured during the ENSO events. Research on the effects of seawater temperature oscillations on *O. mimus* along with its distribution range should be carried out to know how ENSO and other environmental perturbations modulate octopus distribution and fisheries.

Easy acclimation to captivity and acceptance of diverse types of diets make *O. mimus* a promising species for aquaculture. However, metabolic and physiological studies are needed in order to formulate an optimal diet for paralarvae and juveniles, which yields a higher growth rate than in the wild. In addition, the industrial farming of this species will not be possible until rearing technology and feeding physiology are known.

21.7 Conclusion

Nowadays, the culture of *O. mimus* is still at a research level. As other octopus species, *O. mimus* has a great potential to be cultured, but paralarvae rearing should be solved before it becomes a real and profitable species to diversify aquaculture in Chile and Perú. Moreover, considering its importance to the fishery in the two countries and its apparent sensitivity to temperature changes (Uriarte et al. 2011; Zúñiga et al. 2012), *O. mimus* could be considered as an experimental model species to help understand the possible alterations of marine ecosystem to climate warmings.

References

Baltazar P, Rodríguez P, Rivera W, Valdivieso V (2000) Cultivo experimental de *Octopus mimus*, Gould 1852 en el Perú. Rev Perú Biol 7(2):151–160

Bidder AM (1957) Evidence for an absorptive function in the liver of *Octopus vulgaris* Lam. Pubbl Stn Zool Napoli 29:139–150

Boletzky SV (1975) Le développement d'*Eledone moschata* (Mollusca, Cephalopoda) élevée au laboratoire. Bull Soc Zool Fr 100:361–367

Cardoso F, Villegas P, Estrella C (2004) Observaciones sobre la biología de *Octopus mimus* (Cephalopoda: Octopoda) en la costa peruana. Rev Perú Biol 11:45–50

Carrasco S, Guisado Ch (2010) Effects of alimentary regime on feeding, growth, and proximal composition of *Octopus mimus* Gould, 1852 (Cephalopoda: Octopodidae). J Shellfish Res 29:455–461

Castro H, Olivares A, Quintana A, Zúñiga O (2002) Descripción del desarrollo embrionario y paralarvas de *Octopus mimus* Gould 1852 (Mollusca: Cephalopoda) en cautiverio. Estud Oceanol 21:13–25

Clarke A (2004) Is there universal temperature dependence of metabolism? Funct Ecol 18:252–256

Clarke A, Johnston NM (1999) Scaling of metabolic rate with body mass and temperature in teleost fish. J Anim Ecol 68:893–905

Cortez T, Castro BG, Guerra A (1995a) Feeding dynamics of *Octopus mimus* (Mollusca: Cephalopoda) in northern Chile waters. Mar Biol 123:497–503

Cortez T, Castro B, Guerra A (1995b) Reproduction and condition of female *Octopus mimus* (Mollusca: Cephalopoda). Mar Biol 123:505–510

Cortez T, González A, Guerra A (1999) Growth of cultured *Octopus mimus* (Cephalopoda, Octopodidae). Fish Res 40:81–89

Defeo O, Castilla JC (1998) Harvesting and economic patterns in the artisanal *Octopus mimus* (Cephalopoda) fishery in a northern Chile cove. Fish Res 38:121–130

Farías A, Uriarte I, Hernández J, Pino S, Pascual C, Caamal C, Domíngues P, Rosas C (2009) How size relates to oxygen consumption, ammonia excretion, and ingestion rates in cold (*Enteroctopus megalocyathus*) and tropical (*Octopus maya*) octopus species. Mar Biol 156:1547–1558

Forsythe JW, De Rusha RH, Hanlon RT (1983) Notes on the laboratory culture of *Octopus bimaculoides* the California mudflat octopus. Am Malacol Bull 2:92–93

Galleguillos R, Canales-Aguirre CB, Ferrada S, Barrera A (2011) SSRs in *Octopus mimus*: development and characterization of nine microsatellite loci. Rev Biol Mar Oceanogr 46(3):491–494

Guerra A, Cortez T, Rocha F (1999) Redescripción del pulpo de los Changos, *Octopus mimus* Gould, 1852, del litoral chileno-peruano (Mollusca, Cephalopoda). Iberus 17:37–57

Hixon RF (1983) *Loligo opalescens*. In: Boyle PR (ed) Cephalopod Life Cycles, Vol 1. Species Accounts. Academic Press, London, p 95–114

Hochberg FG, Nixon M, Toll RB (1992) Orden Octopoda Leach 1818. In: Sweeney MJ, Roper CF, Mangold KM, Clarke MR, Boletzky SV (eds) Larval and juvenil cephalopods: a manual for their identification, vol 513. Smithsonian Institution press, Washington, D.C., pp. 213–279

Hunsicker ME, Essington TE (2010) The contribution of cephalopods to global marine fisheries: can we have our squid and eat them too? Fish Fish 11:421–438

Ibáñez CM, Camus PA, Rocha F (2009) Diversity and distribution of cephalopod species of the coast off Chile. Mar Biol Res 5:374–384

Ikeda Y, Sakurai Y, Shimazaki K (1993) Maturation process of Japanese common squid *Todarodes pacificus* in captivity. In: Okutani T, O'Dor RK, Kuboreda T (eds) Recent advances in cephalopod fisheries biology, Tokai University, Tokyo, pp 179–187

Ishiyama V, Siga B, Talledo C (1999) Biología reproductiva del pulpo *Octupus mimus* (mollusca: cephalopoda) de la región de Matarani, Arequipa, Perú en la costa Peruana. Rev Perú Biol 11(1):110–122

Joll LM (1983) *Octopus tetricus*. In: Boyle PR (ed) Cephalopod Life Cycles, Vol 1, Species Accounts, Academic Press, London, pp 325–334

Koueta N, Boucad-Camou E, Renou AM (1995) Gonadotropic mitogenic activity of the optic gland of the cuttlefish, *Sepia officinalis*, during sexual maturation. J Mar Biol Assoc UK 75:391–404

Mangold K (1987) Reproduction. In: Boyle PR (ed) Cephalopod Life Cycles, Vol II, Comparative Reviews. Academic Press Inc, London, pp 157–200

Mangold K, Boletzky SV (1971) Reproductive biology and embryonic development of *Eledone cirrosa* (Cephalopoda: Octopoda). Mar Biol 8(2):109–117

Minakata H, Shigeno S, Kano N, Haraguchi S, Osugi T, Tsutsui K (2009) *Octopus* gonadotrophin-releasing hormone: a multifunctional peptide in the endocrine and nervous systems of the cephalopod. J Neuroendocrinol 21:322–326

Moguel C, Mascaró M, Avila-Poveda OH, Caamal-Monsreal C, Sanchez A, Pascual C, Rosas C (2010) Morphological, physiological and behavioral changes during post-hatching development of *Octopus maya* (Mollusca: Cephalopoda) with special focus on the digestive system. Aquat Biol 9:35–48

Morales E (1973) Contribución al estudio de la morfología, estructura y anatomía microscópica de la región visceral de *Pteroctopus tetracirrhus* D. Ch. (Octopoda: Incirrata). Inv Pesq 37(2):355–518

Naef A (1928) Die Cephalopoden (Embryologie). Fauna e flora del Golfo di Napoli 35:1–375

Olivares A, Zúñiga O, Retamales E (1994) Escala de madurez sexual de *Octopus mimus* Gould 1852 (Cephalopoda: Octopodidae) de la II Región, Chile. Estud Oceanol 13:90–93

Olivares A, Zúñiga O, Castro G, Segura C, Sánchez J (1996) Bases biológicas para el manejo de *Octopus mimus*: Reproducción y crecimiento. Estud Oceanol 15:61–74

Olivares A, Zamora M, Portilla P, Zúñiga O (2001) Estudio histológico de la ovogénesis y maduración ovárica en *Octopus mimus* (Cephalopoda: Octopodidae) de la II Región de Chile. Estud Oceanol 20:13–22

Olivares A, Bustos-Obregón E, Castillo V, Zúñiga O (2003) Variaciones del funcionamiento testicular en *Octopus mimus* adultos. Int J Morphol 21(4):315–323

Osorio C, Atria J, Mann S (1979) Moluscos marinos de importancia económica en Chile. Biol Pesq (Chile) 11:3–47

Overath H, Boletzky SV (1974) Laboratory observations of spawning and embryonic development of a blue-ringed octopus. Mar Biol 27(4):333–337

Pérez-Losada MP, Guerra A, Sanjuan A (2002) Allozyme divergence supporting the taxonomic separation of *Octopus mimus* and *Octopus maya* from *Octopus vulgaris* (Cephalopoda: Octopoda). Bull Mar Sci 71(2):653–664

Porta J de (2003) La formación del Istmo de Panamá. Su incidencia en Colombia. Rev Acad Colomb Cienc 27(103):191–216

Rocha F, Vega MA (2003) Overview of cephalopod fisheries in Chilean waters. Fish Res 60:151–159

Solís-Ramírez MJ (1967) Aspectos biológicos del pulpo *Octopus maya* Voss y Solís. Publ Inst Nac Inv Pesq 18:1–190

Söller R, Warnke K, Saint-Paul U, Blohm D (2000) Sequence divergence of mitochondrial DNA indicates cryptic biodiversity in *Octopus vulgaris* and supports the taxonomic distinctiveness of *Octopus mimus* (Cephalopoda: Octopodidae). Mar Biol 136:29–35

Thompson JT, Voight JR (2003) Erectile tissue in an invertebrate animal: the octopus copulatory organ. J Zool 261:101–108

Uriarte I, Iglesias J, Domíngues P, Rosas C, Viana MT, Navarro JC, Seixas P, Vidal E, Ausburger A, Pereda S, Godoy F, Paschke K, Farías A, Olivares A, Zúñiga O (2011) Current status and bottle neck of octopod aquaculture: The case of american species. J World Aquacult Soc 42(6):735–752

Uriarte I, Espinoza V, Herrera M, Zúñiga O, Olivares A, Carbonell P, Pino S, Farías A, Rosas C (2012) Effect of temperature on embryonic development of Octopus mimus under controlled conditions. J Exp Mar Biol Ecol 416–417:168–175

van Heukelem WF (1983) *Octopus cyanea*. In: Boyle PR (ed) Cephalopod Life Cycles, Vol 1. Species Accounts, Academic Press, London, pp 267–276

Vidal E, Fuentes L, Da Silva LB (2010) Defining *Octopus vulgaris* populations: A comparative study of the morphology and chromatophore pattern of paralarvae from Northeastern and Southwestern Atlantic. Fish Res 106:117–248

Villanueva R (1995) Experimental rearing and growth of planktonic *Octopus vulgaris* from hatching to settlement. Can J Fish Aquat Sci 52:2639–2650

Villanueva R, Norman MD (2008) Biology of the planktonic stages of benthic octopuses. Oceanogr Mar Biol Ann Rev 46:105–202

Villegas P, Tafur R (2000) Aspectos reproductivos del pulpo (*Octopus mimus*) en el área de Callao. Informes Progresivos, Instituto del Mar de Perú 121:3–15

Warnke K (1999) Observations on the embryonic development of *Octopus mimus* (Mollusca: Cephalopoda) from Northern Chile. Veliger 42(3):211–217

Warnke K, Söller R, Blohm D, Saint-Paul U (2000) Rapid differentiation between *Octopus vulgaris* Cuvier (1797) and *Octopus mimus* Gould (1852), using randomly amplified polymorphic DNA. J Zool Syst Evol Research 38:119–122

Warnke K, Söller R, Blohm D, Saint-Paul U (2004) A new look at geographic and phylogenetic relationships within the species group surrounding *Octopus vulgaris* (Mollusca: Cephalopoda): indications from mitochondrial DNA nucleotide sequences. J Zool Syst Evol Res 42:306–312

Wells MJ, Wells J (1977) Optic glands and the endocrinology of reproduction. Symp Zool Soc Lond 38:525–540

Wodinsky J (1977) Hormonal inhibition, feeding and death in *Octopus*: control by optic gland secretion. Science 198:948–951

Wolff M, Pérez H (1992) Population dynamics, food consumption and gross conversion efficiency of *Octopus mimus* Gould, from Antofagasta (northern Chile). ICES Council Meeting Papers 1992:1–12

Zamora M, Olivares A (2004) Variaciones bioquímicas e histólogicas asociadas al evento reproductivo de la hembra de *Octopus mimus* (Mollusca: Cephalopoda). Int J Morphol 22(3):207–216

Zúñiga O, Olivares A, Ossandón L (1995) Efecto de la luz en la maduración sexual de hembras de *Octopus mimus*. Estud Oceanol 14:75–76

Zúñiga O, Olivares A, Muñoz L (1997) Resultados preliminares de la producción de larvas de pulpos *Octopus mimus* (II región, Chile). VII Congreso Latinoamericano sobre Ciencias del Mar, Libro de Resúmenes, Vol II:567–569

Zúñiga O, Olivares-Paz A, Torres I (2011) Evaluación del crecimiento del pulpo común *Octopus mimus* del norte de Chile alimentados con dietas formuladas. Lat Am J Aquat Res 39(3):584–592

Zúñiga O, Olivares A, Rojo M, Chimal ME, Díaz F, Uriarte I, Rosas C (2012) Thermoregulatory behaviour and oxygen consumption of *Octopus mimus* paralarvae: The effect of age. J Therm Biol 38:86–91

Chapter 22
Octopus minor

Xiao-Dong Zheng, Yao-Sen Qian, Chang Liu and Qi Li

Abstract *Octopus minor* (Sasaki 1920) is widely distributed along the coastal waters of China, Korean Peninsula, as well as south of Sakhalien to Japan. As an important economic cephalopod, culture of *O. minor* has been attempted in recent years. After capture and broodstock acclimatization to captivity, spawning induction and mating, eggs are spawned at the artificial substratum. Females are responsible for protecting the eggs. The embryonic development lasts for 72–89 days before hatching under the conditions of a seawater temperature of 21–25 °C and a salinity of 28–31 psu. The mantle length and total length of new hatchlings range from 8.5 to 11.5 mm and from 25 to 31 mm, respectively. Hatchlings are benthic, going directly to the tank bottom. Two different types of shelters are provided for this species: ceramic pots of 8–12 mm inner diameter or 10-mm-diameter polyvinyl chloride (PVC) tubes. Cladocerans, copepods and enriched *Artemia* nauplii are adequate initial feeds for the hatchling rearing. *Hemigrapsus sanguineus* of less than 4 mm body width are also used to feed 10-day-old hatchlings. Using these prey, the survival rate is 75% after 1 month of culture. After that, a mixed fresh diet such as juvenile crab, shellfish and shrimp becomes the main feed. After 6–7 months of culture indoors, juveniles of about 100 g can be considered as commercial specification, and they are then transferred to outdoor ponds to continue the ongrowing process, or to be released to the sea for stocking or enhancement programmes. Under indoor culture conditions, the average weights of males and females at 250 days are 122.9 and 197.1 g, respectively.

Keywords *Octopus minor* · Embryonic development · Hatchling · Ongrowing · Culture

X. D. Zheng (✉) · Y. S. Qian · C. Liu · Q. Li
Fisheries College, Ocean University of China, Qingdao 266003, China
e-mail: xdzheng@ouc.edu.cn

Y. S. Qian
Ocean Fishery Station of Ganyu, Lianyungang 222100, China

J. Iglesias et al. (eds.), *Cephalopod Culture*, DOI 10.1007/978-94-017-8648-5_22,

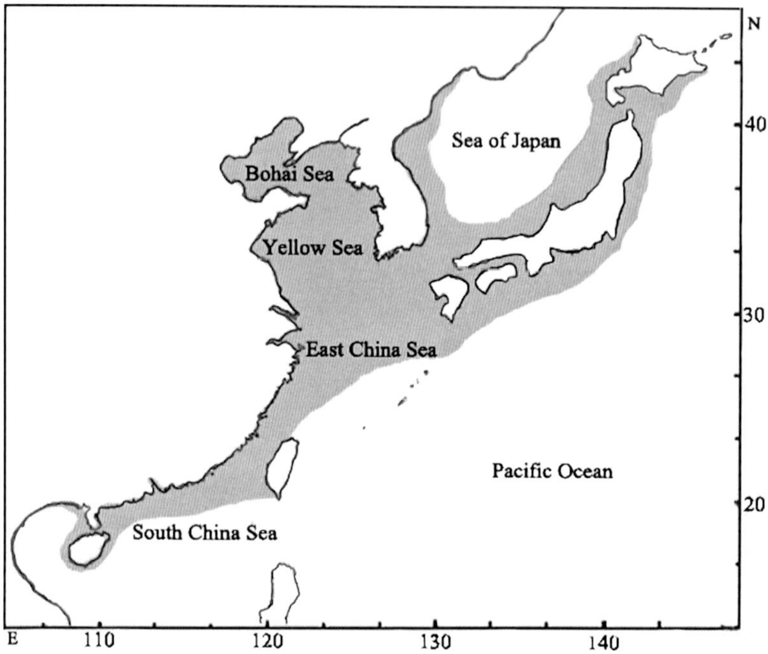

Fig. 22.1 Distribution of *Octopus minor* (*light grey* zone). (Map modified from Roper et al. 1984)

22.1 Importance of this Species in the Market

Octopus minor (Sasaki 1920) is widely distributed along the coastal waters of China, Korean Peninsula, as well as south of Sakhalien to the whole of Japan (Okutani et al. 1987; Fig. 22.1), with *O. variabilis* as a synonym used in Japan and China (Yamamoto 1942; Dong 1988; Lu et al. 2012).

It is a benthic, littoral species occurring down to a depth of 200 m (Roper et al. 1984). *O. minor* is a popular seafood in East Asia, not only with high protein content and high n-3 HUFA (EPA and DHA) levels but also rich in minerals and vitamins (Qian et al. 2010). The overall catch per year is much more than 10,000 t in the north of China only. As a species of great economic importance, it is exported mainly to Korea and Japan. In recent years, the commercial catches have decreased rapidly (Kim et al. 2008).

As other species, such as *O. ocellatus* in Japan (Segawa and Nomoto 2002) and China (Wang et al. 2010; Dong et al. 2013) or *O. maya* in México (Domingues et al. 2007), *O. minor* hatchlings undergo a direct development. The species belongs to the group of "long arm octopus". The first arm is the longest and represents about 80 % of the total length and six to seven times the mantle length. The ligula is large, spoon shaped and about 10–20 % of the length of the third hectocotylized right arm (Okutani 2000).

Fig. 22.2 *Octopus minor* and its nest (*arrows* show holes)

As a potential aquaculture species, some researches on its ecology, reproductive biology, physiology as well as genetics have been done (Taki 1944; Iwakoshi et al. 2000; Seol et al. 2007; Zuo et al. 2011; Cheng et al. 2012; Kang et al. 2012; Qian et al. 2013), but very few documents have reported on its life history or culture in captivity. The culture of the species has been developed in China in the current decade. Meanwhile, the artificial releasing of hatchery-produced octopus has become the main objective of resource protection and stock enhancement programmes especially in the coastal waters of Shandong Province.

22.2 Capture Methods

The main way to capture mature individuals of *O. minor* is by the bottom trawl in the northeast coastal waters of China during the period from May to June. In shallow areas, because of their burrowing behaviour, fishermen usually catch them directly from the mud holes (Fig. 22.2). Young octopuses are also caught by fishermen with a kind of spade or a small hand hook, digging the muddy bottom at low tide, mainly in spring and autumn. They are then transferred to the indoor ponds for culture.

Besides the artificial production of juveniles and their culture indoors, collection of wild eggs and juveniles of *O. minor* is also possible in China. The species lives in shallow calm waters with a muddy bottom, burrowing a deep tunnel to hide in. The collection process of wild eggs involves the placing of artificial egg collectors (Fig. 22.3) in the muddy bottoms of coastal waters in mid-May, which is the main reproduction season. The plastic collector has two holes (Fig. 22.3a, b), which are open and 30–40 mm above the bottom surface. These artificial egg collectors are picked up together at the end of July. After checking, artificial egg collectors with eggs and/or their parent octopuses are transported to the indoor pools to continue the breeding process.

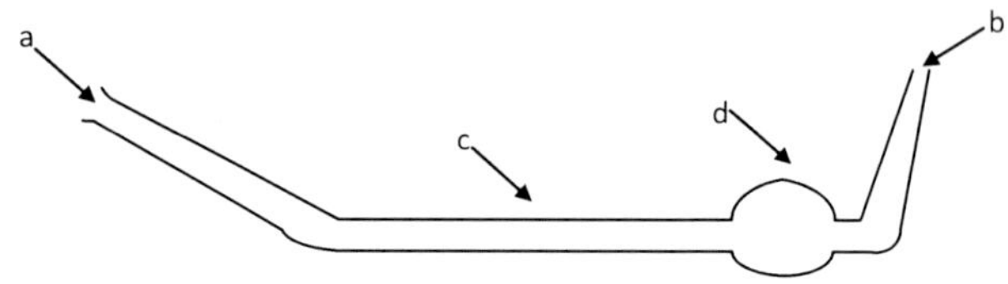

Fig. 22.3 Artificial egg collector. **a** Submarine hole. **b** Breathing hole. **c** Tunnel. **d** Spawning room

22.3 Broodstock Acclimatization to Captivity

22.3.1 Physical and Chemical Parameters

After capture, wild individuals must be chosen carefully, and those with no wounds are selected as broodstock and transported indoors for adaptation to captivity. The broodstock are introduced into 8 × 4 × 1.2 m rectangular concrete pools, whereas nursery pools are much smaller measuring 3 × 1 × 0.6 m. The recommended light intensity ranges from 200 to 800 Lx under a seawater temperature of 20–24 °C, salinity 29–31 psu and dissolved oxygen > 4 mg L^{-1}.

Two kinds of shelter nests are provided in advance after sterilizing: 20–30-cm-long polyvinyl chloride (PVC) pipes or eel culture cages. The broodstock density is controlled to approximately 3–5 individuals m^{-2} and a sex ratio of 1:1.

Small crabs of various species are supplied daily as the main food, in a ratio of 1–1.5 times the total number of broodstock individuals. Other live prey species, such as small-sized marine bivalves (*Ruditapes philippinarum*, *Potamocorbula* sp.), or polychaetes (*Nereis*), etc. are also used. The leftover food should be cleaned up and seawater also changed the next morning, due to the fact that this species preys overnight.

22.3.2 Mating

During the period of adaptation to captivity, mating is usually observed in the early morning or at night under indoor conditions. The male lies on the back of the female and introduces its hectocotylized arm directly into the female mantle. The spermatophores are transferred near the oviduct. The mating behaviour lasts for 10–30 min. After mating, most males die gradually, and females can store spermatophores up to at least 2 weeks before spawning. Reproductive females always hide alone in the PVC tubes or eel cages. If more than two individuals are found in the same shelter, one of them should be immediately transferred to an empty one.

22.3.3 Spawning and Egg Handing

Octopus shelters are usually checked daily and confirmed whether or not eggs appear in the internal walls. As happens in other species, food intake of broodstock decreases sharply once they start to spawn. Females show the behaviour of protecting the egg masses. To avoid other individuals disturbing the spawning female, both ends of the shelter pipes are firmly sealed with a 0.85 mm-diameter fine mesh, so that a female would be able to stay safely with her eggs. Most of the eggs are attached one by one to the inner walls of the shelter. Each egg looks like a long eggplant in shape, with a long diameter of 13–20 mm, a short diameter of 4–6 mm and with a wet weight of 0.15–0.26 g. The number of spawned eggs per female ranges from 9 to 125, depending on the weight among individuals. In general, the spawning time lasts for 1 week per female. Once an individual starts to spawn, it is able to stimulate the others to do the same. The entire spawning period can be extended to at least 1 month, because more than 100 females can be used for the same broodstock.

It is well known that during the spawning period, females hardly eat food in the wild. But they can eat a small amount of food soon after being captured from the sea and taken into the indoor culture conditions. Consequently, some fresh food such as one to two half-shell clams, or crabs without claws, is still supplied to the shelters every 3–5 days. Any leftover food must be retrieved the next morning, to prevent the deterioration of the water quality. Meanwhile, seawater keeps flowing continuously. Under these conditions, most females protect their eggs until hatching.

The embryonic development of *O. minor* lasts for 72–89 days before hatching under the conditions of a seawater temperature of 21–25 °C and a salinity of 28–31 psu. At this temperature range, the process of embryonic development is divided into 20 stages (Fig. 22.4) differentiated by characteristic changes, such as cleavage, blastula formation, gastrula formation, organ rudiment formation as well as two embryo reverses (Qian et al. 2013). The first embryonic reversal takes place at stages VI–VII on day 21 (Fig. 22.4), and embryos turn from the animal pole to the vegetal pole. The second embryonic inversion happens at stage XIX on day 65 (Fig. 22.4). Sporadically, more than two reverses were also observed. The mantle length and total length of hatchlings range from 8.5 to 11.5 mm and from 25 to 31 mm, respectively. And after hatching, they already show benthic behaviour.

22.4 Hatchlings Rearing

New hatchlings are very sensitive to the environment and they can eject ink easily in response to any external stimuli, which may trigger a major injury. To function as a den, it is effective and feasible to place in advance some green macroalgae or seaweed (e.g. *Ulva pertusa, Zostera marina*) in the breeding pools. In the case of

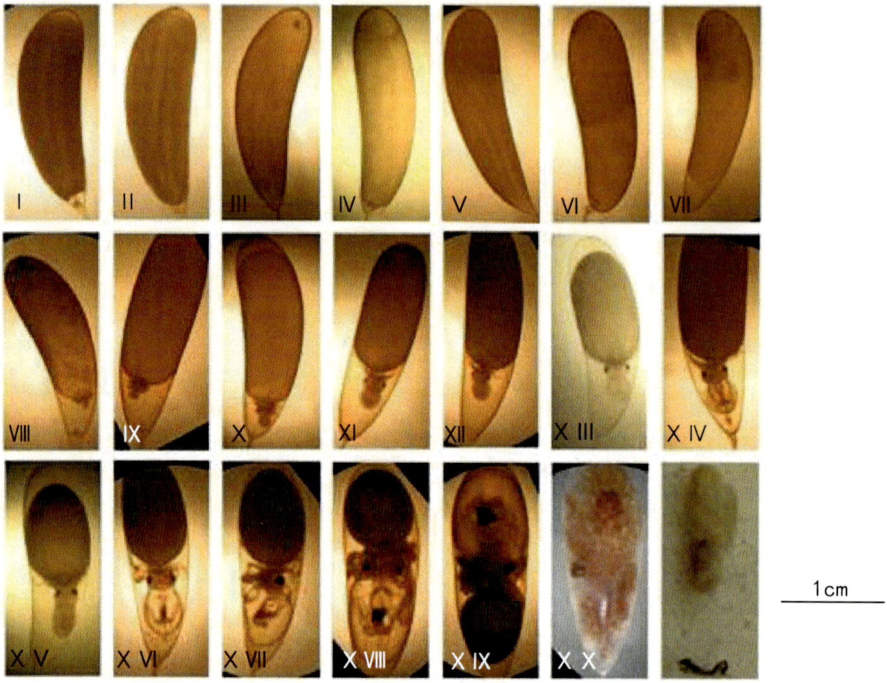

Fig. 22.4 Embryonic development of *O. minor*. (Stages from Qian et al. 2013)

a stress situation and ink ejection, hatchlings will hide in the den surface to avoid being harmed.

As nurseries, concrete pools of around 400 mm depth are used, where the behaviour of hatchlings can be easily observed. In addition to the previously mentioned seaweeds, ceramic pots of 8–12 mm inner diameter (Fig. 22.5a) and grey-green PVC pipes of 10 mm diameter (Fig. 22.5b) should also be supplied in the pool bottom as hatchling shelters. The number of these shelters must be 1.5–2 times the total number of hatchlings. The inner diameter should become a bit larger after 2 months of rearing. Light intensity is important in hatchling culture and should be controlled under the condition of 200–800 Lx in the pool surface.

Live enriched *Artemia* nauplii (density: 0.5–1 individuals mL^{-1}), cladocerans and copepods (5–10 individuals cm^{-2}) are an adequate initial feed for the new hatchlings (Fig. 22.5c). Instead of *Artemia* nauplii, *Hemigrapsus sanguineus* of less than 4 mm body width are used to feed the 10-day-old hatchlings (Fig. 22.5d). The prey density should be three to four times the total number of hatchlings. At this age, the density of cladocerans and copepods will be decreased to 1–3 individuals cm^{-2}.

During the period of hatchling culture, the water must be kept clean and half of the whole volume must be changed every day. Physical and chemical parameters must also be kept in the following ranges: water temperature of 10–25 °C, pH 7.6–8.6, dissolved oxygen >5 mg L^{-1} and ammonia nitrogen <0.2 mg L^{-1}.

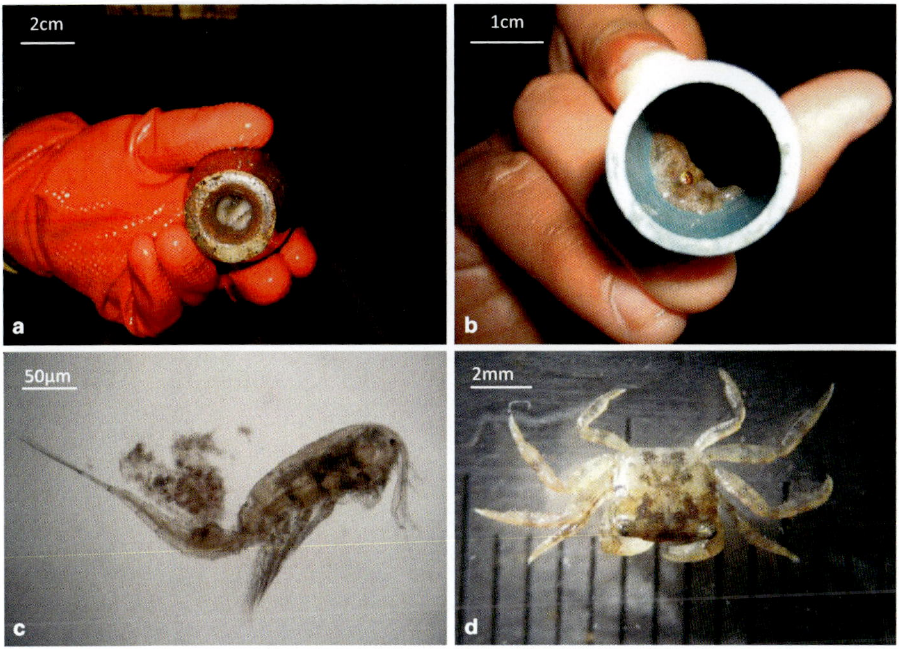

Fig. 22.5 Shelters and food of hatchlings. **a** Ceramic pot. **b** PVC pipe. **c** Copepod. **d** *Hemigrapsus sanguineus*

Due to the mortality, hatchling density decreases progressively from 500 to 100 individuals m^{-3} during the 1st month of culture. When food is offered in sufficient amount, very few hatchlings fight and compete with each other. The mean hatchling weight increases to 2.2 g, and survival rate reaches 75 % after 1 month of rearing.

22.5 Ongrowing

After a month of hatchling rearing, *O. minor* juveniles have reached the ongrowing stage. At this period, seawater temperature should range from 15 to 20 °C and the other physical and chemical parameters should be similar to those used in the previous hatchling process. Octopus juveniles have many characteristics similar to adults, such as aggressiveness, territoriality and cannibalism. When they are fed in the evening, the phenomenon of fierce fighting is observed, especially between the bigger and smaller individuals. Therefore, they must be separated based on their size, but in any case the survival rate during this growing period decreases to 50 %.

Some live small bivalves (genus *Potamocorbula*) are used as diet due to their affordable price and easy availability from the fishery market. The seaweed *U. pertusa* is not only a kind of essential shelter to juveniles but also a suitable food to cultivate gammarids which are also a favourite prey of octopus juveniles. At

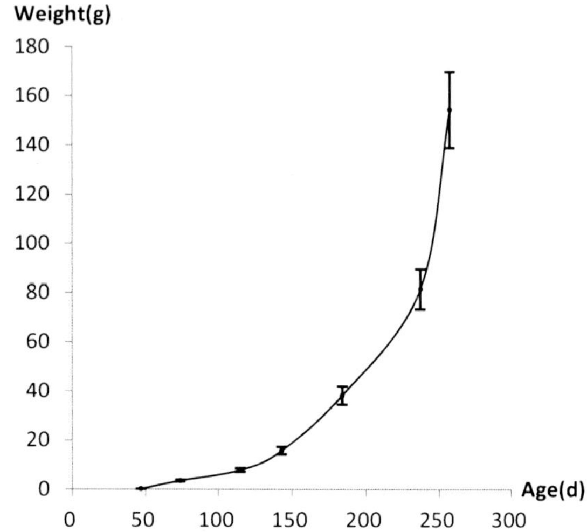

Fig. 22.6 Growth curve of *Octopus minor*

the same time, fresh and frozen feeds become acceptable substitutes for live ones. Small pieces of fresh muscles of *R. philippinarum* are well accepted as an inert diet. Generally, the feeding period of juveniles is within 2 h after being fed at night. The leftover food must be cleaned the next morning in order to maintain good water quality. The juvenile density gradually decreases to 30–40 individuals m^{-3} after 2 months of growth.

Based on the experiments of 2010–2012, the sex of 115-day-old octopus could be identified, when their mean fresh weight reached 15 g. The survival rate remained stable from that time on. Later on, at 250 days, the average weights of males and females were 122.9 and 197.1 g, respectively. The growth curve for this period is shown in Fig. 22.6 which is based on data of four to ten individuals of different stages. During this period, mating behaviour was observed, followed by the death of males. The latest spawn took place at 346 days under indoor culture conditions and it was performed sequentially. The weight of the female decreased to 117.3 g after spawning.

22.6 Harvesting and Resource Enhancement

Most juveniles reach a weight of 100 g in 6–7 months under indoor culture conditions, which corresponds to the commercial marketable size and therefore can be harvested for marketing.

Otherwise, these individuals can also be transferred to outdoor ponds to grow further or to be released into the wild for stock enhancement purposes. For example, more than 100,000 juveniles have been released into the Moon Lake National Aquatic Resources Conservation Area of China, located in Rongcheng (Shandong

Province), for the past 4 years. The detailed effect of resource restoration is being evaluated.

22.7 Trends in Research and Industrial Level

One of the main subjects studied in China at present is to obtain an indoor method to develop the rearing process of *O. minor* all year round. The researches on biological zero temperature and effective accumulated temperature for embryonic development of *O. minor* have been carried out under artificial conditions (Liu 2013). The embryonic development could stop when the temperature decreased below the biological zero (3.79 °C) and the embryo could tolerate down to approximately 0 °C. Once above the biological zero temperature, the embryonic development could resume and also hatch as usual. These recent results indicate that the incubation of *O. minor* will be feasible indoors all year round in the north of China. Though artificial culture can greatly enhance the hatching and the survival rate, the number of spawned eggs per *O. minor* female is only 9–125, and besides this the embryonic development lasts for at least 70 days (Qian et al. 2013). It is obvious that these results restrict severely the industrial culture of *O. minor* due to such a low fecundity and long embryonic development period. Certainly, optimization of feed formulation is also another key factor for its commercial culture which still requires further research.

The artificial nursery of *O. minor* in the Moon Lake National Aquatic Resources Conservation Area, China, has been working since 2009, and the releasing of hatchery-produced octopus has become the main approach of resource protection and stock enhancement programmes.

22.8 Conclusions

As an important economic cephalopod, *O. minor* has been cultured in the northeast coastal waters of China in recent years. The whole culture process includes broodstock capture, acclimatization to captivity, spawning, hatchling culture, ongrowing and overwintering culture and harvesting. The main conclusions are the following:

1. Eggs are spawned at the artificial substratum, and females are responsible for protecting the eggs (Fig. 22.7a). The number of spawned eggs ranges from 9 to 125 under the artificial conditions. The embryonic development lasts for 72–89 days before hatching under the conditions of a seawater temperature of 21–25 °C and a salinity of 28–31 psu. The mantle length and total length of new hatchlings range from 8.5 to 11.5 mm and from 25 to 31 mm, respectively.
2. Hatchlings are benthic. The shelters provided are ceramic pots of 8–12 mm inner diameter or 10-mm-diameter PVC tubes. Cladocerans, copepods and enriched

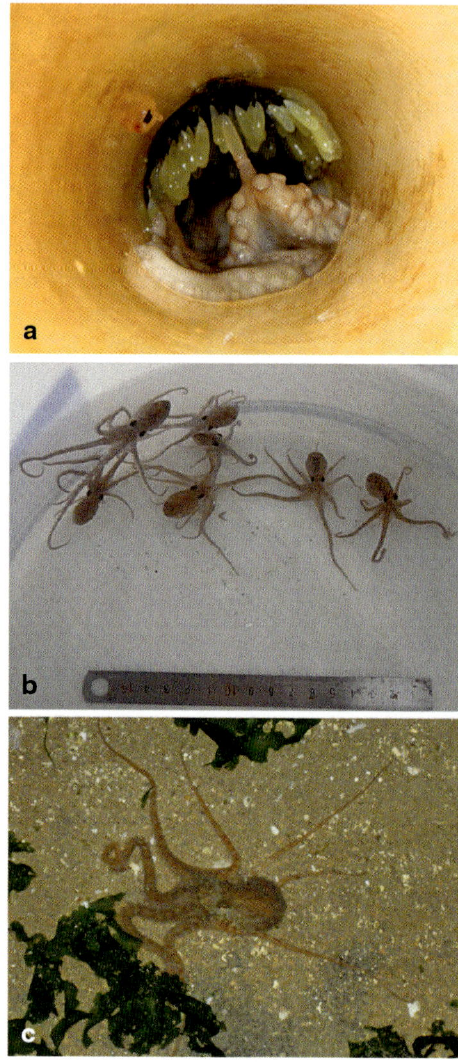

Fig. 22.7 Culture process of *Octopus minor*. **a** Spawning and female protection. **b** 1-month-old hatchlings. **c** 6–7-month-old octopus

Artemia nauplii are adequate initial feeds for the hatchling rearing. *H. sanguineus* of less than 4 mm body width is also used to feed 10-day-old hatchlings. The survival rate is 75 % after 1 month of rearing (Fig. 22.7b).

3. A mixed fresh diet such as juvenile crab, shellfish and shrimp becomes the main feed for juveniles of *O. minor*. After 6–7 months under indoor culture conditions (Fig. 22.7c), *O. minor* bigger than 100 g can be harvested for marketing, or transferred to outdoor ponds to continue the ongrowing process, or released to the sea for stocking or enhancement programmes.
4. Under indoor culture conditions, the longest period of survival for males is 250 days; the female lives 346 days, with a weight of 117.3 g after spawning.

5. The low fecundity and the long embryonic development period of *O. minor* severely restrict industrial cultivation. The species, however, is considered a very good candidate for restocking and enhancement programmes of marine invertebrates.

Acknowledgements The authors thank Dr. Chung-Cheng Lu and Mrs. Mairi Lu from National Chung Hsing University for checking the language and Hai-Bin Zhang from Monterey Bay Aquarium Research Institute for his comments. The study was supported by the grants from the National Marine Public Welfare Research Project (200805069, 201305043).

References

Cheng RB, Zheng XD, Lin XZ et al (2012) Determination of the complete mitochondrial DNA sequence of *Octopus minor*. Mol Biol Rep 39(4):3461–3470

Domingues PM, López N, Muñoz JA et al (2007) Effects of a dry pelleted diet on growth and survival of the Yucatan octopus, *Octopus maya*. Aquac Nutr 13:273–280

Dong G, Yang JM, Wang WJ et al (2013) Studies on biological zero and effective accumulated temperature for embryonic development of *Octopus ocellatus*. [In Chinese with English abstract.] Oceanol Limnol Sin 44:476–481

Dong ZZ (1988) Fauna Sinica. Phylum Mollusca. Class Cephalopoda. Science Press, Beijing, pp 181–812

Iwakoshi E, Hisada M, Minakata H (2000) Cardioactive peptides isolated from the brain of a Japanese octopus, *Octopus minor*. Peptides 21(5):623–630

Kang JH, Kim YK, Park JY et al (2012) Development of microsatellite markers to genetically differentiate populations of *Octopus minor* from Korea and China. Mol Biol Rep 39(8):8277–8286

Kim DH, An HC, Lee KH et al (2008) Optimal economic fishing efforts in Korean common octopus *Octopus minor* trap fishery. Fish Sci 74:1215–1221

Liu C (2013) Studies on culture of the life cycle of *Octopus minor* (Sasaki, 1920). Thesis, Ocean University of China, pp 27–32

Lu CC, Zheng XD, Lin XZ (2012) Diversity of Cephalopoda from the waters of the Chinese mainland and Taiwan. In: Lin M and Wang CG (eds) Proceeding of the 1st Mainland and Taiwan symposium of marine biodiversity studies. Ocean Press, Beijing, pp 76–87

Okutani T (2000) Marine mollusks in Japan. Tokai University Press, Tokyo, pp 1078–1079

Okutani T, Tagawa M, Horikawa H (1987) Cephalopods from continental shelf and slope round Japan. Japan Fisheries Resource Conservation Association, Tokyo, pp 164–165

Qian YS, Zheng XD, Li Q (2010) Analysis and evaluation of nutritive composition of *Octopus minor* in Lake Swan. [In Chinese with English abstract.] Mar Sci 34(12):14–18

Qian YS, Zheng XD, Liu C et al (2013) Studies on the reproductive habit and embryonic development of *Octopus minor* under the artificial conditions. [In Chinese with English abstract.] Oceanol Limnol Sin 44:165–170

Roper CF, Sweeney MJ, Nauen CE (1984) FAO Fisheries Synopsis, no 125, vol 3. FAO species catalogue, vol 3: Cephalopods of the world. An annotated and illustrated catalogue to species of interest to fisheries. FAO, Rome, p 210

Sasaki M (1920) Report of Cephalopods collected during 1906 by the United States Bureau of Fisheries Steamer "Albatross" in the NOIthwestern Pacific. Proc US Nat Mus 57:163–203

Segawa S, Nomoto A (2002) Laboratory growth, feeding, oxygen consumption and ammonia excretion of *Octopus ocellatus*. Bull Mar Sci 71:801–813

Seol DW, Lee JW, Im SY et al (2007) Clove oil as an anesthetic for common Octopus (*Octopus minor*). Aquac Res 38:45–49

Taki I (1944) Studies on the octopus (2) the sexes and the reproductive organs. [In Japanese with English abstract.] Jap J Malacol (The Venus) 13(5–8):267–310

Wang WJ, Yang JM, Zhou QL et al (2010) Reproductive behavior and process of embryonic development of *Octopus ocellatus*. [In Chinese with English abstract]. J Fish Sci Chn 17:1157–1165

Yamamoto T (1942) On the ecology of *Octopus variabilis typicus* (Sasaki), with special reference to its breading habits. [In Japanese with English abstract.] Jap J Malacol 12(1–2):9–20

Zuo ZR, Zheng XD, Yuan Y et al (2011) Development and characterization of 12 polymorphic microsatellite loci in *Octopus minor* (Sasaki, 1920). Conserv Genet Resour 3(3):489–491

Chapter 23
Octopus vulgaris. Paralarval Culture

José Iglesias and Lidia Fuentes

Abstract There have been many attempts worldwide to produce in captivity juveniles of *Octopus vulgaris,* one the most studied cephalopod species in the world because of its very strong market interest. This chapter reviews the different methods used to obtain and maintain broodstocks and the rearing technologies applied to the paralarvae. The main parameters and culture methods to rear the planktonic stage are discussed and a protocol for the rearing of the paralarvae is suggested. The main bottlenecks in the cultivation of this species are emphasized, and further research topics are suggested, including both technical and biological aspects.

In laboratory trials, the best growth and survival of the paralarval phase is currently achieved by feeding a mixed live diet composed of enriched *Artemia* and crustacean zoeae. However, this method is not transferable to a commercial scale as there is limited availability of live zoeae. In order to advance from a research to an industrial scale, it is essential to develop an inert diet with the appropriate nutritional composition to be supplied from an age of 1 month onwards. Another option would be to develop an appropriate enrichment protocol for *Artemia* so that its composition simulates more closely that of crustacean larvae or wild zooplankton.

A protocol for the first month of *O. vulgaris* paralarvae culture, which allows the production of good-quality individuals (in terms of dry weight and survival) to start the settlement process, is proposed. Relatively high survival rates and paralarvae dry weights of 1.3–1.8 mg can be attained after 1 month on a sole diet of *Artemia*. These weights are increased to 2.5–3.5 mg when that diet is supplemented with zoeae.

Keywords *Octopus vulgaris* · Common octopus · Paralarvae rearing · Paralarvae feeding · Culture conditions

J. Iglesias (✉) · L. Fuentes
Oceanographic Center of Vigo, Instituto Español de Oceanografía (IEO),
Subida a Radio Faro 50, 36390 Vigo, Spain
e-mail: jose.iglesias@vi.ieo.es

L. Fuentes
e-mail: lidia.fuentes.moledo@gmail.com

J. Iglesias et al. (eds.), *Cephalopod Culture,* DOI 10.1007/978-94-017-8648-5_23,

23.1 Introduction

23.1.1 Importance of this Species in the Market and Capture Methods

The common octopus, *Octopus vulgaris,* is a species with a substantial global market and consequent capture importance. It is widely distributed with a high market price due to its high demand as a food source. It is appreciated and used in food preparations in Asiatic, Mediterranean and Latin-American countries and can be found in the market as fresh, frozen, whole or sliced products. The edible part of octopus is very high (over 90 % of its body weight), resulting in a highly interesting product concerning yield.

It is a species that supports both industrial and artisanal fisheries. Local fishermen catch them using mainly hooks, lures and pots. In contrast, industrial fishermen catch large quantities in the oceanic sublittoral areas using trawls operated from large fishing boats. The declared annual world catches attributed to *O. vulgaris* declined from more than 100,000 t in the late 1970s to around 40,000–50,000 t during the period 2000–2010 (FAO 2012).

The common octopus is a highly valued species of great commercial interest in Spain. The octopus fishery has been overexploited in the past decades, which has forced the administration to regulate the fishery and, through research, to evaluate aquaculture techniques as an alternative source of supply.

Many of the species' biological features, such as direct embryological development, short life cycle, rapid growth and elevated food conversion index (Vaz-Pires et al. 2004), make it an excellent candidate for aquaculture. In addition, its high fecundity, rapid growth and high food conversion index, together with its high market value, make the common octopus a promising species for diversifying the aquaculture industry.

23.1.2 State of the Art

Research on rearing *O. vulgaris* paralarvae started in Japan when Itami et al. (1963) obtained the first benthic juveniles after 33 days at a mean temperature of 24.7 °C using shrimp (*Palaemon serrifer*) zoeae as prey. Later, Imamura (1990) and Hamasaki et al. (1991) reported successful rearing in 20 m^3 tanks to which *Artemia* and *Nannochloropsis* sp. were added, suggesting for the first time the possibility of mass-producing octopus paralarvae. Subsequent research in Japan focused on the production of juveniles for enhancement programmes (Okumura et al. 2005; Kurihara et al. 2006; Arai et al. 2008), using *Artemia* supplemented with frozen slices of Pacific sandeel (*Ammodytes personatus*) as food for the paralarvae.

In Europe, the first successful rearing trials were carried out by Villanueva (1994, 1995). He obtained benthic juveniles after 47 days using decapod crab zoeae as prey and a rearing temperature of 21.2 °C. Subsequently, Moxica et al. (2002)

increased survival and dry weight of the paralarvae at 1 month by using larger prey (adult *Artemia* of up to 2 mm and spider crab zoeae). Settlement, however, was not attained. The complete culture cycle at an experimental level was first achieved in 2001 (Iglesias et al. 2004), using both *Artemia* and spider crab zoeae as live prey. Later, Carrasco et al. (2005) achieved similar results using the same prey, but different cultivation systems with regard to water circulation, volume, colour and shape of tanks (see Table 23.1). Recently, research on *O. vulgaris* paralarval rearing has spread to many regions of Spain including Andalusia, Asturias (Carrasco et al. 2006), Canary Islands (Hormiga et al. 2010; Feyjoo et al. 2011; Almansa et al. 2012), Catalonia (Estévez et al. 2009), Galicia (Seixas 2009; Seixas et al. 2010; Fuentes et al. 2011) and Valencia (Viciano et al. 2011). An Italian group, Maricoltura Di Rosignano Solvay (Livorno), has also researched this field (Lenzi et al. 2006; De Wolf et al. 2011). In 2007, they successfully reared paralarvae through to 160-day-old juveniles using enriched *Artemia*. More detailed information on growth and survival rates of trials cited above is given in Sects. 23.5.3 and 23.5.4.

In 2005, an international workshop on *O. vulgaris* reproduction and paralarvae rearing was held in Vigo (Spain) in order to discuss the different methods used, identify the causes of larval mortality and establish future research priorities (Iglesias et al. 2007). A review of the biology of the planktonic stages of benthic octopuses was later published (Villanueva and Norman 2008). An international workshop on Latin-American cephalopod culture was held during 2008 in Puerto Montt (Chile), the conclusions of which were published by Uriarte et al. (2011). Recently, another workshop on cephalopod culture, organised by CIAC 2012 (Cephalopod International Advisory Council), took place in Florianópolis, Brazil, with the aim of defining the current status and research priorities of four cultured cephalopod species (*Sepia officinalis, Sepioteuthis lessoniana, Octopus maya* and *O. vulgaris*; Vidal et al. manuscript in elaboration).

23.2 Broodstock Conditioning

High fecundity is one of the characteristics that encourages *O. vulgaris* to be considered as a serious candidate for diversification in aquaculture. According to Mangold (1983), wild females can lay up to 500,000 eggs and, in captivity, an output of approximately 100,000 eggs per kg has been achieved (Iglesias et al. 1997).

After several decades of work on the culture of this species, there is now little difficulty in capturing wild subadult and adult individuals, acclimatizing them to captive conditions and obtaining viable egg masses. When keeping wild males and females together under suitable environmental conditions and providing them with shelters, nearly 100% of females can mature and lay egg strings (Iglesias et al. 2007).

Welfare considerations need to be taken into account when maintaining octopus in captivity. It has been argued that the European Union Directive 2010/63/EU on animal welfare should be applied to cephalopod breeding and experimentation in aquaculture research (Sykes et al. 2012). These authors suggested revisions to the

Table 23.1 Summary of paralarvae rearing conditions of *Octopus vulgaris* carried out by different research groups. (Adapted from Iglesias et al. 2007)

	Andalusia PA	Asturias CEP	Catalonia ICM-CSIC	Catalonia IRTA	Canary Islands ICCM	Galicia IEO	Galicia USC	Italia MRS	Brazil FURG	Japan YS
Reference	Iglesias et al. (2007)	Carrasco et al. (2006)	Villanueva (1994, 1995)	Estévez et al. (2009)	Iglesias et al. (2007)	Iglesias et al. (2007)	Seixas et al. (2010)	De Wolf et al (2011)	Iglesias et al. (2007)	Okumura et al. (2005)
Tank volume (L)	400	30	25–50	500	100	1,000	50	100–6,000	100	500
Tank colour	Black Grey	White	Black	Black	Grey	Black	White	Light-grey Black walls and white bottoms	Black	Orange
Tank shape	Cylindrical Rectangular	Parabolic	Cylindrical Parabolic	Cylindro-conical	Cylindrical	Cylindrical	Conical	Circular Slightly cylindro-conical	Cylindrical	Cylindrical
Water system	Open	Open (recirculation)	Open	Recirculation First week stagnant then open (0.7–1 L min-1)	Open 25 % day^{-1}	First week stagnant then semi-open (3–4 h = 100 % day^{-1})	10 % day^{-1}	100–200 % day^{-1}	Open (recirculation)	First 5 days stagnant then open
Aeration	Yes, gentle	Yes, gentle	No	–	Yes, gentle	Yes, intermediate	–	Yes, gentle	No	Yes, gentle
Light	Natural photoperiod	12 h L–12 h D 1 fluorescent 40 W	24 h Bulb 60w 900 Lx	16 h L–8 h D 500 Lx	Natural photoperiod	24 h 2 fluorescents 36 W 2,000 Lx	14 h L–10 h D Fluorescent daylight lamp	14 h L–10 h D Artificial light 60–250 Lx	10 h L–14 h D Natural + cold light	1 fluorescent 36W
Temperature (°C)	19–22	20–22	19–23	18	21.5–22.5	20–22	19–20	18.5–25	19–24	25
Clear/green water	Green *Tetraselmis* + *Isochrysis*	Clear	Clear	Green during first week	Clear	Green *Isochrysis* + *Nannochloropsis*	Clear	Green *Nannochloropsis* sp., *Isochrysis galbana* and a mix	Clear	Green Freshwater *Chlorella* sp.

Table 23.1 (continued)

	Andalusia IFAPA	Asturias CEP	Catalonia ICM-CSIC	Catalonia IRTA	Canary Islands ICCM	Galicia IEO	Galicia USC	Italia MRS	Brazil FURG	Japan YS
Paralarvae density (ind L^{-1})	20	25	13–48	20	25	5	10	1–15	5–30	3
Type and prey density (ind mL^{-1})	Zoeae (*Carcinus*, *Palaemon* and *Maja*) (<0.1) + *Artemia* + Moina (4–5 day old; 1.0)	Zoeae *Maja* (0.7–1) + *Artemia* (3 times week^{-1}) (0.5–0.7)	Zoeae (*Liocarcinus* and *Pagurus*), nauplii *Artemia* (2–6) and *Artemia* biomass	*Artemia* metanauplii (1–3) alone or mixed with zooplankton *Palaemon* sp. zoeae and copepods (0.05–1)	Zoeae *Grapsus* (15) + *Artemia* 72 h (2)	Zoeae *Maja* (0.01–0.1 when available) + *Artemia* (0.05–0.1)	Enriched *Artemia* juveniles (0.05)	Rotifers (5) and *Artemia* nauplii (1–2) and adult (0.05–0.1)	Crustacean zoeae, copepods, mysids, nauplii and adult *Artemia* (0.15–0.3)	*Artemia* nauplii + fish flakes from fifth day
Prey size [most cases TL (mm)]	Zoeae: 0.8–1.0 Moina: 1.0–1.2 *Artemia*: 1–3	Zoeae: 1 *Artemia* retained in 300 μm sieve	Zoea: 1.3–3.1 *Artemia* nauplii to 1–3 mm *Artemia* biomass	Zooplankton fraction < 0.4 mm 5 days *Artemia*	Zoeae: 1.5 *Artemia*: 0.85	Zoeae: 1 *Artemia*: 2–3	*Artemia*: 1.5–2.8	*Artemia* nauplii: 0.75–0.85 Adult *Artemia*: 12–20	0.4–8	*Artemia:* 0.650 Fish flakes: 10–20 mm diameter, 0.5–1 mm thickness
Artemia enrichment	Reared and enriched with *Tetraselmis* + *Isochrysis*, SuperSelco Prolon	Reared and enriched with *Tetraselmis*	DCSuperSelco, Methionine	Reared for 5 days with *Tetraselmis suecica* and *Isochrysis galbana*	*Artemia* enrichment (A_1 Selco Inve)	Reared in commercial cereal flour, enriched with *Nannochloropsis* (5×10^6 cells mL^{-1})	Reared with *Rhodomonas lens* and *Isochrysis galbana* and then enriched with different procedures	A1DHASelco® by Inve; *Isochrysis* and Prolon®	Super SELCO and DHA SELCO Inve	Fish egg powder (Plus Aquaran, BASF Japan)
Sampling	Every 7–10 days	Every 10 days	Every 7–10 days	Every 10 days	Every 7 days	Every 7 days	Days 15, 25 and 35	Every 10 days	Daily up to day 7 and every 5 days thereafter	Every 5 days

Table 23.1 (continued)

	Andalusia IFAPA	Asturias CEP	Catalonia ICM-CSIC	Catalonia IRTA	Canary Islands ICCM	Galicia IEO	Galicia USC	Italia MRS	Brazil FURG	Japan YS
Survival (%)	5–15 (day 35)	89.6–93.5 (day 20) and 3.4 (day 60)	0.8 (day 60) with zoeae, and 54 (day 20) with *Artemia* nauplii (with poor growth)	–	11–27 (day 30)	31.5 (day 40)	35–53 (day 15) 7–20 (day 25)	8 (day 45)	1–20 (day 40) with *Artemia* and 20–39 (day 40) with *Artemia* and copepods	10–30 (day 30)
Cleaning	Daily tank bottom siphoning	Every 20 days changing tank by pipetting and checking the survival	Daily tank bottom siphoning	–	No bottom cleaning	–	Siphoning	–	Bottom siphoning daily or on alternate days	Daily tank bottom siphoning after fifth day

TL Total Length, *IFAPA* Instituto de Investigación y Formación Agraria y Pesquera, *CEP* Centro de Experimentación Pesquera, *ICM-CSIC* Instituto de Ciencias del Mar-Consejo Superior de Investigaciones Científicas, *IRTA* Institut de Recerca i tecnología Agroalimentàries, *ICCM* Instituto Canario de Ciencias Marinas, *IEO* Instituto Español de Oceanografía, *USC* Universidad de Santiago de Compostela, *MRS* Maricoltura di Rosignano Solvay, *FURG* Universidade Federal do Río Grande, *YS* Yashima Station, *L* light, *D* darkness, *ind* individuals,—no information animal welfare legislation, the definition of live cephalopods, stress, pain and suffering. Although some recent reports have been published on this topic, research on appropriate culture conditions, behaviour and stress responses must be carried out in the near future. For more information on this topic, see Chap. 6.

Iglesias et al. (2007) made a comparative analysis of the different broodstock conditioning systems used in the world and concluded that whereas methods of capture, transport, food supply and light intensity are similar among different research groups, male to female ratios and the holding temperatures (14–25 °C) may differ widely. With regard to this subject, it is possible to establish the following recommendations.

23.2.1 Broodstock Capture and Transport

Trawl nets enable the capture of large numbers of octopus, but can both harm the individuals and cause a negative environmental impact. Consequently, selective fishing methods like creels or individual traps are recommended for the capture of spawners.

Individual mesh bags or separate containers can be used during and after transportation in order to avoid attacks between individuals, stress and subsequent mortality (Fuentes et al. 2005). Oxygen supply is recommended, particularly when stocking densities are high.

23.2.2 Food Supply

Broodstock diet can influence the biochemical composition and biometrical relationships of the newly hatched paralarvae (Quintana et al. 2007, 2009; Márquez et al. 2013). Frozen and fresh crustaceans and fish of low commercial value are usually used as food with optimal results (Iglesias et al. 2007; Quintana 2009; Estefanell 2012). Adult females refuse feed when spawning is imminent (see Chap. 1); in consequence, the quantity of food should be reduced during this period to maintain good water quality.

23.2.3 Sex Ratio

Some authors (Villanueva 1995; Okumura et al. 2005) capture only wild mature females during the natural spawning season to obtain eggs in captivity; these females are usually already fertilised from previous matings and can preserve viable sperm in their oviducts for long periods. In these cases, males are not needed.

Nevertheless, most authors obtain viable eggs by keeping males and females together in a ratio of 1:3 (Iglesias et al. 2000, 2007).

23.2.4 Physical and Chemical Parameters

Physical and chemical parameters in broodstock tanks are important determinants of successful egg laying. Density should not exceed 5 kg m^{-3} and tanks should be provided with filtered seawater and a minimal renewal rate of 400–800% day^{-1} in order to maintain abiotic water parameters at optimal levels (Iglesias, personal communication). Dissolved oxygen levels should be kept around 100% saturation. Temperature should follow that of the natural seawater temperature cycle as much as possible and should not drop below 14 °C nor exceed 25 °C (Iglesias et al. 2007). Similarly, water salinity should simulate local seawater values. Semi-dark conditions are commonly used, but natural photoperiod with shaded natural light is also utilized (De Wolf et al. 2011). Individual shelters (terracotta or polyvinyl chloride (PVC) tubes) should be provided to facilitate egg laying.

23.3 Spawning Process

23.3.1 Female Conditions

Each spawning female with its strings of eggs should be transferred to an individual tank to avoid being disturbed by conspecifics and to facilitate the counting of eggs laid and hatched paralarvae. Tank volume should be 200–500 L, and temperature must be the same as the broodstock tank.

Females take care of egg masses by oxygenating and cleaning them throughout the embryonic development process. They do not feed during this period and consequently reduce their weight; a 4-kg female, for example, can lose 30% of its weight during this process (Fig. 23.1). Under suboptimal or stressful circumstances, females may leave the shelter and abandon the egg clusters. In this case, hatching percentage can fall to as little as 50% due to egg detachment, fungal infection, etc. (Lenzi et al. 2002). Exceptionally, another different female from the broodstock may take over the egg-caring role (Iglesias, personal communication). A special incubator to maintain eggs without a female was patented by Rosas et al. (2010). It was originally developed for *O. maya,* a large-egg species, but is currently being tested for *O. vulgaris* in the Austral University of Chile. Simple PVC tube systems with continuous water flow and aeration have also been successfully tested at the Spanish Institute of Oceanography.

23.3.2 Egg Handling

A weekly check of each egg cluster should be established to estimate the time in days necessary to hatch according to the incubation temperature. Using this method, the hatching date can be predetermined in order to estimate in advance the needs for

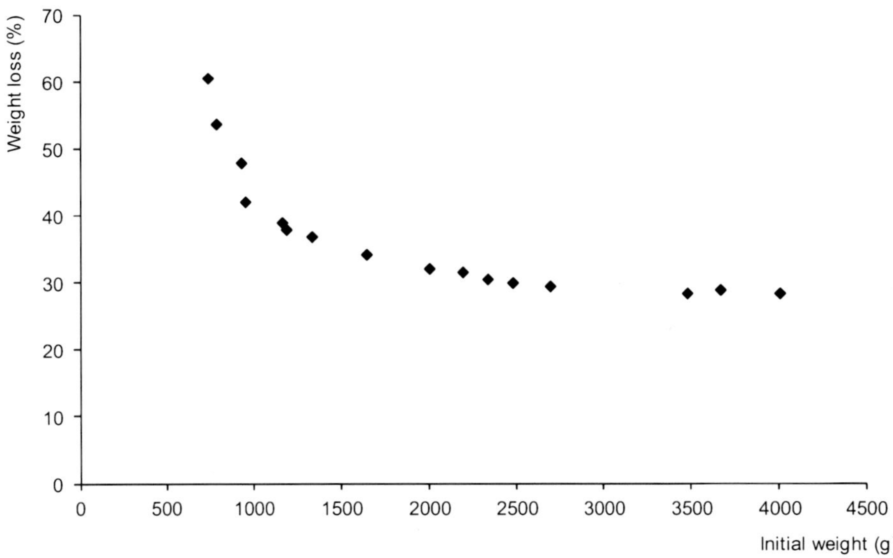

Fig. 23.1 Weight loss (%) of *Octopus vulgaris* breeding females as a percentage of their initial weight (after egg laying) during the egg-care period

larval rearing (phytoplankton and *Artemia*; Boletzky 1987). Figure 23.2 shows the evolution of the embryonic development of *O. vulgaris* at 18 °C, which takes around 38 days to hatch; the different stages have been identified according to Naef (1928).

Transportation of egg strings should be performed together with the female in the original spawning shelter, and swinging movements that would cause damage to the egg mass should be avoided. In order to avoid premature hatching, this process must be performed before the second embryonic inversion occurs (Fig. 23.2f). Due to high oxygen consumption by eggs (Parra et al. 2000), additional oxygenation should be used during transportation. Drastic fluctuations in other parameters (e.g. temperature, salinity, pH, light intensity) should be avoided to prevent premature hatching.

In cases when egg strings need to be transported without the female, they should be placed in plastic containers filled with oxygen supersaturated seawater and a low temperature be maintained. Strings should be held in a vertical position by hanging them from the container cap.

23.4 Paralarvae Culture Conditions

Since the 1990s, many attempts have been made by several research groups to determine the feasibility of paralarval culture, and a wide range of rearing conditions have been tested (reviewed by Iglesias et al. 2007; De Wolf et al. 2011). Table 23.1

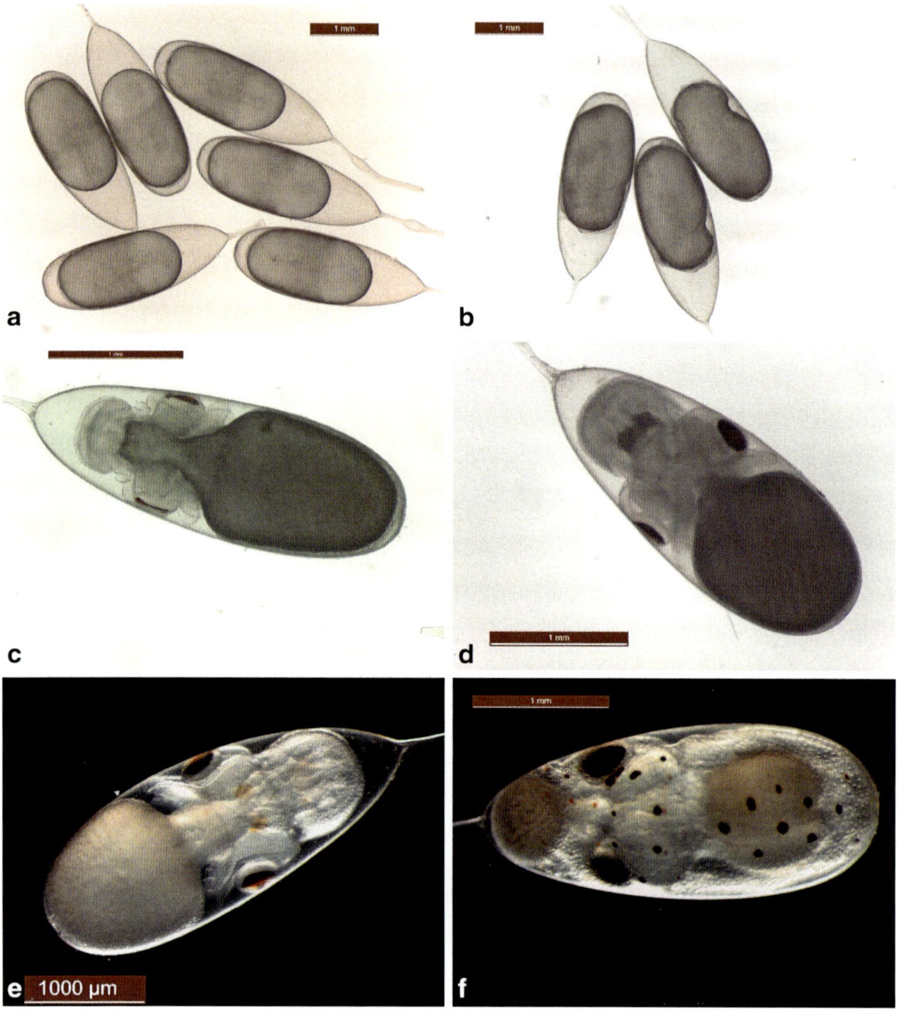

Fig. 23.2 Embryonic development of the common octopus *Octopus vulgaris* at 18 °C. **a** Prior to first inversion (Naef IV–VI stage). **b** Post first inversion (Naef VIII stage). **c** Embryo differentiation (Naef XI stage). **d** Naef XII–XIV stage. **e** Chromatophores and internal yolk (Naef XV stage). **f** Post second inversion (Naef XIX stage). (Photographs by M. Nande)

shows the culture conditions used, including the different feeding strategies, prey concentrations and larval densities.

The main problem that hampers the success of commercial culture of octopus is the high mortality rate observed during the first 2 months of paralarval rearing. The lack of a standardised culture system and the absence of appropriate food sources that fulfil nutritional requirements have been identified as two possible factors responsible for this mortality (Iglesias et al. 2007). Even if survival rates during the

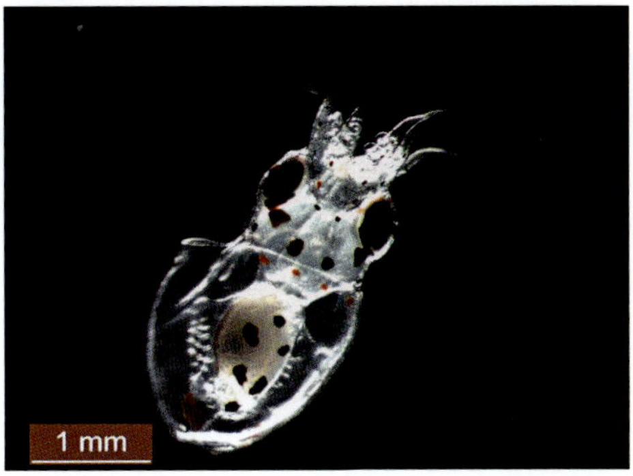

Fig. 23.3 Ventral view of recently hatched *Octopus vulgaris* paralarvae (total length: approximately 3 mm; 3 suckers per arm). (Photograph by M. Nande)

planktonic stage have increased considerably in the past decades, successful juvenile settlement is still very difficult to achieve.

23.4.1 Paralarvae Collection and Transfer

Recently hatched *O. vulgaris* paralarvae (Young and Harman 1988) measure around 3 mm TL, have three suckers per arm and an individual dry weight of 0.20–0.35 mg (Villanueva and Norman 2008; Iglesias et al. 2007; Fig. 23.3).

The transfer of paralarvae to rearing tanks should be accomplished carefully in order to avoid stress or damage. When concentrations of paralarvae are high, they can be collected with 1–2 L plastic beakers, but when concentrations are low a 30 cm diameter PVC collector with 0.5 mm mesh can be used to collect them. Paralarvae can be counted either individually or by volumetric estimation (Iglesias et al. 2006; Fuentes et al. 2011).

23.4.2 Tank Colour and Volume

Dark tanks are used by most authors, even though observing paralarvae in these tanks is quite difficult. Other authors have also reported very good results for growth and survival using completely white tanks (Carrasco et al. 2005) or tanks with black walls and white bottom (De Wolf et al. 2011).

There is a general agreement that good growth and survival can be achieved in 500 and 1,000 L cylindro-conical tanks. Sánchez et al. (2010) concluded that growth is positively related to tank volume; paralarvae reared in 1,000 L tanks attained a dry weight of 1.73 ± 0.27 mg after 21 days, significantly greater than the 1.44 ± 0.33 mg obtained in 100 L tanks. De Wolf et al. (2011) also obtained better results using larger tanks attributing the differences to the less dramatic fluctuations in physical conditions in larger tanks. Seixas (2009) and Villanueva et al. (2002), using small rearing tanks of 25–50 L, obtained dry weights of 0.83 and 0.90 mg, respectively, for 1-month-old paralarvae. On the other hand, Moxica et al. (2006) and Viciano et al. (2011), working with 1,000 L tanks and using enriched *Artemia* (cultured with *Isochrysis* sp. and further enriched with *Nannochloropsis* sp.), obtained dry weights of 1.76 mg and 1.88 mg, respectively, for the same-aged paralarvae.

23.4.3 Physical and Chemical Culture Parameters

23.4.3.1 Water Circulation

Stagnant water is commonly used for the first week to maintain a green-water system; thereafter, some authors exchange water for 4 h day^{-1}, corresponding to a 100% day^{-1} renewal rate (Iglesias et al. 2007; Viciano et al. 2011), whereas others recommend maintaining a constant water flow to give an exchange rate of at least 150% day^{-1} (De Wolf et al. 2011). The water inlet should produce a mild tangential circulation of the surface water. However, both lateral and centrally bottom-placed water outlets have been used by some investigators.

In order to obtain a homogeneous distribution of paralarvae and live prey in the culture tanks, gentle aeration can be supplied by air stones (Iglesias et al. 2004; Okumura et al. 2005) or open tube aeration can be done (De Wolf et al. 2011). Currents that result in the accumulation of paralarvae in very small areas should be avoided.

It is also important to prevent the production of small air bubbles that can be trapped in the paralarvae mantle. For this reason, special devices for the distribution of the incoming water, such as multiple superficial inlets (Villanueva 1995) or bottom water inlets combined with a superficial water outlet (Carrasco et al. 2005, 2006), have been used. Nevertheless, the use of a high water flow can increase the risk of skin damage and arm erosion, resulting in an increased mortality of the paralarvae (Vidal et al. 2002).

23.4.3.2 Light

A wide range of light conditions (natural light, incandescent bulbs, fluorescent tubes) has been used in paralarval culture. Optimal light conditions seem to depend on the type of tanks that are used. In general, higher light intensities and longer photoperiods are used in black tanks than in light-coloured tanks or white-bottom tanks. Surface light intensity in the rearing tank should be 500–700 Lx when using black tanks (Iglesias and Fuentes 2013), whereas other authors recommend using lower light intensities (60–250 Lx for a 14 h L (light):10 h D (dark) photoperiod) in black-walled and white-bottom tanks (De Wolf et al. 2011). Garrido et al. (2012) studied the effect of different types of lighting on the growth and survival of *O. vulgaris* paralarvae.

23.4.3.3 Temperature

The change of temperature between the broodstock tank and the paralarvae rearing tank must be gradual (an increase of approximately 1 °C per day). Temperature is the most important determinant of development and growth of paralarvae (Mangold and Boletzky 1973). Paralarval rearing temperatures cited in the literature range between 19 and 25 °C (Iglesias et al. 2007), but temperatures of 20–22 °C are recommended to obtain optimal growth and survival. High daily temperature fluctuations should be avoided during paralarval culture.

23.4.3.4 Chemical Parameters

The most important water quality parameters in the larval rearing phase are salinity, pH and concentrations of dissolved oxygen, ammonia, nitrite and nitrate.

Octopuses show very low tolerance to low salinity; therefore, seawater of around 32–35 psu should be used. Salinity should be as constant as possible because sudden changes are not tolerated by paralarvae. Rearing places close to rivers and freshwater sources need to take into account this parameter.

Dissolved oxygen is crucial for gas exchanges and depends on temperature. Optimal levels in the paralarvae rearing process should be kept between 6 and 8 mg L^{-1}, but should not be allowed to fall below 4 mg L^{-1}. Cerezo-Valverde and García-García (2005) determined optimal oxygen saturation levels for subadult and adult individuals to be between 100 and 65 % and suboptimal saturation levels between 65 and 35 % (dangerous below 35 %) for temperatures between 17 and 20 °C.

pH, nitrite and ammonia levels need to be monitored according to water renewal but at least once a week. Feyjoo et al. (2011) determined the acute toxicity of unionized ammonia and nitrite on newly hatched *O. vulgaris* paralarvae. The lethal concentration 50 (LC50) value after 24 h of exposure was 10.7 ppm for ammonia and 19.9 ppm for nitrite. This suggests that paralarvae are quite resistant to free ammonia compared to marine fish larvae, but much less resistant to nitrite. At concentrations much lower than the LC50 values, negative effects are observed on both prey intake and chromatophore activity (Feyjoo et al. 2011).

Gas supersaturation can explain the extensive mortality that sometimes happens during the early life stages in intensive production of marine species (Gunnarsli et al. 2008). Air bubbles can be formed in gills, fins, skin and blood of fishes (gas bubble trauma), whilst in the case of *O. vulgaris* paralarvae, bubbles usually appear inside the mantle cavity. This phenomenon can be related to water circulation, temperature and chemical parameters; for example, it can appear through the mechanical process of heating the water, long pipe runs, pump cavitation, etc. In order to prevent water supersaturation (mainly nitrogen gas), it is recommended to have a method of trickling water over a large surface area, as in a packed-column aerator (Hargreaves and Tucker 1999).

23.5 Paralarvae Feeding

23.5.1 *Prey Size*

Paralarvae can start feeding from the first day of life, but usually a greater number of attacks are recorded 2 days after hatching at temperatures of 18–20 °C (Iglesias et al. 2006). These authors also reported that larger *Artemia* (1.4±0.4 mm) were clearly preferred to smaller *Artemia* (0.8±0.1 mm) at first feeding. Nevertheless, a wide range of prey sizes has been used in research on paralarvae rearing. Navarro and Villanueva (2003) used *Artemia* nauplii of 450–750 µm, in the first few weeks of culture, while others (Moxica et al. 2002; Iglesias et al. 2004; Carrasco et al. 2005; Estévez et al. 2009) used *Artemia* of 2 mm length or bigger.

In order to investigate the effect of prey size on growth, Fuentes et al. (2009) compared the use of small *Artemia* enriched for 24 h with *Isochrysis galbana* (0.7 mm TL) with that of larger *Artemia* grown for 4 days with the same microalga (1.5 mm TL) during a 30-day larval rearing experiment. There were no significant differences in growth during the first 15 days, but growth was significantly faster when larger *Artemia was* used as diet during the next 15 days of trial (Table 23.1). Considering these results, for experimental larval rearing, it is recommended that *Artemia* of 0.5–0.7 mm length are used for the first 15 days followed by *Artemia* of 1.5–2 mm length.

23.5.2 *Paralarval Nutritional Requirements*

Navarro and Villanueva (2000, 2003) considered that a lack of balance in lipid and fatty acid composition and a deficiency in polyunsaturated fatty acids (PUFAs) of the food supplied could be responsible for the high mortality and low growth during *O. vulgaris* paralarvae rearing. Okumura et al. (2005) used *Artemia* supplemented with flakes of *A. personatus* to increase the proportion of PUFAs in the diet. Seixas (2009) and Seixas et al. (2010) fed *Artemia* enriched with different microalgae species rich in PUFAs, whereas Bersano (2003) and Estévez et al. (2009) used live copepods and other zooplankton groups, respectively. Besides PUFAs, amino acids seem to be another important component in the diet. Villanueva et al. (2004) found that lysine, arginine and leucine represent about half of the total essential amino acids in cephalopod hatchlings. In addition, Villanueva and Bustamante (2006) studied the importance of essential elements in the diet and detected a higher content of copper in *O. vulgaris* hatchlings and wild juveniles than in *Artemia*-fed paralarvae. An approach to meeting their vitamin A and E requirements in culture has been published by Villanueva et al. (2009). Further information on nutritional requirements can be found in Chap. 5.

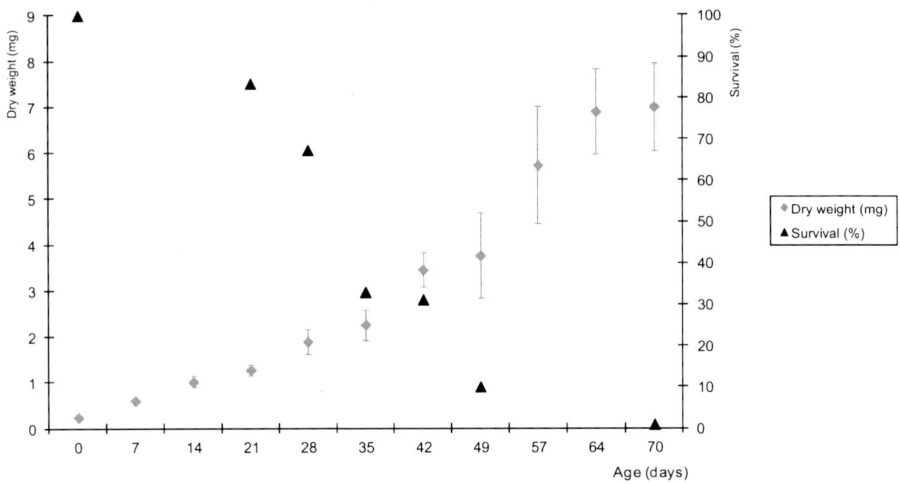

Fig. 23.4 Growth of *Octopus vulgaris* during the first month obtained using *Artemia* cultured for 4 days with *Isochrysis galbana* and further enriched with *Nannochloropsis* sp. as food. (From Moxica et al. 2006)

23.5.3 *Artemia* Enrichment

Even though *Artemia* is easily available and well accepted by *O. vulgaris* paralarvae, Navarro and Villanueva (2003) have reported that *Artemia* per se has an inadequate lipid composition, the docosahexaenoic acid (DHA) and eicosapentaenoic acid (EPA) content being especially low.

The only experiment in which rearing to the adult stage has been reached on a sole diet of *Artemia* was carried out by Moxica et al. (2006). They obtained 67% survival of paralarvae (dry weight of 1.89 mg) after 1 month of culture (Fig. 23.4), 0.9% after 70 days, three individuals at 3 months and finally one adult octopus. In that experiment, a green-water system based on *Nannochloropsis* sp. was used, and the *Artemia* was cultured for 4 days with *I. galbana* and further enriched with *Nannochloropsis* sp. Fuentes et al. (2011) suggested that the combination of the DHA provided by *I. galbana* and the high EPA content present in *Nannochloropsis* sp. could cover the basic PUFA requirements of octopus paralarvae.

Hamasaki and Takeuchi (2000) and Hamasaki and Morioka (2002) also reported successful results when adding *Nannochloropsis* sp. to the culture tank of planktonic larvae and as food for *Artemia*. Hamasaki and Takeuchi (2000), using *Artemia* biomass (1–2.7 mm) with or without *Nannochloropsis* sp. in the rearing tank, attained 88% survival at day 24; Hamasaki and Morioka (2002) obtained 62.5% survival rate at day 40, using *Artemia* biomass (1.5–2 mm) fed with *Nannochloropsis* sp. Another possible beneficial aspect of adding *Nannochloropsis* sp., previously cited for other marine species (Skiftesvik et al. 2003), is its inhibitory effect on harmful microflora in the culture tank. Besides the nutritional role, De Wolf et al. (2011) suggested that microalgae may generate a shading effect resulting in better prey visualisation and a more homogeneous distribution of the paralarvae in the water column.

A different *Artemia* enrichment process has been proposed by Seixas et al. (2010). They suggested that a microalgae combination of *I. galbana* and *Rhodomonas lens* was the best because it provides a high level of PUFAs (*I. galbana*) and a very high level of protein (*R. lens*).

The higher protein to lipid ratio present in *Artemia* can also positively affect paralarvae growth (Fuentes et al. 2011). Seixas et al. (2010) reached a similar conclusion and suggested that in order to sustain good paralarvae growth, a minimum dietary protein to lipid ratio should be maintained.

Other authors have used enrichments other than microalgae. Kurihara et al. (2006) enriched *Artemia* with fish egg powder (Plus Aquaran, BASF Japan), which was supplemented with frozen *A. personatus* flakes (improving DHA content), and obtained 10 % survival rate at day 42. Hamasaki and Takeuchi (2001) used *Artemia* biomass (2 mm), enriched or not with yeast or shark egg powder, with *Nannochloropsis* sp. in the tank and obtained 24 % survival rate at day 20.

23.5.4 Crustacean Zoeae as Optimal Prey

Very few studies of the natural prey of *O. vulgaris* paralarvae have been published because almost no early stages of this species have been found in nature. Roura et al. (2012) used a polymerase chain reaction (PCR)-based method with group-specific primers selected to identify prey consumed by *O. vulgaris* paralarvae in the wild; they identified 12 families of crustaceans (11 belonging to the order Decapoda and 1 to the order Euphausiacea) and two families of fishes (Gobiidae and Carangidae). Additionally, Couto (2012) found that a high number of attacks on cirripede larvae were registered when live wild zooplankton was provided to 2-day-old *O. vulgaris* paralarvae in first feeding laboratory experiences.

As an approach to determining the hypothetical optimal composition of prey for *O. vulgaris* paralarvae, Navarro and Villanueva (2000, 2003), Villanueva and Bustamante (2006) and Villanueva et al. (2004, 2009) studied the lipids, fatty acids, amino acids, essential and nonessential elements and vitamins A and E of mature ovaries, eggs, early stages and juveniles of wild *O. vulgaris*.

In general, the best growth and survival of paralarvae in culture conditions were attained using crustacean larvae alone or as a complement to enriched *Artemia*. Villanueva and Norman (2008) provided a detailed review of prey offered, survival and duration of the planktonic period. Itami et al. (1963), using shrimp (*P. serrifer*) zoeae of 2–4 mm body length as prey, obtained the first benthic juveniles (10–15 mm TL) after 33 days, with survival rates of 9 % at 40 days and 4 % after 3 months of culture. Villanueva (1994, 1995) obtained benthic juveniles with settlement starting after 47 days with a survival rate of 9 %. Wet weight after 60 days was 173.2 mg. They used crustacean zoeae (*Liocarcinus depuratus* and *Pagurus prideaux*) as prey. Afterwards, Moxica et al. (2002), using adult *Artemia* up to 2 mm and spider crab zoeae (*Maja brachydactyla*) as food, obtained 8.3 % survival rate after 1 month and 0.2 % after 56 days when the mean dry weight of the paralarvae was 9.21 ± 0.92 mg.

The complete culture cycle at an experimental level was attained for the first time in 2001 (Iglesias et al. 2004) using both *Artemia* and spider crab (*Maja*

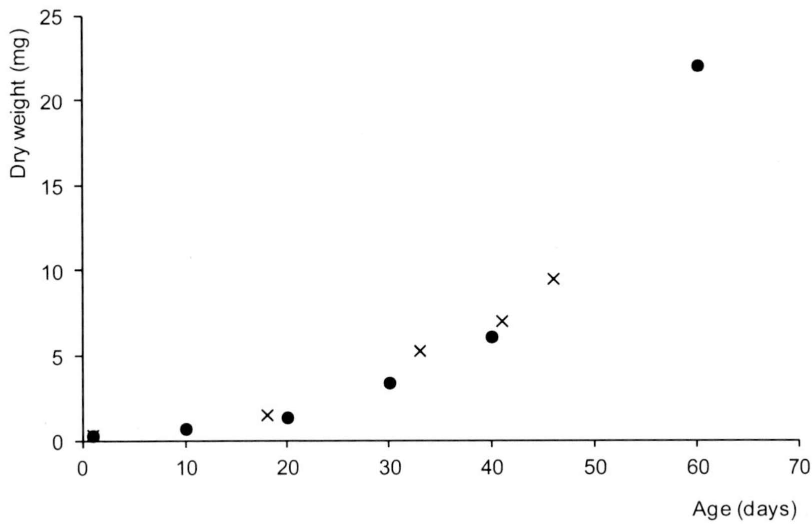

Fig. 23.5 Comparison of growth obtained by (*x*) Iglesias et al. (2004) at 22.5 °C and (•) Carrasco et al. (2006) at 21.2 °C using similar preys (*Artemia* and zoeae crustacean) and different rearing systems

brachydactyla) zoeae as live prey, and obtaining a paralarvae dry weight of 9.5 ± 1.9 mg and a survival rate of 31.5 % after 40–45 days. Carrasco et al. (2006) using the same prey but different rearing systems (Fig. 23.5) achieved 13 % survival rate at day 40 (5.40 ± 0.20 mm dorsal mantle length and 6.10 mg dry weight). Settlement began from day 52 and survival at day 60 was 3.4 % when the paralarvae were 6.54 ± 0.13 mm long and weighed 22.02 ± 3.08 mg. Almansa et al. (2012) used a combination of *Artemia* juveniles and *Grapsus adcensionis* zoeae as live prey for first feeding and obtained one *O. vulgaris* juvenile of 7 months old (ventral mantle length 23 mm, 10.4 g).

Therefore, a mixed live diet, composed of enriched *Artemia* and crustacean zoeae, is currently the most balanced diet for achieving the best growth and survival results in the paralarval phase (Fig. 23.5). However, this methodology is not readily transferable to a commercial level since there is limited availability of live zoeae, and it is therefore difficult to expand the culture methodology to an industrial scale.

23.6 Paralarvae Rearing Protocol

In accordance with data reported in this chapter, a protocol to attain good-quality 30-day-old *O. vulgaris* paralarvae is proposed.

Use a minimum of three replicate 500–1,000 L cylindro-conical tanks. Use stagnant water for the first week, after which open the circuit for 4 h day^{-1} (100 % day^{-1} renewal rate). The number of replicates will ensure a suitable statistical treatment of the results. The recommended colour for tank walls and bottom is black, although a white (or clear) bottom can be used in order to improve observation of the paralarvae. Temperature should be kept at 20–22 °C and salinity 32–35 psu. Tangential surface water input and drainage consisting of a central cylindrical pipe with 250 µm mesh are suggested. Another option is to use stagnant water during the day and open the flow during the night with a 500-µm outlet mesh in order to keep a more homogeneous enriched *Artemia*. Surface cleaners and moderate central aeration are recommended. Surface light intensity should be 500–700 Lx for 24 h photoperiod when using black wall and black bottom tanks, whereas light intensity can be 60–250 Lx for a 14 h:10 h (L:D) photoperiod when a clear bottom tank is used. A concentration of 1×10^6 cells mL^{-1} of *Nannochloropsis* sp. should be used in the culture medium (green-water system), and paralarvae concentration should be 10 individuals L^{-1}. Paralarval prey should consist of 24 h *Artemia* (0.5 *Artemia* mL^{-1}) enriched with *I. galbana* at a concentration of 0.75×10^6 cells mL^{-1} for the first 15 days, followed by larger *Artemia* (1.5–2 mm TL) cultured for 4–5 days with *I. galbana* and further enriched with *Nannochloropsis* sp., at a concentration of 1×10^7 cells mL^{-1}, keeping a prey concentration of 0.3 *Artemia* mL^{-1}. When zoeae of spider crab are used in co-feeding with *Artemia,* they should be added at a concentration of 0.05–0.1 individuals mL^{-1}, at least 3–4 days per week but preferably every day.

For comparative purposes, total length and dry weight of 20 paralarvae should be recorded at the beginning of experiments and periodically (fortnightly is recommended) in each rearing trial.

23.7 Settlement Process

Around 65–75 days after hatching (at 20 °C), when paralarvae have between 17 and 20 suckers per arm, they change their pelagic behaviour to a benthic life. At this stage, paralarvae start crawling on the walls and bottom of the tank, moving by attaching their arms to the substrate. This period, which lasts for 2 weeks before they fully adapt to benthic stage, is known as settlement.

During this phase, it is necessary to change feeding and habitat conditions. *Artemia* concentration should be gradually reduced while the benthic juveniles are increasingly fed with mussel, crab and sea urchin muscle and gonad (Iglesias et al. 2004) and/or frozen mysidaceans (Carrasco et al. 2005). In both cases, the authors placed small-angle or T-shaped PVC pieces on the bottom of the tanks as shelters and the bottom substrate was modified by the inclusion of pebbles (Fig. 23.6).

23.8 Conclusions

Reproduction in captivity has proved feasible on an experimental scale, but the high mortality observed during the paralarval and settlement stages are still the main constraints to the industrial cultivation of the common octopus.

Fig. 23.6 *Octopus vulgaris* 100-day-old juvenile obtained in captivity (Iglesias et al. 2004) in the artificial habitat (PVC tubes and pebbles) used during the settlement period

A mixed live diet of enriched *Artemia* and crustacean zoeae is currently the most successful in terms of growth rate and survival during the paralarval phase. However, this feeding protocol is not transferable to a commercial level due to the limited availability of live zoeae.

A protocol for the first 30 days of *O. vulgaris* paralarvae culture is proposed that supports the development of individuals with good fitness (in terms of dry weight and survival) through to the settlement process. Using this method, relatively high survival rates and paralarvae dry weights between 1.3 and 1.8 mg can be attained after 1 month when enriched *Artemia* is used as the sole diet, but these values increase up to 2.5–3.5 mg when zoeae are used in a co-feeding regimen with *Artemia*.

23.9 Trends in Research and Industrial Level

The following research topics are recommended.

23.9.1 Wild Paralarvae and Natural Prey Studies

It is strongly recommended to study further the biochemical composition of wild paralarvae and their natural prey preferences and behaviour. It is also of interest to complete sequence information included in molecular databases (GenBank and

Barcode of Life) of zooplankton species that are potential natural prey of *O. vulgaris* paralarvae in order to allow its molecular identification and traceability.

23.9.2 Nutrition

In order to move from the actual research situation to the next phase of industrial cultivation of the common octopus, it is essential to develop an inert diet to be supplied from 1 month of age onwards, with an appropriate nutritional composition that meets the requirements of the octopus paralarvae (PUFAs, lipids, proteins, protein to lipid ratio, amino acids, essential elements, vitamins, etc.), as well as a good level of acceptance, buoyancy, leaching, etc. Another issue to be addressed would be that of developing appropriate enrichment protocols for *Artemia* that would make its composition to more closely match that of crustacean zoea or natural wild zooplankton.

23.9.3 Zootechnical Improvements

Standardized methods should be established for the whole rearing process including the live prey period, weaning and settlement.

The settlement process needs to be further studied in order to better define requirements for food, shelter and type of substrate through a better understanding of juvenile behaviour.

23.9.4 Water Quality, Microbiology and Pathology

Further investigations of the chronic toxicity of nitrite compounds and other metabolites are strongly recommended in order to further improve water treatment systems. Gas supersaturation effects should be analysed in more detail. Specific pathogens and bacterial growth in the rearing systems should be also taken into account and identified. The effect of the application of probiotics in this context should also be evaluated in *O. vulgaris* paralarvae rearing.

23.9.5 Histology and Paralarvae Development

Studies of the anatomical changes of internal and external organs during the process of paralarval rearing are essential in defining welfare and fitness issues. Other aspects related to paralarval morphology, physiology, immunology, growth and internal rhythms are required to more fully understand paralarvae quality.

23.9.6 Welfare and Ethical Considerations

Although some recent reports have been published on this topic, research on appropriate culture conditions, behaviour and stress responses should be carried out in the near future.

References

Almansa E, Shcherbakova A, Jiménez P, Rodríguez C, Riera R, Felipe BC, Martín MV, Andrade JP, Sykes A (2012) Effects of different tank volumes and feeding regimes on growth, survival and lipid composition of *Octopus vulgaris* paralarvae. Book of abstracts, Aqua 2012. Czech Republic, Prague, p 52

Arai D, Kurihara A, Komi R, Iwamoto A, Takeuchi T (2008) Effect of feeding various amounts of Pacific sandeel flakes on growth survival and carcass fatty acid composition of common octopus *Octopus vulgaris* paralarvae. Aquac Sci 56(4):595–600

Bersano JGF (2003) Intensive cultivation of the calanoid copepod *Acartia tonsa*. A potential source of live food for aquaculture. In: Book of abstracts, proceedings: word aquaculture, p 95, Salvador, BA, Brazil, 19–23 May 2003

Boletzky SV (1987) Embryonic phase. In: Boyle PR (ed) Cephalopod life cycles, vol 2. Academic Press, London, pp 5–31

Carrasco JF, Rodríguez C, Rodríguez M (2005) Cultivo intensivo de paralarvas de pulpo (*Octopus vulgaris*, Cuvier) utilizando como base de la alimentación zoeas vivas de crustáceos. Libro de actas, IX Congreso Nacional de Acuicultura, Sevilla, pp 219–222

Carrasco JF, Arronte JC, Rodríguez C (2006) Paralarval rearing of the common octopus, *Octopus vulgaris* (Cuvier). Aquac Res 37:1601–1605

Cerezo Valverde J, García GB (2005) Suitable dissolved oxygen for common octopus (*Octopus vulgaris*, cuvier, 1797) at different weights and temperatures: analysis of respiratory behaviour. Aquac 244:303–314

Couto A (2012) Study on the wild diet of *Octopus vulgaris* paralarvae in Ría de Vigo (NW Spain). Dissertation application to the Master's degree in marine sciences and marine resources submitted to Institute of Biomedical Sciences Abel Salazar, University of Porto. pp 1–90

De Wolf T, Lenzi S, Lenzi F (2011) Paralarval rearing of *Octopus vulgaris* (Cuvier) in Tuscany, Italy. Aquac Res 42:1406–1414

Estefanell JA (2012) Optimización de las condiciones de engorde y avances en el conocimiento de los requerimientos nutricionales del pulpo común *Octopus vulgaris* (Cuvier, 1797). Doctoral Thesis, Universidad de Las Palmas de Gran Canaria, Gran Canaria, p 292

Estévez A, Gairin I, Berger E (2009) Wild zooplancton for *Octopus vulgaris* larval rearing. In: Hendry CI, Van Stappen G, Wille M, Sorgeloos P (eds) LARVI 09, Fish & shellfish larviculture symposium, special publication No. 38. European Aquaculture Society, Oostende, pp 88–91

FAO (2012). FAO yearbook. Fishery and aquaculture statistics. 2010. Rome, pp 609

Feyjoo P, Riera R, Felipe CB, Skalli A, Almansa E (2011) Tolerance response to ammonia and nitrite in paralarvae of *Octopus vulgaris* and its toxic effects on prey consumption rate and chromatophores activity. Aquac Int 19(1):193–204

Fuentes L, Iglesias J, Sánchez FJ, Otero JJ, Moxica C, Lago MJ (2005) Técnicas de transporte de paralarvas y adultos de pulpo (*Octopus vulgaris*). Bol Inst Esp Oceanogr 21(1–4):155–162

Fuentes L, Sánchez FJ, Otero JJ, Lago MJ, Iglesias J (2009) Utilización de zooplancton natural y *Artemia* en el cultivo de paralarvas de pulpo *Octopus vulgaris*. Libro de resúmenes, Congreso Nacional de Acuicultura. Madrid, pp 122–123

Fuentes L, Sánchez FJ, Lago MJ, Iglesias J, Pazos G, Linares F (2011) Growth and survival of *Octopus vulgaris* (Cuvier 1797) paralarvae fed on 3 *Artemia*-based diets complemented with frozen fish flakes, crushed zooplankton and marine microalgae. Sci Mar 75(4):771–777

Garrido D, Reis D, Orol D, Gonçalves R, Martín MV, Sykes AV, Rodríguez C, Felipe BC, Santamaría FJ, Zheng X, Almansa E (2012) Efecto de distintos tipos de iluminación en el cultivo de las paralarvas del pulpo común (*Octopus vulgaris* Cuvier, 1797). In: V Foro Iberoamericano de los Recursos Marinos y la Acuicultura, Cádiz, 26–29 Nov 2012

Gunnarsli K, Toften H, Mortensen A (2008) Effects of nitrogen gas supersaturation on growth and survival in juvenile Atlantic cod (*Gadus morhua* L.). Aquac 283:175–179. http://dx.doi.org/10.1016/j.Aquac. Accessed 29 June 2008

Hamasaki K, Morioka T (2002) Effects of temperature on egg incubation period, paralarval survival and growth of common octopus, *Octopus vulgaris* reared in the laboratory. Suisanzoshoku 50:407–413

Hamasaki K, Takeuchi T (2000) Effects of the addition of *Nannochloropsis* to the rearing water on survival and growth of planktonic larvae of *Octopus vulgaris*. Saibai- Giken 28:65–68

Hamasaki K, Takeuchi T (2001) Dietary value of *Artemia* enriched with ω-yeast or shark eggs as feed for planktonic larvae of *Octopus vulgaris*. Saibai-Giken 28:13–16

Hamasaki H, Fukunaga K, Yoshida Y, Maruyama K (1991) Effects of marine microalgae *Nannochloropsis* sp. on survival and growth on rearing pelagic paralarvae of *Octopus vulgaris*, and results of mass culture in the tank of 20 metric tons. Saibai-Giken 19:75–84

Hargreaves JA, Tucker CS (1999) Design and construction of degassing units for Catfish hatcheries. Southern Regional Aquaculture Center, publication No. 191. Mississippi State University

Hormiga JA, Almansa E, Sykes AV, Torres NV (2010) Model based optimization of feeding regimens in aquaculture: application to the improvement of *Octopus vulgaris* viability in captivity. J Biotechnol 149:209–214. doi: 10.1016/j.jbiotec.2009.12.008

Iglesias J, Fuentes L (2013) Cephalopods paralarval rearing with special reference to the common octopus (*Octopus vulgaris*, Cuvier 1797). In: Allan G, Burnell G (eds) Advances in aquaculture hatchery technology, Number 242. Woodhead Publishing Series in Food Science, Technology and Nutrition, Oxford, pp 374–403

Iglesias J, Sánchez FJ, Otero JJ (1997) Primeras experiencias sobre el cultivo integral del pulpo (*Octopus vulgaris*) en el Instituto Español de Oceanografía. In Costa J, Abellan E, García B, Ortega A, Zamara S (eds) Libro de actas, VII Congreso Nacional de Acuicultura. Cartagena, pp 221–226, ISBN: 84-491-0323

Iglesias J, Sánchez FJ, Otero JJ, Moxica C (2000) Culture of octopus (*Octopus vulgaris*, Cuvier). Present knowledge, problems and perspectives. Cah Options Méditer 47:313–321

Iglesias J, Otero JJ, Moxica C, Fuentes L, Sánchez FJ (2004) The completed life cycle of the octopus (*Octopus vulgaris*, Cuvier) under culture conditions: paralarval rearing using *Artemia* and zoeae, and first data on juvenile growth up to 8 months of age. Aquac Int 12:481–487

Iglesias J, Fuentes L, Sánchez FJ, Otero JJ, Moxica C, Lago MJ (2006) First feeding of *Octopus vulgaris* Cuvier, 1797 paralarvae using *Artemia*: effect of prey size, prey density and feeding frequency. Aquac 261(2):817–822

Iglesias J, Sánchez FJ, Bersano JGF, Carrasco JF, Dhont J, Fuentes L, Linares F, Muñoz JL, Okumura S, Roo J, van der Meeren T, Vidal EAG, Villanueva R (2007) Rearing of *Octopus vulgaris* paralarvae: Present status, bottlenecks and trends. Aquac 266:1–15

Imamura S (1990) Larval rearing of octopus (*Octopus vulgaris* Cuvier). The progress of technological development and some problems remained. Collect Breed 52:339–343

Itami K, Izawa Y, Maeda S, Nakai K (1963) Notes on the laboratory culture of the Octopus larvae. Bull Jap Soc Sci Fish 29(6):514–520

Kurihara A, Okumura S, Iwamoto A, Takeuchi T (2006) Feeding Pacific sandeel enhances DHA level in common octopus paralarvae. Aquac Sci 54:413–420

Lenzi F, Cittolin G, Ingle E, Tibaldi E (2002) Allevamento del polpo (*Octopus vulgaris*): riproduzione e allevamento larvale in avannotteria industriale. Ricerca per lo sviluppo dell'Aquacoltura Toscana, risultati conseguiti, pp 73–83

Lenzi F, Capiferri U, De Wolf T (2006) Paralarval rearing of the common Octopus *Octopus vulgaris*: state of the art in Italy. In: Book of abstracts Aqua, Firenze, Florence, Italy, p 523, 9–13 May 2006

Mangold K (1983) *Octopus vulgaris*. In: Boyle PR (ed) Cephalopod life cycles, vol I. Academic Press, London, pp 335–364

Mangold K, Boletzky S (1973) New data on reproductive biology and growth of *Octopus vulgaris*. Mar Biol 19:7–12

Márquez L, Quintana D, Lorenzo A, Almansa E (2013). Biometrical relationships in developing eggs and neonates of *Octopus vulgaris* in relation to parental diet. Helgoland Marine Research

Moxica C, Linares F, Otero JJ, Iglesias J, Sánchez FJ (2002) Cultivo intensivo de pararlarvas de pulpo, *Octopus vulgaris* Cuvier, 1797, en tanques de 9 m^3. Bol Inst Esp Oceanogr 18(1–4): 31–36

Moxica C, Fuentes L, Hernández J, Iglesias J, Lago MJ, Otero JJ, Sánchez FJ (2006) Efecto de *Nannochloropsis* sp. en la supervivencia y crecimiento de paralarvas de pulpo *Octopus vulgaris*. IX Foro dos Recursos Mariños e da Acuicultura das Rías Gallegas 9:255–261

Naef A (1928) Cephalopoda embryology, part I, vol II (final part of monograph no. 35). In: Fauna and flora of the Bay of Naples (translated by the Smithsonian Institution Libraries). Smithsonian Institution Libraries, Washington 35:1–461

Navarro JC, Villanueva R (2000) Lipid and fatty acid composition of early stages of cephalopods: an approach to their lipid requirements. Aquac 183:161–177

Navarro JC, Villanueva R (2003) The fatty acid composition of *Octopus vulgaris* paralarvae reared with live and inert food: deviation from their natural fatty acid profile. Aquac 219: 613–631

Okumura S, Kurihara A, Iwamoto A, Takeuchi T (2005) Improved survival and growth in *Octopus vulgaris* paralarvae by feeding large type *Artemia* and Pacific sandeel, *A. personatus*. Aquac 244:144–157

Parra G, Villanueva R, Yúfera M (2000) Respiration rates in late eggs and early hatchlings of the common octopus, *Octopus vulgaris*. J Mar Biol Ass UK 80:557–558

Quintana D (2009) Valoración de los requerimientos nutricionales de reproductores de pulpo común *Octopus vulgaris*. PhD Thesis, Universidad de La Laguna, Tenerife, p 225

Quintana D, Márquez L, Arévalo JR, Almansa E, Lorenzo A (2007) Aplicación de análisis multivariante al estudio de la influencia de la dieta de los reproductores de *Octopus vulgaris* sobre la composición lipídica de huevos y paralarvas y su relación con la calidad de puesta. Poster. XI Congreso Nacional de Acuicultura, Vigo, Sept 2007

Quintana D, Márquez L, Suárez H, Rodríguez E, Jerez S, Almansa E (2009) Efecto de la dieta de los reproductores de pulpo común (*Octopus vulgaris*) sobre la composición de aminoácidos de huevos y paralarvas: Relación con la calidad de puesta. XII Congreso Nacional de Acuicultura, Madrid

Rosas C, Caamal C, Cazares RJ (2010) Incubation process for octopuses and incubator. Universidad Autónoma de México. Organización Mundial de la Propiedad Intelectual. WO 2010/030155 A1

Roura A, González AF, Redd K, Guerra A (2012) Molecular prey identification in wild *Octopus vulgaris* paralarvae. Mar Biol 159(6):1335–1345

Sánchez FJ, Fuentes L, Otero JJ, Lago MJ, Linares F, Pazos G, Iglesias J (2010) Effect of tank volume on the growth and survival of reared *Octopus vulgaris* paralarvae. In: Aquaculture Europe, CD abstracts, Porto, pp 1170–1171

Seixas P (2009) Composición bioquímica y crecimiento de paralarvas de pulpo (*Octopus vulgaris* Cuvier, 1797), alimentadas con juveniles de *Artemia* enriquecidos con microalgas y otros suplementos nutricionales. PhD Thesis, University of Santiago de Compostela, p 279. ISBN: 978-84-9887-253-8

Seixas P, Rey-Méndez M, Valente LMP, Otero A (2010) High DHA content in *Artemia* is ineffective to improve *Octopus vulgaris* paralarvae rearing. Aquac 300:156–162

Skiftesvik AB, Browman HI, St-Pierre JF (2003) Life in green water: the effect of microalgae on the behaviour of Atlantic cod (*Gadus morhua*) larvae. In: 26th annual larval fish conference, Norway OS, 22–26 July 2002. In: Browman HI, Skiftesvik AB (eds) Big fish bang, pp 97–103

Sykes AV, Baptista FD, Gonçalves RA, Andrade JP (2012) Directive 2010/63/EU on animal welfare: a review on the existing scientific knowledge and implications in cephalopod aquaculture research. Rev Aquac 4:142–162

Uriarte I, Iglesias J, Domingues P, Rosas C, Viana MT, Navarro JC, Seixas P, Vidal E, Ausburger A, Pereda S, Godoy F, Paschke K, Farías A, Olivares A, Zúñiga O (2011) Current status and bottleneck of octopod aquaculture: the case of American species. J World Aquac Soc 42(6):735–752

Vaz-Pires P, Seixas P, Barbosa A (2004) Aquaculture potential of the common octopus (*Octopus vulgaris Cuvier*, 1797): a review. Aquac 238(1–4):221–238

Viciano E, Iglesias J, Lago MJ, Sánchez FJ, Otero JJ, Navarro JC (2011) Fatty acid composition of polar and neutral lipid fractions of *Octopus vulgaris* Cuvier, 1797 paralarvae reared with enriched on-grown *Artemia*. Aquac Res 42:704–709

Vidal EAG, DiMarco FP, Wormuth JH, Lee PG (2002) Optimizing rearing conditions of hatchling loliginid squid. Mar Biol 140:117–127

Vidal et al (manuscript in elaboration) In: Vidal E, Villanueva R (eds) Cephalopod culture: current status on main biological models and research priorities. Adv Mar Biol

Villanueva R (1994) Decapod crab zoeae as food for rearing cephalopod paralarvae. Aquac 128:143–152

Villanueva R (1995) Experimental rearing and growth of planktonic *Octopus vulgaris* from hatching to settlement. Can J Fish Aquat Sci 52:2639–2650

Villanueva R, Bustamante P (2006) Composition in essential and non-essential elements of early stages of cephalopods and dietary effects on the elemental profiles of *Octopus vulgaris*. Aquac 261:225–240

Villanueva R, Norman MD (2008) Biology of the planktonic stages of benthic octopuses. In: Gibson RN, Gordon JDM (eds) Oceanography and marine biology: an annual review, vol 46. CRC Press, pp 105–202

Villanueva R, Koueta N, Riba J, Boucaud-Camou E (2002) Growth and proteolytic activity of *Octopus vulgaris* paralarvae with different food rations during first-feeding, using *Artemia* nauplii and compound diets. Aquac 205:269–286

Villanueva R, Riba J, Ruíz-Capillas C, González AV, Baeta M (2004) Amino acid composition of early stages of cephalopods and effect of amino acid dietary treatments on *Octopus vulgaris* paralarvae. Aquac 242:455–478

Villanueva R, Escudero JM, Deulofeu R, Bozzano A, Casoliva C (2009) Vitamin A and E content in early stages of cephalopods and their dietary effects in *Octopus vulgaris* paralarvae. Aquac 286:277–282

Young RE, Harman RF (1988) Larva, paralarvae and subadult in cephalopod terminology. Malacologia 29:201–207

Chapter 24
Octopus vulgaris: Ongrowing

Francisco Javier Sánchez, Jesús Cerezo Valverde and Benjamín García García

Abstract This chapter covers the most important topics related to the common octopus *Octopus vulgaris* ongrowing from an experimental and commercial standpoint, citing the most outstanding works of the last decade on the subject. It first describes the current situation and then lists the parameters that are most important in the rearing process, from the purely environmental (salinity, temperature, oxygen, nitrogen compounds) to those related to the biology of the species studied such as growth rate and survival in relation to sex, initial body weight and stocking density. It also provides information on ongrowing performance depending on the type of food, natural or formulated and the degree of acceptance of this with respect to its texture. Finally, we present results of growth in tanks and sea cages and give recommendations on the transport of live octopus.

Keywords *Octopus vulgaris* · Ongrowing · Environmental parameters · Culture parameters · Diet

24.1 Introduction

The common octopus (*Octopus vulgaris*) has aroused much interest in recent years as a species that might be considered suitable for large-scale production, given its commercial value (Vaz-Pires et al. 2004); its fecundity, each female producing

F. J. Sánchez (✉)
Instituto Español de Oceanografía (IEO), Oceanographic Center of Vigo,
Subida a Radio Faro 50, 36390 Vigo, Spain
e-mail: javier.sanchez@vi.ieo.es

J. Cerezo Valverde · B. García García
IMIDA-Acuicultura, Consejería de Agricultura y Agua de la Región de Murcia,
Puerto de San Pedro del Pinatar, San Pedro del Pinatar, Murcia, Spain
e-mail: jesus.cerezo@carm.es

B. G. García
e-mail: benjamin.garcia@carm.es

J. Iglesias et al. (eds.), *Cephalopod Culture*, DOI 10.1007/978-94-017-8648-5_24,

100,000–500,000 eggs (Mangold 1983; Iglesias et al. 2000); rapid growth, exceeding 16–26 g day^{-1} (Mangold 1983; Aguado-Giménez and García García 2002); high protein content, representing 70–90 % of its body weight (BW) as dry matter (Lee 1994); and high feed conversion factor, incorporating 30–60 % of ingested food into its own body mass (Mangold and Boletzky 1973; Mangold 1983; Aguado-Giménez and García García 2002).

In Spain, the country where most research is being carried out on rearing of *O. vulgaris,* interest arose primarily from the results obtained in studies of its culture in captivity by Villanueva (1995), Rama-Villar et al. (1997) and Iglesias et al. (2000). From 1996 onwards, a few private companies started the small-scale production of octopus, with varying results, which, however, were sufficiently promising to suggest the great potential of this species for commercial culture. At the same time, several research groups, some in conjunction with the above companies, began research into octopus ongrowing in floating sea cages moored in areas with flat water conditions (inside harbours or estuaries; Rama-Villar et al. 1997; Luaces Canosa and Rey Méndez 1999; Tuñón et al. 2001, 2002; Chapela et al. 2006; Rodríguez et al. 2006; Oltra et al. 2005; Socorro et al. 2005; Iglesias et al 2007), in benthic cages (Estefanell et al. 2012a) and also in offshore conditions (García García et al. 2009). The results suggested that there were several parameters that could optimise production and identified some factors of interest for developing these types of facilities.

Specific studies of feeding and nutrition of octopus (Cagnetta and Sublimi 2000; García García and Aguado Giménez 2002; García García and Cerezo Valverde 2006; Domingues et al. 2009a; Prato et al. 2010), their physiology (Caruso et al. 2004; Cerezo Valverde and García García 2004, 2005; Katsanevakis et al. 2005; Petza et al. 2006; Sieiro et al. 2006; Mazón et al. 2007; García-Garrido et al. 2010a, b; García García et al. 2011; Morillo-Velarde et al. 2011) and the influence of a variety of factors on growth and survival (Aguado Giménez and García García 2002; Miliou et al. 2005; Domingues et al. 2010; Delgado et al. 2011) have also been considered and these factors have contributed to the identification of the optimal conditions for the production of this species in intensive cultures.

24.2 Environmental Parameters

24.2.1 Salinity

In general, cephalopods are stenohaline species that live in a salinity range of 27–38 psu (Boletzky and Hanlon 1983). In an ongrowing experiment conducted in sea cages in the Galician Rías, Chapela et al. (2006) recorded substantial mortality when salinity fell below 32 psu, as a result of run-off from the land following heavy rainfall, whilst in tanks Delgado et al. (2011) found no significant differences in growth or survival in the salinity range of 29–34 psu.

24.2.2 Temperature

Aguado-Giménez and García García (2002) developed models of growth and feeding rates as a function of BW (0.18–3.50 kg), temperature (13–28 °C) and feed type (bogue *Boops boops* and crab *Carcinus maenas*) based on 236 observations made in tanks containing isolated animals. Maximum growth was obtained at 17.5 °C, falling as the temperature rose to reach negative values at 28 °C. The maximum feeding rate was obtained at 20 °C and as a consequence of these two findings, the maximum feed efficiency would be around 16.5 °C. Mortality rose sharply above 22 °C, so that the authors established that the optimal range for octopus ongrowing is between 16 and 22 °C. Subsequently, Miliou et al. (2005), who obtained similar results, identified a significant interaction between weight and temperature, probably because they used smaller specimens than did the above authors. This interaction meant that the temperature at which maximum growth was achieved varied with the weight; this temperature for weights of 50–150 g was 25 °C.

24.2.3 Dissolved Oxygen

Cerezo Valverde and García García (2005) studied the octopus respiratory behaviour in relation to decreasing saturation level, measuring oxygen consumption and ventilation frequency for a wide range of BWs and temperatures, establishing three oxygen concentrations: optimal, suboptimal and dangerous. In the optimal range, neither oxygen consumption nor ventilation frequency varied, while in the suboptimal range ventilation frequency was modified and in the dangerous range both parameters were modified. These ranges were temperature dependent: At 16 °C, the minimum saturation level in the optimal range was 55 %, while at 25 °C it was 85 %. The lethal oxygen concentration was found to be below 15 % and directly dependent on both BW and temperature, varying from 3 % saturation (15 °C and 0.2 kg) to 9 % (27 °C and 2 kg).

24.2.4 Nitrogenated Compounds

Little attention has been paid to the effect of nitrogenated compounds (ammonia/ammonium, nitrates and nitrites) in cephalopods so that the maximum levels assumed are those that have been established in a general way for fish (Tucker 1998).

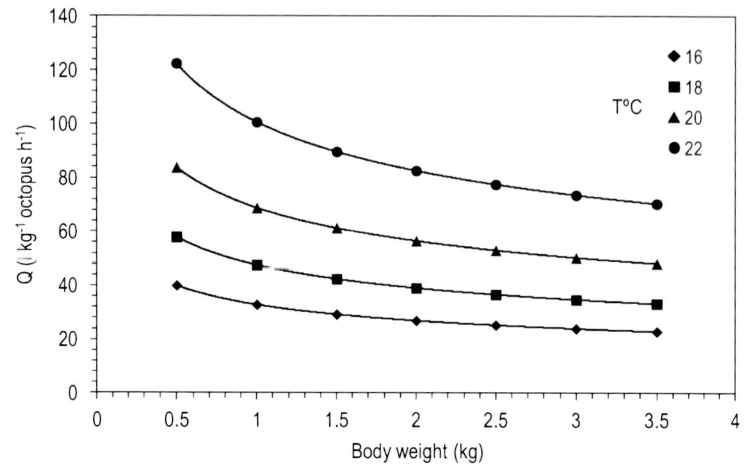

Fig. 24.1 Water flow rate (Q) as a function of octopus body weight and temperature water (16, 18, 20 and 22 °C), expressed in litres of water to 100 % saturation of oxygen per octopus kg and hour. Estimates made from the oxygen consumption equations developed by Cerezo Valverde and García García (2004) and the optimal level of dissolved oxygen for this species. (Cerezo Valverde and García García 2005)

However, results obtained by Gómez et al. (2005) point to a decrease of the ammonia excretion rate above 1.2 mg total ammonia nitrogen (TAN) L^{-1}at 19 °C, suggesting that higher values may be toxic.

24.3 Culture Parameters

24.3.1 Oxygen Consumption and Water Flow

The water flowing through a culture system (e.g. a tank) must ensure that the oxygen supply is sufficient to cover the metabolic needs of the animals in question, but also maintains optimal levels of oxygen, which must never be a limiting factor for production. The water also must ensure the removal of nitrogen compounds, but the oxygen requirement is demanding a greater flow rate. That is why the oxygen consumption models developed for the octopus (Cerezo Valverde and García García 2004) are necessary in order to estimate the flow water requirements in different situations (Fig. 24.1), but for calculation purposes we must also consider optimal ranges of dissolved oxygen in this species (Cerezo Valverde and García García 2005).

It is important to emphasise that a rise in temperature, at a constant salinity value, reduces oxygen solubility, so that each litre of water that enters the tank supplies less oxygen, and the oxygen levels at the inlet must be increased to maintain optimal levels at the outlet. Consequently, the oxygen available per litre of water for cultivated octopus falls as the temperature increases, so that the flow must be increased. Figure 24.1 shows the flows, expressed as L h^{-1} and per kg of octopus, as a function of the mean weight of the latter and at different temperatures. As can be seen, and based on the above, the water flow must be increased dramatically as the temperature rises, especially at 21–24 °C.

24.3.2 Generation of Solid Wastes and Nitrogenated Products

In general, natural diets generate a large amount of wastes, and this is particularly true in the case of cephalopods. The common octopus uses 77 % of the bogue supplied, leaving as solid waste the head, spine and tail, while it uses 50 % of crab when given as food, leaving the carapace (Mazón et al. 2007). These authors also estimated that of the nitrogen intake, 33–37 % is incorporated into the body (growth), and 1–2 % is eliminated in particulate form (faeces) and 62–65 % in dissolved form (excretion). The latter is the most pernicious in land-based ongrowing systems, particularly recirculating water supply systems, since the nitrogenated products may become a limiting factor. In cephalopods, the major final product of protein catabolism is ammonium (Potts 1965; Boucher-Rodoni and Mangold 1985), while urea represents a low percentage (Hoeger et al. 1987), which is rapidly hydrolyzed to ammonium and carbon dioxide. The absolute excretion rate of ammonium in octopus depends on the weight and amount of protein ingested and can be expressed by an equation (García García et al. 2011), while, in relative terms (related with BW), it has a value of 84 mg TAN kg^{-1} BW day^{-1} when the octopus is not fed and 265 mg TAN kg^{-1} BW day^{-1} when protein ingestion is 9 g day^{-1}. The excretion rate as a function of quantity of food supplied when the level of protein is not limiting varies linearly as a function of BW from 10 to 57 g TAN kg^{-1} feed dry matter day^{-1} when protein ingestion is 9 g day^{-1}.

24.3.3 Transport

The need to use live cephalopods has led to the development of a variety of transportation systems. Adult octopuses can tolerate quite extreme transport conditions, as long as the water temperature does not exceed 20 °C (Aguado-Giménez et al. 2001) and the level of dissolved oxygen is suitable. In Galicia, fishermen transport octopus to the ongrowing cages in tubes wrapped in nets, thus avoiding aggressive behaviour among the animals. At the experimental level, simulating transport for 24 h, Fuentes et al. (2005) obtained mortality rates of less than 5 % under extreme density and temperature conditions (165 kg m^{-3}, 19.5 °C). Estefanell (2012) used 500-L tanks and an oxygenation system to maintain the levels of dissolved oxygen at 10–12 mg L^{-1}; in this way, they transported adult octopuses (1 kg) at a maximum density of 40 kg m^{-3} for 60–70 min.

24.3.4 Octopus Ongrowing as a Function of Sex, Initial weight and Stocking Density

First commercial attempts to rear *O. vulgaris* used circulating water tanks and floating cages in the late 1990s. These experiments were designed to determine the optimal rearing conditions to reach a commercial size (2.5–3.5 kg). It was found that males grow slightly more rapidly than females (Sánchez et al. 1998), although the final weights of both sexes in ongrowing cages did not differ if they were cultivated separately. Studies on growth, oxygen consumption and ammonia production carried out excluding the period of gonadal maturation in females showed no significant differences between males and females (Aguado-Giménez and García García 2002; Cerezo Valverde and García García 2004; García García et al. 2011). Moreover, the best performance in ongrowing trials was observed with no or low state of sexual maturation in females (Estefanell et al. 2012b, 2013). Indeed, maturing of females during the rearing period has been a problem because from the moment egg laying occurs stop eating and grow. In this sense, Chapela et al. (2006) suggest separation of sexes prior to ongrowing process and prevent loss of commercial value. However, companies engaged in floating sea cages rearing octopus, mixed male and female in the same cage for practical reasons.

Another important factor to take into account is the initial weight of octopuses, since in many markets animals weighing more than 2 kg reach a higher price. Indeed, one of the advantages of cephalopods is the rate at which they grow: starting ongrowing with individuals of 330 g, they can reach market size (>2 kg) in 4 months (Iglesias et al. 2000). However, in the current situation, the companies prefer to start ongrowing with octopuses of 1 kg because the survival and absolute growth are higher and lower rearing time is necessary, thus obtaining a higher profit.

For the ongrowing of both juveniles and adults, the initial stocking density in tanks and cages is of great importance since the growth rate is so high that the initial biomass can triplicate in a few weeks, generating problems related to oxygen consumption and detritus production and even the creation of hierarchies as a result of size variations. This leads to stress and may end in cases of cannibalism and autophagy. In experiments using juveniles cultivated at different initial densities (10 kg m^{-3} and 20 kg m^{-3}) in tanks, Iglesias et al. (2000) found no differences in growth between the groups after 56 days but found that survival was greater in the lower density group. Chapela et al. (2006) recommended using similar-sized octopuses in floating cages (eight animals at 14 kg m^{-3}). The same authors used food of low value or from by-catch, composed of fish, crustaceans and mussels. Under these conditions, juveniles weighing 700 1,000 g reached a weight of 3 kg in 3 or 4 months, and it was possible to have two or three ongrowing cycles per year, avoiding winter, when growth is poorer. Working with the same cages, Iglesias et al. (2007) also recorded higher weights in summer, while in experiments made in floating cages measuring 4 × 3 x 1 m, Rodríguez et al. (2006) used octopuses at different concentrations and at different times of the year in northern Spain, finding the best results at initial concentrations of 10–12 kg m^{-3} and temperatures of 15–21 °C in spring–summer. Domingues et al. (2010) found no significant differences in growth to initial densities of 4, 8 and 15 kg m^{-3} in tanks; nevertheless, mortality was significantly lower for the low density.

24.4 Feeding

24.4.1 Ongrowing with Natural Diets

Both growth and food conversion rates vary widely as a function of the food supplied. The best growth (18–20 g day^{-1}; 1.8 % BW day^{-1}) and survival rates were obtained using a diet composed entirely of crustaceans or when crustaceans made up an important part of mixed diets (Cagnetta and Sublimi 2000; Tuñón et al. 2001; Aguado Giménez and García García 2002; García García and Cerezo Valverde 2004). García García and Cerezo Valverde (2006) obtained similar results with a diet based solely on crustaceans (*C. mediterraneus*) and another one in which they were partially replaced by fish (2 days *Boops boops* and 1 day *C. mediterraneus*). This pattern was subsequently applied successfully to ongrowing in the open sea (García García et al. 2009). Similarly, the partial replacement of crustaceans by *Diplodus vulgaris* provided excellent results (Prato et al. 2010). Growth rates similar to those obtained using crustaceans, and even higher than those obtained with mixed diets including crustaceans, were obtained with diets based on *B. boops* with a medium–high lipid content (7–17 % lipids wet weight; Prato et al. 2010; Biandolino et al. 2010; Estefanell et al. 2011, 2012b). García García and Aguado-Giménez (2002) also tested the influence of the lipid content, obtaining better growth and conversion index in octopus fed *B. boops* (6 % lipids; 1.14 % BW day^{-1}) than with *Sardina pilchardus* (20 % lipids; 0.84 % BW day^{-1}). Petza et al. (2006) obtained poorer growth rates including in the smallest specimens (0.43–0.95 % BW day^{-1}) the use of anchovy, *Engralius encrasicolus*. Other natural diets low in lipids (<2 % lipids) such as squid (*Loligo gahi* or *Illex coindetii*) or hake have also been seen to generate good growth (1.7–1.9 % BW day^{-1}) and show good feed efficiency (Cagnetta and Sublimi 2000; Domingues et al. 2009a). Poorer results were obtained with mussel (0.3–0.9 % BW day^{-1}) and crayfish (1.1 % BW day^{-1}), attributable in the first case to low acceptability (López et al. 2009; Prato et al. 2010) and in the second case to a poor conversion index (Domingues et al. 2009a). In general, diets based on fish or squid, or mixed diets have been notable for their high feed efficiencies (40–64 %) compared with monodiets of crustaceans (25–40 % in all studies). Estefanell et al. (2013) attributed the better feed efficiency of fish and its protein to the efficient use of lipids, while other studies have demonstrated the capacity of the octopus to mobilise its reserves of triglycerides in the digestive gland, involving up to 30 % of its total energy costs during starvation (García-Garrido et al. 2010b; Morillo-Velarde et al. 2012a).

24.4.2 Ongrowing with Formulated Diets

From the beginning of the 1990s until 2008, much effort in cephalopod culture was invested in preparing formulated diets, especially in cultures of *Sepia officinalis* and *O. maya* (Lee et al. 1991; Castro 1991; Castro et al. 1993; Castro and Lee 1994;

Domingues et al. 2005, 2007, 2008; Aguila et al. 2007; Rosas et al. 2007). All these works had one outcome in common: poor ingestion and growth rates, whether wet or formulated dry feeds were used, the results being markedly poorer than when natural diets were provided. Some of these results were attributed to poor palatability or, if the feeds were accepted, to their deficient nutritional content. The influence of certain chemical attractants on the acceptance or rejection of the feeds became evident (Lee et al. 1991; Domingues et al. 2007). It had previously been demonstrated that octopuses have a high chemoreceptor capacity towards many stimuli, including acids, sugars, crab extracts, amino acids and nucleotides (Wells 1963; Wells et al. 1965; Boyle 1983; Chase and Wells 1986; Lee 1992). Studies on the biochemical composition of cephalopods (Lee 1994; Rosa et al. 2005; Sieiro et al. 2006), their digestive enzymes (Boucher-Rodoni 1982; Caruso et al. 2004), digestibility (O'Dor et al. 1984; Hernández and García García 2004; Mazón et al. 2007) and bioenergetic subjects (Katsanevakis et al. 2005; Petza et al. 2006; García García et al. 2011) demonstrated the predominant role of the protein metabolism and of amino acids, emphasising the importance that these would have in the design of diets.

Between 2007 and 2010, two separate but complementary research lines were encouraged by the Spanish National Plans for Aquaculture, JACUMAR (2007–2009). One was dedicated to the study of acceptability and performance of different forms of feed and the other focused on a biochemical study of cephalopods, their natural diets and possible prime materials for formulating feeds. It was deduced that such feeds should be stable in water and with a firm and homogeneous (non-granular) texture, and have a degree of flexibility so as not to disaggregate during manipulation, as in the case of fish feed. Also, the presence of chemical substances that might lead to rejection had to be avoided. As a result, feeds were developed in formats that were acceptable to the octopuses and produced good growth (0.7–1.5 % BW day^{-1}; Cerezo Valverde et al. 2008; Quintana et al. 2008; Estefanell et al. 2013). All these feeds used a fish, crustacean or mollusc paste mixed with different binders (alginates or gelatines), which gave a moist feed (> 70 % water) with a texture that was suitable for manipulation and ingestion.

Gelatine was found to be more suitable than alginates for elaborating feeds because of its better digestibility and acceptability (Rosas et al. 2008; Seiça Neves et al. 2010; García-Garrido et al. 2010a), although alginate is more stable in water. Heating the mixtures gave them greater stability although it also led to a loss in nutritive value (Cerezo Valverde et al. 2008; Domingues et al. 2009b). The biochemical studies carried out showed that phospholipids and polyunsaturated fatty acids (docosahexaenoic acid (DHA) and eicosapentaneoic acid (EPA)) were the key nutrients (Navarro and Villanueva 2003; Estefanell et al. 2013; Cerezo Valverde et al. 2012a). Supplementation with arginine and the complementation of crustacean or bivalve protein with fish or animal meals were proposed as alternatives to increasing the protein yield (Cerezo Valverde et al. 2013).

Great advances have been made regarding knowledge of the nutritional requirements of cephalopods as a result of developing feed with a known nutritional composition. The incorporation of free amino acids, such as glutamate or arginine, had no effect on the feed rates but was associated with increased growth and protein productive value (Cerezo Valverde et al. 2012b). The better feed efficiency of diets with a moderate lipid content (6–9% wet weight) compared with low lipid diets (<2%) has been demonstrated (Estefanell et al. 2011; Cerezo Valverde et al. 2012b). Krill, fish and crab meals did not give good results as regards acceptability or growth (Estefanell et al. 2009; López et al. 2009; García-Garrido et al. 2010a). The tendency now is to work without fresh ingredients in favour of dry, preferably freeze-dried ones (Morillo-Velarde et al. 2012b, 2013) and to manufacture extruded feeds with a composition that differs from granulated fish feed (Querol et al. 2012a, b).

24.5 Culture Yield: Growth and Survival

The growth and survival results obtained in different ongrowing experiments in floating cages and tanks (Table 24.1) and their variability have been related, as mentioned above, with factors such as water temperature, culture density, size dispersion, the reproduction season and the type of feed supplied. However, erring on the conservative side, it can be concluded that under optimal culture conditions a production cycle in which the octopuses grow from 750 to 3,500 g can be carried out in 90–120 days with a survival rate of 80–90%. This means that up to three production cycles per year are possible although this depends on the geographical position of the installations and, in particular, on the temperature. In the Mediterranean, sea temperatures are very high for this species (exceeding 20–24 °C) so that production is only viable between mid-October and mid-June, as demonstrated experimentally (García García et al. 2009) in offshore cages (Fig. 24.2). In Galicia, on the other hand, it is the low temperatures of winter that limit culture, so that here, too, only two cycles are possible. In areas like the Canary Islands where the water temperature varies very little around 18–20 °C, three cycles are possible although this has not been demonstrated experimentally. Three cycles are also possible in tanks equipped with recirculation systems and temperature control.

24.6 Conclusions

There is enough information to establish in a general way and with certain degree of reliability optimum conditions for ongrowing process to *O. vulgaris*. Unfortunately, much of that information is available only for conference papers. Thus, specific and rigorous studies are still needed both for the influence of environmental parameters as to establish the optimum parameters of the culture in each geographical area and for different production systems (land tanks or sea cages: protected areas, offshore and benthic). Exhaustive studies on the biochemistry of cephalopods, their natural diets and different raw material have led to great advances in our understanding of their nutritional requirements. With semi-humid formulated diets moderate growth rates have been achieved, similar to when a fish-based diet is supplied.

Table 24.1 Results of some trials in *Octopus vulgaris* ongrowing in floating cages and tanks

	System	Time (days)	Volume (m^3)	T (°C)	Sex	Initial density (kg m^{-3})	Final density (kg m^{-3})	Initial weight (g)	Final weight (g)	RFR (% day^{-1})	AGR (g day^{-1})	Survival (%)	Food
Rama Villar et al. 1997	FC	127	16.5	[a]	M, F	21.1	6.04	990	3,000	5	15.83	94.3	P, MU
Tuñón et al. 2001	FC	55	12.4	[a]	M, F	10.9	19.79	841	2,020	7	21.44	76.50	C, P
Tuñón et al. 2001	FC	44	12.4	[a]	M, F	19.3	21.39	1497	2,296	7	18.03	83	C, P
Rey Méndez et al. 2003	FC	115	13.5	[a]	M, F	5.0	16.9	761	3,074	-	20.11	82.22	-
García García and Cerezo Valverde 2004	T	77	3.5	18	M	3.15	10.25	929	3,260		30.4	92	C, P
García García and Cerezo Valverde 2004	T	58	1.4	18	M	3.89	11.1	908	3,098	6,9	37.8	92	C, P
Rodríguez et al. 2006	FC	86	4	14	M	24.5	22.5	1,090	1,694	5–10	7.02	59	C, P, MU
Rodríguez et al. 2006	FC	96	4	16	M	10.7	36.5	917	3,739	5–10	29.39	83	C, P, MU
Chapela et al. 2006	FC	75	27	15	M	7	15	790	2,430	5	21.86	71.2	P, C, MU
Chapela et al. 2006	FC	75	27	15	H	6	13	810	2,440	5	21.73	65.3	
Iglesias et al. 2007	FC	120	8.25	15	M, H	15.0	26.3	1,122.7	2,427	3–7	10.87	81.1	P, MU
Iglesias et al. 2007	FC	120	8.25	15	M, H	20.0	31.61	1,081.5	2,269	3–7	9.90	75	P, MU
Domingues et al. 2010	T	70	2	20	M, H	4	8	1,175	2,660	5	21.2	98	S
Domingues et al. 2010	T	70	2	20	M, H	8	14	1175	2,419	5	17.5	80	S
Domingues et al. 2010	T	70	2	20	M, H	15	24	1,175	2,310	5	15.8	78	S
Delgado et al. 2011	T	60	0.65	19	M, H	12.4	44.4	803	2,690	6	31.9	96.6	C, P
Delgado et al. 2011	T	60	0.65	19	M, H	24.1	67.8	784	2,416	10	27.4	91.6	C, P
Estefanell et al. 2012a	FC	63	2.5	20	M, H	10.1	29.6	873	2,615	8	27.7	97	B
Estefanell et al. 2012a	BC	63	2	20	M, H	10.3	30.7	932	3,067	8	33.9	91	B

RFR relative feeding rate, *AGR* absolute growth rate, *M* males, *F* females, *FC* floating cage, *BC* benthic cage, *T* tank; *P* fish; *MU* mussel, *C* crab, *B* bogue, *S* squid

[a] No available data

Fig. 24.2 Offshore experimental cage in the Mediterranean Sea. (García García et al. 2009)

Noteworthy are the high rates of feed efficiency and digestibility of formulated diets, with food conversion rates close to 1, both dry and semi-moist feeds. Future research should be aimed primarily at improving the formats to achieve better feeding rates. Obviously, the environmental impact from inedible portion of natural diets (carapaces from crustacean or spines, heads and tails from fish) for octopus rearing in sea cages would be reduced with the formulated feed developed to date.

Acknowledgments This work was financed by the European Regional Development Fund (POI07-001 FEDER project).

References

Aguado-Giménez F, García García B (2002) Growth and food intake models in *Octopus vulgaris* Cuvier 1797): influence of body weight, temperature, sex and diet. Aquac Int 10(5):361–377

Aguado-Giménez F, Rueda FM, Egea MA, Hernández MD, Faraco F, García García B (2001) Efecto de la temperatura sobre el transporte, estabulación y engorde de *Octopus vulgaris* Cuvier en el Mediterráneo Occidental. VII Congreso Nacional de Acuicultura, Las Palmas (Gran Canaria), 19–21 May 1999. Serie Monografías del ICCM Instituto Canario de Ciencias Marinas ed(.) nº 4, pp 174–179

Aguila J, Cuzon G, Pascual C, Domingues PM, Gaxiola G, Sánchez A, Maldonado T, Rosas C (2007) The effects of fish hydrolysate (CPSP) level on *Octopus maya* (Voss and Solis) diet: Digestive enzyme, blood metabolites, and energy balance. Aquaculture 273:641–655

Biandolino F, Portacci G, Prato E (2010) Influence of natural diets on growth and biochemical composition of *Octopus vulgaris*, Cuvier 1797. Aquac Int 18:1163–1175

Boletzky S, Hanlon RT (1983) A review of the laboratory maintenance, rearing and culture of cephalopod molluscs. Memoirs of the National Museum Victoria 44:147–187

Boucher-Rodoni R (1982) La glande digestive des céphalopodes, organe de synthèse et de sécrétion d'enzymes digestives. Cah Biol Mar 23:299–318

Boucher-Rodoni R, Mangold K (1985) Ammonia excretion during feeding and starvation in *Octopus vulgaris*. Mar Biol 86:193–197

Boyle PR (1983) Ventilation rate and arousal in the octopus. J Exp Mar Biol Ecol 69:129–136

Cagnetta P, Sublimi A (2000) Productive performance on the common octopus (*Octopus vulgaris* C.) when fed on a monodiet. CIHEAM Cah Options Méditerr 47:331–336

Caruso G, Giordano D, Mancuso M, Genovese L (2004) Studio preliminare degli enzimi digestivi di Sepia officinalis Linnaeus, 1758 ed *Octopus vulgaris* Cuvier, 1797 (Mollusca: Cephalopoda). Biol Mar Med 11:367–369

Castro BG (1991) Can *Sepia officinalis* L. be reared on artificial food? Mar Behav Physiol 19:35–38

Castro BG, Lee PG (1994) The effects of semi-purified diets on growth and condition of *Sepia officinalis* L. (Mollusca: Cephalopoda). Comp Biochem Physiol 109A:1007–1016

Castro BG, Dimarco FP, DeRusha RH, Lee PG (1993) The effects of surimi and pelleted diets on the laboratory survival, growth, and feeding rate of the cuttlefish *Sepia officinalis* L. J Exp Mar Biol Ecol 170:241–252

Cerezo Valverde J, García García B (2004) Influence of body weight and temperature on post-prandial oxygen consumption of common octopus (*Octopus vulgaris*). Aquaculture 233:599–613

Cerezo Valverde J, García García B (2005) Suitable dissolved oxygen levels for common octopus (*Octopus vulgaris* Cuvier, 1797) at different weights and temperatures: analysis of respiratory behavior. Aquaculture 244(1–4):303–314

Cerezo Valverde J, Hernández MD, Aguado-Giménez F, García García B (2008) Growth, feed efficiency and condition of common octopus (*Octopus vulgaris*) fed on two formulated moist diets. Aquaculture 275:266–273

Cerezo Valverde J, Hernández MD, García-Garrido S, Rodríguez C, Estefanell J, Gairín JI, Rodríguez CJ, Tomás A, García García B (2012a) Lipid classes from marine species and meals intended for cephalopod feeding. Aquac Int 20:71–89

Cerezo Valverde J, Hernández MD, Aguado F, Morillo-Velarde PS, García GB (2012b) Performance of formulated diets with different level of lipids and glutamate supplementation in *Octopus vulgaris*. Aquacu Res. doi:10.1111/j.1365–2109.2012.03201.x

Cerezo Valverde J, Tomás A, Martínez-Llorens S, Jover M, García S, Rodríguez C, Estefanell J, Gairín I, Domingues PM, Rodríguez CJ, García García B (2013) Amino acids composition and protein quality evaluation of marine species and meals for feed formulations in cephalopods. Aquac Int 21:413–433

Chapela A, González AF, Dawe EG, Rocha F, Guerra A (2006) Growth of common octopus (*Octopus vulgaris*) in cages suspended from rafts. Sci Mar 70:121–129

Chase R, Wells MJ (1986) Chemotactic behaviour in Octopus. J Comp Physiol A 158:375–381

Delgado M, Gairín JI, Carbó R, Aguilera C (2011) Crecimiento de *Octopus vulgaris* en tanques en El Delta Del Ebro: efecto de la densidad de cultivo y de la salinidad. In: Villarroel M, Reig L, Almansa C (eds) Book of abstracts, XIII Congreso Nacional de Acuicultura. Castelldefels, Barcelona

Domingues PM, DiMarco FP, Andrade JP, Lee PG (2005) The effects of diets with amino acid supplementation on the survival, growth and body composition of the cuttlefish *Sepia officinalis*. Aquac Int 13:423–440

Domingues PM, López N, Muñoz JA, Maldonado T, Gaxiola G, Rosas C (2007) Effects of a dry pelleted diet on growth and survival of the Yucatan octopus, *Octopus maya*. Aquac Nutr 13:273–280

Domingues PM, Ferreira A, Márquez L, Andrade JP, López N, Rosas C (2008) Growth, absorption and assimilation efficiency by mature cuttlefish (*Sepia officinalis*) fed with alternative and artificial diets. Aquac Int 3:215–229

Domingues PM, García S, Hachero-Cruzado I, Lopez N, Rosas C (2009a) The use of alternative prey (crayfish, *Procambarus clarki*, and hake, *Merlucius gayi*) to culture *Octopus vulgaris* (Cuvier 1797). Aquac Int 18:487–499

Domingues PM, Marquez L, López N, Rosas C (2009b) Effects of food thermal treatment on growth, absorption, and assimilation efficiency of juvenile cuttlefish, *Sepia officinalis*. Aquac Int 17:283–299

Dominges P, García S, Garrido D (2010) Effects of three culture densities on growth and survival of *Octopus vulgaris* (Cuvier, 1797). Aquac Int 18:165–174

Estefanell JA (2012) Optimización de las condiciones de engorde y avances en el conocimiento de los requerimientos nutricionales del pulpo común *Octopus vulgaris* (Cuvier, 1797). Tesis doctoral. Dpto de Biología, Universidad de las Palmas de Gran Canaria

Estefanell J, Socorro J, Roo J, Naranjo D, Martín A, Fernández-Palacios H, Izquierdo M (2009) Utilización de cangrejo y harina en el pienso para pulpo común *Octopus vulgaris*. In: Beaz D, Villarroel M, Cárdenas S (eds) Book of abstracts. XII Congreso Nacional de Acuicultura. 24–26 Nov, Madrid, pp 675–678

Estefanell J, Socorro J, Tuya F, Izquierdo M, Roo J (2011) Growth, protein retention and biochemical composition in *Octopus vulgaris* fed on different diets based on crustaceans and aquaculture by-products. Aquaculture 322–323:91–98

Estefanell J, Roo J, Guirao R, Izquierdo M, Socorro J (2012a) Benthic cages versus floating cages in *Octopus vulgaris*: biological performance and biochemical composition feeding on *Boops boops* discarded from fish farmas. Aquac Eng. doi:10.1016/j.aquaeng.2012.02.001

Estefanell J, Roo J, Guirao R, Izquierdo M, Socorro J (2012b) Biological performance of *Octopus vulgaris* in integrated aquaculture models: effect of two potential fresh diets, sex and sexual maturation during the grow-out phase. Aquac Res. doi: 10.1111/are.12094

Estefanell J, Roo J, Guirao R, Afonso JM, Fernández-Palacios H, Izquierdo M, Socorro J (2013) Efficient utilization of dietary lipids in *Octopus vulgaris* (Cuvier 1797) fed fresh and agglutinated moist diets based on aquaculture by-products and low price trash species. Aquac Res 44:93–105. doi: 10.1111/j.1365–2109.2011.03014.x.

Fuentes L, Iglesias J, Sánchez FJ, Otero JJ, Moxica C, Lago MJ (2005) Métodos de transporte de paralarvas y adultos de pulpo *Octopus vulgaris* Cuvier, 1797. Bol Inst Esp Oceanogr 21(1–4):155–162

García García B, Aguado-Giménez F (2002) Influence of diet on ongrowing and nutrient utilization in the common octopus (*Octopus vulgaris*). Aquaculture 211:171–182

García García B, Cerezo Valverde J (2004) Influencia del número de días de ayuno a la semana sobre el crecimiento, el índice de conversión y la supervivencia en el pulpo de roca (*Octopus vulgaris* Cuvier, 1797). AcuaTIC 21:34–41

García García B, Cerezo Valverde J (2006) Optimal proportions of crabs and fish in diet for common octopus (*Octopus vulgaris*) ongrowing. Aquaculture 253:502–511

García García B, Cerezo Valverde J, Aguado-Giménez F, García García J (2009) Growth and mortality of common octopus *Octopus vulgaris* reared at different stocking densities in Mediterranean offshore cages. Aquac Res 40:1202–1212

García García B, Cerezo Valverde J, Gómez E, Hernández MD, Aguado-Giménez F (2011) Ammonia excretion of Octopus (*Octopus vulgaris*) in relation to body weight and protein intake. Aquaculture 319:162–167

García-Garrido S, Domingues P, Navarro JC, Hachero I, Garrido D, Rosas C (2010a) Growth, partial energy balance, mantle and digestive gland lipid composition of *Octopus vulgaris* (Cuvier, 1797) fed with two artificial diets. Aquac Nutr 17(2):174–187

García-Garrido S, Hachero-Cruzado I, Garrido D, Rosas C, Domingues PM (2010b) Lipid composition of mantle and digestive gland of *Octopus vulgaris* juveniles (Cuvier, 1797) exposed to prolonged starvation. Aquac Int 18:1223–1241

Gómez E, Mazón MJ, Piedecausa MA, García García B (2005) Efecto de la concentración de amoniaco total sobre la tasa de excreción en el pulpo de roca (*Octopus vulgaris* L): resultados preliminares. In: Jover M, de la GF (eds) Book of Abstracts, X Congreso Nacional de Acuicultura. Gandía, 17–21 Oct, pp 386–387

Hernández MD, García García B (2004) Digestibility of natural foods in common octopus (Octopus vulgaris). In: Biotechnologies for Quality, Aquaculture Europe 2004, Barcelona Spain, pp 414–415

Hoeger U, Mommsen TP, O'Dor R, Webber D (1987) Oxygen uptake and nitrogen excretion in two cephalopods, octopus and squid. Comp Biochem Physiol 87A:63–67

Iglesias J, Sánchez FJ, Otero JJ, Moxica C (2000) Culture of octopus (*Octopus vulgaris*): present knowledge, problems and perspectives. Recent advances in Mediterranean aquaculture finfish species diversification. Proceedings of the Seminar of the CIHEAM Network Options Méditerranéennes 47:313–322

Iglesias J, Otero JJ, Moxica C, Fuentes L, Sánchez FJ (2004) The complete life cycle of octopus (*Octopus vulgaris* Cuvier) under culture conditions: paralarvae rearing using Artemia and zoeae, and first data on juvenile growth up to eight months of age. Aquac Int 12:481–487

Iglesias J, Sánchez FJ, Otero JJ, Lago MJ, Moxica C, Fuentes L, Martínez FJ (2007) Engorde industrial de pulpo *Octopus vulgaris* en batea en la Ría de Vigo. Actas del XI Congreso Nacional de Acuicultura, Tomo I, 683–686. ISBN 978-84-611-9086-7

Katsanevakis S, Stephanopoulou S, Miliou H, Moratiou-Apostolopoulou M, Verriopoulos G (2005) Oxygen consumption and ammonia excretion of *Octopus vulgaris* (Cephalopoda) in relation to body mass and temperature. Mar Biol 146:725–732

Lee PG (1992) Chemotaxis by *Octopus maya* Voss et Solis in a Y-maze. J Exp Mar Biol Ecol 153:53–67

Lee PG (1994) Metabolic substrates in cephalopods. In: Pörtner HO, O'Dor RK, MacMillan DL (eds) Physiology of cephalopod mollusc. Lyfestyle and performance adaptations. Gordon and Breach Publishers, Basel, pp 35–51

Lee PG, Forsythe JW, DiMarco FP, DeRusha R, Hanlon RT (1991) Initial palatability and growth trials on pelleted diets for cephalopods. Bull Mar Sci 49:362–372

López M, Rodríguez C, Carrasco JF (2009) Engorde de juveniles de pulpo (*Octopus vulgaris* Cuvier, 1797) con distintas dietas naturales y artificiales. In: Beaz D, Villarroel M, Cárdenas S (eds) Book of abstracts. XII Congreso Nacional de Acuicultura. 24–26 Nov, Madrid, pp 170–171

Luaces-Canosa M, Rey-Méndez M (1999) El engorde industrial del pulpo (*Octopus vulgaris*) en jaulas: análisis de dos años de cultivo en la Ría de Camariñas. VII Congreso Nacional de Acuicultura (1999). In: Fernández-Palacios H, Izquierdo M (eds) Convergencia entre Investigación y Empresa: un Reto para el Siglo XXI. Monografías del ICCM, Las Palmas de Gran Canaria, p 184–189

Mangold KM (1983) *Octopus vulgaris*. In: Boyle, PR (ed) Cephalopod Life Cycles, vol. 1. Academic Press, London, pp 335–364

Mangold KM, von Boletzky SV (1973) New data on reproductive biology and growth of *Octopus vulgaris*. Mar Biol 19:7–12

Mazón MJ, Piedecausa MA, Hernández MD, García García B (2007) Evaluation of environmental nitrogen and phosphorus contributions as a result of intensive ongrowing of common octopus (*Octopus vulgaris*). Aquaculture 266:226–235

Miliou H, Fintikaki M, Kountouris T, Verriopoulos G (2005) Combined effects of temperature and body weight on growth and protein utilization of the common octopus, *Octopus vulgaris*. Aquaculture 249:245–256

Morillo-Velarde PS, Cerezo Valverde J, Serra Llinares RM, García García B (2011) Energetic contribution of carbohydrates during starvation in common octopus (*Octopus vulgaris*). J Molluscan Stud 77:318–320

Morillo-Velarde PS, Cerezo Valverde J, Serra Llinares RM, García García B (2012a) Changes in lipid composition of different tissues of common octopus (*Octopus vulgaris*) during short-term starvation. Aquac Res. doi:10.1111/j.1365–2109.2012.03131.x.

Morillo-Velarde PS, Cerezo Valverde J, Hernández MD, Aguado-Giménez F, García García B (2012b) Growth and digestibility of formulated diets based on dry and freeze-dried ingredients in the common octopus (*Octopus vulgaris*). Aquaculture 368–369:139–144

Morillo-Velarde PS, Cerezo Valverde J, García García B (2013) A simple format feed to test the acceptability of ingredients for common octopus (*Octopus vulgaris* Cuvier, 1797). Aquac Res. doi:10.1111/are.12233

Navarro JC, Villanueva R (2003) The fatty acid composition of *Octopus vulgaris* paralarvae reared with live and inert food: deviation from their natural fatty acid profile. Aquaculture 219:613–631

O'Dor RK, Mangold K, Boucher-Rodoni R, Wells MJ, Wells J (1984) Nutrient Absorption, Storage and Remobilization in *Octopus vulgaris*. Mar Behav Physiol 11:239–258

Oltra R, Alemany F, Roig M, Mezquita F (2005) Engorde pulpo (*Octopus vulgaris*) en jaulas flotantes en la Comunidad Valenciana. In: Jover M, de la GF (eds) Book of Abstract, X Congreso Nacional de Acuicultura. Gandía. 17–21 Oct, Spain, pp 408–409

Petza D, Katsanevakis S, Verriopoulos G (2006) Experimental evaluation of the energy balance in *Octopus vulgaris*, fed ad libitum on a high-lipid diet. Mar Biol 148:827–832

Potts WTW (1965) Ammonia excretion in *Octopus dofleini*. Comp Biochem Physiol 14:339–355

Prato E, Portacci G, Biandolino F (2010) Effect of diet on growth performance, feed efficiency and nutritional composition of *Octopus vulgaris*. Aquaculture 309:203–211

Querol P, Morillo-Velarde PS, Cerezo Valeverde J, Martinez Llorens S, Moñino AV, Jover M, Tomás A (2012a). First assessment of acceptance of dry extruded diets for Octopus vulgaris (Cuvier, 1797). Aquac Res. doi:10.1111/are.12006

Querol P, Morillo-Velarde PS, Cerezo Valverde J, Martinez Llorens S, Moñino AV, Jover M, Tomás A (2012b). Inclusion of fish and krill meal in extruded diets for Octopus vulgaris (Cuvier, 1797): assessment of acceptance. Aquac Res. doi:10.1111/are.12093

Quintana D, Domingues PM, García S (2008) Effect of two artificial wet diets agglutinated with gelatin on feed and growth performance of common octopus (*Octopus vulgaris*) sub-adults. Aquaculture 280:161–164

Rama-Villar A, Faya-Angueira V, Moxica C, Rey-Méndez M (1997) Engorde de pulpo (*Octopus vulgaris*) en batea. In: Costa J, Abellán E, García García B, Ortega A, Zamora S (eds), Book of abstracts, VI Congreso Nacional de Acuicultura, Cartagena 9–11 de Julio de 1997. MAPA, Madrid, pp 245–250

Rey-Mendez M, Tuñón EG, Amor R (2001) Estudio comparativo del tipo de jaula y disposición de refugios sobre la mortalidad en el engorde del pulpo (*Octopus vulgaris* Cuvier, 1797). COLACMAR 2001, San Andrés, Colombia

Rey-Méndez M, Tuñón E, Luaces-Canosa M (2003) Estudio de los efectos de peso inicial y el sexo sobre el comportamiento, la mortalidad y el crecimiento del pulpo (*Octopus vulgaris* Cuvier, 1797) en cultivo industrial. Actas IX Congreso Nacional de Acuicultura, pp 276–277. Cádiz, Mayo 2003

Rodríguez C, Carrasco JF, Arronte JC, Rodríguez M (2006) Common octopus (*Octopus vulgaris* Cuvier, 1797) juvenile ongrowing in floating cages. Aquaculture 254:293–300

Rosa R, Pereira J, Nunes ML (2005) Biochemical composition of cephalopods with different life strategies, with special reference to a giant squid, *Architeuthis* sp. Mar Biol 146:739–751

Rosas C, Cuzon G, Pascual C, Gaxiola G, Chay D, López N, Maldonado T, Domingues PM (2007) Energy balance of *Octopus maya* fed crab or an artificial diet. Mar Biol 152:371–381

Rosas C, Tut J, Baeza J, Sánchez A, Sosa V, Pascual C, Arena L, Domingues PM, Cuzon G (2008) Effect of type of binder on growth, digestibility, and energetic balance of *Octopus maya*. Aquaculture 275:291–297

Sánchez FJ, Iglesias J, Moxica C, Otero JJ (1998) Growth of octopus (*Octopus vulgaris*) males and females under culture conditions. Europ Aquac Soc. Spec Public 26:244–245

Seiça Neves MM, Cerezo Valverde J, García García B (2010) Digestibility of a formulated diet with alginate as binder in octopus. In: Seafarming tomorrow, EAS Aquaculture Europe 2010. Porto, Portugal, pp 500–501

Sieiro MP, Aubourg SP, Rocha F (2006) Seasonal study of the lipid composition in different tissues of the common octopus (*Octopus vulgaris*). Eur J Lipid Sci Technol 108:479–487

Socorro J, Roo J, Fernández-López A, Guirao R, Reyes T, Izquierdo M (2005) Engorde de pulpo (*Octopus vulgaris*) alimentado exclusivamente con boga (*Boops boops*) de descarte de la acuicultura. In: Jover M, de la GF (Eds.) Book of Abstract, X Congreso Nacional de Acuicultura. Gandía 17–21 Oct, pp 428–429

Tucker JW (1998) Marine Fish Culture. Kluwer Academic Publishers, London

Tuñón E, Parada A, Caeiro C, Rey MM (2001) Efecto del tamaño inicial en el engorde industrial del pulpo (*Octopus vulgaris*, Cuvier 1797) en jaulas suspendidas en bateas. In: Rey-Méndez M, Fernández Casal J, Izquierdo Rodríguez M (eds) III Foro dos Recursos Mariños e da Acuicultura das Rías Galegas. O Grove, Spain pp 181–199

Tuñón E, Parada A, Caeiro C, Rey-Méndez M (2002) Estudio comparativo basado en la dieta diferenciada para el engorde de pulpo *Octopus vulgaris*, Cuvier 1797; en una explotación industrial. In: Rey-Méndez M, Fernández Casal J, Izquierdo Rodríguez M (eds), IV Foro dos Recursos Mariños e da Acuicultura das Rías Galegas. O Grove, Spain pp 255–269

Vaz-Pires P, Seixas P, Barbosa A (2004) Aquaculture potential of the common octopus (*Octopus vulgaris* Cuvier, 1797): a review. Aquaculture 238:221–238

Villanueva R (1995) Experimental rearing and growth of planktonic *Octopus vulgaris* from hatching to settlement. Can J Fish Aquat Sci 52:2639–2650

Wells MJ (1963) Taste by touch; some experiments with Octopus. J Exp Biol 40:187–193

Wells MJ, Freeman NH, Ashburner M (1965) Some experiments on the chemotactile sense of octopuses. J Exp Biol 43:553–563

Chapter 25
Robsonella fontaniana

Íker Uriarte and Ana Farías

Abstract *Robsonella fontaniana* (d'Orbigny 1834) is a moderate-sized octopus that inhabits the subantarctic region of South America. This benthic resource has been considered as a potential species for aquaculture diversification, mainly focused on the Asian "baby octopus" market. Most of the information about this species comes from the natural populations in Argentina and Chile, and from its experimental rearing in Chile. This chapter describes the major advances achieved in reproductive biology, ontogenetic development and hatchery cultivation under laboratory conditions of *R. fontaniana*.

Keywords Reproductive conditioning · Baby octopus larviculture · Ontogenetic development · *Robsonella fontaniana*

25.1 Introduction

25.1.1 State of the Art

Robsonella fontaniana (d'Orbigny 1834) is a small-sized octopus found off the Chilean coast and southern Argentina. It can be distinguished from other sympatric octopus species by its dorsal white spot anterior to the head on fresh and live specimens (Ibáñez et al. 2008). This cold-water benthic species is distributed along almost the entire coastal tip of South America, from Peru to Cabo de Hornos in Chile (Pacific Ocean) and Puerto Madryn in Argentina (Atlantic Ocean). Main reproductive aspects related to survival and feeding aspects of *R. fontaniana* paralarvae under controlled conditions have been reported in southern Chile (González et al. 2008; Uriarte et al. 2011a, b). Morphometric changes have been described for *R. fontaniana* embryos, paralarvae and juveniles up to 160 days after hatching

Í. Uriarte (✉) · A. Farías
Instituto de Acuicultura, Sede Puerto Montt, Universidad Austral de Chile,
1327, Puerto Montt, Chile
e-mail: iuriarte@uach.cl

A. Farías
e-mail: afarias@uach.cl

J. Iglesias et al. (eds.), *Cephalopod Culture,* DOI 10.1007/978-94-017-8648-5_25,

(DAH) under controlled conditions (Uriarte et al. 2009, 2010). Cannibalism among paralarvae siblings has recently shown a shift mainly by managing rearing density among other factors (Miranda et al. 2011). Ontogenic changes on the digestive system of paralarvae have also been studied (Pereda et al. 2009).

R. fontaniana is not currently a fishery resource in South America (Ibáñez et al. 2009). However, its potential for being exported to the Asian "baby octopus" market has been analyzed with interesting commercialization levels and prices (González et al. 2008).

Juvenile individuals of the Patagonian red octopus *Enteroctopus megalocyathus* have been captured jointly with adults of *R. fontaniana* for fisheries research, given that both are sympatric species along the Patagonian coastline of the Pacific Ocean, and also the fishermen do not necessarily discriminate between them. The abundance of paralarvae of this species could represent 85 % of octopus paralarvae sampled in the far south channels of southern Chile (43°S), ranging between 4 and 164 paralarvae 1,000 m (Vega et al. 2000), so it may be more abundant than *E. megalocyathus*. Recent advances in the biology of *R. fontaniana* planktonic phase (González et al. 2008; Pereda et al. 2009; Uriarte et al. 2010, 2011b) have been well documented, with promising advances successfully reaching the benthic phase (Uriarte et al. 2010) for a possible artificial juvenile production. Because of the difficulty in developing the controlled culture of *E. megalocyathus* (see Chap. 19), *R. fontaniana* was considered as an alternative to develop the juvenile production of a merobenthic octopus species (Uriarte et al. 2008, 2011b).

25.1.2 Geographic Distribution

R. fontaniana is an endemic benthic resource from the subantarctic region of South America; it is distributed along the Pacific Ocean from Peru (6°S) to Tierra del Fuego (55°S) and along the Atlantic Ocean from the Gulf of Saint Matias (41°S) to the South. It bathymetric distribution is from coastal zone and up to 225 m, mainly over hard substrates and in caves (Ibáñez et al. 2008; Ortiz and Ré 2011).

25.1.3 Catching Methods

Based on field and experimental evidence, Ibáñez et al. (2009) suggested that *R. fontaniana* has a selective hunting behaviour and a specialized diet with a narrow trophic niche breadth, showing a high preference for crustaceans as food items when compared with their null preference for molluscs. This characteristic allows for baits to be chosen to capture them effectively without the use of hooks, the most common method among intertidal shellfish gatherers. *R. fontaniana* is caught along with *Octopus tehuelchus* (similar in size) in the Atlantic zone of Argentina (Ré 1998). Among the techniques used to catch baby octopuses in the Argentine fishing areas, the most prominent are subtidal traps with and without bait, and the intertidal harvesting with hooks or by hand (Narvarte et al. 1996; Ré 1998).

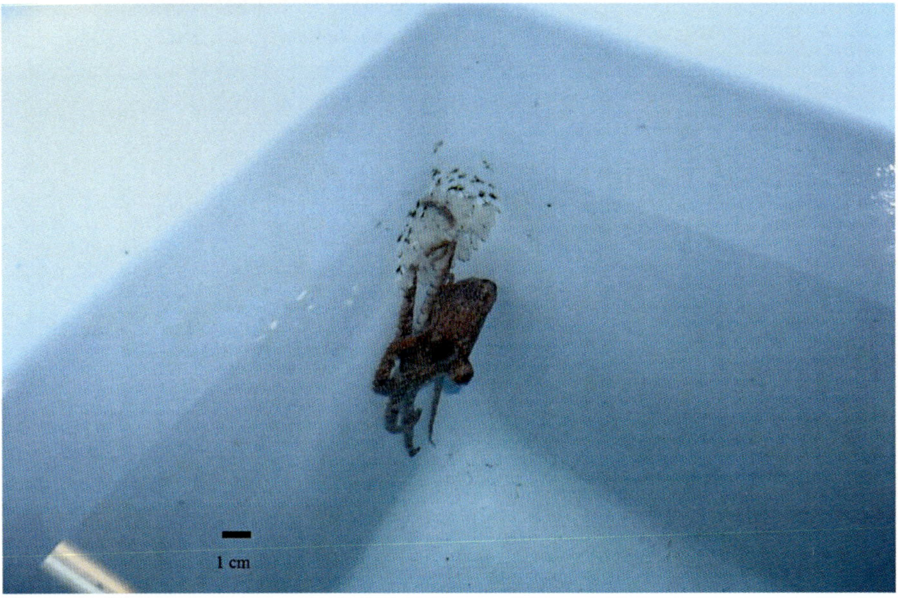

Fig. 25.1 *Robsonella fontaniana*. Female octopus with spontaneous clutches observed at Hatchery of Marine Invertebrates of the Universidad Austral de Chile (HIM-UACh). The measuring slide indicates 1 cm.

25.1.4 Morphological Characteristics

Ibáñez et al. (2008) provide a description of the moderate-sized octopus. It can be distinguished from other sympatric octopus species by a dorsal white spot anterior to the head on fresh and live specimens. The same author reports maximum sizes of 273 mm total length for males and 209 mm for females, dorsal mantle length (ML) reaching up to 69 mm in females and 68.8 mm in males, with arms length accounting for about 70 % of the total length (Fig. 25.1). It has a saccular mantle, bumpy skin with intraocular papilla, arms with two sets of suckers, third arm hectocotylized in males, enlarged suckers in males, siphon organ in a W shape, ink sac present, *radula* with seven teeth by transversal row, marginal plaques present. The hectocotylus has 47–60 suckers, a small ligula and seven copulatory lamellae.

25.1.5 Reproductive Cycle

A total of 21 different clutches of eggs under the care of females were obtained from natural subtidal refuges between November 2007 and November 2008 by Uriarte et al. (2009). The length of the embryonic development period calculated according to Uriarte et al. (2009) indicated that the onset of egg laying had taken place in September 2007, October 2007, January 2008 and April 2008, suggesting that in the

Xth Region, southern Chile (41°52'S, 73°51'W), the spawning starts in spring and ends in autumn, in accordance with González et al. (2008) who observed individuals spawning under controlled conditions between October and March. There is not much information in the literature about the reproductive cycle of the species, but Ortiz and Ré (2011) reported egg batches in April 2005, June 2005 and April 2007 in the coastal waters of the San Jose Gulf (42°15'S, 66°14'W). Although knowledge on the reproductive biology of *R. fontaniana* is limited, laboratory studies show that this species can easily spawn up to 2,500 eggs (Rocha et al. 2001; Briceño-Jacques 2004; González et al. 2008).

25.2 Larviculture Advances and Production of Juvenile Octopus

25.2.1 Broodstock Conditioning

Until now, broodstock conditioning has not been documented for this species. Therefore, eggs must be collected from the wild in order to study the ontogenetic development of embryos as well as the incubation of eggs under controlled conditions. Besides, spontaneous egg laying of this species has been documented under laboratory conditions (Uriarte et al. 2011b). Females can show a fecundity of $1{,}798 \pm 823$ eggs per female (González et al. 2008) and the eggs are not placed within dens but over the surface of rocks. After egg laying, *R. fontaniana* females attend to the eggs throughout the embryonic development. This may produce differences in dietary habits between the sexes (Ibáñez et al. 2008) and determine the beginning of the senescence period of the females, finishing in death.

25.2.2 Egg Incubation

The embryonic development of *R. fontaniana* takes between 39 and 103 days (González et al. 2008) but this depends on the incubation temperature because 74 days have been reported at 12 °C by Uriarte et al. (2009), while 39 days at 14 °C and 91 days at 8 °C, with a high range between embryos within the same batch, were reported by Ortiz and Ré (2011). This period also shows a slower development rate when compared to tropical species of the *Octopus* genus, probably because of the temperature, but still it is faster than the development rate of cold-water species *E. megalocyathus*, probably due to the smaller size of moderate-sized octopus eggs when compared to those of the Patagonian red octopus observed at the Hatchery of Marine Invertebrate of Universidad Austral de Chile (HIM-UACh).

The size of the eggs ranges between 3.9 and 5.2 mm which is lower than 10% of the ML; therefore, its development corresponds to a merobenthic octopus with a paralarval stage. The growth rate of embryo length is exponential, reaching an

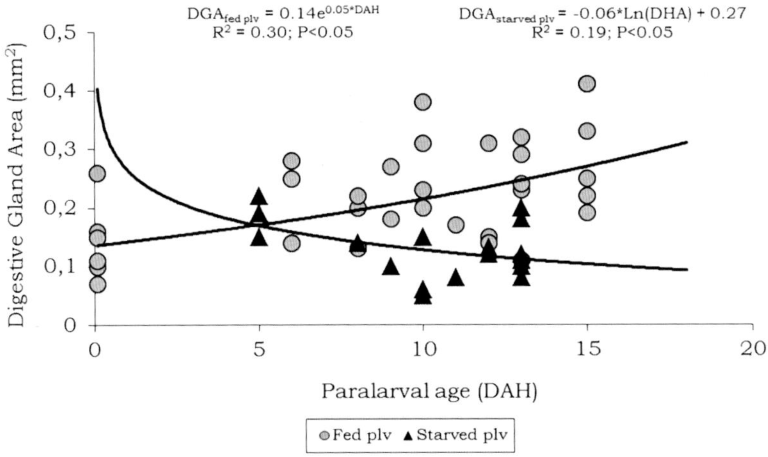

Fig. 25.2 *Robsonella fontaniana*. Variation in the size of the digestive gland in paralarvae subjected to feed or fasting. (Data published with permission of Espinoza V, Viana MT, Rosas C, Uriarte I and Farías A)

average of 0.35 ± 0.11 % day at 12 °C. The embryo is conspicuous at day 22 at 12 °C (Uriarte et al. 2009), when reaching stage IX (according to Naef 1928). At the end of the embryonic period, paralarvae ML is 4.4 ± 0.1 mm, with an arm length (AL)/ML ratio of 0.61 ± 0.01 (Uriarte et al. 2010).

25.2.3 Paralarvae Culture

When the paralarvae of *R. fontaniana* hatch, they have a total length of 3.4–6.0 mm (González et al. 2008; Uriarte et al. 2010; Ortiz and Ré 2011). According to Ortiz and Ré (2011), the chromatophore pattern, in terms of shape and distribution of paralarvae of this species, allows its easy distinction from other octopods from the Patagonian Atlantic and the Patagonian Pacific. Carrasco et al. (2012) differentiated paralarvae of both *E. megalocyathus* and *R. fontaniana* in oceanographic zooplankton samples collected in the Chilean Fjord region (41–43°S).

Uriarte et al. (2010) demonstrated that the total length of the paralarvae of *R. fontaniana* increases exponentially from hatching to benthic settlement. The AL and eye diameter increase proportionally to the total length of the paralarvae, while the mantle increases logarithmically. According to these authors, this demonstrates an important increase of the paralarvae predatory abilities throughout their ontogenetic development. Fasting affects the morphometry of the paralarvae, noting that starved paralarvae had reduced AL and AL/ML ratio (Espinoza 2009). In paralarvae subjected to prolonged fasting at HIM-UACh, the digestive gland suffered a detectable atrophy from fasting over 8 days (Fig. 25.2), reaching only 1/3 of its size during fasting periods of 13 days.

The study of enzyme activity in the digestive system of the paralarvae shows that until 10 DAH at 12 °C, the paralarvae show less enzyme activity of trypsin and chymotrypsin, independent of the diet. After this age, when they are fed with *Artemia* or maintained in a fasting state (without food) the paralarvae present little trypsin and chymotrypsin activity, while the enzymes are significantly activated in paralarvae when they feed on zoea of king crab, *Lithodes santolla* (Pereda et al. 2009).

Even though the zoea of king crab is a diet that allows a normal development throughout the paralarval life until benthic settlement, the paralarvae show a high level of mortality during the first 15 DAH with an absence of growth. The exponential growth of paralarvae starts only after 22 DAH. Consequently, this early period is characterized by a high level of mortality and zero growth could be related with two phenomena: (1) the arms do not reach enough length to manipulate the zoeae of king crab according to Uriarte et al. (2010) and (2) the digestive system is immature according to Pereda et al. (2009). Therefore, only paralarvae with enough energy reserve to manage surviving the first 2 or 3 weeks post hatching could capture the zoeae and survive.

The complete development of *R. fontaniana* is documented for paralarvae fed with *L. santolla* (Uriarte et al. 2010), but the paralarvae fed only with Artemia at 12 °C survived but did not grow during 70 days of culture (Uriarte et al. 2011a).

Amphipod prey was not consumed by the paralarvae, and the consumption of zoea of *Pagurus sp.* and *Petrolisthes laevigatus* did not allow the growth and survival of *R. fontaniana*. This was due to the zoeae behaviour and the lower organic content in these species in comparison to the king crab zoeae (Uriarte et al. 2011a, b).

Furthermore, *R. fontaniana* paralarvae presented a high mortality due to cannibalism. According to Miranda et al. (2011), this behaviour increases significantly along with paralarvae cultivation density. This prevents cultivation of this species at densities higher than 4 paralarvae L^{-1}. Below this density no cannibalism is observed, even though the paralarvae remain in fasting. According to the authors, cannibalism causes between 12.5 and 65.6% mortality when cultivation has a density between 8 and 56 paralarvae L^{-1}, respectively.

The tendency of a linear growth of AL and eye diameter with regard to the total length is maintained after finishing the paralarval period and throughout the early juvenile period.

The juvenile growth rate reached only 3.19% day^{-1} in *R. fontaniana* during 90 days after settlement; this value was lower than the specific growth rate observed during the previous paralarval stage that reached a value of 4.43% day^{-1} (Uriarte et al. 2010). Settlement and transition to juvenile stage mark the beginning of the consumption of benthic preys such as small crabs from the *Petrolisthes* genus (Uriarte et al. 2010).

At the HIM-UACh, females with egg laying in the subtidal zone are caught and transported to the laboratory. They are kept in a recirculating system at 11 ± 1 °C until paralarvae hatching. Once they hatch, paralarvae are cultured in 3 L cylindrical tanks with mild and constant aeration, and total replacement of seawater filtered to 1 µm and UV sterilized every 24 h. The cultures are performed at 11 ± 1°C. Culture density is 5 paralarvae L^{-1}, ruling out daily mortality. The supplied food is zoeae of king crab (*L. santolla*) and *Artemia* nauplii, at a density of 1–2 prey $paralarva^{-1}$

day^{-1}. They are moved to rectangular 15 L tanks with PVC dens once they show signs of settlement behaviour.

25.3 Ongrowing

Studies on natural food preferences of *R. fontaniana* carried out using collected samples in Concepcion, Chile (36°44'S, 73°10'W), showed that Brachyuran crabs are the most common part of their diet. Gastropods are not consumed, and they prefer highly mobile shellfish. These findings indicate that this species displays a hunter behaviour and has a narrow trophic niche (Ibáñez et al. 2009).

Recent studies concerning the defence of *R. fontaniana* against predators show that they have two main strategies: The first is to reduce the probabilities of being detected by a predator such as *Schroederichthys chilensis* (pintarroja) by using camouflage, while the second focuses on avoiding being captured by increasing their inactivity period as a defence mechanism (Ruíz et al. 2012).

There are no cases of grow-out neither in tanks nor in cages reported for this species; therefore, appropriate growth rate during grow-out of juvenile is not documented.

25.4 Trends in Research and Industrial Level

This species of baby octopus is distributed along the Chilean Pacific Coast and part of the Argentinian Atlantic Coast. It covers a wide range of temperatures and salinity levels. Its commercial value as a "baby octopus" is recently being explored in Chile, which means that its catches are not significant yet, although they are in Argentina.

Considering its wide geographical distribution, its reproduction potential should be studied throughout its distribution, both to establish population differences in the reproductive, paralarval and juvenile characteristics and to select those populations that best support controlled rearing conditions with productive purposes.

Studies on the composition of toxic (Hg, Pb and Cd) and essential elements (Cu, Zn, Co, Ni, Cr, Mn, Fe, Mg, Ca, K, Na and Se) found in seafood, including common Mediterranean and Philippines octopus (1–4 kg) and baby octopus from Thailand, have shown high levels of Cd in baby octopus, as all of its parts are consumed including viscera, while those common octopuses with arms as their only edible part show very low Cd levels (Kwoczek et al. 2006). With regard to essential elements, baby octopus show high amounts of Se and Mg (Kwoczek et al. 2006) similar to other seafood. Studies looking to avoid the presence of toxic elements as well as to characterise the functional nutrients are very important to position baby octopus such as *R. fontaniana* in global markets as healthy and safe foods. A major positioning will promote research for the aquaculture production of this resource.

25.5 Conclusions

Future tendencies are determined by the market but the results of controlled cultivation indicate that juvenile production is possible, and at 12 °C, it takes 70 days of egg incubation, 70 days of paralarval cultivation and a further 90 days to obtain juveniles of approximately 2 g, starting from eggs collected from the subtidal zone. However, there are no studies reporting growth rate of juveniles in grow-out until the harvest size of the octopus. Based on the specific growth rate of early juveniles documented by Uriarte et al. (2010), the specimens could reach a weight of 50 g in 260 days. Therefore, the period of total production could reach up to 14 months considering the collection of eggs from the wild as the beginning of the productive cycle.

References

Briceño-Jacques F (2004) Acondicionamiento y crecimiento de reproductores de Robsonella fontaniana (d' Orbigny, 1835) (Cephalopoda, Octopodidae) bajo condiciones de ambiente semi-controlado en el litoral central de Chile, Quintay (V Región). (Broodstock conditioning of Robsonella fontaniana under semi controlled conditions in Central Chile). Universidad Andrés Bello, Santiago

Carrasco SA, Maltrain R, Villenas F, Vega MA (2012) New records of early life-stages of cephalopods in the Chiloé Interior Sea. Lat Am J Aquat Res 40:229–235

Espinoza V (2009) Punto de no retorno en paralarvas de pulpo (Robsonella fontaniana Orbigny, 1834). (Starving effects in paralarvae of Robsonella fontaniana). Universidad Austral de Chile, Puerto Montt

González ML, Arriagada S, López DA, Pérez MC (2008) Reproductive aspects, eggs and paralarvae of Robsonella fontanianus (d'Orbigny, 1834). Aquac Res 39:1569–1573

Ibáñez CM, Sepúlveda RD, Guerrero J, Chong J (2008) Redescription of Robsonella fontaniana (Cephalopoda: Octopodidae). J Mar Biol Ass UK 88:617–624

Ibáñez CM, Sepúlveda S, Sanhueza EJ, Ruiz F, Chong J (2009) Feeding strategies of Robsonella fontaniana (d'Orbigny, 1834) (Cephalopoda: Octopodidae). Rev biol mar oceanogr 44(2):277–283

Kwoczek M, Szefer P, Hac E, Grembecka M (2006) Essential and toxic elements in seafood available in Poland from different geographical regions. J Agric Food Chem 54:3015–3024

Miranda RM, Espinoza V, Dörner J, Farías A, Uriarte I (2011) Sibling cannibalism on the small octopus Robsonella fontaniana (d'Orbigny,1834) paralarvae. Mar Biol Res 7:746–756

Naef A (1928) Fauna and Flora of the Bay of Naples: Cephalopoda, Embryology. Part I, Vol II (Final part of Monograph No. 35). 35. Smithsonian Institution Libraries, Washington D.C., pp 1–461

Narvarte A, González R, Sica I (1996) Estado actual de la pesquería del pulpito patagónico Octopus tehuelchus en el golfo San Matías. Fundación Patagonia Natural, Argentina, p 1–37

Ortiz N, Re ME (2011) The eggs and hatchlings of the octopus Robsonella fontaniana (Cephalopoda: Octopodidae). J Mar Biol Ass UK 91(3):705–713

Pereda S, Uriarte I, Cabrera JC (2009) Effect of diet and paralarval development on digestive enzyme activity in the cephalopod Robsonella fontaniana. Mar Biol 156:2121–2128

Ré ME (1998) Pulpos octopódidos (Cephalopoda: Octopodidae). In: Boschi EE (ed) El Mar Argentino y sus Recursos Pesqueros. Tomo 2: Los moluscos de interés pesquero. Cultivos y estrategias reproductivas de bivalvos y equinoideos. Instituto Nacional de Investigación y Desarrollo Pesquero, Mar del Plata, p 69–98

Rocha F, Guerra A, González A (2001) A review of the reproductive strategies in cephalopods. Biol Rev 76:291–304

Ruíz JF, Sepúlveda R, Ibáñez C (2012) Behaviour of Robsonella fontaniana in response to a potential predator. Lat Am J Aquat Res 40:253–258

Uriarte I, Farías A, Paschcke K, Ausburger A, Romo C (2008) Informe Término Proyecto FONDEF D04 I1401 (Technical report of Project "Desarrollo científico-tecnológico de la producción de juveniles de cefalópodos"). Universidad Austral de Chile, Puerto Montt

Uriarte I, Zúñiga O, Olivares A, Espinoza V, Cerna V, Farías A, Rosas C (2009) Morphometric changes and growth during embryonic development of Robsonella fontaniana. Vie Milieu 59:315–323

Uriarte I, Hernández J, Dörner J, Paschke K, Farías A, Crovetto E, Rosas C (2010) Rearing and growth of the octopus Robsonella fontaniana (Cephalopoda: Octopodidae) from planktonic hatchlings to benthic juveniles. Biol Bull 218:200–210

Uriarte I, Farías A, Paschke K, Navarro JC, Rosas C (2011a) Observations on feeding and biochemical characteristics to improve larviculture of Robsonella fontaniana (Cephalopoda: Octopodidae). Aquaculture 315:121–124

Uriarte I, Iglesias J, Rosas C, Viana MT, Navarro JC, Domíngues P, Seixas P, Vidal E, Ausgburger A, Pereda S, Godoy F, Paschke K, Farías A (2011b) Current status, potential and prospects of cephalopod larviculture in Latin American. J World Aquac Soc 42:735–752

Vega MA, Rocha FJ, Osorio C (2000) Variaciones espaciales y temporales de paralarvas de cefalópodos en el Canal Moraleda (43°S), sur de Chile. Cienc Tecnol Mar 23:69–82

Part III
Conclusions and Future Trends

Chapter 26
Current Status and Future Challenges in Cephalopod Culture

Roger Villanueva, António V. Sykes, Érica A.G. Vidal, Carlos Rosas, Jaruwat Nabhitabhata, Lidia Fuentes and José Iglesias

Abstract This chapter presents an overall perspective on the current status of cephalopod culture, its bottlenecks and future challenges. It focuses on the species that have received more research effort and consequently accumulated more scientific literature during the present century, namely *Sepia officinalis, Sepioteuthis lessoniana, Octopus maya* and *Octopus vulgaris*. Knowledge regarding physiology, metabolism and nutrition of different species is still lacking. Two main challenges are identified: the development of a sustainable artificial diet and the control of reproduction. Understanding cephalopod physiology and nutrition will probably be the biggest challenge in developing the large-scale culture of this group of molluscs

R. Villanueva (✉)
Institut de Ciències del Mar (CSIC), Passeig Marítim de la Barceloneta 37-49,
08003, Barcelona, Spain
e-mail: roger@icm.csic.es

A. V. Sykes
CCMAR-CIMAR L.A., Centro de Ciências do Mar do Algarve, Universidade do Algarve,
Campus de Gambelas, 8005-139, Faro, Portugal
e-mail: asykes@ualg.pt

É. A. G. Vidal
Centro de Estudos do Mar (CEM), Universidade Federal do Paraná (UFPR),
Cx. P. 61, Pontal do Paraná PR, 83255-976, Brazil
e-mail: ericavidal2000@yahoo.com.br

C. Rosas
Unidad Multidisciplinaria de Docencia e Investigación, Facultad de Ciencias,
Universidad Nacional Autónoma de México (UNAM), Puerto de Abrigo s/n Sisal,
Yucatán, Mexico
e-mail: crv@ciencias.unam.mx

J. Nabhitabhata
Excellence Centre for Biodiversity of Peninsular Thailand (CBIPT), Faculty of Science, Prince of Songkla University, Hatyai, Songkhla 90112, Thailand
e-mail: jaruwat.n5@gmail.com, jaruwat.n@psu.ac.th

L. Fuentes · J. Iglesias
Oceanographic Center of Vigo, Instituto Español de Oceanografía (IEO),
Subida a Radio Faro 50, 36390, Vigo, Spain
e-mail: lidia.fuentes.moledo@gmail.com

J. Iglesias
jose.iglesias@vi.ieo.es

J. Iglesias et al. (eds.), *Cephalopod Culture,* DOI 10.1007/978-94-017-8648-5_26,

on a medium to long term. In addition, zootechnical parameters need future research and improvement. The performance of an ethical experimentation with cephalopods is strongly encouraged and any zootechnical development should be performed and adapted accordingly. The potential of cephalopod culture extends far beyond its use for research and human consumption and probably it will be translated in a remarkable production in the coming years.

Keywords Artificial feed development · Sustainable aquaculture · Cephalopod culture bottlenecks · Control of reproduction · Embryo · Hatchling · Paralarvae · Juvenile · Subadult and adult life phases · Metabolism and nutrition

26.1 Introduction

The study of live cephalopods under controlled laboratory conditions has provided considerable scientific information to understand their life cycles, otherwise difficult to obtain. This experimental approach has facilitated their use as species models in neurobiology and behaviour and also provided the basis for the pilot commercial culture of some species. The chapters of this book show the state of the art of key aspects related to cephalopod culture and the recent advances on the culture of the main species studied in different laboratories around the world during the present century. From the beginning of cephalopod culture (see Chap. 4), new species have been added to the list of cephalopods maintained, reared or cultured in captivity. The experimental and pilot culture of a number of cephalopod species has been explored for different research purposes and with different degrees of success. Researchers from all over the world have contributed to the extended knowledge that is summarized in this book, which includes information on existing culture technology for nautiluses, six sepioid species (*Sepia officinalis, Sepia pharaonis, Sepiella inermis, Sepiella japonica, Euprymna hyllebergi* and *Euprymna tasmanica*), three squid species (*Loligo vulgaris, Doryteuthis opalescens* and *Sepioteuthis lessoniana*) and seven octopus species (*Amphioctopus aegina, Enteroctopus megalocyathus, Octopus maya, Octopus mimus, Octopus minor, Octopus vulgaris* and *Robsonella fontaniana*). This list contains species from all the oceans, with the exception of polar areas.

However, only the culture of few species seems to persist through time, being used as target species by research laboratories in different countries that have managed to accumulate considerable scientific knowledge; they are *S. officinalis, S. lessoniana, O. maya* and *O. vulgaris*. These species have attained the status of cephalopod culture models and are used not only as the preferential cephalopod aquaculture candidates but also as the most common cephalopod model organism in other fields of research, such as neurobiology. The advances in culture technology obtained through these species have been partially shared among them and employed to other subtropical and tropical cuttlefish, squid and octopus, which are distributed mainly in littoral waters (with the exception of the nautiluses). The reasons why these culture cephalopod models have been selected to receive more attention from researchers around the world probably helped us to define the present status of cephalopod culture and to identify the main bottlenecks that need further research. These four species have a biochemical profile common to most coastal cephalopods: They are very rich in protein and essential amino acids but have low carbohydrate and lipid content. Overall, this composition is translated to the status of delicacy in certain regions of the world, where its consumption is part of a well-balanced diet and of cultural habits. This is particularly true in countries of the Mediterranean basin, Gulf of Mexico and Asia, where these four species attain high commercial interest with market channels historically well established from the fishery industry.

From a biological point of view, it makes more sense to culture these four cephalopod species because they display fast growth, high survival rates under high stocking densities and food conversion rates similar or higher to those of most cultured finfishes. All of which is translated in obtaining a marketable size and high biomass in just a few months or less than a year, depending on the size preference of a given market. All of these species mate and spawn in captivity and lay eggs that will hatch as large hatchlings, with the exception of *O. vulgaris*. Eggs and adults are resistant to handling and able to endure shipping to other facilities if properly packaged. These are just some of the general qualities that make them excellent candidates as experimental laboratory animals as well for commercial culture.

Experiments and pilot cultures of these species have identified their main qualities and the problems to solve before the development of their aquaculture (see Chaps. 11, 17, 20, 23 and 24 for details). At this point, it should be noted that cephalopod culture is still in its infancy in comparison with finfish culture. For that reason, nearly all zootechnical aspects still need improvement and research. However, the main identified areas that need urgent research in cephalopod culture are related to the development of sustainable artificial foods and to the control of reproduction. These main bottlenecks are discussed in the following.

26.2 Future Challenges in Cephalopod Culture

26.2.1 Development of a Sustainable Artificial Diet

Understanding cephalopod physiology and nutrition probably is the main challenge in developing large-scale culture for this group of molluscs to medium or long term. Species such as *S. officinalis, S. lessoniana* and *O. maya* are already being cultured for the complete life cycle on consecutive bred generations, and commercial on-growing for *O. vulgaris* subadults has been developed. However, all of them still depend on the use of live or frozen natural food, signalling the development of a sustainable artificial diet as the main step to solve in the near future. Formulated feeds are needed for hatchlings, juveniles and adults, as animals at these life phases will surely have different nutritional requirements. Cephalopods are carnivorous and, unfortunately, protein is expensive from both an economical and a sustainable point of view. Likewise, as most marine carnivores, coastal cephalopods have particular lipid requirements with relatively high content in phospholipids and polyunsaturated fatty acids. In consequence, cephalopods can be clustered with the carnivorous marine finfish species cultured worldwide, which use marine raw materials, such as fishmeal, fish oil and other fishery products for formulated feeds. Fishmeal includes marine species of high trophic levels on its formulation that are partially obtained from extractive fisheries, therefore contravening the primary goal of a sustainable aquaculture, i.e. reducing human pressure on the wild resources and the environment. Cephalopods are poikilothermic and display many physiological adaptations that are homologous to vertebrates (Lee 1994). They are fast growers and mainly composed of protein, with low lipid and carbohydrate content. In addition to protein and lipids, recent studies showed that carbohydrate, mainly metabolised from glycogen, can contribute up to 10% of the daily energy costs in *O. vulgaris* during starvation (Morillo-Velarde et al. 2011), indicating that the role of carbohydrates as glycogen and its relationship with protein content in artificial diets should be strengthened in future research.

The future of carnivorous fish and cephalopod farming probably will rely on feeding them mostly with local discard fish and vegetal raw materials with high-quality protein, a goal that has not been yet reached, although considerable research is being conducted on this aspect for carnivorous fishes (see among others: Ceulemans et al. 2003; Hansen et al. 2007; Benedito-Palos et al. 2008; Díaz-López et al. 2009; Dias et al. 2010; Enes et al. 2010; Pratoomyot et al. 2010). Although, at present, there are evidences that demonstrate that the vegetable meal produces the worst amino acid balance for *O. vulgaris* (Cerezo-Valverde et al. 2013), the use of those ingredients in diets for cephalopods should be investigated in order to make a sustainable culture. Multidisciplinary approaches and the ability to integrate a full range of skills, seldom found in a single group of researchers, may be able to provide solutions to obtain sustainable artificial feeds for cephalopods. These feeds need to be visually attractive and have suitable palatability and texture to be well handled, accepted and ingested, as well as appropriate digestibility to be well absorbed and metabolised to fit nutritional requirements. This goal will be particularly difficult to achieve for the paralarvae and for juveniles of some species. The feeding of the early stages of cephalopods has been historically one of the main bottlenecks in the development of their culture technology. Due to species-specific biological characteristics, this problem has been partially solved in *S. officinalis, S. lessoniana* and *O. maya*, which have large hatchlings directly called juveniles. For benthic hatchlings of *O. maya,* a semi-humid squid paste bound with gelatin has recently been developed and can be used as the main food, opening a promising standpoint for other species (Rosas et al. 2008, 2013). In *S. officinalis,* hatchlings were early fed on frozen food leading to the perspective of acceptance of a prepared food from the first day (Sykes et al. 2013). The well-developed nervous system, responsible for the sensory world of cephalopods and their learning capabilities, may be a valuable tool in training these animals to feed on suitable artificial feeds. Recent studies have shown the ability for embryonic learning in the cuttlefish *S. officinalis* (Darmaillacq et al. 2008), a characteristic that may be used in the future for the juvenile rearing by conditioning hatchling prey preferences, an unexplored field of research in cephalopod culture. In addition, greater effort is being made to understand the digestive capability of cephalopods, and recent results in octopuses may be able to clarify the assimilation process and the feeding requirements during the juvenile phase (see Chaps. 5 and 20). The acidic environment and the presence of cathepsins (enzyme of the lineage of pepsin) in the gastric juice of *O. maya* juveniles have shown that artificial food for octopus, and possibly for cephalopods in general, should have characteristics very different from those used to feed other aquatic organisms (Martínez et al. 2011).

On the other hand, in species displaying a small-sized paralarvae after hatching and being planktonic during this life stage, the delicate first feeding period represents a major source of mortality, where the suitability of food is suspected to be the main factor influencing the poor survival such as for *O. vulgaris* (Iglesias et al. 2007; Villanueva and Norman 2008). In these cases, to go beyond the experimental rearing that relies on natural prey, as decapod crustacean zoeae (Villanueva 1994), a suitable enriched *Artemia* protocol is needed to feed the planktonic paralarvae, and recent studies have provided encouraging results to this respect (Guinot et al. 2013, Chap. 23). Although nutritional approaches will play an important role in improving rearing conditions, future efforts also should focus on improving feeding rates of paralarvae and thus their survival and growth. The mechanisms by which paralarvae select their prey and that determine prey-capture success remain as an area of much needed research.

26.2.2 The Control of Reproduction

After nutrition, the second main bottleneck in cephalopod culture probably is the control of reproduction. Although most cultured species reproduce in captivity and eggs are obtained during the spawning season, overall control over reproduction is lacking. Cephalopods are semelparous and individuals of all coastal species tested for culture die after spawning; thus, the broodstock must be renewed every culture cycle. Further investigation to describe methods and establish protocols for accelerating and retarding the collection of egg masses is needed. Until now, cephalopod egg masses are obtained by: (a) spontaneous spawning of broodstocks, (b) eggs collected from the field and (c) *in vitro* fertilization. As coastal cephalopods spawn at the end of their life cycle, experienced workers can tentatively expect the collection of egg masses in laboratory according to the existing knowledge of a given species life cycle for a particular geographical location. This includes information on the spawning season, behaviour and external body characteristics. However, this method is relatively basic and sometimes subjected to high individual variation. Studies that allow the ability to control sexual maturation and spawning entirely are necessary to accurately obtain egg masses according to the planned culture design. Knowledge of the influence of natural variables on sexual maturation and spawning should be clearly defined. The effect of light intensity and photoperiod on sexual maturity has been studied in few species (Richard 1971; Zúñiga et al. 1995) and need further research. Despite the functional neuroendocrine analogy with vertebrates (Legall and Feral 1985), no induction or hormonal methods have been used to obtain cephalopod maturation, and the role of olfaction and chemical messaging and its effect on sexual behaviour have started to be recently understood. The reproductive behaviour of *S. officinalis* and its relation to chemical cues have been described (Boal and Marsh 1998; Boal and Golden 1999; Boal et al. 2010). In addition, sex pheromones expressed from precursors in the oviducal gland during egg laying have been identified, and induce hyperventilation as well as stimulate mating (Enault et al. 2012). Despite the identification of these substances, studies that combine their effects at reproductive and behavioural levels in cultured cephalopods are still needed. Research should also be focused on closing the life cycle under controlled conditions, which will allow obtaining a sufficient number of progenitors to avoid deleterious effects on the natural populations, renewing the broodstock. Some experimental studies used broodstocks to obtain animals for the multiple generations as for *S. lessoniana* (Lee et al. 1994; Walsh et al. 2002) and *S. officinalis* (Forsythe et al. 1994; Sykes et al. 2006). Such closed-cycle practice with captive breeders may have led to reproductive isolation from wild populations, resulting in a loss of genetic variability due to the low effective breeding population size and inbreeding. This eventual inbreeding in consecutive generations needs to be addressed by determining the effective number of breeders contributing for reproduction by using behavioural analysis and paternity studies, and quantifying the loss of genetic variation in consecutive cultured generations at given culture conditions.

Reproduction in captivity needs interdisciplinary studies, integrating theoretical with experimental methods to understand the magnitude of different factors. Studies should consider zootechnical factors (tank type, light type and intensity, quantity and quality of food) as well as the chronobiology, sex ratios, inbreeding increase, maternal effects and male contribution to offspring production, systematic line crossing for the reduction of the rate of inbreeding accumulation, chemical communication, mating behaviour and manipulation of reproduction using pheromones. Controlled production of sufficient quantities of quality offsprings remains a major constraint for development of cephalopod culture. In fact, the viability of eggs and hatchlings will, to a large extent, depend on the environmental influences during egg development. Nonetheless, little progress has been made in elucidating the connections between the influence of environmental factors and how they should be manipulated during embryonic development for the production of high-quality and competent hatchlings for rearing. How air bubbles in oversaturated air water affect the viability of *O. vulgaris* paralarvae may be an example of this (see Chap. 23). In this respect, improvements of broodstock conditioning and determination of maternal effects and their impact on hatchling quality should, therefore, be categorized as a high-priority research topic.

In addition to nutrition and reproduction, additional fields of research and zootechnical techniques need future studies and improvement. Cephalopods have delicate skin and the available knowledge on infections and general pathologies is still very scarce and needs further research (see Chap. 6). In some countries, cephalopods are under the scope of animal welfare legislation. In any case, the performance of an ethical experimentation with cephalopods is strongly encouraged and any zootechnical development should be performed and adapted according to the adherence to the 3Rs principles: replacement, refinement and reduction (Mather and Anderson 2007; Moltschaniwskyj et al. 2007; Andrews et al. 2013, Chap. 6). Zootechniques are in constant improvement and should be adapted to each species and areas. For example, in developing countries located in tropical zones with relatively high temperatures (associated with fast growth in some cephalopod species), the open seawater system, as the floating net cages or earthen ponds used for *S. lessoniana* and *S. inermis* (Nabhitabhata et al. 2005, Chaps. 13 and 17), seems more appropriate because of the low cost for construction and maintenance.

Finally, cephalopod culture is completely justified since cephalopods are used as biological models in neuroscience (Williamson and Chrachri 2004; Sio 2011), behaviour (Wells 1978; Hanlon and Messenger 1996; Tricarico et al. 2011; Gherardi et al. 2012), evolution (Budelmann 1995; Strugnell et al. 2011) and climate change (Pörtner and Farrell 2008; Gutowska et al. 2010; Uriarte et al. 2012; Dorey et al. 2013; Noyola et al. 2013; Zúñiga et al. 2013). They have been recently introduced as models for mechatronics (Laschi et al. 2012), biological adhesive systems (Byern and Klepal 2006; Cyran et al. 2010) and tissue regeneration (Feral 1978, 1979, 1988; Rohrbach and Schmidtberg 2006).

Moreover, some cephalopods are among a short number of species with aquaculture production potential where nearly the whole animal may be used or recycled (see Chap. 8). For instance, after making use of the muscle as human food, cephalopod by-products may be used for feeds and natural products, such as viscera for the fish-feed industry (Le Bihan et al. 2006, 2007); the cuttlebone (made of calcium carbonate—99 % aragonite) has application for medicinal and pharmacological industry (Rocha et al. 2005; Kannan et al. 2007; Cadman et al. 2012; Kim et al. 2012); and ink can be used for human food industry (sepia spaghetti). In addition, secondary metabolites from cephalopods, particularly from their ink, have showed their properties as promoters of immune function in vertebrates (Sundaram 2009; Liu et al. 2011), as well as antibacterial (Mochizuki 1979; Benkendorff 2010; Nithya et al. 2011; Gomathi et al. 2010; Mohanraju et al. 2013), antimutagenic (Liu et al. 2008) and antitumoral activity (Chen et al. 2010; Senan et al. 2013), with potential use in biomedicine.

Cephalopods are also among some of the most charismatic marine animals that seduce the general public in aquaria and their culture will play an important role in providing animals already adapted to captive conditions. So, their production potential should probably expand beyond their use as human food. Hence, there is an overall recognition of the potential of cephalopods that need to be translated into actual production in the coming years, as a way of promoting diversification of culture outputs in a sustainable and economical form.

Acknowledgements RV was funded by the research project CALOCEAN-2 (AGL2012-39077) from the Ministry of Economy and Competitiveness of Spain (MINECO). AS was funded by a Post-Doc grant (SFRH/BPD/36100/2007) and projects SEPIAMETA (PTDC/MAR/102348/2008), SEPIABREED (PTDC/MAR/120876/2010), SEPIATECH (31-03-05-FEP-2) and ASSEMBLE (227799 FP7) from the Fundação para a Ciência e a Tecnologia and DGPA PROMAR Programme of the Portuguese Government as well as from the European FP7 Infrastructure Programme. EAGV was funded by the Brazilian National Research Council (CNPq- Pro 307204/2011-1). CR was funded by DGAPA-UNAM project IN212012 and CONACYT- CB201001 project No. 150810, México. LF was funded by the Subprogram of Technical Support Staff (PTA2010-3326-T) and JI by the research project OCTOPHYS (AGL2010-22120-CO3-01), both from MINECO, Spain.

References

Andrews PLR, Darmaillacq AS, Dennison N, Gleadall IG, Hawkins P, Messenger JB, Osorio D, Smith VJ, Smith JA (2013) The identification and management of pain, suffering and distress in cephalopods, including anaesthesia, analgesia and humane killing. J Exp Mar Biol Ecol 447:46–64

Benedito-Palos L, Navarro JC, Sitjà-Bobadilla A, Bell G, Kaushik S, Pérez-Sánchez J (2008) High levels of vegetable oils in plant protein-rich diets fed to gilthead sea bream (*Sparus aurata* L.): growth performance, muscle fatty acid profiles and histological alterations of target tissues. Br J Nutr 100:992–1003

Benkendorff K (2010) Molluscan biological and chemical diversity: secondary metabolites and medicinal resources produced by marine molluscs. Biol Rev 85:757–775

Boal JG, Golden DK (1999) Distance chemoreception in the common cuttlefish, *Sepia officinalis* (Mollusca, Cephalopoda). J Exp Mar Biol Ecol 235:307–317

Boal JG, Marsh SE (1998) Social recognition using chemical cues in cuttlefish (*Sepia officinalis* Linnaeus, 1758). J Exp Mar Biol Ecol 230:183–192

Boal JG, Prosser KN, Holm JB, Simmons TL, Haas RE, Nagle GT (2010) Sexually mature cuttlefish are attracted to the eggs of conspecifics. J Chem Ecol 36:834–836

Budelmann BU (1995) The cephalopod nervous system: what evolutions has made of the molluscan design. In: Breidbach O, Kutsuch W (eds) The nervous system of invertebrates. An evolutionary and comparative approach. Birkhauser Verlag, Basel, pp 115–138

Byern J, Klepal W (2006) Adhesive mechanisms in cephalopods: a review. Biofouling 22:329–338

Cadman J, Zhou S, Chen Y, Li Q (2012) Cuttlebone: characterisation, application and development of biomimetic materials. J Bionic Eng 9:367–376

Cerezo-Valverde J, Martínez-Llorens S, Vidal A, Jover M, Rodríguez C, Estefanell J, Gairín J, Domingues P, Rodríguez C, García B (2013) Amino acids composition and protein quality evaluation of marine species and meals for feed formulations in cephalopods. Aquac Int 21:413–433

Ceulemans S, Coutteau P, Robles-Arozarena R (2003) Fishmeal, fish oil replacements in sea bream, sea bass diets need nutritional compensation. Global Aquac Advocate 6:46–51

Chen SG, Wang JF, Xue CH, Li H, Sun BB, Xue Y, Chai WG (2010) Sulfation of a squid ink polysaccharide and its inhibitory effect on tumor cell metastasis. Carbohydr Polym 81:560–566

Cyran N, Klinger L, Scott R, Griffiths C, Schwaha T, Zheden V, Ploszczanski L, Byern J (2010) Characterization of the adhesive systems in cephalopods. In: Byern J, Grunwald I (eds) Biological adhesive systems. Springer, Vienna, pp 53–86

Darmaillacq AS, Lesimple C, Dickel L (2008) Embryonic visual learning in the cuttlefish, *Sepia officinalis*. Anim Behav 76:131–134

Dias J, Yúfera M, Valente LMP, Rema P (2010) Feed transit and apparent protein, phosphorus and energy digestibility of practical feed ingredients by Senegalese sole (*Solea senegalensis*). Aquaculture 302:94–99

Díaz-López M, Pérez MJ, Acosta NG, Tocher DR, Jerez S, Lorenzo A, Rodriguez C (2009) Effect of dietary substitution of fish oil by Echium oil on growth, plasma parameters and body lipid composition in gilthead seabream (*Sparus aurata* L.). Aquac Nutr 15:500–512

Dorey N, Melzner F, Martin S, Oberhänsli F, Teyssié JL, Bustamante P, Gattuso JP, Lacoue-Labarthe T (2013) Ocean acidification and temperature rise: effects on calcification during early development of the cuttlefish *Sepia officinalis*. Mar Biol 160:2007–2022

Enault J, Zatylny-Gaudin C, Bernay B, Lefranc B, Leprince J, Baudy-Floc'h M, Henry J (2012) A complex set of sex pheromones identified in the cuttlefish *Sepia officinalis*. PLoS ONE 7(10):e46531

Enes P, Peres H, Couto A, Oliva-Teles A (2010) Growth performance and metabolic utilization of diets including starch, dextrin, maltose or glucose as carbohydrate source by gilthead sea bream (*Sparus aurata*) juveniles. Fish Physiol Biochem 36:903–910

Feral JP (1978) Regeneration of arms of cuttlefish *Sepia officinalis* L (Cephalopoda, Sepioidea): 1—Morphological study. Cahiers Biol Marine 19:355–361

Feral JP (1979) Regeneration of the arms of *Sepia officinalis* L (Cephalopoda, Sepioidea): 2—Histologic and cytologic study. Cahiers Biol Marine 20:29–42

Feral JP (1988) Wound-healing after arm amputation in *Sepia officinalis* (Cephalopoda, Sepioidea). J Invert Pathol 52:380–388

Forsythe JW, Derusha RH, Hanlon RT (1994) Growth, reproduction and life-span of *Sepia officinalis* (Cephalopoda, Mollusca) cultured through 7 consecutive generations. J Zool 233:175–192

Gherardi F, Aquiloni L, Tricarico E (2012) Revisiting social recognition systems in invertebrates. Anim Cognit 15:745–762

Gomathi P, Nair JR, Sherief PM (2010) Antibacterial activity in the accessory nidamental gland extracts of the Indian squid, *Loligo duvauceli* Orbigny. Indian J Mar Sci 39:100–104

Guinot D, Monroig O, Navarro JC, Varó I, Amat F, Hontoria F (2013) Enrichment of *Artemia* metanauplii in phospholipids and essential fatty acids as a diet for common octopus (*Octopus vulgaris*) paralarvae. Aquac Nutr 19:837–844

Gutowska M, Melzner F, Pörtner H, Meier S (2010) Cuttlebone calcification increases during exposure to elevated seawater pCO2 in the cephalopod *Sepia officinalis*. Mar Biol 157:1653–1663

Hanlon RT, Messenger JB (1996) Cephalopod behaviour. Cambridge University Press, New York

Hansen AC, Rosenlund G, Karlsen O, Koppe W, Hemre GI, Karlsen Ø (2007) Total replacement of fish meal with plant proteins in diets for Atlantic cod (*Gadus morhua* L.) I. Effects on growth and protein retention. Aquaculture 272:599–611

Iglesias J, Sánchez FJ, Bersano JGF, Carrasco JF, Dhont J, Fuentes L, Linares F, Muñoz JL, Okumura S, Roo J, van der Meeren T, Vidal EAG, Villanueva R (2007) Rearing of *Octopus vulgaris* paralarvae: Present status, bottlenecks and trends. Aquaculture 266:1–15

Kannan S, Rocha JHG, Agathopoulos S, Ferreira JMF (2007) Fluorine-substituted hydroxyapatite scaffolds hydrothermally grown from aragonitic cuttlefish bones. Acta Biomater 3:243–249

Kim B-S, Kim JS, Sung H-M, You H-K, Lee J (2012) Cellular attachment and osteoblast differentiation of mesenchymal stem cells on natural cuttlefish bone. J Biomed Mater Res A 100A:1673–1679

Laschi C, Cianchetti M, Mazzolai B, Margheri L, Follador M, Dario P (2012) Soft robotic arm inspired by the octopus. Advanced Robotics 26:709–727

Le Bihan E, Zatylny C, Perrin A, Koueta N (2006) Post-mortem changes in viscera of cuttlefish *Sepia officinalis* L. during storage at two different temperatures. Food Chem 98:39–51

Le Bihan E, Perrin A, Koueta N (2007) Effect of different treatments on the quality of cuttlefish (*Sepia officinalis* L.) viscera. Food Chem 104:345–352

Lee PG (1994) Nutrition of cephalopods: fueling the system. Mar Fresh Behav Physiol 25:35–51

Lee PG, Turk PE, Yang WT, Hanlon RT (1994) Biological characteristics and biomedical applications of the squid *Sepioteuthis lessoniana* cultured through multiple generations. Biol Bull 186:328–341

Legall S, Feral C (1985) Neuroendocrine metabolism of reproduction in a marine invertebrate, *Sepia officinalis*—functional analogy with the hypothalmo-hypophyseal axis of vertebrates. Ann Endocrinol-Paris 46:N22–N22

Liu CH, Li XD, Li YH, Feng Y, Zhou S, Wang FS (2008) Structural characterisation and antimutagenic activity of a novel polysaccharide isolated from *Sepiella maindroni* ink. Food Chem 110:807–813

Liu H, Luo P, Chen S, Shang J (2011) Effects of squid ink on growth performance, antioxidant functions and immunity in growing broiler chickens. Asian-Australasian J Anim Sci 24:1752–1756

Martínez R, Santos R, Álvarez A, Cuzon G, Arena L, Mascaró M, Pascual C, Rosas C (2011) Partial characterization of hepatopancreatic and extracellular digestive proteinases of wild and cultivated *Octopus maya*. Aquac Int 19:445–457

Mather JA, Anderson RC (2007) Ethics and invertebrates: a cephalopod perspective. Dis Aquat Org 75:119–129

Mochizuki A (1979) An antiseptic effect of cuttlefish ink. Bull Jap Soc Sci Fish 45:1401–1403

Mohanraju R, Marri DB, Karthick P, Narayana K, Murthy KN, Ramesh Ch (2013) Antibacterial activity of certain cephalopods from Andamans, India. Indian J Phar Biol Sci 3:450–455

Moltschaniwskyj NA, Hall K, Marian J, Nishiguchi M, Sakai M, Shulman DJ, Sinclair B, Sinn DL, Staudinger M, Van Gelderen R, Villanueva R, Warnke K (2007) Ethical and welfare considerations when using cephalopods as experimental animals. Rev Fish Biol Fish 17:455–476

Morillo-Velarde PS, Cerezo-Valverde J, Serra Llinares RM, Garcia BG (2011) Energetic contribution of carbohydrates during starvation in common octopus (*Octopus vulgaris*). J Mollusc Stud 77:318–320

Nabhitabhata J, Nilaphat P, Promboon P, Jaroongpattananon C, Nilaphat G, Reunreng A (2005) Performance of simple large-scale cephalopod culture system in Thailand. Phuket Mar Biol Cent Res Bull 66:337–350

Nithya M, Ambikapathy V, Panneerselvam A (2011) Effect of pharaoh's cuttlefish ink against bacterial pathogens. Asian J Plant Sci Res 4:49–55

Noyola J, Mascaró M, Caamal C, Noreña-Barroso E, Díaz F, Re AD, Sanchez A, Rosas C (2013) Effect of temperature on energetic balance and fatty acid composition of early juveniles of *Octopus maya*. J Exp Mar Biol Ecol 445:156–165

Pörtner HO, Farrell AP (2008) Physiology and climate change. Science 322:690–692

Pratoomyot J, Bell JG, Tocher DR, Bendiksen E (2010) Effects of increasing replacement of dietary fishmeal with plant protein sources on growth performance and body lipid composition of Atlantic salmon (*Salmo salar* L.). Aquaculture 305:124–132

Richard A (1971) Contribuition à l'étude expérimentale de la croissance et de la maturation sexuelle de *Sepia officinalis* L. (Mollusque, Céphalopode). Université de Lille, Lille, p 264

Rocha JHG, Lemos AF, Agathopoulos S et al (2005) Scaffolds for bone restoration from cuttlefish. Bone 37:850–857

Rohrbach B, Schmidtberg H (2006) Sepia arms and tentacles: Model systems for studying the regeneration of brachial appendages. Vie Milieu 56:175–190

Rosas C, Tut J, Baeza J, Sánchez A, Sosa V, Pascual C, Arena L, Domingues P, Cuzon G (2008) Effect of type of binder on growth, digestibility, and energetic balance of *Octopus maya*. Aquaculture 275:291–297

Rosas C, Valero A, Caamal-Monsreal C, Uriarte I, Farias A, Gallardo P, Sánchez A, Domingues P (2013) Effects of dietary protein sources on growth, survival and digestive capacity of *Octopus maya* juveniles (Mollusca: Cephalopoda). Aquac Res 44:1029–1044

Senan VP, Sherief PM, Nair JR (2013) Cytotoxic effect of ink extracts of cuttlefish and squid on chick embryo fibroblasts'. Int J Phar Sci Res 4:1893–1896

Sio FD (2011) Leviathan and the soft animal: Medical humanism and the invertebrate models for higher nervous functions, 1950s–90s. Med Hist 55:369–374

Strugnell JM, Cherel Y, Cooke IR, Gleadall IG, Hochberg FG, Ibáñez CM, Jorgensen E, Laptikhovsky VV, Linse K, Norman M, Vecchione M, Voight JR, Allcock AL (2011) The Southern Ocean: Source and sink? Deep Sea Res Part 2 Tropical Stud Oceanogr 58:196–204

Sundaram S (2009) The various uses of cephalopods. Fish Chimes 29:23–25

Sykes AV, Domingues PM, Andrade JP (2006) Effects of using live grass shrimp (*Palaemonetes varians*) as the only source of food for the culture of cuttlefish, *Sepia officinalis* (Linnaeus, 1758). Aquac Int 14:551–568

Sykes AV, Gonçalves RA, Andrade JP (2013) Early weaning of cuttlefish (*Sepia officinalis*) with frozen grass shrimp (*Palaemonetes varians*) from the first day after hatching. Aquac Res 44:1815–1823 doi:10.1111/j.1365-2109.2012.03186.x

Tricarico E, Borrelli L, Gherardi F, Fiorito G (2011) I know my neighbour: Individual recognition in *Octopus vulgaris*. Plos One 6(4):e18710

Uriarte I, Espinoza V, Herrera M, Zúñiga O, Olivares A, Carbonell P, Pino S, Farias A, Rosas C (2012) Effect of temperature on embryonic development of *Octopus mimus* under controlled conditions. J Exp Mar Biol Ecol 416-417:168–175

Villanueva R (1994) Decapod crab zoeae as food for rearing cephalopod paralarvae. Aquaculture 128:143–152

Villanueva R, Norman MD (2008) Biology of the planktonic stages of benthic octopuses. Oceanogr Mar Biol Ann Rev 46:105–202

Walsh LS, Turk PE, Forsythe JW, Lee PG (2002) Mariculture of the loliginid squid *Sepioteuthis lessoniana* through seven successive generations. Aquaculture 212:245–262

Wells MJ (1978) Octopus—physiology and behaviour of an advanced invertebrate. Chapman & Hall, London

Williamson R, Chrachri A (2004) Cephalopod neural networks. Neurosignals 13:87–98

Zúñiga O, Olivares A, Rojo M, Chimal ME, Díaz F, Uriarte I, Rosas C (2013) Thermoregulatory behavior and oxygen consumption of *Octopus mimus* paralarvae: the effect of age. J Therm Biol 38:86–91

Zúñiga OR, Olivares AP, Ossadón LR (1995) Influencia de la luz en la maduración sexual de hembras *Octopus mimus*. Estudios Oceanol 14:75–76

Index

J. Iglesias et al. (eds.), *Cephalopod Culture*, DOI 10.1007/978-94-017-8648-5,

Printed by Publishers' Graphics LLC
LMO140411.15.20.89